Neuroscience

The INSTANT NOTES series

Series editor
B.D. Hames
School of Biochemistry and Molecular Biology, University of Leeds, Leeds, UK

Animal Biology
Ecology
Microbiology
Genetics
Chemistry for Biologists
Immunology
Biochemistry 2nd edition
Molecular Biology 2nd edition
Neuroscience

Forthcoming titles
Psychology
Developmental Biology
Plant Biology

The INSTANT NOTES Chemistry series
Consulting editor: Howard Stanbury

Organic Chemistry
Inorganic Chemistry

Forthcoming titles
Physical Chemistry
Analytical Chemistry

Neuroscience

A. Longstaff

Science writer and freelance lecturer in neuroscience

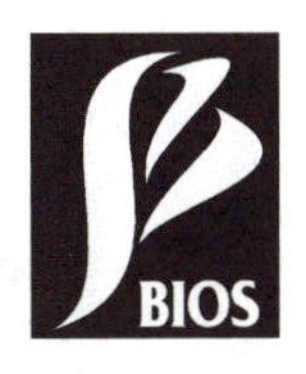

First published 2000

A CIP catalogue record for this book is available from the British Library.

ISBN 1 85996 082 0

BIOS Scientific Publishers Ltd
9 Newtec Place, Magdalen Road, Oxford OX4 1RE, UK
Tel. +44 (0)1865 726286. Fax +44 (0)1865 246823
World Wide Web home page: http://www.bios.co.uk/

Published in the United States of America, its dependent territories and Canada by Springer-Verlag New York Inc., 175 Fifth Avenue, New York, NY 10010-7858, in association with BIOS Scientific Publishers Ltd.

Published in Hong Kong, Taiwan, Cambodia, Korea, The Philippines, Brunei, Laos and Macau by Springer-Verlag Hong Kong Ltd, Unit 1702, Tower 1, Enterprise Square, 9 Sheung Yuet Road, Kowloon Bay, Kowloon, Hong Kong, in association with BIOS Scientific Publishers Ltd.

Production Editor: Fran Kingston
Typeset and illustrated by Phoenix Photosetting, Chatham, Kent, UK
Printed by Biddles Ltd, Guildford, UK, www.biddles.co.uk

Contents

Abbreviations

ACh	acetylcholine
AChE	acetylcholinesterase
ACTH	adrenocorticotrophic hormone
AD	Alzheimer's disease
AII	angiotensin II
AMPA	α-amino-3-hydroxy-5-methyl-4-isoxazole-proprionic acid
ANS	autonomic nervous system
AP	action potential
apoE	apolipoprotein E
APP	amyloid precursor protein
APV	D-2-amino-5-phosphonovalerate
ATN	anterior thalamic nuclei
ATP	adenosine 5′-triphosphate
AVP	arginine vasopressin
βA	β-amyloid
βAR	β adrenoceptors
BAT	brown adipose tissue
BDNF	brain derived neurotrophic factor
bl	basal lamina
BMP	bone morphogenetic protein
α-BTX	α-bungarotoxin
CA	cornu ammonis
CaM	calmodulin
CAM	cell adhesion molecule
CaMKII	calcium–calmodulin-dependent protein kinase II
cAMP	cyclic adenosine monophosphate
CAT	computer assisted tomography
cbf	cerebral blood flow
CC	cingulate cortex
CCK	cholecystokinin
CF	characteristic frequency
cGMP	3′,5′-cyclic guanosine monophosphate
ChAT	choline acetylesterase
CL	central laminar nucleus (of thalamus)
CNG	cyclic-nucleotide-gated channel
CNS	central nervous system
CoA	coenzyme A
CPG	central pattern generators
CRH	corticotrophin releasing hormone
CRO	cathode ray oscilloscope
CS	conditioned stimulus
CSF	cerebrospinal fluid
CVA	cerebrovascular accident
CVLM	caudal ventrolateral medulla
CVO	circumventricular organ
DAG	diacylglycerol
DBL	dorsal blastopore lip
DCML	dorsal column–medial lemniscal system
DCN	dorsal column nuclei
2-DG	2-deoxyglucose
DHC	dorsal horn cell
DI	diabetes insipidus
DLPN	dorsolateral pontine nucleus
DOPAC	dihydroxyphenyl acetic acid
DRG	dorsal root ganglion
DYN	dynorphin
ECT	electroconvulsive therapy
EEG	electroencephalography
EGF	epidermal growth factor
EGL	external granular layer
EMG	electromyography
ENK	encephalin
ENS	enteric nervous system
epp	endplate potential
epsp	excitatory postsynaptic potential
ER	endoplasmic reticulum
F-actin	filamentous actin
FEF	frontal eye field
FF	fast fatiguing
FM	frequency modulation
fMRI	functional magnetic resonance imaging
FR	fatigue resistant
FRA	flexor reflex afferents
FSH	follicle stimulating hormone
G_i	inhibitory G protein
G_q	G protein coupled to phospholipase
G_s	stimulatory G protein
GABA	γ-aminobutyrate
GC	guanylyl cyclase
GDP	guanosine 5′-diphosphate
GFAP	glial fibrillary acidic protein
GH	growth hormone
GHRH	growth hormone releasing hormone
GnRH	gonadotrophin releasing hormone
GPe	globus pallidus pars externa
GPi	globus pallidus pars interna

GR	glucocorticoid receptor
GTO	Golgi tendon organs
GTP	guanosine 5′-triphosphate
5-HIAA	5-hydroxyindoleacetic acid
HPA	hypothalamic–pituitary–adrenal (axis)
HPG	hypothalamic–pituitary–gonadal (axis)
HPT	hypothalamic–pituitary–thyroid (axis)
HRP	horseradish peroxidase
5-HT	5-hydroxytryptamine (serotonin)
5-HTP	5-hydroxytryptophan
HVA	high voltage activated
IaIN	Ia inhibitory interneurons
IbIN	Ib inhibitory neurons
IC	inferior colliculus
ICSS	intracranial self-stimulation
Ig	immunoglobulin
IGF-1	insulin-like growth factor 1
IGL	internal granular layer
iGluR	ionotrophic glutamate receptor
ILD	interaural level differences
IP_3	inositol 1,4,5-trisphosphate
ipsps	inhibitory postsynaptic potential
IT	inferotemporal cortex
JGA	juxtoglomerular apparatus
L-DOPA	L-3,4-dihydroxyphenylalanine
LC	locus cerulus
LCN	lateral cervical nucleus
LDCV	large dense-core vesicle
LGN	lateral geniculate nucleus
LH	luteinizing hormone
LSO	lateral superior olivary nucleus
LTD	long-term depression
LTM	long-term memory
LTN	lateral tegmental nucleus
LTP	long-term potentiation
LVA	low voltage activated
M	magnocellular pathway
M/T	mitral/tufted cells
mAChR	muscarinic cholinergic receptor
MAO	monoamine oxidase
MAP	mean arterial (blood) pressure
MB	mammillary bodies
mepp	miniature endplate potential
MFB	medial forebrain bundle
MFS	mossy fiber sprouting
mGluR1	type 1 metabotropic glutamate receptor
MGN	medial geniculate nucleus
MI	primary motor cortex
MII	secondary motor cortex
MLCK	myosin light chain kinase
MLR	mesencephalic locomotor region
MOPEG	3-methoxy,4-hydroxy phenylglycol
MPOA	medial preoptic area
MPP^+	1-methyl-4-phenyl pyridinium
mpsp	miniature postsynaptic potential
MPTP	1-methyl-4-phenyl-1,2,3,6-tetrahydropyridin
MR	mineralocorticoid receptor
MRI	magnetic resonance imaging
MSO	medial superior olivary complex
MST	medial superior temporal cortex
NA	noradrenaline
nAChR	nicotinic cholinergic receptor
NGF	nerve growth factor
NMDA	*N*-methyl-D-aspartate
NMDAR	*N*-methyl-D-aspartate receptor
nmj	neuromuscular junction
NMR	nuclear magnetic resonance
NPY	neuropeptide Y
NREM	nonrapid eye movement sleep
NRM	nucleus raphe magnus
NST	nucleus of the solitary tract
NT3-6	neurotrophins 3-6
OC	olivocochlear
OCD	obsessive–compulsive disorder
OHC	outer hair cells
6-OHDA	6-hydroxydopamine
ORN	olfactory receptor neurons
OVLT	vascular organ of the lamina terminalis
P	parvocellular pathway
PAD	primary afferent depolarization
PAG	periaqueductal gray matter
Pc	Purkinje cells
PD	Parkinson's disease
PDE	phosphodiesterase
PDS	paroxysmal depolarizing shifts
PET	positron emission tomography
pf	parallel fibers
PFC	prefrontal cortex
PGO	pontine–geniculate–occipital spikes
PHF	paired helical filaments
PIP_2	phosphatidyl inositol-4,5-bisphosphate
PKA	protein kinase A
PM	premotor cortex

PNS	peripheral nervous system
POA	preoptic area
POM	posterior complex (medial nucleus) of thalamus
POMC	pro-opiomelanocortin
PP	posterior parietal complex
PRL	prolactin
PSNS	parasympathetic nervous system
psp	postsynaptic potential
PVN	paraventricular nucleus
RA	retinoic acid
REM	rapid eye movement sleep
RF	receptive field
RHT	retinohypothalamic tract
S	slow twitch fiber
SCG	superior cervical ganglion
SCN	suprachiasmatic nucleus
Sc	Schaffer collateral
SDN–POA	sexually dimorphic nucleus of the preoptic area
SER	smooth endoplasmic reticulum
SH2	src homology domain 2
SHH	sonic hedgehog protein
SMA	supplementary motor area
SNpc	substantia nigra pars compacta
SNzc	substantia nigra zona compacta
SNS	sympathetic nervous system
SON	supraoptic nucleus
SP	substance P
SPL	sound pressure level
SR	sarcoplasmic reticulum
SSRI	selective serotonin reuptake inhibitors
SSV	small clear synaptic vesicle
STM	short-term memory
STN	subthalamic nucleus
STT	spinothalamic tract
TB	trapezoid body
TCA	tricyclic antidepressants
TEA	tetraethylammonium
TENS	transcutaneous electrical nerve stimulation
TH	tyrosine hydroxylase
TM	transmembrane
TRH	thyrotropin releasing hormone
trk	tyrosine kinase receptors
TSH	thyroid releasing hormone
TTX	tetrodotoxin
UR	unconditioned response
US	unconditioned stimulus
VDCC	voltage-dependent calcium channel
VDKC	voltage-dependent potassium channel
VDSC	voltage-dependent sodium channel
VIP	vasoactive intestinal peptide
VLH	ventrolateral hypothalamus
VLPO	ventrolateral preoptic area
VMAT	vesicular monoamine transporter
VMH	ventromedial hypothalamus
VOR	vestibulo-ocular reflexes
VPL	ventroposterolateral nucleus (of thalamus)
VPM	ventroposteromedial nucleus (of thalamus)
VRG	ventral respiratory group
VST	ventral spinocerebellar tract
VZ	ventricular zone

PREFACE

Neuroscience is one of the most rapidly advancing areas of science and, as a consequence, spawns a literature which is growing dramatically. At one level it attempts to provide a mechanistic account of the most complex 'device' in the known Universe, the human brain. Moreover, neuroscience is multidisciplinary, having contributions from biochemistry and molecular biology, physiology, anatomy, psychology and clinical medicine, to name the most obvious. For these reasons, it is becoming increasingly difficult for lecturers and textbook authors to present neuroscience in a way that manages to be comprehensive, up-to-date and accessible, while still being sufficiently rigorous to prepare students to be successful explorers of the literature for themselves. *Instant Notes Neuroscience* is not intended as a replacement for lectures or the standard textbooks, but as an affordable text to supplement them, which is of a manageable size and in a format which aids learning.

The text is designed to provide the core of the subject in 18 sections containing 93 topics. When coming to a new subject, it is my experience that students commonly express two concerns: first, how to sort out the important ideas and facts from the wealth of detail, and second, how to get to grips with the unfamiliar terminology. Lecturers, in addition, will want students (especially later in their studies) to be able to integrate their knowledge across the subject. *Instant Notes Neuroscience* attempts to address each of these issues. Each topic is supported by a 'Key Notes' panel which gives a concise summary of the crucial points. Whenever a term appears for the first time it is in bold and immediately followed by a definition or explanation. Extensive cross-references are provided between topics so that students can forge the links that are important for integration.

This is a much slimmer volume than most neuroscience texts, which can be dauntingly large. A number of features contribute to this. First, I have tried to minimize the amount of detail without compromising the need for a database for further study. Second, while many of the methods used by neuroscientists are included, *individual* experiments or items of evidence are included only where I thought it essential to illustrate a point, or on matters that would need some justification to be convincing. Third, with a few exceptions, I have restricted examples to those most appropriate to the human condition. In so doing I have always qualified the species, since species differences matter. If not, then rats and cats would behave as humans do, which clearly they do not!

Section A introduces the cells of the nervous system, showing how they are specialized for the functions they serve. The next three sections are essentially cellular neuroscience. Section B is concerned mostly with action potentials, Section C with synapses, while Section D deals with how nerve cells act as information processors. These sections provide an introduction to the electrophysiological techniques used to study nerve cells, and say something about the molecular biology of the ion channels and receptors that govern their behavior. Section E takes a broad view of neuroanatomy and summarizes techniques, such as brain imaging, used to investigate nervous system structure. How information is encoded by the firing and connectivity of neurons is considered in Section F. All the material thus far might reasonably be found in first-year courses.

The next seven sections (G–M) form the core of systems neuroscience. Section G reviews the body senses, touch, pain and balance. Sections H and I deal with vision and hearing, respectively, while Section J looks at the chemical senses, smell and taste. The properties of skeletal muscle, motor reflexes, and the cortical control of voluntary movement are the subject of Section K, while the involvement of the cerebellum (including proprioception) and the basal ganglia in movement is covered in Section L. Neuroendocrinology, and both peripheral and central aspects of the autonomic nervous system appear in Section M, which also (and unusually for standard neuroscience texts) includes the functions of smooth and cardiac muscle, and the enteric nervous system. A short Section N describes the essential features of amine transmission, the basis for much neuropharmacology, and paves the way for understanding aspects of behavior, such as motivation and sleep that are included in Section O. Section P is an overview of how the embryonic nervous system develops, ranging from how the basic plan is genetically specified, to how differences between male and female brains might arise. Section Q addresses how the nervous system continues to rewire itself on the basis of experience (i.e., learning and memory). Finally, although quite a number of nervous system disorders are considered at appropriate places throughout the book, section R takes the four most common neuropathologies, stroke, epilepsy, Parkinson's disease and Alzheimer's dementia and looks in some detail at what has gone amiss and what current and future treatments may do. Space precluded the inclusion of topics on two major psychiatric disorders (schizophrenia and depression) in the text; it is intended that these topics will be made available on the BIOS website free of charge. At the end of the book is a reading list for those who wish to take their studies further.

As a student, how should you use this book? Restrict your reading only to the sections and topics covered by your current course. That said, Sections A–F are likely to appear in, or be required knowledge for, just about any neuroscience program; you will probably need to work through these first. Later sections can be dipped into in any order. Read the main sections thoroughly first, making sure that you *understand* the ideas, and use the 'Related topics' to make links, just as you would if you were surfing the internet. Where appropriate, reference is made to areas that are covered in more detail in the companion volumes, *Instant Notes in Biochemistry*, 2nd edn, and *Instant Notes in Molecular Biology*, 2nd edn. At this stage you can incorporate additional material from lectures or other textbooks in the gaps at the end of topics, or highlight things which seem to be particularly important for your course. Studying *Instant Notes Neuroscience* 'little but often' is a good strategy. The information density in the text is high, so many short, concentrated, bursts are much more effective than a few eight-hour stints. The more times you work through a topic, the better your understanding, and the more likely you will be to remember it clearly. When it comes to revision, use the 'Key Notes' as a prompt. In addition, you should aim to be able to write, from memory, a few sentences about each of the terms that appears in bold in the main text. Being able to reproduce the simpler diagrams is also an effective way of getting your point across in an exam. Neuroscience is an extraordinary endeavor because it aims to reveal what, in essence, it is to be human; how we behave, think and feel as we do. At the moment we are a long way from being able to give a coherent account of any of these faculties; that there is so much still to be done is one reason that this science is so exciting. This book is an account of the remarkable progress made so far. I hope you find that it serves your needs well and that, like me, you enjoy discovering neuroscience.

Acknowledgements

A number of colleagues – Barry Hunt, Vasanta Raman and John Wilkinson, all at the University of Hertfordshire – kindly read through some individual topics and made very helpful suggestions. David Hames (University of Leeds, UK), Kevin Alloway (Penn State University, USA), and Patricia Revest (Queen Mary and Westfield College, London University, UK), were each brave enough to read through the entire text and their thought-provoking comments have been important in shaping the final version. I am very grateful to all of these people for their time and expertise. Finally, I thank Jonathan Ray, Rachel Offord, Will Sansom and Fran Kingston at BIOS Scientific Publishers for their encouragement and patience.

Alan Longstaff

A1 Neuron structure

Key Notes

Cell body

The neuron cell body contains all the subcellular organelles found in a typical animal cell but it is specialized to maintain high rates of protein synthesis, as shown by the ribosome-packed Nissl bodies.

Neurites

Neurites are long projections from the cell body. There are two types of neurite, dendrites and axons. Dendrites are large extensions of the cell body and receive most of the synaptic inputs impinging onto the cell. Neurons may have one or many dendrites. Neurons have a single axon arising from the axon hillock. Axon terminals form the presynaptic components of synapses.

Axon or dendrite?

The two neurites can be distinguished on structural grounds. Dendrites contain many organelles and are capable of protein synthesis. By contrast, axons cannot synthesize protein, so axonal proteins are derived from the cell body. Axons and dendrites both have mitochondria. Organelles are transported into neurites via microtubules.

Related topic

Morphology of chemical synapses (A3)

Cell body

The cell body (**soma** and **perikaryon** are synonyms) of a neuron (see *Fig. 1*) contains the nucleus, Golgi apparatus, ribosomes and other subcellular organelles, and is responsible for most of its routine metabolic 'housekeeping' functions. The neuronal perikaryon is not so different from non neuronal cells although structurally it is specialized to maintain high levels of biosynthetic activity. The rough endoplasmic reticulum, for example, is so densely packed as to produce distinct structures called **Nissl bodies** which are extremely rich in ribosomes. This reflects the high rates of protein synthesis of which neurons are capable.

Because there is a great variety of types of neurons, their cell bodies vary in size considerably. The smallest are some 5–8 μm in diameter, the largest 120 μm across.

Neurites

Neurons are distinguished from other cells by **neurites**, long (relative to the cell body) cylindrical processes that come in two varieties, dendrites and axons. **Dendrites** are highly branched extensions of the cell body, may be up to 1 mm in length and account for up to 90% of the surface area of many neurons. Dendrites on some neurons are covered with hundreds of tiny projections termed **dendritic spines** on which synapses (see below) are made. Nerve cells with spines are sometimes called **spiny neurons**, those lacking them **aspiny neurons**. A neuron may have one or many dendrites, arranged in a pattern which is cell typical and collectively referred to as the **dendritic tree.** The majority of synaptic inputs from other neurons are made on dendrites.

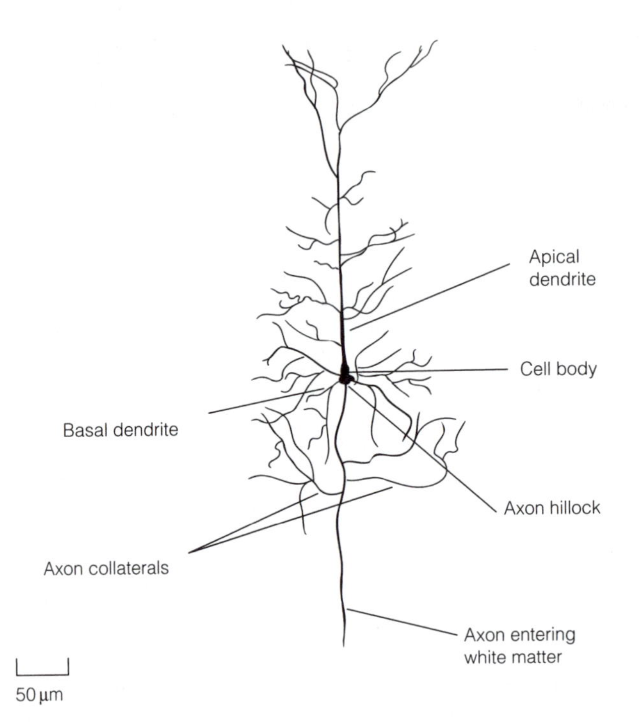

Fig. 1. Key features of a neuron. A drawing of a pyramidal cell showing the distribution of neurites (dendrites and axon).

Nerve cells usually have only one **axon** which arises typically from the cell body but may emerge from a proximal dendrite (the end of a dendrite closest to the soma). In either case, the site of origin is termed the **axon hillock**. Axons have diameters ranging from 0.2 to 20 μm in humans (though axons of invertebrates can reach 1 mm) and vary in length from a few μm to over a meter. They may be encapsulated in a **myelin sheath**. Axons usually branch, particularly at their distal end (furthest from the soma). These branches are referred to as **axon collaterals.** The ends of an axon are swollen **terminals** (or **boutons**) and usually contain mitochondria and vesicles. Some axons have a tuft of branches (a **terminal arbor**) at their tip, each with its terminal bouton, some have boutons along their length where they are described as **varicosities**. Axon terminals form the presynaptic component of chemical synapses.

Axon or dendrite?

Axons can be distinguished from dendrites on structural grounds. Axons tend to be long, untapered, less highly branched, never spiny and may have a myelin sheath, whereas dendrites are shorter, tapered, highly branched and may bear spines. Dendrites are extensions of the cell body in that they contain Golgi apparatus, rough endoplasmic reticulum and ribosomes – organelles not seen in axons. By contrast both axons and dendrites have mitochondria. Since axons do not possess protein synthetic machinery, proteins in axons must be made in the cell body and subsequently moved into and along the axon by a mechanism called axoplasmic transport. Axon terminals are often rich in mitochondria which indicates their high requirement for metabolic energy.

Differences in organelle composition are thought to result from the differing arrangements of **microtubules** in the two types of neurite. Microtubules are long protein polymers that are part of the internal scaffolding of cells called the **cytoskeleton**. Microtubules act as 'tramlines' along which organelles are moved within the cell. The two ends of a microtubule are different, designated + and – ; thus microtubules have a distinct polarity which means that organelles move along them in a specific direction. Mitochondria move in the – to + direction whereas other organelles move in the + to – direction. Both dendrites and axons have microtubules, but whereas dendrites have them orientated in either direction, those in axons are always arranged with their + ends away from the cell body. Thus axonal microtubules can transport mitochondria out of the cell body into the axon but cannot transport other organelles. Because microtubules are orientated in both directions in dendrites, all organelles are transported into these neurites.

When a neuron develops, it first grows a number of processes which are indistinguishable. It is not yet known how one of these is subsequently selected to differentiate into an axon. The first sign that a process will become an axon is that it grows at a much faster rate than those which are destined to be dendrites.

A2 Classes and numbers of neurons

Key Notes

Neuron classification Neurons may be classified by their morphology, function or by the neurotransmitter they secrete. Cells with one, two or three or more neurites are classed as unipolar, bipolar or multipolar respectively. The shape of the dendritic tree, or whether the dendrites have dendritic spines and the length of the axon have proved useful in categorizing neurons. Functional classification distinguishes between sensory neurons which respond directly to physiological stimuli and motor neurons which synapse with effectors.

Neuron numbers The human nervous system may contain 300–500 billion neurons. Neuron density is quite constant across the cerebral cortex and between cerebral cortices of different mammals. Smaller brains have fewer neurons.

Related topics Organization of the peripheral nervous system (E1)
Organization of the central nervous system (E2)

Neuron classification

There is no such thing as a 'typical' neuron. Nerve cells come in a wide variety of shapes and sizes with widely differing numbers and patterns of synaptic contacts using distinct neurotransmitters. Hence neurons are classified either according to morphology, function or by their neurotransmitters, and the assumption is that all neurons falling into a single class have similar functions.

Structural considerations to classifying a given cell include the size of the cell body, the number of neurites it has, the pattern of its dendritic tree, axon length and the nature of the connections it makes. A neuron with a single neurite is **unipolar**, one with two neurites is **bipolar** while a neuron with three or more neurites is said to be **multipolar** (*Fig. 1*). The majority of neurons in the vertebrate nervous system are multipolar but there are important exceptions. For example, a population of neurons in the retina which synapse with photoreceptors are bipolar and sensory neurons in the dorsal root ganglion are described as **pseudounipolar**; technically they are bipolar because they start life with two processes but these subsequently fuse. Invertebrate nervous systems are dominated by unipolar neurons.

Dendrites are used to classify neurons on the basis of whether or not they have dendritic spines and the overall pattern of their dendritic tree. The shape of any dendritic tree helps to determine the efficacy of its synaptic connections and so to the functioning of the cell. **Pyramidal** cells, so-called because of the shape of their cell bodies, comprise some 60% of neurons in the cerebral cortex and have dendrites which extend to fill a pyramidal space. A second population of cortical cells are termed stellate cells because of the star-like appearance of

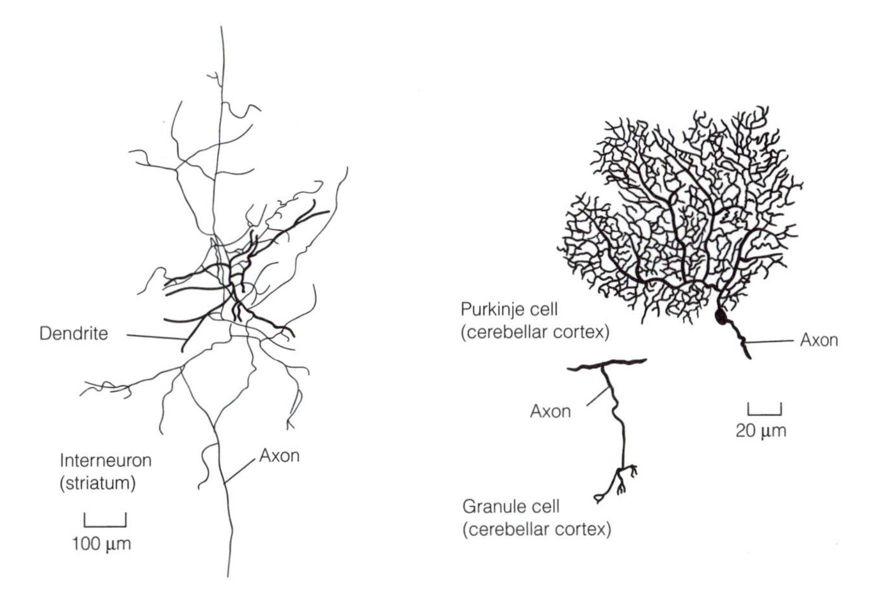

Fig. 1. The morphologies of three common types of neuron. The full length of the axons is not shown. The bifurcating axon of the granule cell extends for several millimeters in each direction. Note how the axon of the interneuron branches extensively.

their dendritic trees. **Purkinje** cells of the cerebellar cortex have the unique feature that their extensive network of dendrites forms a two-dimensional array.

Neurons can also be classified on the basis of the lengths of their axons. **Projection** neurons (**principal**, **relay** or **Golgi type I** are synonyms) have long axons which extend way beyond the region of the nervous system in which their cell body resides. Pyramidal and Purkinje cells fall into this category. In contrast, **interneurons** (**intrinsic** or **Golgi type II** neurons) have short axons. These local circuit neurons, such as stellate cells, produce direct effects only in their immediate neighborhood.

By examining the connections that a neuron makes it is possible to classify neurons (crudely) by function. Any given region of the nervous system receives inputs from **afferent** neurons and projects by **efferent** neurons to other regions of the nervous system or an effector organ (such as a muscle or gland). Afferent neurons that synapse with sensory receptors or which are themselves capable of responding directly to physiological stimuli are **sensory** neurons. Efferent neurons which synapse with skeletal muscles are termed **motor** neurons. Sometimes the term motor neuron is applied to projection neurons in motor pathways even if they do not synapse directly with a muscle.

Finally, neurons can be classified according to the neurotransmitters which they secrete. Moreover, there is often a clear correlation between neuron morphology and neurotransmitter. In other words, the shape of a neuron allows an intelligent guess to be made about which transmitter it secretes. For example, pyramidal cells release glutamate whereas stellate and Purkinje cells secrete γ-aminobutyrate. This, in turn, provides very strong circumstantial evidence of function. This illustrates a general point, namely that these apparently separate ways of classifying neurons overlap.

Neuron numbers

Estimates of the number of neurons in the nervous system can be made by statistical analysis of cell counts in thin tissue sections viewed under light microscope. This shows that the number of neurons per unit area of the cerebral cortex is remarkably constant from one area of the cortex to another in humans, and across mammalian species, at around 80 000 mm^{-2}. The exception is the primary visual cortex where the neuron density rises to 200 000 mm^{-2}. Assuming a total surface area for the human cerebral cortex of 2000 mm^2, these figures suggest that there are some 1.6×10^{11} neurons in the cerebral cortex alone. The most populous cells in the mammalian nervous system are small granule cells of the cerebellum; in humans they may number 10^{11}. Hence the human nervous system contains at least 2.5×10^{11} neurons; the total is likely to be between 300 and 500 billion! Smaller mammals have smaller brains because they have fewer neurons, not because their neurons are smaller.

A3 MORPHOLOGY OF CHEMICAL SYNAPSES

Key Notes

Synapse location

Synapses may be electrical or chemical. Chemical synapses can be classified according to where they are located on the receiving neuron. Axodendritic synapses are made on dendrites, axosomatic on the cell body, and axoaxonal on the axon. The majority of synapses are axodendritic.

Synapse structure

An axodendritic synapse is formed between an axon terminal presynaptically and a dendrite postsynaptically. The synaptic cleft between these elements is 30 nm wide. The axon terminal contains mitochondria, spherical synaptic vesicles and dense projections on the presynaptic membrane. The cleft contains proteins which link pre- and postsynaptic membranes. The postsynaptic membrane is thickened to form the postsynaptic density.

Synapse diversity

Two major types of synapse have been identified. Type I are the axodendritic synapses described above and are usually excitatory. Type II synapses have less well developed dense projections, cleft material and postsynaptic density. They have ovoidal vesicles, an axosomatic location and are usually inhibitory. Synapses that secrete catecholamines or peptides have large dense-core vesicles; some of these have little discernible pre- or postsynaptic specialization and a wide cleft. Many synapses contain both small clear vesicles and large dense-core vesicles, evidence that many neurons secrete more than one transmitter.

Related topics

Fast neurotransmission (C2)
Slow neurotransmission (C3)
Retinal processing (H5)
Olfactory pathways (J2)
Norepinephrine neurotransmission (N2)

Synapse location

Signaling between nerve cells occurs via synapses. There are two types of synapse, **electrical** and **chemical**, but chemical synapses far outnumber electrical ones. A chemical synapse is formed from the close approximation of an axon terminal, which is the presynaptic component, with some region of the postsynaptic cell. The gap between the presynaptic terminal and postsynaptic cell, the **synaptic cleft**, varies between 20 nm and 500 nm depending on the precise nature of the synapse. Synapses are made on all regions of a receiving cell and may be classified on the basis of where they are located. The majority of synapses are formed on dendrites. On spiny dendrites each spine is the target of an axon terminal and comprises the postsynaptic component of a single synapse. Synapses between axons and dendrites are called **axodendritic**. Particularly powerful synapses are made between axons and the cell body of a postsynaptic cell. These are called **axosomatic** synapses, the word somatic coming from the alternative term for the cell body, **soma**. Synapses between axon terminals and axons of postsynaptic neurons are said to be **axoaxonal**.

Synapse structure

Electron microscopy must be used to resolve synapses because of their small size. This has revealed numerous morphologically distinct types of synapse, but all share common features. *Fig. 1* shows a typical axodendritic synapse. The **small clear synaptic vesicles** (SSVs) which store neurotransmitter, are spherical with a diameter of about 50 nm, and are scattered throughout the terminal apparently in close association with microtubules which might be involved in transporting them to the presynaptic membrane. The presynaptic membrane is thickened and may show inwardly directed **dense projections** which are involved in the docking of synaptic vesicles at the **active zone**, the region from which transmitter release occurs. Note that the axon terminal harbors several mitochondria.

The synaptic cleft of axodendritic synapses is 30 nm wide and contains protein filaments that stretch transversely from the pre- to the postsynaptic side that may serve to keep the two membranes in close apposition.

The cell membrane of the dendrite in the region of the synapse appears thickened to form the **postsynaptic density**. This is caused by the accumulation of proteins, receptors, enzymes and the like, involved in the postsynaptic cells' response to transmitter.

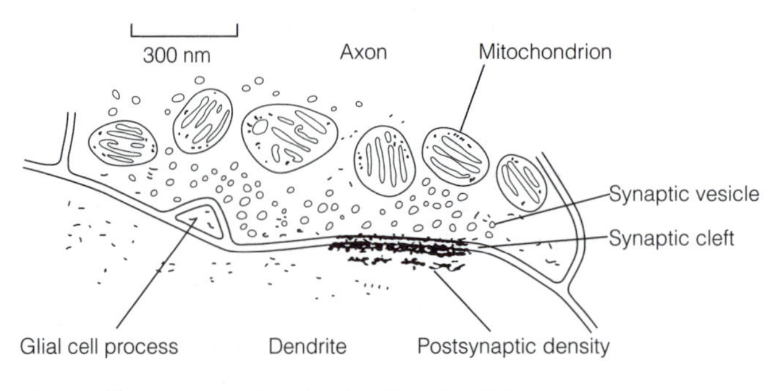

Fig. 1. The structure of a chemical (axodendritic) synapse.

Synapse diversity

Studies of cerebral cortex and cerebellar cortex have revealed that the majority of synapses fall into one of two types. **Type I** are those described above. **Type II** synapses have little or no presynaptic dense projections, poorly defined material in a cleft only 20 nm across and a thin postsynaptic density. These synapses are frequently axosomatic. Type II synapses contain ovoidal vesicles. Physiological studies of these two types of synapse shows that type I are generally excitatory whereas type II are usually inhibitory.

Although most synapses contain SSVs, some (*Fig. 2*) have spherical vesicles with electron-dense centers called **large dense-core vesicles** (LDCVs) which fall into two populations according to size; 40–60 nm vesicles are found in neurons that release catecholamine transmitters, whereas 120–200 nm vesicles are associated with neurosecretory neurons that liberate peptide hormones in the posterior pituitary. Some synapses lack obvious specialized contact zones on both pre- and postsynaptic sides and have extremely wide synaptic clefts (100–500 nm). These are often catecholaminergic (i.e. they secrete one or other of the catecholamines), have dense-core vesicles and can be found both in the central nervous system and in the peripheral nervous system.

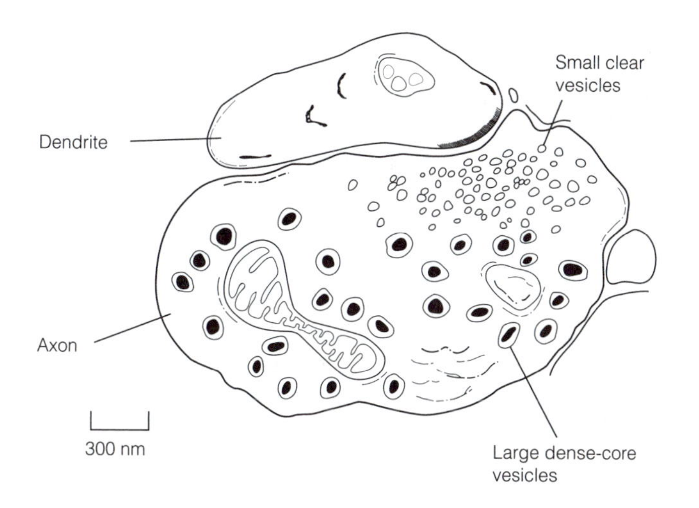

Fig. 2. A type I synapse containing both small clear vesicles and large dense-core vesicles. From Revest, P.A. and Longstaff, A. (1998) Molecular Neuroscience *© BIOS Scientific Publishers Ltd, Oxford.*

Many synapses contain more than one population of vesicles. Commonly, SSVs are found together with LDCVs. This is clear structural evidence that many neurons secrete more than one transmitter.

Synaptic diversity is greater than implied above. In some specialized regions of the nervous system synapses can be found which are quite distinct from the typical, for example, the triad ribbon synapse in the retina (see Topic H5), reciprocal synapses in the olfactory bulb (Topic J2) have some monoamine-secreting terminals (Topic N2). These exceptions will be described later where appropriate.

A4 GLIAL CELLS AND MYELINATION

Key Notes

Classes of glial cells

Glial cells perform a number of functions which support neurons. There are many more glial cells than neurons. Glial cells can be assigned to one of three major populations: astrocytes, oligodendrocytes (including peripheral Schwann cells) and microglia.

Astrocytes

Astrocytes are large, numerous, star-shaped glia which have elongated processes tipped with endfeet. They cover synapses, form contacts with capillary endothelial cells and with the pia mater where they form a limiting glial membrane. The functions of astrocytes include homeostatic regulation of the extracellular K^+ concentration, removal of neurotransmitters from the synaptic cleft, regulating the supply of glucose to neurons, and generating the formation of a tight blood–brain barrier.

Oligodendrocytes and Schwann cells

Oligodendrocytes in the central nervous system (CNS) and Schwann cells in the peripheral nervous system are responsible for forming the insulating myelin sheath that surrounds many axons. The sheath is produced by the mesaxon of the glial cell spiralling around the axon a number of times. The sheath is interrupted at regular intervals by nodes of Ranvier, a tiny gap where the axon membrane is naked.

Microglia

Microglia are small phagocytic immune cells derived from macrophages which serve repair functions and proliferate during inflammatory conditions of the nervous system.

Related topics

Blood–brain barrier (A5)
Resting potentials (B1)
Action potential conduction (B5)
Neurotransmitter inactivation (C7)

Classes of glial cells

As well as neurons, the nervous system contains **glial cells** (Greek: *glia*; glue). These are thought not to be directly involved in information processing but instead perform a variety of supporting functions without which neurons could not operate. Estimates suggest that glial cells outnumber neurons perhaps by as much as 10-fold. It is hardly surprising then that the total cell density in nervous tissue is extremely high and the brain has the lowest extracellular space of any organ in the body. Glial cells are divided into **macroglia** and **microglia**. Several distinct populations of macroglia are recognized; **astrocytes**, **oligodendrocytes** and **Schwann cells**.

Astrocytes

Astrocytes are the largest and most numerous of glial cells. They have irregularly shaped cell bodies and many have long processes which superficially resemble the dendrites of neurons. Astrocytes can readily be distinguished from neurons; they do not have Nissl bodies and can be stained using immunocytochemistry for a specific astrocyte marker, **glial fibrillary acidic protein** (GFAP). Astrocytes fill most of the space between neurons leaving gaps only about 20 nm across. Astrocytes processes surround synapses and some form **endfeet** which butt onto capillaries or onto the pia mater (the innermost layer of meninges, see Topic E5) to produce a layer covering the surface of peripheral nerves and CNS called the **glial membrane.**

Astrocytes have a wide variety of functions:

- When neurons are highly active, significant amounts of K^+ accumulate in the extracellular space. Astrocytes take up the excess K^+ and extrude it in regions with low K^+. Because neighboring astrocytes are coupled to each other via gap junctions, they form interconnected networks able to move K^+ over quite long distances. Much of the excess K^+ is dumped from the endfeet across the glial membrane into capillaries. This spatial buffering of potassium maintains appropriate K^+ concentrations in the vicinity of neurons.
- By virtue of the fact that astrocytes surround synapses, they influence neurotransmission in two ways. First, they serve as a barrier to the diffusion of transmitter away from the synaptic cleft. Second, the astrocyte plasma membrane contains specific transport proteins which enable high affinity reuptake of transmitter from the cleft into the astrocyte. These two processes have opposite effects on the length of time that a transmitter molecule remains in the synaptic cleft.
- Astrocytes may play a role in regulating the supply of glucose to neurons. They have glucose transporters that allow facilitated diffusion of glucose into them where it can be stored as glycogen. It is possible that glucose can be released from astrocytes to supply neurons when they are highly active and require greater quantities of glucose than can be supplied across the blood–brain barrier.
- Astrocyte endfeet on capillaries cause the endothelial cells to form the extremely tight junctions that are the hallmark of the blood–brain barrier (Topic A5).

Oligodendrocytes and Schwann cells

Oligodendrocytes in the CNS and **Schwann cells** in the peripheral nervous system have the common function of providing the **myelin sheath**, an electrically insulating covering around many axons. Those axons with a myelin sheath are said to be **myelinated**, those without are termed **unmyelinated**. The myelin sheath is formed in the peripheral nervous system in the following way. Schwann cells line up along the axon and surround the axon with a pseudopodium-like structure, the **mesaxon**. For unmyelinated axons the process stops at this point. For myelinated axons the mesaxon spirals around the axon some 8–12 times. During this ensheathing most of the cytoplasm gets left behind (except for the innermost layer) so the majority of layers simply consists of a double thickness of plasma membrane (see *Fig. 1*). Each Schwann cell myelinates between 0.15 and 1.5 mm of axon. In general, the thicker the axon the longer the region myelinated by a single glial cell. Between adjacent ensheathed regions is a tiny (0.5 μm) gap of naked axon called the **node of Ranvier**. Here the axon membrane is directly exposed to the extracellular space. Since a peripheral nerve may be quite long a few hundred Schwann cells might be required to generate the sheath. Myelinated axons vary in total nerve fibre diameter between 3 and 15 μm but across this range the proportion of the diameter contributed by the myelin sheath is roughly constant.

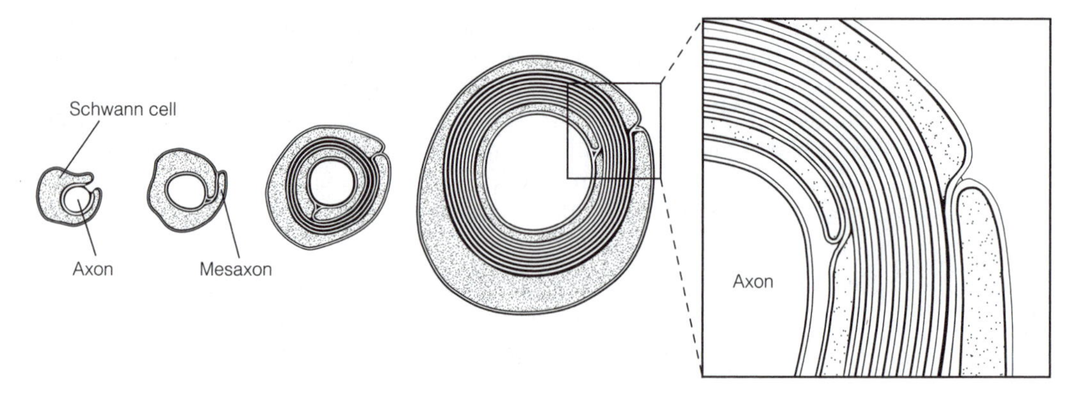

Fig. 1. Myelination of a peripheral axon. The myelin sheath is generated by the growth of the mesaxon which wraps itself around the axon. Redrawn from Gray's Anatomy, *37th edn, 1989, with permission from Harcourt Publishers Ltd.*

Myelination proceeds in a similar way in the CNS except that each oligodendrocyte extends several processes so that it can contribute to the myelination of several adjacent axons. This ensures that fewer glial cells are needed for CNS myelination and, therefore, saves space.

Multiple sclerosis is a progressive disorder in which stretches of myelin sheath, both in the peripheral and central nervous system, are subjected periodically to destruction. It is thought to be an autoimmune disease in which, inappropriately, the immune system mounts an attack on one or more of the proteins of which myelin is made. This results in failure of propagation of action potentials. Another disease which seems to have a related etiology, **Guillain-Barré syndrome**, is confined to damage to the myelin of peripheral sensory and motor neurons and, fortunately, spontaneous recovery is usual in this condition.

Microglia

The smallest of the glial cells are the **microglia.** These are components of the immune system and are derived from macrophages. They are phagocytic, and proliferate in a wide variety of conditions that produce inflammation of the nervous system, including infections, trauma and tumors. Scar tissue formation in the CNS as a result of the activities of microglia is called **gliosis**.

A5 Blood–brain barrier

Key Notes

Structure of the blood–brain barrier

The blood–brain barrier is formed by capillary endothelial cells, which are coupled by tight junctions with extraordinarily high electrical resistance. Astrocytes are required to produce the barrier. A few regions of the brain, the circumventricular organs, do not have a blood–brain barrier and are able to secrete substances directly into the blood, or monitor the concentrations of materials in the blood. These regions are isolated from the rest of the brain by tanycytes, which are coupled together by tight junctions.

Functions of the blood–brain barrier

The blood–brain barrier is a highly selective permeability barrier which allows the passage of water, some gases, and lipid soluble molecules by passive diffusion, and contains specific carrier-mediated transporters for the selective transport of molecules crucial to neural function (such as glucose and amino acids). It prevents the entry of circulating neuroactive compounds and is able to exclude lipophilic, potential neurotoxins via an active transport mechanism mediated by P-glycoprotein. Cerebral edema is the accumulation of excess water in the extracellular space of the brain, and results when hypoxia causes the blood–brain barrier to open.

Related topics

Organization of the central nervous system (E2)
Meninges and cerebrospinal fluid (E5)
Posterior pituitary function (M2)

Structure of the blood–brain barrier

The blood–brain barrier governs strictly what is allowed to cross into the brain extracellular fluid from the blood. The physical barrier is provided by brain capillary **endothelial cells** which are coupled to each other by tight junctions with a very high electrical resistance (at least 1000 Ohm cm^{-2}; some 100-fold higher than is typically the case for tight junctions in other capillaries). This means that even small ions will not permeate between endothelial cells in brain capillaries. Furthermore, brain capillary endothelial cells have a relative lack of two major transport mechanisms possessed by other endothelial cells: **pinocytotic vesicles**, which normally allow the bulk transfer of fluid across the cell, and **receptor-mediated endocytosis**, by which a variety of substrates, for example lipoproteins, are normally specifically transported. Brain capillaries are entirely covered by the endfeet of neighboring astrocytes which secrete as yet undefined factors which promote the formation of the very tight junctions between the endothelial cells (*Fig. 1*).

In a few regions of the brain the capillaries are fenestrated, so there is no blood–brain barrier. These regions are the **circumventricular organs** (CVOs; including the posterior pituitary and choroid plexus) situated around the ventricles of the brain. The location of CVOs is shown in *Fig. 1*, Topic M2. These areas are isolated from the rest of the brain by specialized **ependymal cells** (epithelial cells lining the ventricles) called **tanycytes**. Tanycytes are coupled together by tight

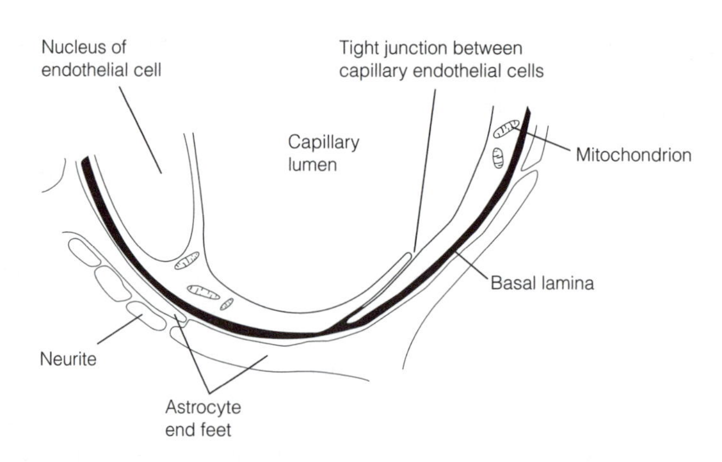

Fig. 1. Structural features of the blood–brain barrier. The barrier is made by the tight junctions between the endothelial cells.

junctions effectively sealing off the CVOs from the rest of the brain. The lack of a blood–brain barrier at the posterior pituitary permits oxytocin and vasopressin to be secreted directly into the systemic circulation, and at other sites it allows the brain to monitor the concentrations of water, ions and selected molecules for homeostatic functions. The choroid plexus is dealt with elsewhere (Topic E5).

Functions of the blood–brain barrier

The plasma membranes of endothelial cells, like those of any cell, consist of phospholipid bilayers into which are inserted numerous protein species. The lipid component will exclude ions or charged molecules and all but the smallest of polar molecules. Only water, gases which are water or lipid soluble (e.g. O_2 or volatile general anesthetics respectively) and lipophilic molecules (e.g. steroids) will be permeant to any extent. The transport of ions, charged or polar molecules must be by carrier-mediated mechanisms. Many of the proteins in the endothelial cell plasma membrane are transporters or ion channels serving this function.

By means of its selective permeability, the blood–brain barrier protects neurons from the actions of neuroactive molecules in the blood, such as circulating catecholamines or glutamate, and ensures that crucial molecules (e.g., glucose and amino acids) are taken across into the brain. The blood–brain barrier is able to actively exclude a wide range of lipophilic compounds that are potentially neurotoxic, many of which are ingested as part of a natural diet. This is achieved by a transport protein, **P-glycoprotein**, expressed in high levels in the plasma membrane of endothelial cells. Lipophilic toxins that diffuse into the endothelial cell are rapidly pumped back out into the blood, by P-glycoprotein. Unfortunately, many brain tumors also express P-glycoprotein and so are able to exclude a variety of chemically unrelated chemotherapeutic agents, a phenomenon known as **multi-drug resistance**. It explains why chemotherapy is often not successful in treating brain tumors.

The blood–brain barrier opens in cerebral ischemia causing **cytotoxic cerebral edema**, which is a medical emergency. The lack of oxygen causes a decline in endothelial cell ATP which compromises the function of the Na^+/K^+-ATPase (see *Instant Notes in Biochemistry*, 2nd edn). Consequently, Na^+ accumulates inside the cell and water then enters osmotically so that the cell swells; this causes tight junctions to open allowing influx of ions and water into the brain extracellular space.

B1 Resting potentials

Key Notes

Excitable properties

When stimulated, excitable cells are able to produce action potentials – brief reversals of the electrical potential across their plasma membrane. Excitable cells include neurons and muscle cells.

Intracellular recording

This is a technique for measuring transmembrane potentials. It uses a fine, electrolyte-filled glass microelectrode to impale the cell. The microelectrode output goes to an amplifier, is compared with a reference electrode signal, and then sent to a cathode ray oscilloscope or computer for display, storage and analysis.

Resting potentials

The resting potential is the voltage across the plasma membrane of an unstimulated excitable cell. All transmembrane potentials are expressed as inside relative to outside. Resting potentials are inside-negative, and range from about –65 mV to –90 mV in neurons. The resting potential is caused largely by the tendency for potassium ions to leak down their concentration gradient, so unmasking a tiny excess of negative charge on the inside of the cell membrane. Other ions (e.g. sodium) make a small contribution to the resting potential. The electrochemical force tending to drive an ion across a membrane is the difference between the resting potential and the equilibrium potential for the ion. The equilibrium potential for an ion is the potential at which there is no net flow of ions across a membrane. Equilibrium potentials can be calculated for individual ions using the Nernst equation. Resting potentials can be calculated from the Goldman equation that takes account of all contributing ions.

Related topics Action potentials (B2) Fast neurotransmission (C2)

Excitable properties

A modest difference in electrical potential exists across the plasma membrane that surrounds all cells. A few cell types are said to be **excitable** because they are able, when suitably stimulated, to generate rapid, brief, alterations in this voltage that can be actively propagated over the cells' surface, an event termed an **action potential**. Excitable cells include neurons, skeletal, cardiac and smooth muscle cells, some endocrine cells (for example, the insulin secreting B cells) and (for a brief time) the plasma membrane of some oocytes. The transmembrane potential that exists across an excitable cell when in an unstimulated state is called the **resting potential**.

Intracellular recording

Being able to measure directly resting potentials, action potentials and the other potentials that occur in nerve and muscle cells is crucial for understanding how these cells work and interact. A standard technique for measuring transmembrane potentials in individual cells is **intracellular recording**.

To record the potential difference across a membrane it is necessary to have two electrodes, one inside the cell, the other outside, both connected to a voltmeter of some description (*Fig. 1*). Because neurons are small the tip of the intracellular electrode impaling the cell needs to be very fine. To achieve this, glass micropipettes are manufactured to have a tip diameter of less than 1 μm. The micropipette is filled with an electrolyte (commonly KCl at a concentration between 0.15 and 3 M) to carry the current, so forming the **microelectrode**. Typically transmembrane potentials are less than 0.1 V and so must be amplified with an **operational amplifier**. This has inputs from both the intracellular microelectrode that impales the cell and the **reference** (**bath** or **indifferent**) electrode, which is placed in the solution bathing the cell. If no potential difference exists between the microelectrode and the reference electrode, the amplifier output will be zero. If a potential difference exists between the electrodes, however, the amplifier generates a signal, the magnitude of which is proportional to the potential. The output of the amplifier goes to a suitable recording device, traditionally a cathode ray **oscilloscope**, but nowadays it is likely be the analog-to-digital port of a computer running software which emulates an oscilloscope and which allows display, storage and analysis of data.

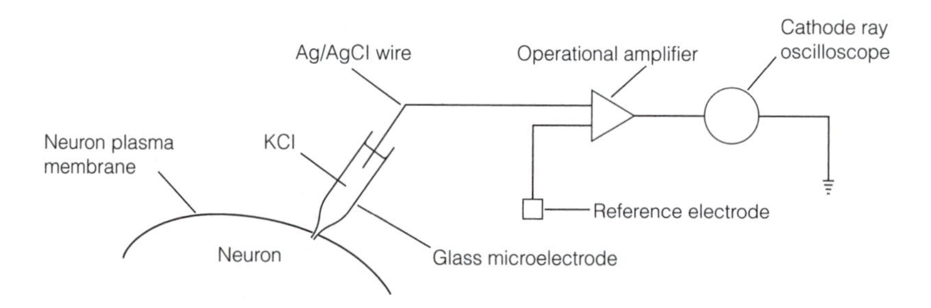

Fig. 1. The circuitry used for intracellular recording.

Resting potentials

Resting potentials (V_m) arise because there is a difference in the concentrations of ions between the inside and outside of the cell and because the cell membrane has different permeabilities for these ions. *Table 1* gives values for concentrations and relative permeabilities for the ions that are most important in relation to the resting potential.

The extracellular fluid that bathes cells is essentially a dilute solution of sodium chloride. The intracellular solution, in contrast, has quite a high concentration of potassium ions that are balanced by a variety of anions to which the cell membrane is completely impermeable (although not listed individually in *Table 1*, these include organic acids, sulfates, phosphates, some amino acids and some proteins). The cell membrane is permeable to K^+ and because there is a

Table 1. Ionic concentrations across mammalian membranes (mmol l^{-1})

Ion	Extracellular fluid	Axoplasm
K^+	2.5	115
Na^+	145	14
Cl^-	90	6

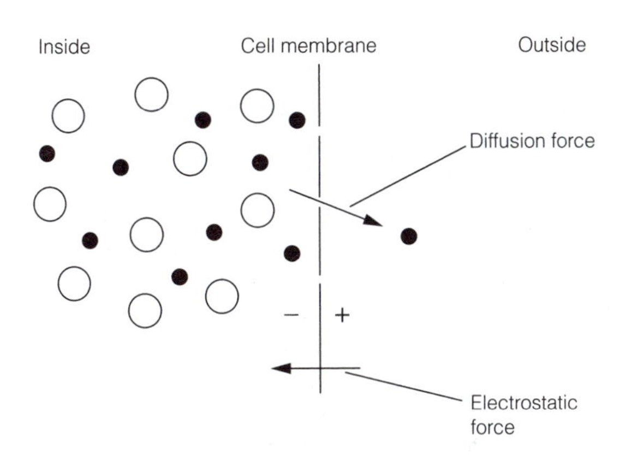

Fig. 2. Illustration of how a potassium equilibrium potential is formed. A small potential exists across the membrane when the diffusional force equals the electrostatic force. Small filled circles represent K^+ ions, large open circles represent anions.

concentration gradient for K^+ across the membrane there is a **diffusional force** acting to drive the K^+ from the inside to the outside of the cell (*Fig. 2*). However the cell membrane is completely impermeable to the much larger anions which therefore remain inside the cell. As the potassium diffuses out a potential difference forms across the membrane because some of the intracellular anions are no longer neutralized by K^+. The potential difference now means that an attractive **coulombic (electrostatic) force** is generated which acts to prevent potassium ions diffusing out. At some point, the diffusional force driving K^+ out is exactly balanced by the electrostatic force preventing K^+ leaving.

At this equilibrium a small potential difference exists, termed an **equilibrium potential** because at this potential there is no net flow of K^+ ions across the membrane. In the case where the potential arises as a result of the distribution of diffusible K^+ it is called a potassium equilibrium potential (E_K). Typically nerve cells have potassium equilibrium potentials around –90 mV. Three important points should be noted:

- Transmembrane potentials are always quoted as inside relative to the outside, which is taken to be zero. So, $E_K = -90$ mV means that the inside of the cell is negative with respect to the outside.
- The number of ions which migrate across the membrane to establish an equilibrium potential is extremely small.
- The potential difference exists only at the plasma membrane, which by storing charge acts as a **capacitor**.

Equilibrium potentials can be calculated using the **Nernst** equation:

$$E = (RT/zF) \ln C_e/C_i$$

where R is the universal gas constant, T is absolute temperature, z is the oxidation state of the ion, F is Faraday's constant and C_e and C_i are extracellular and intracellular concentrations respectively of the ion.

The potassium equilibrium potential is close to the resting potential (V_m) for excitable cells. This suggests that V_m arises largely as a result of the distribution of potassium ions across the cell membrane. Neuron resting potentials range between –65 mV and –80 mV. The discrepancy between E_K and V_m arises because ions other than potassium also make a contribution by virtue of their equilibrium potentials. The most important is sodium (E_{Na} = +55 mV) but since the relative permeability of Na^+ is low it exerts only a modest influence. The effect of sodium is to drag the resting potential away from E_K towards E_{Na} by an amount that reflects the relative permeabilities of the two ions. The difference between the resting potential and the equilibrium potential for any ion, $V_m - E_{ion}$, is termed the **ionic driving force** and is a measure of the **electrochemical force** tending to drive the ion across the cell membrane. At rest the driving force for K^+ is quite small but that for Na^+ is high.

In most excitable cells the ionic driving force for chloride ions is close to zero (that is $E_{Cl} = V_m$). This is because Cl^- ions are passively distributed across the membrane according to the resting potential set up by the combined effects of E_K and E_{Na}. The reason for Cl^- being passively distributed whilst K^+ and Na^+ directly determine the resting potential is because the resting concentration gradients for potassium and sodium are actively maintained by the actions of Na^+/K^+-ATPase, whereas their are no active transport mechanisms acting to maintain a fixed Cl^- concentration gradient.

The resting membrane potential can be calculated by the **Goldman equation**, which takes account of the concentration ratios and relative permeabilities (P) of K^+, Na^+ and Cl^-:

$$E = (RT/zF)\ln \frac{P_K[K^+]_e + P_{Na}[Na^+]_e + P_{Cl}[Cl^-]_i}{P_K[K^+]_i + P_{Na}[Na^+]_i + P_{Cl}[Cl^-]_e}$$

B2 Action potentials

Key Notes

Stimulating neurons

Neurons can be stimulated by using a stimulator which delivers a current to the cell via a microelectrode. The current usually has a square waveform, the frequency, amplitude and pulse width of which can be varied independently. Inward currents cause neurons to depolarize (i.e. the membrane potential becomes smaller) whereas outward currents cause neurons to hyperpolarize.

Action potentials

An action potential or nerve impulse is a short-lived reversal of the membrane potential. The spike lasts less than 1 ms and peaks at about +30 mV. The after-hyperpolarization that follows lasts a few milliseconds.

Action potential properties

Action potentials are triggered at the axon hillock and propagated along the axon. They obey the all-or-none rule: a stimulus must be sufficiently large to depolarize a neuron beyond a threshold voltage before it will fire; all action potentials in a given cell are the same size. There is a short delay, the latent period, between the onset of the stimulus and the beginning of the action potential. Neurons become completely inexcitable to further stimulation during the spike and harder to excite during the after-hyperpolarization. These constitute the absolute and relative refractory periods respectively. Refractory periods limit the maximum rate at which neurons can fire, prevent action potentials from summating, and restrict action potentials to propagation in one direction only along an axon.

Related topics

Resting potentials (B1)
Voltage-dependent ion channels (B3)
Action potential conduction (B5)
Fast neurotransmission (C2)
Properties of neurites (D1)

Stimulating neurons

In vivo, neurons are excited either by the cascade of synaptic inputs onto their dendrites and cell body from other neurons or by receptor potentials generated by sensory organs. Neurophysiologists often stimulate a neuron directly by injecting an electrical current into it via a stimulating microelectrode. The stimulator normally delivers a square wave current pulse. Three variables can be altered at will on most **stimulators**: the **duration** of the pulse, the **amplitude** of the injected current, and the **frequency** of the pulses. The direction of the current (which is defined as the flow of positive charge) determines the response of the neuron. If a small **inward current** is injected into a cell it will become a little more inside-positive. This is a decrease in the membrane potential because V_m gets closer to zero and is called a **depolarization**. If, on the other hand, an **outward current** is injected (that is, if current is withdrawn from the cell) then the membrane potential increases; this is called **hyperpolarization**. The sizes and timecourses of depolarizing and hyperpolarizing potentials seen in nerve cells injected with small currents are determined solely by the passive electrical properties of the neuron.

Action potentials

If a sufficiently large inward current is injected into a neuron its membrane potential will depolarize enough to generate an action potential (nerve impulse). This is defined as a brief reversal of the transmembrane potential which is propagated over the surface of the cell. Intracellular recording of a neuronal action potential shows (see *Fig. 1*) that the membrane potential rapidly depolarizes to zero, overshoots to about +30 mV then repolarizes back towards V_m all in less than 1 ms. This constitutes the **spike** of the action potential. Immediately after the spike the neuron membrane hyperpolarizes. This **after-hyperpolarization** lasts for a few milliseconds, decaying as the membrane potential returns to its resting value.

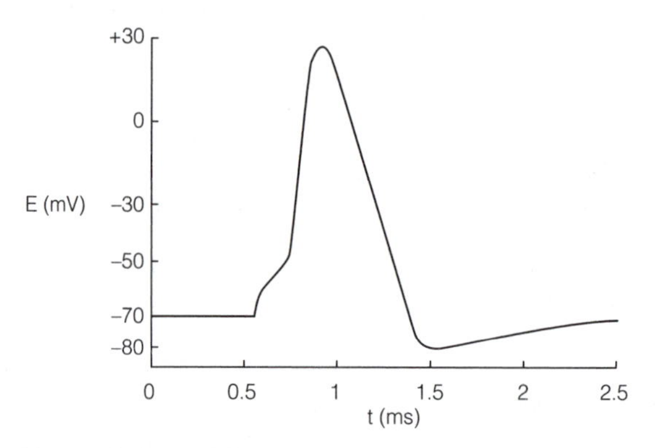

Fig. 1. An intracellular recording of a neuron action potential. The resting membrane potential is –70 mV.

Action potential properties

Action potentials have a number of important properties:

- Under physiological conditions action potentials are triggered at the axon hillock (initial segment) and are propagated along the axon towards its terminals.
- Action potentials are threshold phenomena. A minimum size stimulus current is required to produce an action potential. This is the **threshold stimulus**, and is defined as the current that will cause a neuron to fire on 50% of the occasions it is used. Stimuli smaller than this are called **subthreshold**, those which are larger are termed **suprathreshold**. The crucial thing about a threshold stimulus is that it causes the neuron to depolarize to a critical **threshold voltage**. The size of the threshold stimulus depends on the size of the neuron. In most neurons the threshold voltage is some 15 mV less than V_m. Action potentials are triggered at the axon hillock because this region of the neuron has the lowest threshold.
- All action potentials are about the same size (in a given cell) regardless of the amplitude of the stimulus. The spike of the action potential contains no clue as to the size of the stimulus that produced it. The combined effects of this and the previous property are often paraphrased as the **all-or-none** rule: a neuron either fires or it does not.
- There is a short delay between the onset of the stimulus and the start of the action potential. This is called the **latency** (or **latent period**). The latency gets shorter as the strength of the stimulus increases.

- During the spike a neuron becomes completely inexcitable. This is the **absolute refractory period**, during which time a nerve cell will not fire again no matter how large the stimulus. After the spike, while the neuron remains hyperpolarized, the neuron can be excited only by suprathreshold stimuli. This period is the **relative refractory period** and is explained by the fact that when the cell is hyperpolarized larger stimuli are needed to drive the membrane potential to the threshold voltage. That neurons are temporarily refractory to stimuli during an action potential has three important consequences for nerve cell function. Firstly, it imposes a limit on the maximum rate at which a neuron can fire. Secondly, it means that a second action potential cannot be superimposed on the first; technically put, action potentials do not summate (see synaptic potentials, Topic C2). Thirdly, because the region of neuron membrane that has just fired an action potential is inexcitable, the action potential will not reinvade it and so must be propagated forwards. In other words, action potentials are normally propagated in only one direction along an axon. Refractory periods can be explained by the behavior of the ion channels responsible for the action potential (Topic B3).

B3 Voltage-dependent ion channels

Key Notes

Voltage-dependent ion channels

Voltage-dependent ion channels are transmembrane proteins that are ion selective and voltage sensitive. They are often named for the ion species to which they are most permeable. They can exist in at least two interchangeable states, open or closed, depending on the membrane potential across them.

Voltage-dependent sodium channels

Sodium channels are transmembrane glycoproteins found in most excitable cells. Normally closed, they are opened by depolarization of the membrane potential beyond threshold, allowing sodium ions to permeate into the cell. This causes the depolarization phase of the action potential. After about 0.5–1 ms the channels flip into an inactivated state in which sodium can no longer permeate. This, together with the low ionic driving force for sodium at positive membrane potentials curtails the spike amplitude. Sodium channel inactivation accounts for the absolute refractory period.

Voltage-dependent potassium channels

The delayed outward rectifying potassium channel is responsible for the downstroke of the action potential spike and the subsequent after-hyperpolarization. These transmembrane channels are activated by depolarization, permitting potassium ions to flow out of the cell, which carries the membrane potential to increasingly negative values. The after-hyperpolarization is responsible for the relative refractory period.

Voltage clamping

Voltage clamping is a technique that allows the current that flows across nerve cell membranes to be measured. It does this by holding the membrane potential constant. It provided the crucial evidence that action potentials are caused by an early inward current due to sodium and a late outward current due to potassium. These currents can be blocked independently by tetradotoxin and tetraethylammonium respectively.

Related topics

Resting potentials (B1)
Action potentials (B2)
Molecular biology of sodium and potassium channels (B4)

Voltage-dependent ion channels

Neurons are excitable, i.e. they can produce action potentials, because they have in their plasma membrane a class of transmembrane proteins called voltage-dependent ion channels. They are so called because they have two properties, **ion selectivity** and **voltage sensitivity**. Ion channels will allow only certain small ions to flow through them. Voltage-dependent channels are preferentially selective for one of three ions, Na^+, K^+ or Ca^{2+}. While there are over 30 individual voltage-dependent ion channels, they are usually classified according to the ion for which they are permeable (for example six distinct voltage-dependent sodium channels have been identified to date).

Voltage-dependent ion channels can exist in at least two interchangeable states: **open** (**activated**), when they allow ions to flow through them, or **closed**, when they do not. Whether they are open or closed depends on the voltage across them.

Voltage-dependent sodium channels

Voltage-dependent sodium channels (VDSCs) are large glycoproteins that span the full thickness of the plasma membrane of most excitable cells. At the resting potential they are closed. If a region of membrane is depolarized by only a modest amount (e.g. 10 mV) they remain closed (*Fig. 1a*). However, if the membrane is depolarized to the threshold voltage or beyond, VDSCs change shape so that the channels open, permitting Na^+ ion to flow through. Channel opening, **activation** (*Fig. 1b*), is an extremely rapid event: the transition from the closed to the open state takes only about 10 µs. During an action potential in a neuron a given sodium channel will remain open for about 0.5 to 1 ms. In this time about 6000 Na^+ ions will flow through the channel. The combined effect of sodium influx through a few hundred VDSCs will produce the upstroke of the spike that is the depolarizing phase of the action potential. Only a few voltage-dependent sodium channels need be driven beyond threshold initially to trigger an action potential as the localized influx of Na^+ causes a depolarization which drives other channels to open and so on. It is this **self-regeneration** property of the action potential that causes the explosive rise in Na^+ permeability.

At the top of the spike the increase in sodium permeability is halted for three reasons:

- All of the available VDSCs in the active region of membrane have opened.
- The ionic driving force of sodium becomes smaller as the membrane depolarizes towards the sodium equilibrium potential.
- The voltage-dependent sodium channels flip into the **inactivated** state.

During this state the channel does not permit the flow of any ions; but this is not the same as the closed state, because while in the inactivated state a sodium channel cannot be made to open. In addition to curtailing the spike, it is this inactivation of VDSCs that is responsible for the absolute refractory period. The inactivation wears off after a few milliseconds and the sodium channel enters the closed state from which it can be reactivated by subsequent depolarization.

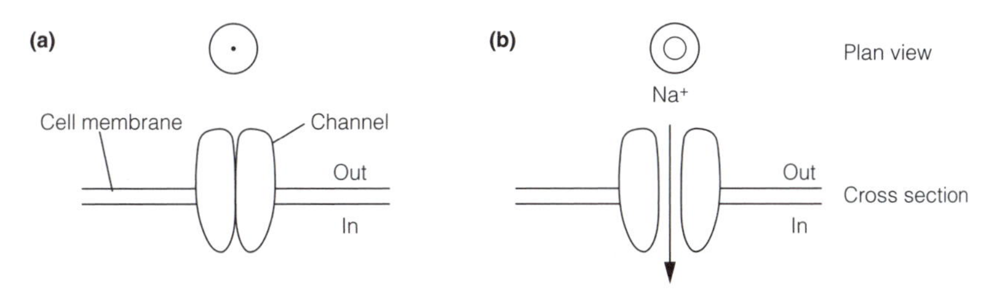

Fig. 1. Behavior of a voltage-dependent sodium channel (a) at rest when it is in the closed state and (b) during the spike of an action potential, when it is activated.

Voltage-dependent potassium channels

Excitable cells have a great variety of distinct types of potassium channel each with their own particular properties. One type, the **delayed outward rectifier** is involved in the repolarization phase of the action potential. These channels are transmembrane glycoproteins with a molecular structure related to that of voltage-dependent sodium channels; like them, they are opened by depolarization, which allows potassium ions to flow down their concentration gradient out of the cell. The inside of the neuron becomes less positive, that is, it repolarizes. This accounts for the downstroke of the spike of the action potential. At the bottom of the downstroke most of the VDSCs are inactivated so there is no flow of sodium into the cell. Delayed outward rectifying voltage-dependent potassium channels (VDKCs) either do not inactivate or inactivate much more slowly than VDSCs (depending on species) so immediately after the spike the neuron remains highly permeable to K^+ whilst being impermeable to Na^+. The consequence is that for a few milliseconds after the spike, potassium ions continue to leave the cell carrying the membrane potential more negative than V_m. This is the **after-hyperpolarization** phase of the action potential. It accounts for the relative refractory period, because during this time, for a stimulus to excite the cell to threshold it must make the neuron depolarize from a more inside-negative state. Finally the membrane potential returns to the resting potential as the potassium channels flip into their closed or inactivated state in a time-dependent manner. The changes in ion conductance during the action potential are summarized in *Fig. 2*.

Two points follow from the above:

- The extent of the hyperpolarization is determined by the potassium equilibrium potential. When the efflux of K^+ is sufficient to bring the membrane potential to E_K the ionic driving force on the potassium ions will be zero. No more K^+ will leave.
- Since depolarization of the neuron to threshold causes the opening of VDSCs (allowing Na^+ in) and opening of VDKCs (allowing K^+ out) how does an action potential arise? The reason is that the sodium channels respond to the depolarization earlier than the potassium channels so the increase in sodium permeability precedes the increase in potassium permeability.

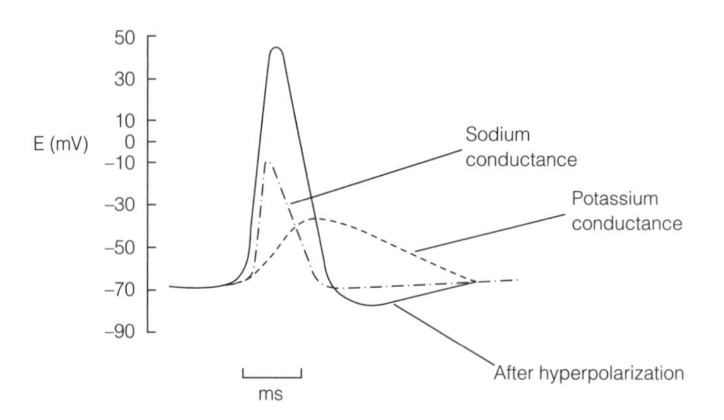

Fig. 2. Changes in ion conductance during the action potential.

Voltage clamping

The ion fluxes that underlie neuron action potentials were discovered in the early 1950s using a technique called voltage clamping, which remains a key technique in electrophysiology. It measures the currents that flow across an excitable cell membrane at a fixed potential. Measuring currents is important because it provides information on which ions might be responsible for changes in membrane potential. Current (I) cannot directly be found from potential (V): it requires, in addition, the membrane resistance (R). Only if both V and R are known can I be calculated from Ohm's law, $V = IR$. Voltage clamping circumvents this problem by measuring the transmembrane potential and having a **feedback amplifier** in the circuit which injects into the cell the current which is needed to keep the membrane potential constant, i.e. the voltage is clamped. The current that must be injected by the circuit to keep the voltage fixed has to be the same size as the current flowing through the ion channels which would normally cause the potential to change. The voltage at which the membrane is clamped is called the **command voltage**. By examining the currents that flow across a membrane over a range of command voltages it is possible to determine which ions carry the currents.

The use of voltage clamping is illustrated in *Fig. 3*, which shows an experiment in which the giant axon of a squid is initially clamped at its resting potential of –60 mV and the command voltage is then switched to zero. This gives rise to a **capacitance current**, I_c. This arises because the neuron plasma membrane is an insulator (a phospholipid bilayer), sandwiched between two conductors (the extracellular fluid and the intracellular fluid), and so acts as a capacitor. Capacitors have the property that they store charge in proportion to the potential difference across them, that is: $q \propto V$, i.e. $q = CV$, where the constant of proportionality is the capacitance, C. Capacitance is determined by the area of the capacitor and the distance between the conductors: C increases as the separation of charges decreases. Neuron plasma membranes are so thin that their capacitance is very high (about 1 $\mu F\ cm^{-2}$). When the voltage across the membrane changes, the amount of stored charge must change and this gives rise to a current flow, the capacitance current.

After the capacitance current comes an early inward current followed by a late outward current. These are the currents that normally flow during an action potential (*Fig. 3*, total current). When the squid axon is bathed in a solution that contains no sodium the early inward current is abolished. The same result is seen if an axon immersed in normal seawater is poisoned with the neurotoxin **tetradotoxin** (TTX). By binding to the external mouth of the VDSC, TTX prevents Na^+ from permeating through the channel. Since sodium carries the early inward current, TTX added to any nerve preparation abolishes action potentials. Similarly, **tetraethylammonium** (TEA) which blocks VDKCs, when added to the bathing medium abolishes the late outward current, showing that it is carried by K^+.

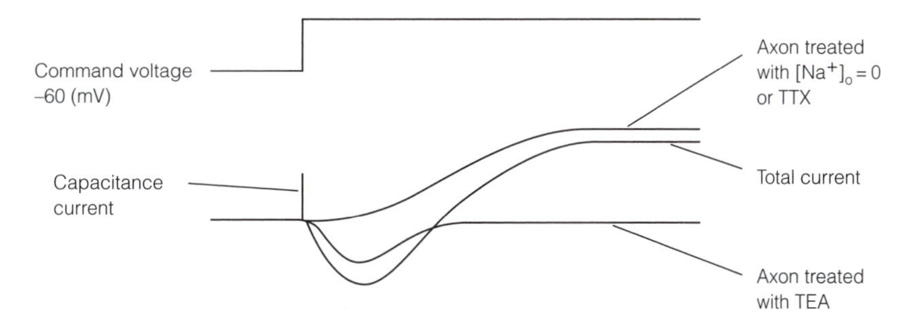

Fig. 3. A voltage clamping experiment to show the currents that flow during an action potential. Outward currents are upward deflections, inward currents downward deflections. Initially the neuron is clamped at –60 mV, and the step to 0 mV generates first a capacitance current, then the ionic currents. The experiment is done with the axon under three different conditions. See text for details. TTX, Tetrodotoxin; TEA, tetraethylammonium.

B4 Molecular biology of sodium and potassium channels

Key Notes

Patch clamping

Patch clamping measures the currents that flow through single ion channels. It depends upon electrically isolating the tiny patch of membrane that lies under the tip of a microelectrode. In the cell-attached mode the patch remains on the cell and single channel currents are recorded. Rupturing the patch gives the whole cell mode, which allows recording of the macroscopic currents that flow through the plasma membrane of the cell. Alternatively the patch can be completely removed to give two other single channel configurations. In outside-out mode the effects of channel ligands applied to the bath can be studied. In inside-out mode agents can be added to the bath to investigate the role of second messengers in modulating channel activity.

Structure of voltage-dependent sodium channels

Cloning and sequencing the DNA encoding voltage-dependent sodium channels (VDSCs) has permitted the primary sequence of these large transmembrane glycoproteins to be deduced, from which much about secondary structure can be inferred. These channels consist of four homologous domains each with six α-helical segments (S1–S6) that traverse the membrane. The positively charged S4 segments are involved in activation. Each domain contains an H5 loop interposed between segments S5 and S6. This loop is thought to line the channel pore. The third cytoplasmic loop linking domains three and four is required for inactivation. The tertiary structure of the channel is formed by the four domains clustering around a central pore.

Structure of voltage-dependent potassium channels

The delayed outward rectifier channels involved in the action potential and a number of related potassium channels are homologous with a single domain of the sodium channel. Functional channels are presumed to be tetrameric homo-oligomers. Channel inactivation (in those channels that do inactivate) is by a ball and chain mechanism. In channels from the fruit fly, *Drosophila melanogaster*, a cluster of amino acids at the N terminus blocks the internal mouth of the channel. In mammalian channels the block is produced by a separate β-subunit associated with the N terminus.

Related topics

Voltage-dependent ion channels (B3)
Voltage-dependent calcium channels (C6)

Patch clamping

Patch clamping is a technique that makes it possible to study the electrophysiology of single ion channels. It works by forming a very high electrical resistance seal between a glass micropipette and the surface of a cell. Only currents flowing through the patch under the electrode will be recorded. This permits the extremely small currents that flow through single ion channels (about 1 pA) to be measured. The electronics allow the voltage of the patch to be clamped so voltage clamping experiments can be performed (*Fig. 1a*).

There are several configurations of patch clamping, each useful for particular types of experiment (*Fig. 1b*).

(1) **Cell attached**. This is used for measuring single channel currents in intact cells. Second messenger-induced modifications of the patched channels can be investigated in response to bathing the cell with specific agents, e.g. neurotransmitters.

(2) **Whole cell**. In this configuration the patch under the microelectrode is ruptured. The current flowing through the electrode represents the sum of all the currents flowing through individual ion channels on the cell surface. Hence whole cell patching measures **macroscopic currents**.

(3) **Outside-out**. This is one of two patch clamp modes used to study single channel currents in which the patch is removed from the cell but remains sealed to the pipette tip. This configuration is used to study the effects of ligands such as neurotransmitters, hormones or externally acting drugs on channels. These ligands are added to the bath because bath solutions can be much more easily and rapidly changed than the pipette solution. This has obvious advantages for performing complicated experiments such as investigating dose–response relationships.

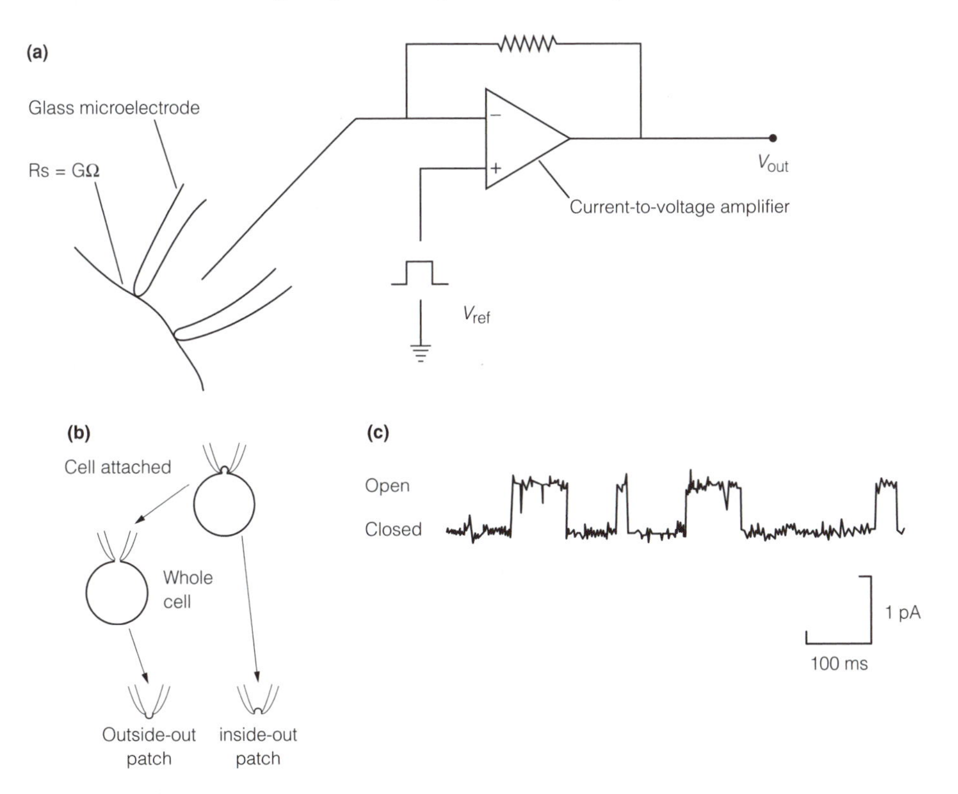

Fig. 1. Patch clamping. (a) The circuitry, V_{ref} is used to clamp the potential of the neuron membrane, the method relies on forming a high resistance seal, R_s, between the pipette tip and the cell membrane. (b) Patch clamp configurations. (c) Single channel currents through a $GABA_A$ receptor.

(4) **Inside-out**. This is the second of the patch only configurations, usually used for the detailed examination of second messengers because these can be applied directly to the inside face of the membrane via the bath solution.

A representative patch clamp recording is shown in *Fig. 1c*. Each of the square wave events is the opening of a single channel. The height of the wave is the **unitary channel current**, its length the time for which it is open. Statistical analysis of large numbers of opening events provides estimates for parameters such as **mean channel open time**, and allows models of channel kinetics to be tested. Such studies are very useful in deducing how neuroactive drugs act at the molecular level. Patch clamping is also proving to be invaluable in verifying the outcome of genetic engineering experiments to clone specific ion channels and receptors.

Structure of voltage-dependent sodium channels

By cloning and sequencing the DNA that encodes a protein it is possible to deduce the primary amino acid sequence of that protein. This provides clues as to the secondary structure of the protein such as the presence of α-helical or β-pleated sheet regions. **Hydropathy plots**, which represent how hydrophobic or hydrophilic particular amino acid stretches are, indicate which regions of the protein might be located in the plasma membrane. Recognizing consensus sequences for glycosylation suggest which parts of a molecule might be extracellular; similarly consensus sequences for phosphorylation imply an intracellular location. Information of this nature provides evidence for how a given protein is orientated in the membrane. **Site-directed mutagenesis**, in which the DNA encoding a protein is altered in precise ways using genetic engineering techniques, and the mutated protein subsequently expressed in some convenient way, can be used to investigate the functions of particular amino acids in a channel. An expression method that has proved very useful involves transcribing the mutant channel DNA *in vitro* and injecting the resulting mRNA into *Xenopus* oocytes. These egg cells will translate the mutant protein and express it in their plasma membrane, where it can be studied by patch clamping.

The structure of voltage-dependent sodium channels (VDSCs) has been deduced using the technologies outlined above. Sodium channels are large transmembrane glycoproteins thought to consist of four domains (I–IV) linked by cytoplasmic loops (*Fig. 2*). The domains are quite similar when judged on amino acid sequence, that is they share a high **homology**. Each domain has six highly hydrophobic segments (S1–S6), about 20 amino acids long, that are thought to be α-helices spanning the membrane. A central pore surrounded by the four domains forms the functional channel.

One of the most striking features of the VDSC is the S4 segment. There is a remarkably high homology between the S4 segments in each domain suggesting that they have hardly altered during the evolution of the channel and so serve an important function. Every third amino acid along the S4 segment is positively charged, either lysine or arginine. Site-directed mutagenesis shows that S4 is required for activation of the channel so it is likely that the positive charges are part of the voltage sensor of the channel.

The large cytoplasmic loop between the third and fourth domains must be crucial for inactivation because injection of proteolytic enzymes intracellularly destroys the ability of VDSCs to inactivate and channels engineered by site-directed mutagenesis to lack the third cytoplasmic loop fail to inactivate.

Between segments S5 and S6 within each domain, is an H5 loop consisting of

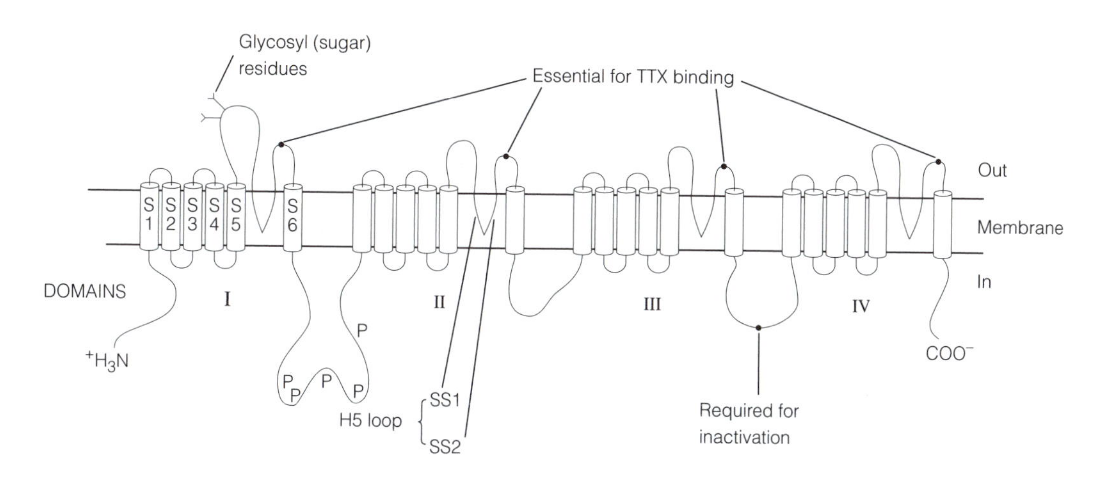

Fig. 2. A cartoon depicting the secondary structure of a VDSC. Segments S1–S6 are labeled only in domain I. The four H5 loops that form the pore are in bold. P, Consensus sequences for phosphorylation.

two short segments called SS1 and SS2. These loops are thought to form the pore by analogy with studies of VDKCs that contain the same motif. In the tertiary structure of the channel all four domains cluster around the central pore which is lined by the four H5 loops.

Structure of voltage-dependent potassium channels

There are numerous types of potassium channel but many, including the delayed outward rectifier that is responsible for the downstroke of the action potential, share a common structure (*Fig. 3*). Potassium channel subunits (α-subunits) closely resemble a single domain of the voltage-dependent sodium channel. A functional channel is a tetrameric homo-oligomer, that is, it consists of four similar subunits arranged around a central pore. Each potassium channel subunit has six transmembrane segments and S4, with its positively charged amino acids, is implicated in activation.

Some potassium channels can be blocked by tetraethylammonium applied either on the outside or inside of the channel. Site-directed mutagenesis has

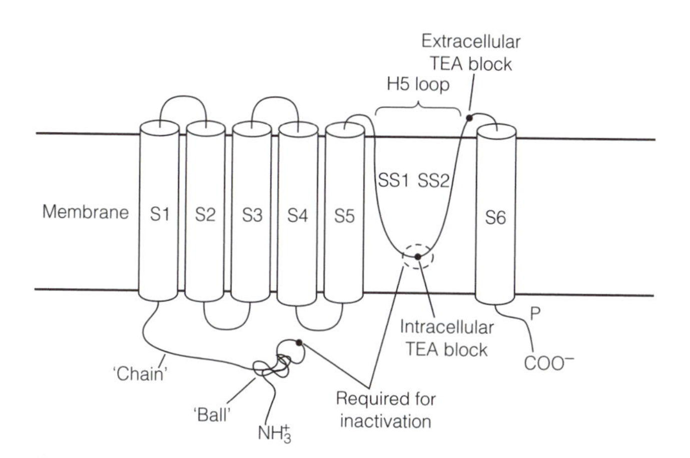

Fig. 3. A cartoon depicting the secondary structure of an outward rectifying potassium channel subunit. Four subunits assemble to form a functioning channel.

shown that the amino acid required for external blockade is only a short distance from the amino acid needed for internal block. This meant that the intervening stretch must span the membrane which is made possible only if a loop exists between segments S5 and S6. This H5 loop is thought to line the channel pore because it contains amino acids that are crucial for conferring potassium ion selectivity on the channel.

Although not all potassium channels involved in action potentials inactivate, many do by the so-called **ball and chain** mechanism. This involves a cluster of amino acids at the intracellular N-terminal end of the molecule swinging up to block the internal mouth of the channel by interacting with an amino acid at the internal tip of the H5 loop. In *Drosophila* channels (the first to be cloned) the ball is part of the channel α-subunit. In mammals the 'ball' is a separate β-subunit associated with the N-terminal end of the α-subunit. *Drosophila* channels engineered to have no N-terminal ball or in which the 'chain' is shortened do not inactivate.

B5 ACTION POTENTIAL CONDUCTION

Key Notes

Propagation of action potentials

Action potentials are generated at the axon hillock (spike initiation zone) and propagate actively, at constant velocity, and without loss of amplitude, down the axon. Because the active zone, the region of the axon at which the action potential sits at a given instant, bears different charges to the axon at rest ahead of it, local circuit currents flow which depolarize the adjacent upstream membrane and so the action potential advances. Local circuit currents also spread backwards but do not allow the action potential to propagate in this direction because the membrane there is refractory.

Conduction velocity in nonmyelinated axons

In nonmyelinated axons the speed of conduction is between 0.5 and $2\,\mathrm{m\,s^{-1}}$. The velocity is proportional to the square root of the axon diameter.

Conduction velocity in myelinated axons

Myelination produces dramatic increases in conduction speed for only modest increases in the overall diameter of axons. Myelinated axons conduct faster because local circuit currents flow around the electrically insulating myelin sheath so that only the axon membrane at the node of Ranvier needs to be depolarized to generate an action potential. The action potential appears to jump from one node to the next. The conduction velocity is proportional to axon diameter and varies from 7 to $100\,\mathrm{m\,s^{-1}}$.

Related topics

Glial cells and myelination (A4) Properties of neurites (D1)

Propagation of action potentials

In a neuron, action potentials are initiated at the axon hillock because this region has the greatest density of voltage-dependent sodium channels (VDSCs) and so the lowest threshold for excitation. For this reason the axon hillock is sometimes referred to as the **spike initiation zone**. Once generated, action potentials are actively propagated (conducted) with constant velocity down the axon without loss of amplitude. Thus, action potentials are undiminished in size even when conducted along peripheral axons that in humans may be up to 1 m in length. This is one of the features that make action potentials reliable signals for information transmission. The details of conduction are a little different depending on whether the neuron is myelinated or not.

In nonmyelinated neurons conduction works as follows (*Fig. 1*). The region of an axon invaded by an action potential at a given time is called the **active zone**. It is a few centimeters long. The part of the active zone occupied by the overshoot of the spike will be inside-positive. Far away from the active zone, ahead of the oncoming action potential or behind it, the membrane potential will be inside-negative. The consequence of this is that a potential difference exists

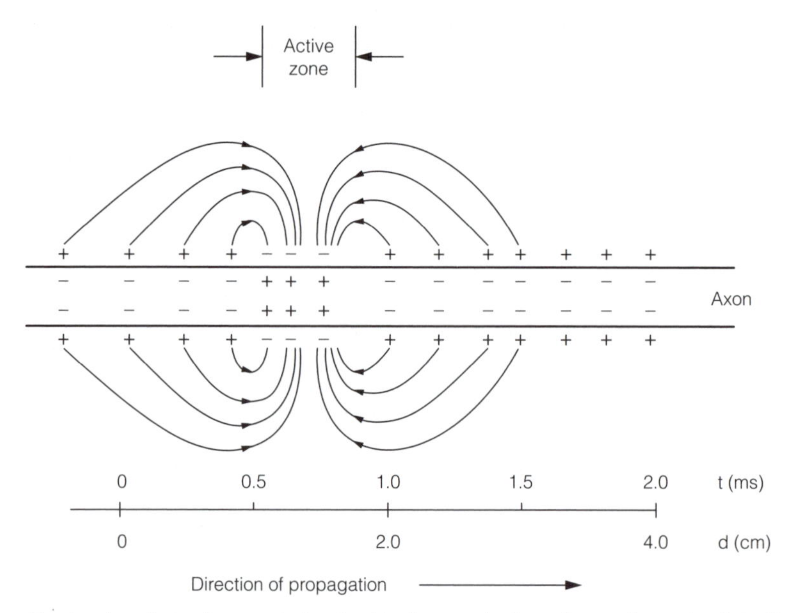

Fig. 1. Local circuit currents involved in the conduction of an action potential. For clarity currents inside the axon are omitted. The action potential is depicted as travelling from left to right along the axon, and the leading edge of the spike (active zone) is 2 cm from the origin (lower scale) after 1 ms (upper scale). t, time; d, distance.

between different regions of the external surface of the axon: the outside of the active zone is more negative than its surroundings. A similar situation exists on the inside surface of the membrane except that here the active zone is more positive than its surroundings. These differences in potential cause currents to flow passively over the axon membrane. By convention, current flows in the positive to negative direction and this is depicted by the arrows in *Fig. 1*.

Current flow across the external surface of the axon is from the regions ahead and behind the action potential into the active zone. These currents are called **local circuit currents**. Just ahead of the axon potential, the local circuit currents drain positive charge from the external surface of the axon and simultaneously dump positive charge on the inside of the axon membrane. The net effect is to depolarize the axon immediately in front of the action potential. When this depolarization becomes suprathreshold, VDSCs in this region activate and the action potential advances. Of course, local circuit currents flow in the same direction along the axon behind the action potential but this region is refractory (VDSCs are inactivated and the membrane is hyperpolarized) and so the currents do not excite here. This explains why action potentials propagate physiologically only in one direction.

Conduction velocity in nonmyelinated axons

The conduction velocity, θ, the speed with which nerve impulses are propagated, is quite slow in nonmyelinated axons. It varies between 0.5 and 2 m s^{-1} depending on the diameter of the axon. Small axons offer a higher resistance to the flow of currents through their cores than large ones, just as thin wires have a higher electrical resistance than thick ones. So, local circuit currents in the axoplasm of small axons spread less well than in larger axons and this is the reason for the slower speed.

Very roughly the relationship is:

$$\theta = ka^{\frac{1}{2}}$$

where a is the axon diameter, k is a constant which depends on the internal resistance of the axon and its membrane capacitance (see Topic D1).

Conduction velocity in myelinated axons

A large number of neurons in the vertebrate nervous system, certainly the majority of those in the peripheral nervous system, have myelinated axons. The function of the myelin sheath is to increase the conduction velocity substantially with relatively little increase in total axon diameter. The evolution of myelination has enabled vertebrates to have a large number of rapidly conducting axons without taking up too much cable space.

Because the myelin sheath consists of plasma membrane, it has a high content of phospholipid with a high electrical resistance. Local circuit currents are forced to take paths of lesser resistance through the electrolyte solution around the sheath. The effect is that local circuits are established, not between adjacent regions of membrane as they are in nonmyelinated axons, but between adjacent nodes of Ranvier, which are relatively far apart. Local circuit currents ahead of an action potential arriving at the next downstream node cause it to depolarize beyond the threshold and trigger an action potential. In this manner action potentials appear to jump from node to node, a mechanism called **saltatory conduction.** The density of VDSCs is about 100-fold greater at nodes than in nonmyelinated axon membrane and the node threshold is consequently much lower. This greatly reduces the risk of nodes not firing in response to local circuit currents weakened by the long distances they must spread.

The rate-limiting factor in determining conduction velocity is not the spread of local circuit currents, which is quite fast, but the time it takes for VDSCs to respond to depolarization. In a nonmyelinated axon each little region of membrane has to be depolarized and respond in succession. For a myelinated axon however only the node membrane needs to be excited. This accounts for the higher speeds of conduction in myelinated axons.

Conduction velocities of myelinated axons vary from about 7 to 100 $m\,s^{-1}$ on average. As with nonmyelinated axons velocity depends on diameter, but the relationship is even simpler:

$$\theta = ka$$

where a is the axon diameter and k is a constant.

C1 OVERVIEW OF SYNAPTIC FUNCTION

Key Notes

Electrical transmission

Electrical synapses are formed by arrays of ion channels, connexons, at gap junctions. Here small ions flow between cells so that they are electrically coupled. Via gap junctions action potentials can spread rapidly between cells without distortion.

Chemical transmission

Neurotransmitter release from the nerve terminal following the arrival of an action potential is triggered by the influx of calcium through voltage-dependent calcium channels. After crossing the cleft, transmitter binds to postsynaptic receptors. These are either ligand-gated ion channels or metabotropic receptors coupled to second messenger systems. Receptor activation either increases or decreases the chance that the postsynaptic cell will fire, responses described as excitatory or inhibitory respectively. Transmission mediated by ligand-gated ion channels is fast, whereas that mediated by metabotropic receptors is slow. Synapses may secrete more than one transmitter.

Related topics

Morphology of chemical synapses (A3)
Fast neurotransmission (C2)
Slow neurotransmission (C3)
Neurotransmitter release (C5)
Neurotransmitter inactivation (C7)

Electrical transmission

Two types of transmission occur in the nervous system, electrical and chemical. Electrical transmission is mediated by electrical synapses – **gap junctions** between adjacent neurons. Gap junctions are arrays of paired hexameric ion channels called **connexons** (*Fig. 1a*). The channel pores are 2–3 nm in diameter, allowing ions and small molecules to permeate between neighboring neurons. By electrically coupling neurons, gap junctions allow any potentials, e.g. action potentials, to spread between cells. Key features of electrical transmission are that it is extremely rapid, it is high fidelity (signals are transmitted with no distortion), and it works in both directions. Gap junctions between cells may close. Each connexon is made up of six subunits called **connexins**. In response to specific chemical signals, such as an increase in intracellular Ca^{2+} concentration, the connexins rotate laterally to close the central pore (*Fig. 1b*). Electrical synapses form only a small proportion of all synapses in adults, but are more numerous during development. Gap junctions are not unique to the nervous system but also commonly couple epithelial and muscle cells.

Chemical transmission

The vast majority of synapses are chemical. At most central synapses chemical neurotransmission happens in the following way: the arrival of an action potential at the axon terminal may result in the release of neurotransmitter from a single presynaptic vesicle. Neurotransmitter release requires a rise in intracellular Ca^{2+} brought about by calcium entry into the axon terminal via **voltage-**

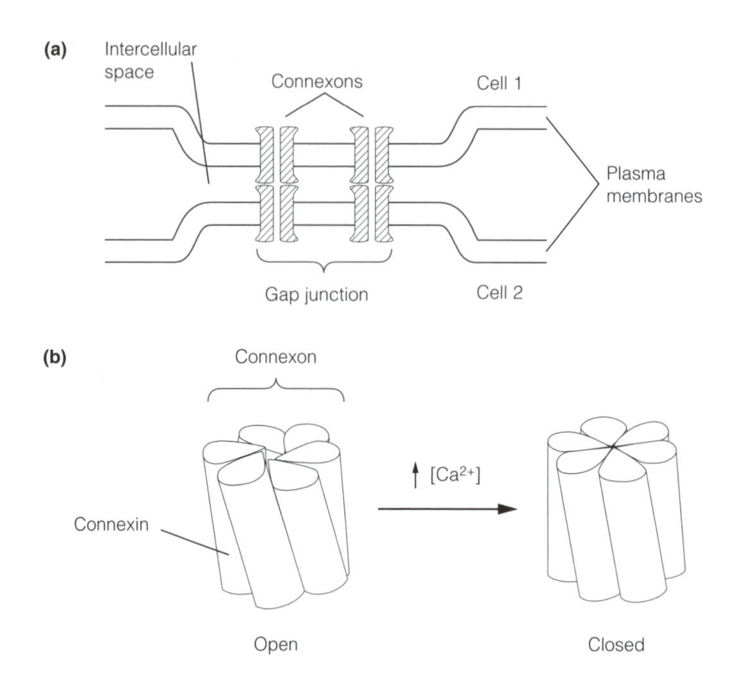

Fig. 1. (a) Gap junction. (b) Change in configuration of the connexons to close a gap junction.

dependent calcium channels. The transmitter diffuses across the synaptic cleft and binds to specific receptors on the postsynaptic membrane. Binding of the transmitter causes a change in the conformation of the receptor. What happens next depends on the receptor, but the overall result is to change the postsynaptic membrane permeability to specific ions.

Neurotransmitter receptors belong to two **superfamilies**. The **ligand-gated ion channel receptors** or **ionotropic receptors** have ion-selective channels as part of the receptor. Binding of the transmitter to the receptor opens the channel, directly increasing its permeability. The second superfamily is the **G protein-linked receptors**, also referred to as **metabotropic receptors**. Binding of transmitter to these receptors activates their associated G-proteins that are capable of diverse and remote effects, both on metabolism and membrane permeability. G-proteins can influence permeability either by binding ion channels directly or by modifying the activity of second messenger system enzymes which phosphorylate ion channels and thereby alter their permeability.

The change in postsynaptic membrane permeability can have essentially one of two effects. It may increase the probability that the postsynaptic neuron fires action potentials, in which case the response is **excitatory**. If the effect is to decrease the probability that the postsynaptic cell might fire, the response is **inhibitory**. Although neuroscientists often describe particular transmitters as excitatory or inhibitory, these actions should more properly be thought of as being produced by a given combination of transmitter and receptor. It is commonly the case that a transmitter is excitatory at one of its receptors, but is inhibitory at another. Any given synapse can be described as excitatory or inhibitory.

Approximately 30 molecules have been unambiguously identified as neurotransmitters and many more are likely candidates. In general they fall into two groups. The classical transmitters are amino acids or amines. Quantitatively by far the most important are glutamate, which is almost invariably excitatory, and γ-aminobutyrate (GABA) which is usually inhibitory. This group also includes acetylcholine, the catecholamines such as dopamine and norepinephrine, and the indoleamine, serotonin. The second, larger, group is an eclectic group of peptides which includes the opioids such as dynorphin and the tachykinins (e.g., substance P). See *Table 1* for a more comprehensive (but far from exhaustive) list.

In general, neurotransmission is thought of as falling into two broad categories based on its time course. **Fast transmission** occurs whenever a neurotransmitter acts via ionotropic receptors whereas **slow transmission** occurs by transmitters acting through metabotropic receptors. Glutamate, GABA and acetylcholine are together responsible for most of the fast transmission. However, each of these molecules also mediates slow transmission by activating their corresponding metabotropic receptors. It is quite common for each of these transmitters to mediate both fast and slow transmission at the same synapse by activating multiple receptor populations. Catecholamine and peptide transmission are invariably slow. Many authors distinguish the two categories above as **classical transmission** and **neuromodulation**.

The term neuromodulation can also be used in a narrower sense to define only those cases in which a neurotransmitter has no measurable effect on membrane permeability *per se*, but simply modulates the responsiveness of a neuron to other inputs.

It is extremely common for a given synapse to release more than one transmitter. This is termed **cotransmission** and usually involves the release of a classical transmitter, coupled with the co-release of one or more peptides at high stimulus frequencies.

Transmitters are rapidly cleared from the synaptic cleft after release by one of three methods: passive diffusion away from the cleft, reuptake into surrounding neurons or glia, or enzyme degradation.

Table 1. Key central nervous system neurotransmitters

Classical	Amino acids	Glutamate	
		Aspartate	
		γ-aminobutyrate	
		Glycine	
	(Mono)amines	Acetylcholine	
		Dopamine	catecholamines
		Norepinephrine	catecholamines
		Epinephrine	catecholamines
		Serotonin (5-hydroxytryptamine) indolamine	
Peptides	Opioids	Dynorphins	
		Endorphins	
		Enkephalins	
	Tachykinins	Substance P	
	Hormones	Cholecystokinin	
		Somatostatin	

C2 FAST NEUROTRANSMISSION

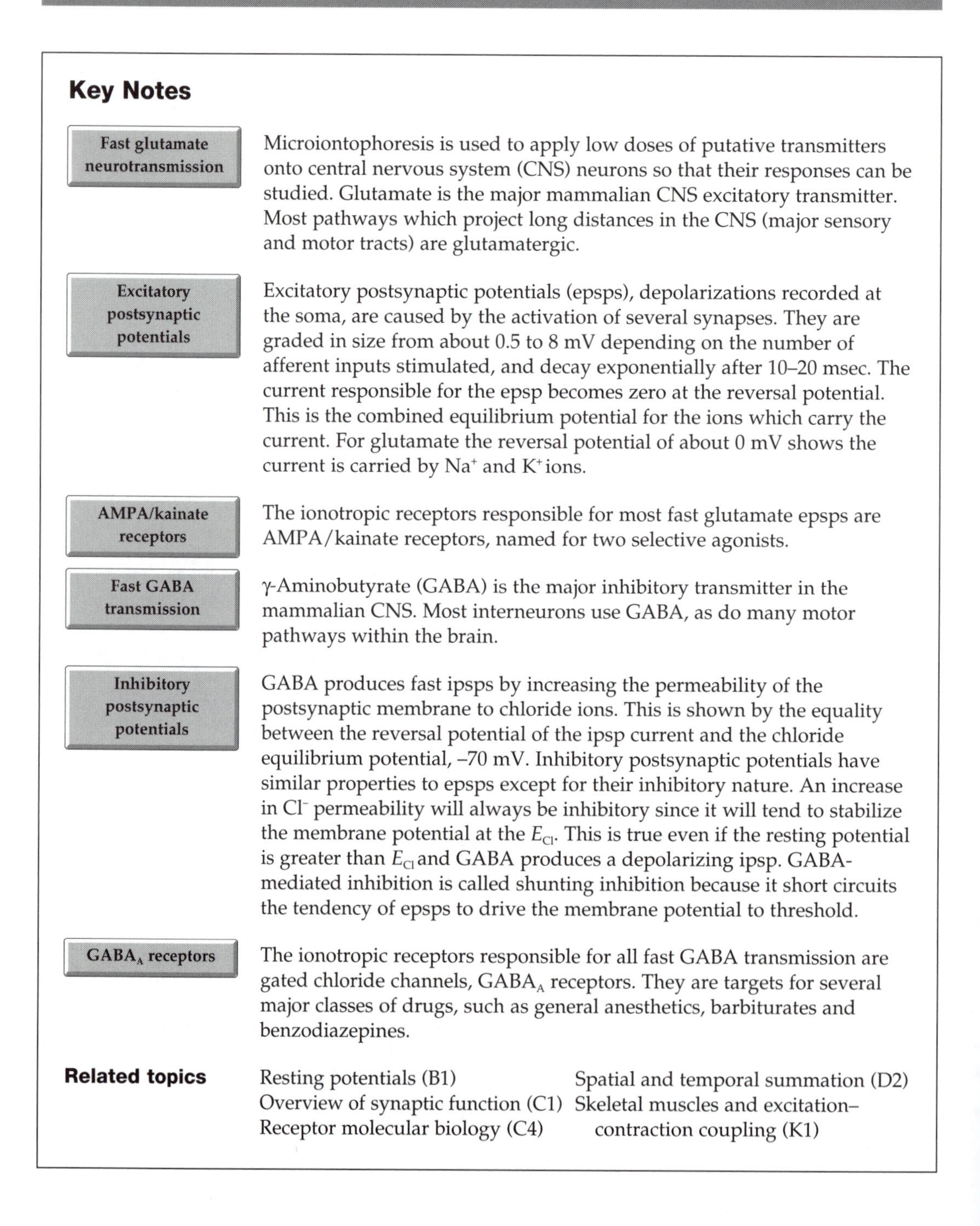

Key Notes

Fast glutamate neurotransmission

Microiontophoresis is used to apply low doses of putative transmitters onto central nervous system (CNS) neurons so that their responses can be studied. Glutamate is the major mammalian CNS excitatory transmitter. Most pathways which project long distances in the CNS (major sensory and motor tracts) are glutamatergic.

Excitatory postsynaptic potentials

Excitatory postsynaptic potentials (epsps), depolarizations recorded at the soma, are caused by the activation of several synapses. They are graded in size from about 0.5 to 8 mV depending on the number of afferent inputs stimulated, and decay exponentially after 10–20 msec. The current responsible for the epsp becomes zero at the reversal potential. This is the combined equilibrium potential for the ions which carry the current. For glutamate the reversal potential of about 0 mV shows the current is carried by Na^+ and K^+ ions.

AMPA/kainate receptors

The ionotropic receptors responsible for most fast glutamate epsps are AMPA/kainate receptors, named for two selective agonists.

Fast GABA transmission

γ-Aminobutyrate (GABA) is the major inhibitory transmitter in the mammalian CNS. Most interneurons use GABA, as do many motor pathways within the brain.

Inhibitory postsynaptic potentials

GABA produces fast ipsps by increasing the permeability of the postsynaptic membrane to chloride ions. This is shown by the equality between the reversal potential of the ipsp current and the chloride equilibrium potential, –70 mV. Inhibitory postsynaptic potentials have similar properties to epsps except for their inhibitory nature. An increase in Cl^- permeability will always be inhibitory since it will tend to stabilize the membrane potential at the E_{Cl}. This is true even if the resting potential is greater than E_{Cl} and GABA produces a depolarizing ipsp. GABA-mediated inhibition is called shunting inhibition because it short circuits the tendency of epsps to drive the membrane potential to threshold.

$GABA_A$ receptors

The ionotropic receptors responsible for all fast GABA transmission are gated chloride channels, $GABA_A$ receptors. They are targets for several major classes of drugs, such as general anesthetics, barbiturates and benzodiazepines.

Related topics

Resting potentials (B1)
Overview of synaptic function (C1)
Receptor molecular biology (C4)
Spatial and temporal summation (D2)
Skeletal muscles and excitation–contraction coupling (K1)

Fast glutamate neurotransmission

Glutamate is the most important excitatory transmitter in the mammalian brain. More than 90% of nerve cells in the cat spinal cord will respond to the application of low doses of glutamate by **microiontophoresis**. This technique allows delivery of precise amounts of a charged molecule onto the surface of a neuron through a micropipette. In the case of glutamate, which carries a net negative charge at physiological pH, initially a current of a few nanoamperes is injected, making the inside of the pipette positively charged so that the glutamate is retained. For delivery of transmitter the current is reversed for a brief period.

It is estimated that 35–40% of synapses use glutamate as a transmitter. Most of the major sensory pathways and some motor pathways are glutamatergic. All pyramidal cells in the cerebral cortex and granule cells in the cerebellar cortex (the most abundant neuron in the mammalian brain) release glutamate.

Excitatory postsynaptic potentials

Glutamate transmission was first studied in the spinal cord where sensory nerve axons from muscles synapse directly with motor neurons (see *Fig. 1*). The synapses are axodendritic, but located on dendrites within about 600 μm of the cell body. Intracellular recording shows that the effect of electrically stimulating the sensory nerve axons is to produce a small depolarization of the motor neuron. This is called an **excitatory postsynaptic potential (epsp)** because it carries the membrane potential closer to the threshold for firing action potentials.

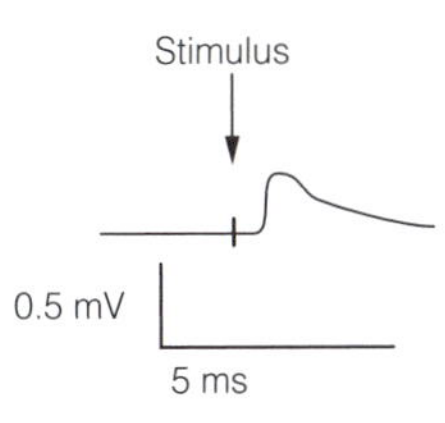

Fig. 1. Excitatory postsynaptic potentials in spinal motor neurons in response to stimulating a single sensory axon.

There are several important points to note about excitatory postsynaptic potentials generally:

(1) Epsps recorded from the cell body are caused by the activation of several synapses. Functional studies of individual central synapses are extremely difficult.
(2) There is a short delay of 0.5–1 msec between stimulating the afferents and the epsp. This is called the **synaptic delay**.
(3) Epsps are small and graded in size, ranging from fractions of 1 mV to about 8 mV, depending on the number of afferent fibres being stimulated. The reason for this is that if more afferent fibres are stimulated, then more synapses are activated (c.f. neuromuscular junction in Topic K1)
(4) Epsps generated by glutamate typically last for about 10–20 msec before decaying exponentially, but those due to slow transmission may last for seconds or even minutes.

Fig. 2 shows synaptic currents rather than synaptic potentials because here the motor neuron is being voltage clamped (Topic B3) at a variety of command potentials while the synapses are being activated.

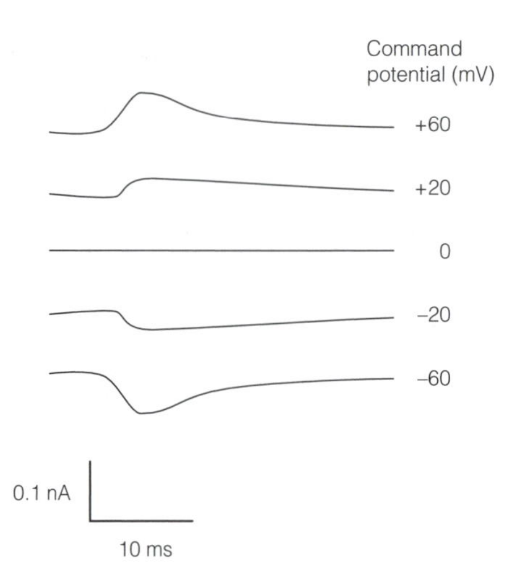

Fig. 2. Reversal potentials of glutamate-evoked currents in voltage-clamped spinal motor neurons.

As the command potential is made progressively less negative the inward current becomes smaller, disappears at about 0 mV, and becomes an outward current at positive potentials. The potential at which there is no net current is the **reversal potential**. It is the equilibrium potential for the ion(s) that carry the current. So, the reversal potential can be a good clue to the ions responsible for generating the synaptic potential. In this case, a reversal potential of 0 mV suggests that the transmitter changes the permeability of the postsynaptic membrane about equally to both Na^+ and K^+. This can be verified by using the Goldman equation (Topic B1). Substituting permeabilities of 1 for both Na^+ and K^+, together with standard values for the outside and inside concentrations of these ions and assuming zero chloride permeability gives a potential close to 0 mV.

AMPA/kainate receptors

A population of receptors called AMPA/kainate receptors mediate most fast glutamate neurotransmission. Widely distributed in the CNS they are named for two compounds which, although not endogenous to the nervous system, act as **agonists** at this class of glutamate receptor. AMPA/kainate receptors are ligand-gated ion channels, transmembrane proteins with an intrinsic ion channel and a recognition site for glutamate on the extracellular face that projects into the synaptic cleft. Binding of glutamate to this site causes the protein to change shape, opening the ion channel. The channel is selective for both Na^+ and K^+. The movement of these ions down their respective concentration gradients simultaneously through several hundred glutamate receptors causes the epsp that can be recorded from the cell body of a neuron.

Fast GABA transmission

Estimates suggest that anything between 17% and 30% of the synapses in the mammalian brain use γ-aminobutyrate (GABA) as a transmitter, making it by far the most important inhibitory neurotransmitter in the CNS. Many pathways involved in motor control are GABAergic, as are most of the interneurons in both cerebral and cerebellar cortices.

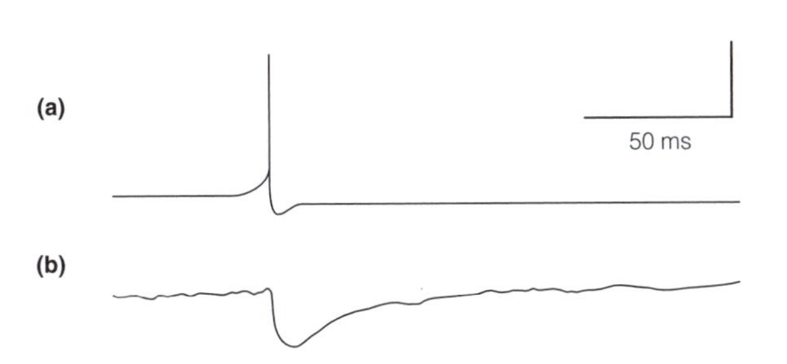

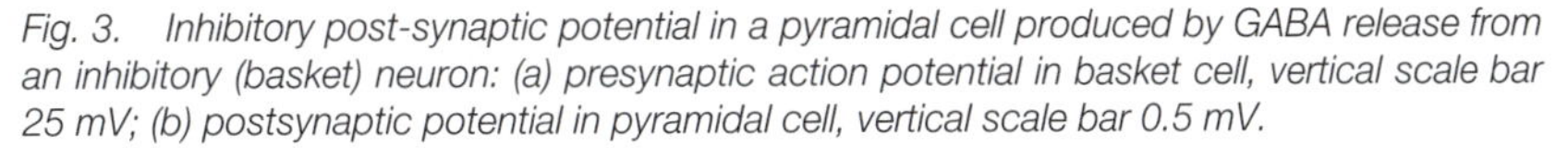

Fig. 3. Inhibitory post-synaptic potential in a pyramidal cell produced by GABA release from an inhibitory (basket) neuron: (a) presynaptic action potential in basket cell, vertical scale bar 25 mV; (b) postsynaptic potential in pyramidal cell, vertical scale bar 0.5 mV.

Inhibitory postsynaptic potentials

The effect of GABA can be seen in the experiment illustrated in *Fig. 3*. Pyramidal cells in the cerebral cortex have many GABAergic synapses impinging on them from interneurons called basket cells. Most of these synapses are axosomatic. By intracellular recording from the pyramidal cell it is possible to see that the effect of activating the basket cell is to produce a modest hyperpolarization. This is an **inhibitory postsynaptic potential (ipsp)** because it carries the membrane potential away from the threshold for firing action potentials.

Inhibitory postsynaptic potentials have very similar properties to epsps. However voltage clamping of neurons shows the reversal potential of the current responsible for fast GABA-evoked ipsps to be about –70 mV (*Fig. 4*). This is the equilibrium potential for Cl^-. Hence in response to the release of GABA the permeability of the pyramidal cell to chloride ions increases. When the membrane potential of the neuron is more positive than the reversal potential Cl^- enters the cell, making it more negative inside, i.e. the cell hyperpolarizes.

Fig. 4 shows that when the membrane potential of the neuron is initially more negative than the reversal potential for chloride, Cl^- leaves the cell, the cell becomes less negative inside (i.e. the cell depolarizes). It is important to note that this is still inhibitory. The effect of increasing Cl^- permeability is to force the membrane potential to remain at the E_{Cl} since whenever the membrane potential is not the same as E_{Cl} there will be an ionic driving force causing chloride ions

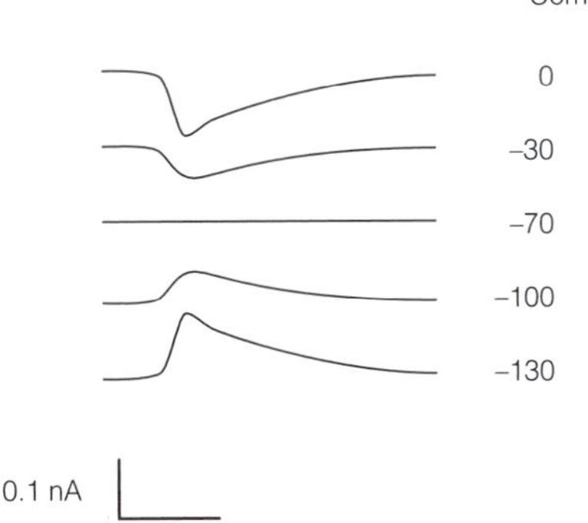

Fig. 4. Reversal potential of fast GABA ipsps, found by voltage clamping. The reversal potential is –70 mV.

either to leave or enter the cell. The increased Cl^- permeability therefore resists any tendency acting to make the membrane potential move away from –70 mV, and so prevents it from being driven towards threshold. Because this inhibition effectively short circuits epsps it is often referred to as **shunting inhibition**.

$GABA_A$ receptors

$GABA_A$ receptors are members of the ligand-gated superfamily of receptors and are responsible for all *fast* GABA transmission. Their activation opens a channel selective for Cl^-. $GABA_A$ receptors are the prime target for several classes of therapeutically useful drugs such as benzodiazepines, barbiturates, steroid anesthetics and volatile gas general anesthetics.

C3 Slow neurotransmission

Key Notes

G proteins

Metabotropic receptors are coupled to ion channels or second messenger enzymes via trimeric guanine nucleotide-binding proteins, G proteins. There are several different G protein families with their own particular targets. Binding of ligand to receptor causes liberation of a guanosine 5′-triphosphate (GTP)-bound form of G protein which activates its targets. The intrinsic GTPase activity of the G protein rapidly hydrolyzes the GTP, hence curtailing its own activity

Activation of adenylyl cyclase

G_S proteins activate adenylyl cyclase, which converts adenosine 5′-triphosphate (ATP) to cyclic adenosine monophosphate (cAMP). This second messenger molecule activates protein kinase A which phosphorylates its target proteins. The cAMP is subsequently degraded by phosphodiesterases. Depletion of cAMP, the action of phosphatases and receptor desensitization all act to curtail the effects of the cAMP second messenger system.

Inhibition of adenylyl cyclase

G_i proteins inhibit adenylyl cyclase. The activity of this enzyme and so the concentration of cAMP in a cell at any time therefore depends on activation of receptors coupled to G_S relative to those coupled to G_i.

Phosphoinositide second messenger system

G_q proteins activate phospholipase C which cleaves a membrane phospholipid to generate two second messenger molecules: diacylglycerol (DAG) activates protein kinase C; inositol trisphosphate (IP_3) mobilizes calcium from internal stores to increase the cytoplasmic Ca^{2+} concentration, which activates calcium-dependent protein kinases. These kinases phosphorylate two overlapping populations of proteins.

Related topics

Overview of synaptic function (C1)
Receptor molecular biology (C4)
Retina (H3)
Olfactory pathways (J2)
Hippocampal learning (Q4)

G proteins

Slow transmission is mediated by G-protein-linked receptors. Responses produced by these receptors may last for seconds or minutes. **G proteins**, trimers consisting of α-, β-, and γ-subunits are so-called because the α-subunit binds guanine nucleotides. Binding of neurotransmitter to metabotropic receptors activates their associated G proteins which may do one or both of the following:

- They may interact directly with ion channels causing them to open or close.
- They may interact with enzymes to switch on or off second messenger cascades that regulate ion channels by phosphorylation, amongst other activities. Two important enzymes are adenylyl cyclase and phospholipase C.

The cycle of events by which a G protein couples metabotropic receptor activation to second messenger modulation is shown in *Fig. 1*.

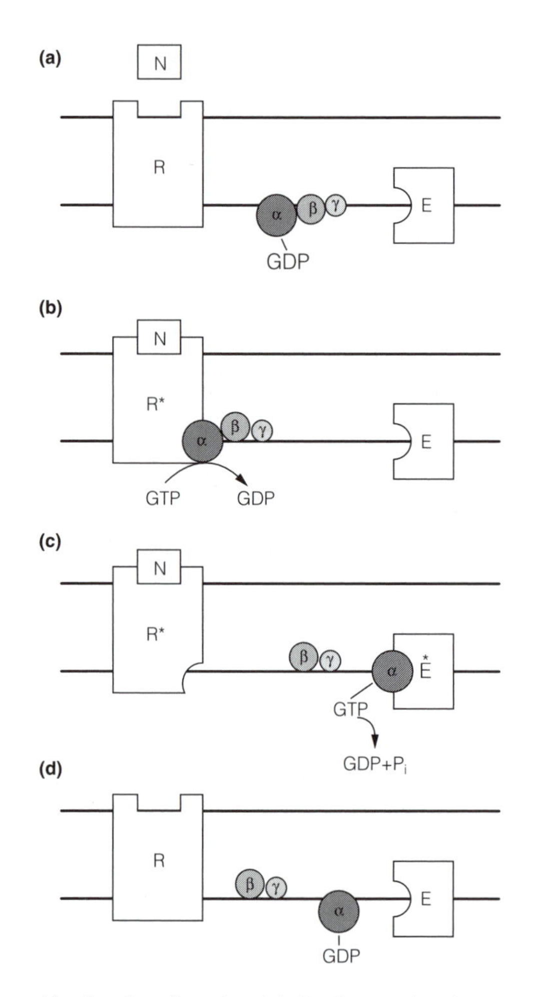

Fig. 1. Coupling of metabotropic receptors to second messenger systems by G proteins. N, Neurotransmitter; R, receptor; E, enzyme.

Binding of the transmitter allows the receptor and G protein to couple. Guanosine 5′-diphosphate (GDP) leaves the α-subunit in exchange for GTP. In its GTP-bound form the G protein dissociates into separate α-, and β/γ-subunits. The α-subunit binds to the enzyme, so activating it. The α-subunit has an intrinsic GTPase activity that cleaves the terminal phosphodiester bond in the GTP converting it to GDP. In its GDP-bound form the α-subunit uncouples from the enzyme, which reverts to its basal activity. One purpose of this cycle is to act as an amplifier. A single transmitter-binding event results in several cycles of G protein shuttling between receptor and enzyme. Furthermore the enzyme will have time to catalyse the synthesis of hundreds of second messenger molecules before it is switched off by the hydrolysis of the G protein-bound GTP. There is evidence for other roles for the β/γ-subunits of G proteins. There are several distinct G proteins, differing largely in their α-subunits. G_S and G_i interact with adenylyl cyclase, G_q with phospholipase C. Despite this multiplicity, G proteins serve as a point for convergence of signals impinging on a neuron because many receptors talk to just a few second messenger systems.

Table 1. Second messenger coupling to selected neurotransmitter receptors

G protein	Second messenger	Receptor
G_s	Increased cAMP	β1, β2, β3 adrenoceptors D1, D5 (dopamine) H2 (histamine)
G_i	Decreased cAMP and/or opening of K^+ channels closing of Ca^{2+} channels	α2 adrenoceptors D2, D4 (dopamine) $GABA_B$ 5-HT1 (serotonin) mGlu, types II and III (glutamate) M2, M4 (muscarinic) μ, δ and κ opioid
G_q	Increased phosphoinositide metabolism	α1 adrenoceptors CCK (cholecystokinin) mGlu, type I (glutamate) 5-HT2 (serotonin) M1, M3, M5 (muscarinic) H1 (histamine) NK (tachykinin)

Table 1 lists some of the major G protein-linked receptors for transmitters, together with the second messenger systems they are coupled to.

Activation of adenylyl cyclase

Adenylyl cyclase is activated by a specific family of G proteins, the **G_S proteins**, so called because their action on adenylyl cyclase is stimulatory. The enzyme catalyses the conversion of ATP to **cyclic adenosine-3′,5′-monophosphate (cAMP)**. This second messenger molecule diffuses freely through the cytoplasm and binds to a kinase enzyme, **protein kinase A (PKA)**, which is thus switched on (*Fig. 2*). The kinase then phosphorylates a variety of target proteins that possess the appropriate amino acid sequence to recognize the kinase, including a number of ion channels. The phosphorylation state of a channel often determines whether it is open or closed. Many channels are opened by phosphorylation whereas others are closed. Clearly a single activated PKA molecule is able to phosphorylate many target proteins thus increasing the amplification achieved by the system. Second messenger cascades must be able to be rapidly turned off so that their signals can be modulated over a time course of tens or hundreds of milliseconds. For the cAMP system this occurs in three ways:

(1) Cyclic AMP is hydrolyzed to AMP by the action of a specific **phosphodiesterase** in the cytoplasm.
(2) There are specific **phosphatases** responsible for dephosphorylating the target proteins. Hence the phosphorylation state of a protein at a given time will depend on the balance of the activities of kinases and phosphatases.
(3) Prolonged occupation of the receptor by the transmitter causes it to **desensitize**. This involves phosphorylation by a specific **kinase** that recognizes the agonist bound form of the receptor followed by the binding of an **arrestin** protein. The resulting complex is unable to recognize the G protein.

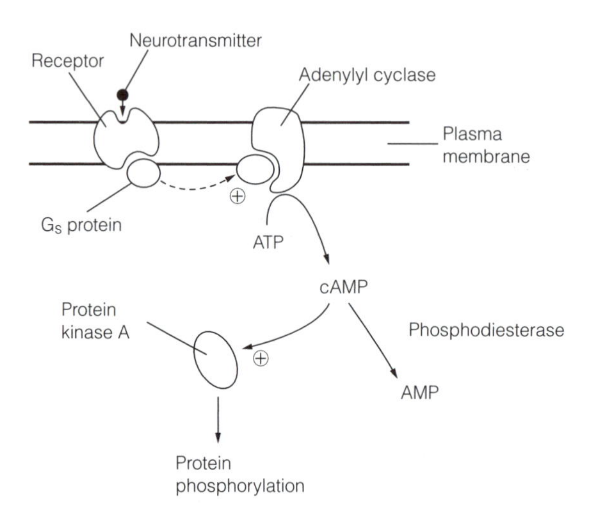

Fig. 2. The adenylyl cyclase-cAMP second messenger system. The activated G_s protein uncouples from the receptor to switch on adenylyl cyclase.

Inhibition of adenylyl cyclase

Some neurotransmitter receptors are negatively coupled to adenylyl cyclase. These receptors associate with **G_i proteins** that inhibit the activity of the enzyme. Exactly how this works is not known but both the α-subunit and β/γ-subunits have been shown to independently block the isoform of adenylyl cyclase common in neurons. The outcome is that the activity of adenylyl cyclase, and so the amount of cAMP in the cell at any given instant will reflect the balance of activation of receptors coupled to G_S and those coupled to G_i.

Phosphoinositide second messenger system

Many receptors are coupled via the **G_q protein** to activation of phospholipase C (*Fig. 3*). This enzyme cleaves a minor phospholipid in the inner leaflet of the plasma membrane, **phosphatidyl inositol-4,5-bisphosphate (PIP_2)**, to give **diacylglycerol (DAG)** and **inositol-1,4,5-trisphosphate (IP_3)**, both of which are second messengers.

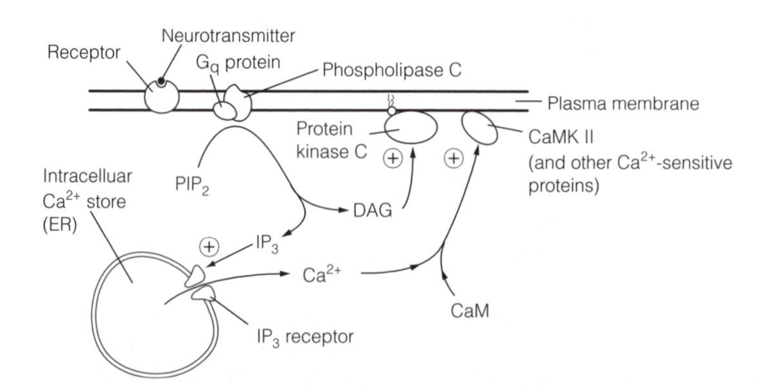

Fig. 3. The phosphoinositide second messenger system. CaM, calmodulin; CaMKII, calcium–calmodulin-dependent protein kinase II; DAG, diacylglycerol; ER, endoplasmic reticulum; IP_3, inositol trisphosphate; PIP_2, phosphatidyl inositol bisphosphate.

DAG, a hydrophobic molecule, diffuses within the lipid where it activates **protein kinase C (PKC)**. In turn this kinase phosphorylates its protein targets, affecting metabolic, receptor and ion channel functions.

IP_3 is water soluble and freely diffusible in the cytosol. Its target is the **IP_3 receptor**, a large IP_3-gated calcium channel located in the membrane of the **smooth endoplasmic reticulum (SER)**. The SER in neurons (and its equivalent, the **sarcoplasmic reticulum** in muscle cells) acts as an intracellular Ca^{2+} store. The binding of IP_3 to its receptors causes the calcium channels to open and Ca^{2+} flows out of the SER into the cytosol. An increase in intracellular calcium concentration has diverse and widespread effects that may be cell typical. An obvious example is that by binding the protein **troponin** in striated muscle, calcium triggers the cascade of biochemical events that leads to muscle contraction. Neurons contain a calcium binding protein called **calmodulin (CaM)** which shares considerable homology with troponin. On binding Ca^{2+}, CaM activates a number of enzymes including **calcium–calmodulin-dependent protein kinase II (CaMKII)**. CaMKII, and the many other calcium-sensitive proteins, mediate the effects of raised intracellular calcium, such as changes in membrane permeability and gene expression.

C4 Receptor molecular biology

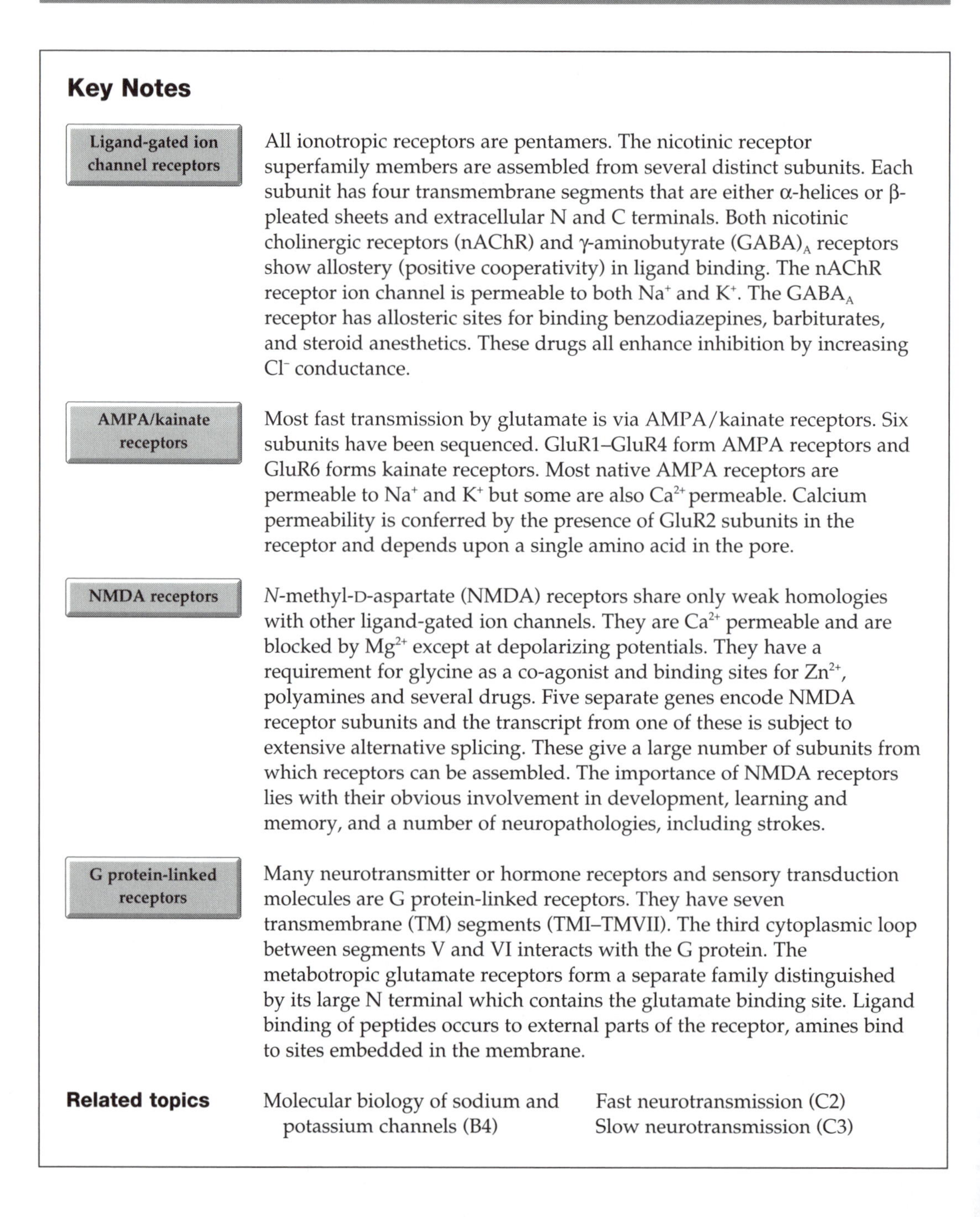

Key Notes

Ligand-gated ion channel receptors

All ionotropic receptors are pentamers. The nicotinic receptor superfamily members are assembled from several distinct subunits. Each subunit has four transmembrane segments that are either α-helices or β-pleated sheets and extracellular N and C terminals. Both nicotinic cholinergic receptors (nAChR) and γ-aminobutyrate $(GABA)_A$ receptors show allostery (positive cooperativity) in ligand binding. The nAChR receptor ion channel is permeable to both Na^+ and K^+. The $GABA_A$ receptor has allosteric sites for binding benzodiazepines, barbiturates, and steroid anesthetics. These drugs all enhance inhibition by increasing Cl^- conductance.

AMPA/kainate receptors

Most fast transmission by glutamate is via AMPA/kainate receptors. Six subunits have been sequenced. GluR1–GluR4 form AMPA receptors and GluR6 forms kainate receptors. Most native AMPA receptors are permeable to Na^+ and K^+ but some are also Ca^{2+} permeable. Calcium permeability is conferred by the presence of GluR2 subunits in the receptor and depends upon a single amino acid in the pore.

NMDA receptors

N-methyl-D-aspartate (NMDA) receptors share only weak homologies with other ligand-gated ion channels. They are Ca^{2+} permeable and are blocked by Mg^{2+} except at depolarizing potentials. They have a requirement for glycine as a co-agonist and binding sites for Zn^{2+}, polyamines and several drugs. Five separate genes encode NMDA receptor subunits and the transcript from one of these is subject to extensive alternative splicing. These give a large number of subunits from which receptors can be assembled. The importance of NMDA receptors lies with their obvious involvement in development, learning and memory, and a number of neuropathologies, including strokes.

G protein-linked receptors

Many neurotransmitter or hormone receptors and sensory transduction molecules are G protein-linked receptors. They have seven transmembrane (TM) segments (TMI–TMVII). The third cytoplasmic loop between segments V and VI interacts with the G protein. The metabotropic glutamate receptors form a separate family distinguished by its large N terminal which contains the glutamate binding site. Ligand binding of peptides occurs to external parts of the receptor, amines bind to sites embedded in the membrane.

Related topics

Molecular biology of sodium and potassium channels (B4)
Fast neurotransmission (C2)
Slow neurotransmission (C3)

Ligand-gated ion channel receptors

Many ligand-gated ion channel receptors have had their primary structures determined by cloning and sequencing their DNA. On the basis of structural features they can be divided into two superfamilies (see *Table 1*), the conventional ligand-gated ion channels typified by nAChR (the first to be characterized) and the ionotropic glutamate receptor family.

Table 1. Ligand-gated ion channels

Nicotinic receptor family	nAChR	
	$GABA_A$	
	$GABA_C$	
	glycine	
	$5\text{-}HT_3$	
Glutamate receptor family	GluR1–GluR4	(AMPA receptors)
	GluR6	(kainate receptors)
	NMDAR	

The nicotinic receptor family members are pentamers, with the five subunits clustered around a central pore. Because several different subunits aggregate to form a given receptor they are called **hetero-oligomers**. For example, nAChR consists of two α-subunits, associated with a β-, a γ- and a δ-subunit. The subunits share modest amino acid homologies both with each other and with corresponding subunits across widely differing species. Each subunit has extracellular N and C terminals and four transmembrane segments (M1–M4) of uncertain secondary structure. Although an α-helical structure has usually been accepted, there are powerful theoretical arguments in favour of a β-pleated sheet configuration for transmembrane segments (*Fig. 1*).

In the case of the nAChR each of the α-subunits contributes to a binding site for acetylcholine (ACh), thus each receptor binds two molecules of ACh. This explains the positive cooperativity displayed in the binding of ACh; binding of one molecule of ACh to a receptor makes binding of the second easier. This property is called **allostery** and is also seen in a number of enzymes (for further information on this topic see *Instant Notes Biochemistry*, 2nd edn). Exactly how the binding of ACh to the nicotinic receptor causes a channel to open has not yet been determined. The channel is a nonselective cation conductance allowing Na^+ influx and K^+ efflux. Since the resulting current has a reversal potential close to

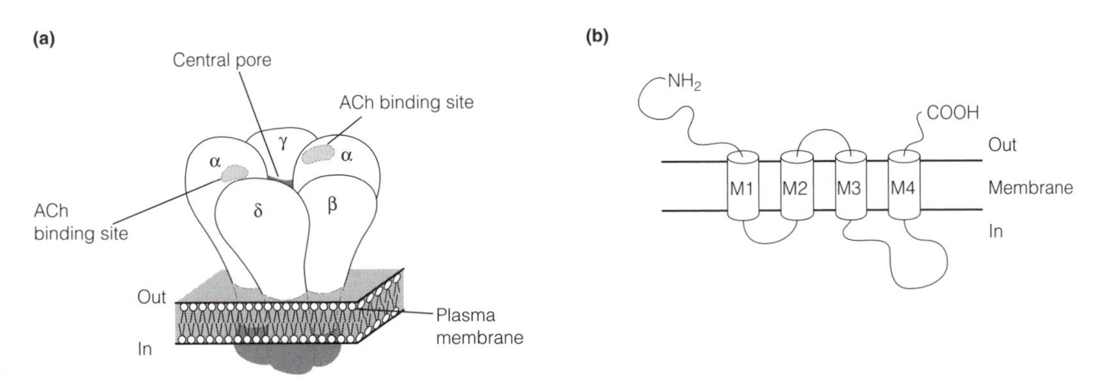

Fig. 1. The nicotinic receptor family: (a) pentameric arrangement of subunits; (b) cartoon of subunit secondary structure.

zero, activating the nAChR causes membrane depolarization: the action of ACh on nicotinic receptors is excitatory.

$GABA_A$ receptors have much in common with nicotinic receptors. They are pentameric, hetero-oligomers, made up from various combinations of subunits designated α, β, γ, δ and ρ, which should not be confused with nAChR subunits with the same designation. $GABA_A$ subunits have the same overall structure as nAChR subunits, with which they have appreciable homology. Like the nAChR, two molecules of transmitter are required to open the ion channel and binding of transmitter shows positive cooperativity. The binding site for GABA even has some homology with the acetylcholine-binding site.

The situation is complicated in the case of $GABA_A$ receptors in two ways. Firstly each subunit comes in a variety of **isoforms** which have 75% homology with each other. Thus, for example, there are six slightly different forms of the α-subunit (α1–α6) each encoded by a separate gene. Hence a large number of distinct $GABA_A$ receptors may exist. Secondly the $GABA_A$ receptor contains binding sites for several major classes of drugs. **Benzodiazepines** (e.g. diazepam) are allosteric modulators in that they bind to a site that is different from the one that binds GABA. The binding of benzodiazepines to $GABA_A$ receptors causes an increase in the affinity of the receptor for GABA. This causes the chloride ion channel in the receptor to open more frequently. The overall effect of benzodiazepines is to potentiate the inhibitory effect of GABA without prolonging it and is thought to account for their antianxiety and anticonvulsant actions. Both α- and γ-subunits are involved in the actions of benzodiazepines.

The **barbiturates** and **steroid anesthetics** also have allosteric actions at the $GABA_A$ receptor but the effect of these drugs is to prolong the length of time that the chloride channel is open so that GABA inhibition lasts longer. The sites of action of these drugs on the receptor molecule are not well defined.

AMPA/kainate receptors

The **ionotropic glutamate receptors (iGluRs)** share only weak homologies with the other types of ligand-gated ion channels such as the nicotinic receptor superfamily. There are two populations of iGluRs, defined by selective agonists: **AMPA/kainate receptors** and **NMDA receptors**. All are probably hetero-oligomers with a pentameric quaternary structure, having five subunits arranged about a central pore.

Six subunits that belong to the AMPA/kainate population have been cloned and sequenced (GluR1–GluR6). The most likely secondary structure for these subunits is shown in *Fig.* 2. The extracellular N-terminal end is large by comparison with the nicotinic receptor family and is glycosylated. The C-terminal end is intracellular and bears amino acid sequences that could be phosphorylated by

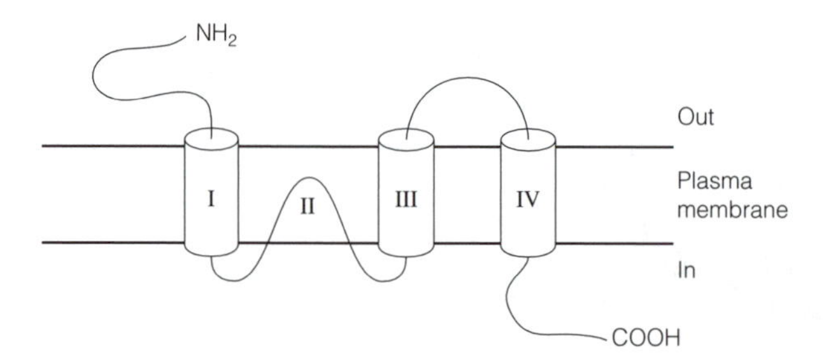

Fig. 2. Ionotropic glutamate receptor subunit structure.

a variety of kinases. Hydropathicity profiles suggest the presence of three TM segments and a loop (TMII) which inserts into the membrane. The region spanning TMI–TMIII bears a striking resemblance to the S5–H5–S6 region of potassium channels albeit inserted into the membrane in the opposite sense. It may be that the TMI–TMIII domain has evolved from some ancient potassium channel. The TMII loop is thought to be the pore region of iGluRs.

By expressing each of the subunits separately in *Xenopus* oocytes, their pharmacology has been studied. GluR1–GluR4 subunits are more sensitive to AMPA than kainate, while GluR6 is a pure kainate receptor. Although weakly activated by glutamate, GluR5 is insensitive to the other agonists. Experiments in which various combinations of subunits are expressed together in *Xenopus* oocytes, **coexpression** studies, have been performed in an attempt to deduce the structure of native AMPA/kainate receptors. These have proved very instructive. Firstly, GluR6 subunits never form channels with any of GluR1–GluR4 suggesting that kainate receptors are distinct from AMPA receptors. Next, GluR1 or GluR3, either alone or expressed together, generates channels permeable to Ca^{2+}. By contrast, coexpression of GluR2 with GluR1 or GluR3 gives rise to channels permeable only to Na^+ and K^+, exactly like the AMPA/kainate receptors identified in cortical pyramidal neurons. The difference in permeability is conferred by just one amino acid located in the TMII pore region. In the GluR2 subunits the amino acid is arginine, whereas in all other subunits the corresponding amino acid is glutamine. Mutant subunits with the 'wrong' amino acid in this position form channels with the 'wrong' permeability.

NMDA receptors

The NMDA receptor is named after the selective agonist **N-methyl-D-aspartate**. The receptor is important because it is implicated in key aspects of brain function such as development, learning and memory, and in pathologies, e.g. strokes and epilepsy. NMDA receptors are targets for a number of drugs including dissociative anesthetics (ketamine) that produce altered consciousness rather than unconsciousness and the hallucinogenic phencyclidine ('angel dust').

NMDA receptors have some unusual properties that are summarized below:

- They are both ligand- and voltage-gated. At resting membrane potentials glutamate will bind to the receptor but the ion channel is blocked by Mg^{2+} ions. This blockade is lifted only by a large depolarization. In other words the ion channel is only opened if glutamate binds and the receptor experiences depolarization at the same time.
- The ion channel is permeable to Ca^{2+} as well as Na^+ and K^+. Under some circumstances calcium entry through NMDA receptors can be a significant factor in raising intracellular Ca^{2+} concentrations.
- Glycine, normally an inhibitory transmitter that acts via receptors that are very similar to $GABA_A$ receptors, acts as a **co-agonist** at NMDA receptors. Glycine acts allosterically to dramatically potentiate the effects of glutamate. The concentration of glycine available to the receptor *in vivo* is about that needed to give maximum potentiation.
- NMDA receptors can be modulated in complicated fashion by Zn^{2+} ions and by polyamines such as spermine. Generally, *in vivo*, zinc inhibits while spermine potentiates the action of glutamate on NMDA receptors.

The structural complexity of NMDA receptors matches their functional complexity. Like all other ionotropic receptors they are assumed to be pentameric. The subunits from which they are assembled are broadly similar to the AMPA/kainate receptor subunits. Five genes code for NMDA receptor subunits; the *nmdar1* gene codes for the NMDAR1 subunit. These can form homomeric channels with all the properties of native NMDA receptors. The remaining four genes code for NMDAR2 subunits A to D. On their own NMDAR2 subunits cannot form channels. However they will form functional channels together with NMDAR1 subunits. Different combinations of R2 with R1 subunits accounts for some of the diversity of NMDA receptors seen *in vivo*. Added complication arises from the fact that NMDAR1 subunits exist in seven distinct variants, as a result of alternative splicing of mRNA (see *Instant Notes Molecular Biology*, 2nd edn). Hence, the number of possible NMDA receptors is extremely large. The significance of this structural diversity, as with $GABA_A$ receptors, is not understood.

G protein-linked receptors

G protein-linked receptors form a huge superfamily. Its members include receptors for slow neurotransmitters, many hormones, and sensory transduction molecules mediating the responses to light by photoreceptors in the retina, and the chemical senses of smell and taste. Detailed X-ray diffraction studies of one member of the superfamily allowed its structure to be deduced and all other members are assumed to be the same, based on the homologies they share in their primary sequences. This structure is illustrated in *Fig. 3*.

The key features include seven membrane-spanning segments (I–VII) – an alternative name for these is seven transmembrane (7TM) receptors – an extracellular N-terminal end which is glycosylated, and a variable sized cytoplasmic loop between TM segments V and VI. Site-directed mutagenesis reveals that this third cytoplasmic loop is the region of the receptor that couples to the G protein. Although metabotropic glutamate receptors are 7TM receptors they have little homology with the others and have a large N-terminal end, which binds glutamate. The peptide-binding receptors have their binding domains associated with several extracellular regions. In those receptors that bind small amines the ligand binds to membrane-spanning regions embedded quite deeply in the membrane.

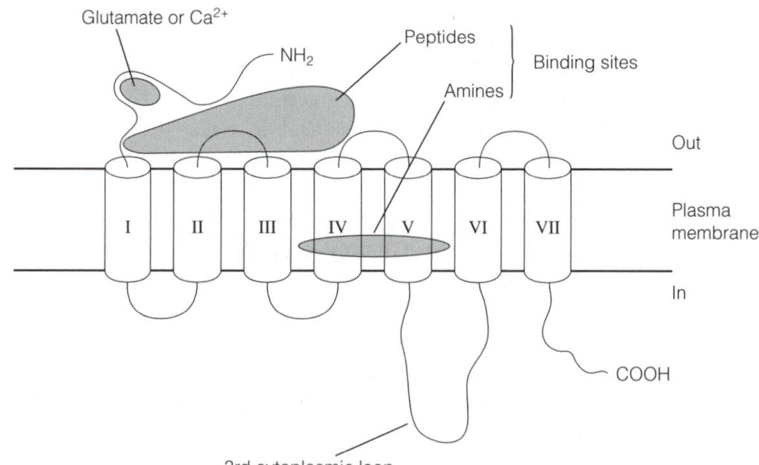

Fig. 3. G protein-linked receptors: cartoon showing transmembrane segments and ligand binding sites; each of the Roman numerals designates a transmembrane segment.

C5 Neurotransmitter release

Key Notes

Vesicular release

Neurotransmitter release occurs most commonly by calcium-dependent exocytosis from vesicles, in response to excitation of the axon terminal by action potentials. Non-vesicular calcium-independent release of glutamate and GABA via transporters can occur under some circumstances.

Release is quantal

Transmitter is released in discrete packets, quanta, that correspond to exocytosis from a single vesicle. The spontaneous, random release of a single quantum causes miniature endplate potentials (at the neuromuscular junction) or miniature postsynaptic potentials (at CNS synapses). Normal postsynaptic potentials arise from the release of several quanta simultaneously. The arrival of an action potential may or may not trigger exocytosis, so neurotransmitter release is probabilistic and can be modeled either as a binomial or as a Poisson process. At the neuromuscular junction postsynaptic receptors are in excess so the size of the postsynaptic response is a marker for the amount of transmitter liberated. At central synapses only one vesicle is available for release at each active zone and there are few postsynaptic receptors so the response is a measure of receptor numbers.

The role of calcium

Calcium imaging shows how Ca^{2+} moves in space and real time through cells. This reveals that following excitation of the nerve terminal calcium influx is restricted to a small region, but the local concentration reaches 200 μM, sufficient to trigger the low affinity exocytosis mechanism for small synaptic vesicles very rapidly.

Exocytosis from large dense-core vesicles

Amines and peptides are released by high frequency stimulation, only after an appreciable delay, because large dense-core vesicles (LDCV) are situated some distance from the active zone. The large vesicle exocytosis mechanism has a high affinity for Ca^{2+}.

Biochemistry of exocytosis

Several linked steps are involved in exocytosis. Recruitment shifts vesicles from a reserve pool into a releasable pool. Binding of vesicle-associated proteins and plasma membrane proteins permits the vesicles to be docked at the active zone in close proximity to voltage-dependent calcium channels (VDCCs). Partial fusion of the vesicle is achieved by priming, mediated by the assembly of a macromolecular complex, and involving the hydrolysis of adenosine 5′-triphosphate (ATP). The final rapid stage of exocytosis occurs when excitation triggers Ca^{2+} influx. Binding of calcium to synaptotagmin permits fusion to go to completion.

Endocytosis

Vesicles are recycled. Vesicle membrane is coated with clathrin so that it invaginates. Fission of coated vesicle is then triggered by hydrolysis of GTP bound to dynamin. Once in the cytoplasm the vesicle loses its clathrin coat.

Refilling

Classical transmitters are imported into vesicles driven by the efflux of H^+ via specific transporters. The proton gradient is generated by a vesicular proton ATPase. Peptides are packaged in the Golgi apparatus from which vesicles bud to be transported to the axon terminal, itself incapable of protein synthesis.

Related topics

Morphology of chemical synapses (A3)
Voltage-dependent calcium channels (C6)
Autoreceptors (C8)
Skeletal muscles and excitation–contraction coupling (K1)

Vesicular release

Most neurotransmitter release occurs by transmitter-loaded synaptic vesicles fusing with the presynaptic membrane so that the contents of the vesicle are discharged into the synaptic cleft. This is an example of **exocytosis**. It is triggered by the arrival at the nerve terminal of an action potential which causes a transient and highly localized influx of Ca^{2+}. Several stages of exocytosis have an absolute requirement for calcium. The link between excitation of the nerve terminal and transmitter release is an example of **excitation-secretion coupling**. After release the vesicle membrane is recycled from the presynaptic membrane to form new vesicles by **endocytosis**. The vesicles are subsequently loaded with transmitter via active transporters localized in the vesicle membrane.

Under some circumstances nonvesicular, Ca^{2+}-independent release of transmitters, particularly GABA and glutamate, can be seen. This is thought to occur by the reversal of transport mechanisms that normally serve to reuptake transmitter from the synaptic cleft back into the nerve terminal (Topic C7).

Release is quantal

In vesicular release, neurotransmitter is secreted in discrete packets or **quanta**. Each quantum represents the release of the contents of a single vesicle, about 4000 molecules of transmitter. At the **neuromuscular junction** (nmj), the rather unusual synapse between motor neurons and skeletal muscle fibers, acetylcholine (ACh) released from a single vesicle diffuses across the cleft in about 2 μs, reaching a peak concentration of around 1 mM, activating 1000–2000 nACh receptors to produce a depolarization of the muscle fiber membrane locally of approximately 0.5 mV. Such events occur randomly and spontaneously under resting conditions and are called **miniature endplate potentials** (mepps). The **endplate potential** produced by a single action potential arriving at the motor neuron terminal results from the summation of about 300 quanta being liberated simultaneously from around 1000 active zones.

The size of the mepp is called the **quantal size**, q. Although the mepp is caused by the activation of postsynaptic receptors, because these are present in large excess at the nmj, quantal size is determined by the amount of ACh released from a single vesicle.

At central synapses **miniature postsynaptic potentials** (mpsps) are seen which are the equivalent of mepps. Mpsps may be excitatory or inhibitory, depending on the transmitter. Quantal size of mpsps is much more variable, ranging from about 100 to 400 μV and depends on the number of postsynaptic receptors available to respond to released transmitter. Thus at central synapses q can provide information about the density and efficacy of postsynaptic receptors. The reason is that at central glutamatergic and GABAergic synapses a single quantum contains more than sufficient transmitter to saturate the 30–100 receptors lying under the **active zone**, the region of the presynaptic membrane where vesicle docking occurs and where transmitter is released. Excitatory and inhibitory postsynaptic potentials represent the summation of multiple mpsps generated by an action potential invading several active zones simultaneously. This is possible either because axons branch to form several discrete terminals or because some terminals appear to have more than one active zone.

Neurotransmitter secretion is stochastic, that is, governed by the laws of probability. The **standard Katz model**, based on studies of the frog nmj, is that a given nerve terminal will have n releasable vesicles. Each quantum can suffer only one of two fates each time an action potential arrives: either it is released or it is not. This is a binomial process, similar to tossing a coin. Hence there is a probability, p, that a quantum will be released during a single action potential. This probability is independent of what is happening to any other quantum. The mean number of quanta released per action potential is the **quantal content** (m), given by:

$$m = np$$

Unfortunately modeling neurotransmitter release by a binomial distribution requires values for the parameters n and p and these can be difficult to determine experimentally. However at low levels of neurotransmitter release (e.g. if $[Ca^{2+}]_o$ is low) p becomes small compared with n. Under this condition the binomial distribution is approximated by a Poisson distribution which requires only a knowledge of m. Now quantal content can be directly measured experimentally from the relationship:

$$m = \log_e N/n_o$$

where N is the number of action potentials and n_o is the number of action potentials which fail to produce a response. So transmitter release at the nmj can be modeled as a Poisson process and m can be used as a measure of the efficiency of presynaptic release.

CNS synapses do not behave according to the standard Katz model. The active zone of many central synapses appears to have only one release site. This is known as the **one vesicle** or **one quantum** hypothesis. Individual active zones behave in an all-or-none fashion because an action potential will either trigger the release of the single quantum or not. The proportion of successes will reflect the probability of release. However at central synapses the probability of release varies between different sites and at least at some synapses p varies with time: that is, p depends on the recent history of the synapse.

The role of calcium

The arrival of an action potential at a nerve terminal causes an influx of Ca^{2+} through calcium channels (see Topic C6). Direct evidence for the role of calcium is provided by **calcium imaging**, a technique which makes visible how Ca^{2+} signals spread in time and space through cells. Fluorescent dyes are used which on binding Ca^{2+} absorb UV light at a different wavelength from that absorbed in the unbound state. Neurons are preloaded with the dye and the emission of UV from the dye is observed in response to its excitation by the two distinct absorption wavelengths. This gives a quantitative measure of how the concentration of Ca^{2+} changes in the neuron in real time.

This technique shows that it takes about 300 μs for calcium channels at the active zone to open in response to an action potential. The driving force for calcium entry is extremely high because of the large concentration gradient. The free Ca^{2+} concentration at rest in a terminal is 100 nM whilst the external concentration is about 1 mM. Despite this huge concentration gradient, the presence of diffusion barriers and calcium buffers in the terminal restrict the rise in calcium concentration to within 50 nm of the channel mouth. This region is called a **calcium microdomain**. The $[Ca^{2+}]$ within 10 nm of the channel mouth rises to between 100 and 200 μM, which matches the half-maximal concentration of Ca^{2+} for glutamate release of 194 μM. Several overlapping microdomains cooperate to trigger the release of a vesicle in close proximity.

Exocytosis from large dense-core vesicles

In contrast to small synaptic vesicles, the mechanism for release from LDCVs has a high affinity for calcium (half-maximal release at about 0.4 μM) but it takes some time for even a small amount of Ca^{2+} to diffuse to the LDCVs which are distant from the active zones. Hence exocytosis of amines and peptides occurs with a delay of about 50 ms and only in response to high frequency stimulation of the neuron which causes high levels of calcium influx.

Biochemistry of exocytosis

Exocytosis from small clear synaptic vesicles (SSVs) involves several linked steps, most of which need calcium. Nerve terminals contain two pools of SSVs. The **releasable pool** is located at the active zone and can take part in repeated cycles of exocytosis and endocytosis at low neuron firing frequencies. The **reserve pool** consists of vesicles tethered to cytoskeletal proteins, and can be mobilized by repetitive stimulation to join the releasable pool: this is called **recruitment**. Liberation of a vesicle from the cytoskeleton requires Ca^{2+}-dependent phosphorylation of **synapsin I**, a protein which anchors vesicles to actin filaments in the terminal.

Vesicles are aligned at specific sites in the active zone by a process termed **docking**, which involves **SNARE** proteins (*Fig. 1*). A vesicle-associated protein, **synaptobrevin** (v-SNARE, VAMP) binds with high affinity to a presynaptic membrane protein, **syntaxin** (t-SNARE). Syntaxin is closely associated with voltage-dependent calcium channels ensuring that vesicles are optimally placed to receive the Ca^{2+} signal. Synaptobrevin and syntaxin, together with a third protein crucial for docking, **SNAP-25**, are targets for **botulinum** and **tetanus toxins**, Zn^{2+} endopeptidases that are powerful inhibitors of neurotransmitter secretion.

After docking comes another calcium dependent step, **priming**, in which a

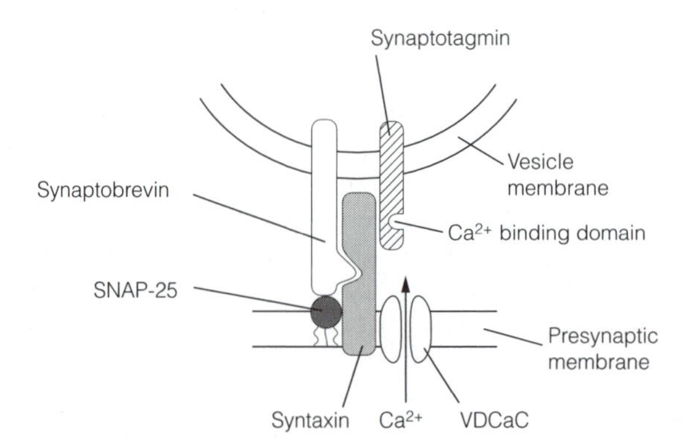

Fig. 1. Proteins involved in the docking of neurotransmitter vesicles. VDCaC, voltage-dependent calcium channel.

number of soluble cytoplasmic proteins form a transient complex with the SNAREs, resulting in partial fusion of vesicle and presynaptic membranes. This step involves the hydrolysis of ATP.

Primed vesicles are poised for exocytosis, requiring only a large pulse of Ca^{2+} to permit complete fusion of the vesicle and presynaptic membranes and opening of the **fusion pore** through which exocytosis occurs. A calcium binding protein located in the vesicle membrane, **synaptotagmin**, is the Ca^{2+} sensor in the exocytotic machinery. It is thought to act as a fusion clamp. In the absence of calcium it prevents complete fusion but when it binds Ca^{2+} it undergoes a conformational change which allows fusion to proceed. This final stage is fast. It must occur within 200 μs.

Endocytosis

Following exocytosis, synaptic vesicles are recycled within 30–60 s by endocytosis. Firstly vesicle membrane acquires a clathrin coat distorting it so that it invaginates into the terminal. Next a GTP-binding protein, **dynamin**, forms a collar around the neck of the invagination. Dynamin has an intrinsic GTPase activity which hydrolyses the bound GTP so triggering the fission of the coated vesicle from the presynaptic membrane. The GTP-bound form of dynamin requires calcium, so the same increase in nerve terminal Ca^{2+} concentration responsible for exocytosis also enables endocytosis. Once free in the terminal, the vesicle loses its clathrin coat (*Fig. 2*).

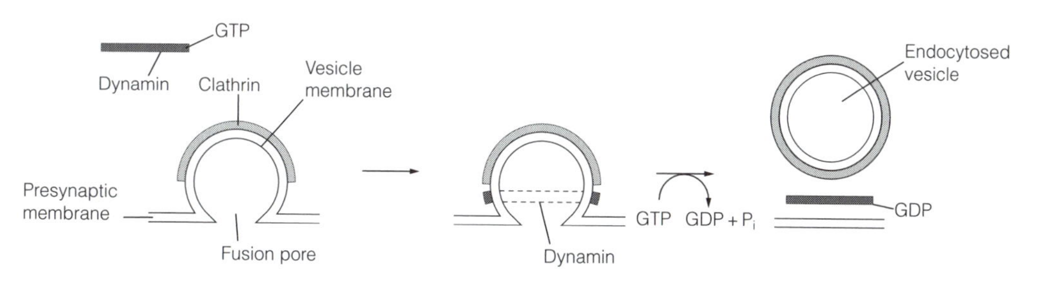

Fig. 2. Vesicle endocytosis. Reprinted from Revest, P.A. and Longstaff, A. (1998) Molecular Neuroscience. *© BIOS Scientific Publishers Ltd, Oxford.*

Refilling

SSVs are reloaded with neurotransmitter in the nerve terminals. The vesicles are acidified by the action of a proton ATPase. The transport of transmitter into vesicles is then driven by secondary active transport with H^+ efflux providing the energy (*Fig. 3*). Vesicle transporters have been identified for a number of

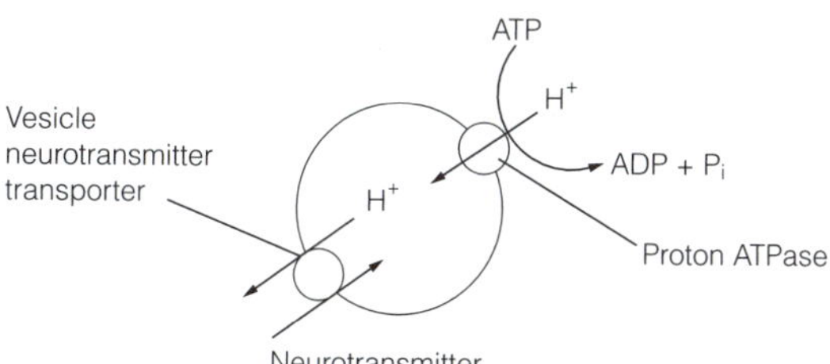

Fig. 3. Vesicle refilling. Reprinted from Revest, P.A. and Longstaff, A. (1998) Molecular Neuroscience. *© BIOS Scientific Publishers Ltd, Oxford.*

transmitters including glutamate, ACh, and catecholamines but not yet for GABA. They are large glycoproteins with 12 transmembrane segments. Surprisingly, they seem unrelated to neurotransmitter transporters located in plasma membranes of neurons or glia (Topic C7). Peptide transmitters, after synthesis on ribosomes in the cell body, are secreted into the lumen of the rough endoplasmic reticulum (RER), and packaged for export by the Golgi apparatus, from which the loaded vesicles are budded. These are then moved to the terminal by fast axoplasmic transport. This is necessary because nerve terminals are devoid of ribosomes and incapable of protein synthesis.

C6 VOLTAGE-DEPENDENT CALCIUM CHANNELS

Key Notes

Channel characterization

Calcium channels are responsible for excitation–secretion coupling in neurons, dendritic action potentials and excitation–contraction coupling in muscles. There are several types of calcium channel that can be characterized by their electrophysiology (activation voltage, conductance, time course of inactivation), by their susceptibility to blockade by specific drugs or toxins, and by their distribution.

Channel types

Most calcium channel types are activated by quite large depolarization. L-type channels are responsible for excitation–contraction coupling in muscle. They are blocked selectively by calcium channel antagonists. N-, P-, and Q-type channels are all implicated in neurotransmitter release. They can each be blocked selectively by toxins. These channels may coexist at some synapses, each contributing to the calcium influx required for exocytosis. T-type channels are activated by small depolarizations. This property underlies burst firing of many neurons.

Molecular biology of voltage-dependent calcium channels

A functional calcium channel consists of an α1 subunit which closely resembles a voltage-dependent sodium channel, together with auxiliary proteins which modify channel properties. That there are many different isoforms of α1 subunits is the cause of the diversity of channel types.

Related topics

Voltage-dependent ion channels (B3)
Properties of neurites (D1)
Skeletal muscles and excitation–contraction coupling (K1)
Cerebellar cortical circuitry (L2)
Autonomic nervous system function (M6)
Sleep (O4)
Epilepsy (R2)

Channel characterization

Voltage-dependent calcium channels control the influx of calcium, which couples excitation to secretion of transmitter. They are also responsible for calcium action potentials in dendrites (see Topic D1) and for excitation–contraction coupling in skeletal, cardiac and smooth muscle. There are several distinct types of calcium channel. They are all Ca^{2+} selective and activated by depolarization but can be differentiated by their electrophysiological properties, their sensitivity to drugs or toxins, and their distributions and functions in the nervous system. The electrophysiological criteria used to distinguish the channels include:

- The size of the depolarization needed to activate them. High voltage activated (HVA), need a large depolarization, and low voltage activated (LVA) require only a small depolarization.
- The conductance of the channel.
- The time course of inactivation.

Channel types

These are summarized in *Table 1*. **L-type** channels are HVA and require depolarization to –20 mV for activation. L-type channels are located in proximal dendrites of pyramidal neurons and contribute to their excitability but are not the presynaptic calcium channels involved in neurotransmitter release. They are the major calcium channels of excitation–contraction coupling. They are the only channel type so far which can be targeted by therapeutically useful agents, the calcium channel antagonists. While these are mostly used in cardiovascular medicine they may also have a role in the treatment of strokes, by reducing neuron excitability.

Three types of HVA calcium channel are implicated in transmitter release by the ability of selective toxins to block release. **N-type** channels, found on a variety of neurons, can be blocked by ω-conotoxin from the cone snail *Conus geographus* and play a more prominent part in GABA than glutamate release. **P-type** channels can be blocked by toxins of the funnel web spider *Agenelopsis aperta*, and are responsible for GABA release from cerebellar Purkinje cells (for which these channels are named), and acetylcholine release at the mammalian neuromuscular junction. Glutamate release from cerebellar granule cells is controlled by **Q-type** channels. P- and Q-type channels coexist in pyramidal cell terminals where each makes a contribution to transmitter release.

T-type channels are LVA channels, activated by depolarizations beyond –65 mV and relatively rapidly inactivating. These properties mean that they can generate burst firing of excitable cells. They are important in the thalamus (see Topics O5 and R2).

Molecular biology of voltage-dependent calcium channels

Calcium channels are macromolecular complexes composed of five different subunits. One, the α1 subunit is the functional channel (the other subunits are auxiliary proteins which can modify channel properties) and resembles the voltage-dependent sodium channel. There are many distinct α1 subunits because there are six genes which code for different versions and each is subjected to alternative splicing (see *Instant Notes Molecular Biology*, 2nd edn and *Instant Notes Biochemistry*, 2nd edn). This accounts for the diversity of channels.

Table 1. Calcium channel types

Type	Named for	Electrophysiology	Location
L	Long-lasting	HVA (–20 mV) Slowly inactivating	Pyramidal cells Skeletal, cardiac and smooth muscle Endocrine cells
T	Transient	LVA (–65 mV) Rapidly inactivating	Cardiac muscle Neurons (e.g. thalamic) Endocrine cells
N	Neuronal	HVA (–20 mV) Moderate inactivation	Neurons
P	Purkinje cell	HVA (–50 mV) Noninactivating	Cerebellar Purkinje cells Mammalian neuromuscular junction
Q	Q after P	HVA	Cerebellar granule cells
R	Remaining	HVA and LVA	

HVA, high voltage activated; LVA, low voltage activated.

C7 Neurotransmitter inactivation

Key Notes

The reason for inactivation

Inactivation of transmitter action occurs so that synapses can be modulated on a fast timescale. Inactivation can occur by enzymic degradation, by transport out of the cleft back into neurons or glia, or by diffusion away from the synapse.

Enzyme degradation

Physiologically, only the hydrolysis of acetylcholine (ACh) by acetylcholinesterase (AChE) is an important example in this category. Choline liberated by the breakdown of ACh is taken up into the nerve terminal via a high affinity Na^+-dependent cotransport system. ACh is synthesized from choline and acetyl CoA by the action of choline acetyltransferase.

Transport

Reuptake from the synaptic cleft into neurons (and glia in the case of the amino acids) is a major mechanism for the inactivation of the classical transmitters. Two major families of transporters are involved. These molecules are unrelated to the vesicular transporters for transmitters. Some of these transporters are targets for antidepressant drugs.

Diffusion

Diffusion away from the synaptic cleft is the major mode of inactivation of the peptides and is also probably important for glutamate and γ-aminobutyrate (GABA). The large size of the peptides makes their diffusion slow and accounts for their protracted actions at synapses.

Related topics

Slow neurotransmission (C3)
Skeletal muscles and excitation–contraction coupling (K1)
Dopamine neurotransmission (N1)

The reason for inactivation

Inactivation is necessary to ensure that the synapse can respond to rapid changes in the firing frequency of the presynaptic neuron. Without it the post-synaptic cell could not be updated on recent changes in incoming signals. Also over a timecourse of a few seconds many neurotransmitter receptors continuously exposed to their ligand (or other agonist) become desensitized. This would reduce the sensitivity of the synapse. There are three ways in which transmitter can be inactivated and they are not mutually exclusive: enzymic degradation, transport out of the synaptic cleft back into neurons or glia, or by passive diffusion away from the synapse.

Enzyme degradation

Many enzymes are involved in the catabolism of both classical and peptide transmitters and pharmacological manipulation (e.g. inhibition) of many of these can have consequences for synaptic transmission. However, only in the case of ACh is enzymic degradation important in the physiological inactivation of transmitter. ACh is hydrolysed by **acetylcholinesterase** (**AChE**), which cleaves the transmitter molecule into choline and acetate. Choline is taken back into the nerve terminal by a Na^+-dependent transporter. AChE has an extremely high catalytic activity and at the neuromuscular junction can reduce the concentration of ACh from about 1 mM immediately after release to virtually zero in about 1 ms. AChE exists in a variety of isoforms. Some (G-forms) are soluble and secreted into the cleft, others (A-forms) have collagen tails by which they are tethered to the plasma membrane with their catalytic domains facing into the cleft.

ACh is resynthesized from choline and acetyl CoA (derived from pyruvate) by choline acetyltransferase (ChAT) and loaded into vesicles by a specific transporter (see *Fig. 3* of Topic C5). Interestingly, the ACh vesicle transporter is coded by part of the ChAT gene. Hence the synthesis of the transporter and the enzyme are coregulated.

Transport

Many classical transmitters are inactivated by their removal from the cleft via high affinity, saturable, secondary active transport. The amino acid transmitters may be transported into both neurons or glia whereas amines are transported only into neurons. Two families of transporter have been identified which serve this function:

(1) *Na^+/K^+ cotransporter family* constitutes the glutamate (and aspartate) transporters. Three have been discovered so far, two present in glia (astrocytes) and one localized in neurons. Glutamate transport is electrogenic, that is, it results in a modest potential difference being set up across the membrane, inside-positive (see *Fig. 1*). A consequence of this is that excessive depolarization of the membrane can reverse the direction of the transport causing glutamate efflux into the cleft. This can have deleterious effects. Glutamate transporters have been cloned and sequenced. There is some doubt about the secondary structures of these molecules but one model gives them 8 transmembrane (TM) segments.

(2) *Na^+/Cl^- cotransporter family* (*Fig. 2*). This is a large family and includes three GABA transporters, the transporters for norepinephrine/epinephrine, dopamine, serotonin, glycine, and the high affinity transporter for choline (see previous subsection). The three GABA transporters are all expressed in both neurons and glia yet pharmacological experiments can distinguish glial from neuronal GABA uptake. It is likely then that more GABA transporters remain to be discovered. The norepinephrine and serotonin transporters are the targets for the tricyclic antidepressant class of drugs. The recent development of drugs acting at the serotonin transporter, selective serotonin reuptake inhibitors, such as fluoxetine (prozac), is proving useful

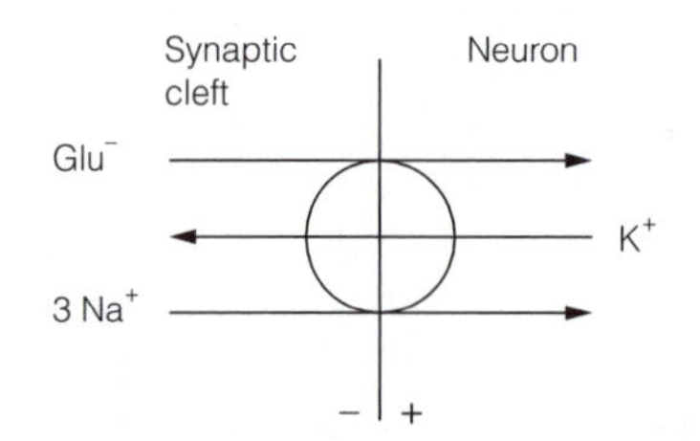

Fig. 1. Glutamate transport by a Na^+/K^+ cotransporter.

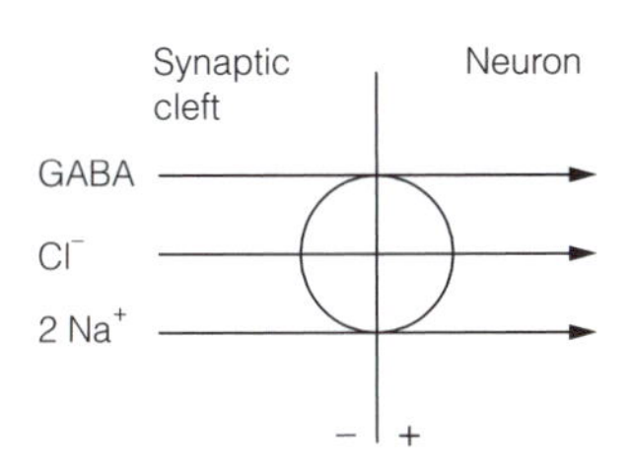

Fig. 2. GABA transport by a Na^+/Cl^- cotransporter.

in the treatment of depression because they have fewer side effects than the tricyclic antidepressants. The dopamine transporter is the target for cocaine. By inhibiting dopamine reuptake, cocaine deranges dopamine transmission in reward pathways in the brain, accounting for its addictive properties (Topic O1). Many members of this family have been sequenced. They are large glycoproteins with 12 TM segments but have no homology with the 12 TM vesicle transporters (*Fig. 2b*).

Diffusion

Despite the existence of transporters, diffusion out of the synapse may be important for the inactivation of glutamate and GABA at synapses in the cerebral cortex. There is no high affinity reuptake for peptides and although peptides may be internalized by neurons via receptor-mediated endocytosis and then degraded by nonspecific peptidases, this is probably not an important mechanism for their inactivation. Hence the major route for terminating the synaptic action of peptides is by diffusion. However, peptides are very much larger than the small classical transmitter molecules and there are significant barriers for free diffusion out of the cleft. This means that peptides clear only slowly from a synapse, which helps to explain why their actions can be so prolonged.

C8 AUTORECEPTORS

Key Notes

Autoreceptor functions
Autoreceptors respond to the transmitter released by the neuron in which they are located. They occur at the presynaptic terminal, the soma and dendrites. They regulate neurotransmitter release, synthesis, and neuron firing rate, usually homeostatically.

Regulation of neurotransmitter release
Most autoreceptors act to reduce transmitter release by reducing the calcium influx into the nerve terminal. At a few sites, autoreceptor activation increases secretion.

Regulation of neurotransmitter synthesis
Autoreceptors on catecholaminergic and serotonergic cells reduce the synthesis of their corresponding transmitter. In the case of dopamine, this is effected by D2 receptor-mediated reduction in cyclic adenosine monophosphate (cAMP) which reduces the activity of tyrosine hydroxylase.

Heteroceptors
Presynaptic receptors that do not respond to the transmitter liberated by the neuron in which they are placed are heteroceptors. They regulate transmitter secretion.

Related topics
Slow neurotransmission (C3)
Neurotransmitter release (C5)
Dopamine neurotransmission (N1)

Autoreceptor functions

Neurotransmitter receptors are not confined to the postsynaptic membrane but also exist in the presynaptic membrane, where they are termed **presynaptic receptors**, and over the cell body and dendrites. If these receptors have, as their ligand, a transmitter released by the neuron in which they are located, they are **autoreceptors**. Autoreceptors have several functions that are normally homeostatic. Those on the presynaptic membrane are involved in regulating neurotransmitter release. In addition, in catecholamine- and serotonin-using neurons presynaptic autoreceptors regulate the synthesis of the transmitter, while somatodendritic autoreceptors regulate the firing rate of the neuron. Autoreceptors are invariably metabotropic receptors.

Regulation of neurotransmitter release

Presynaptic autoreceptors, when activated, usually decrease the release of neurotransmitter. This is a negative feedback mechanism acting to limit the degree of transmitter release, either to avoid excessive excitation, or to curtail postsynaptic receptor desensitization which would reduce the sensitivity of the synapse. Presynaptic autoreceptors reduce transmitter release by reducing calcium influx into the presynaptic terminal. For example, at central γ-aminobutyrate (GABA)ergic synapses the autoreceptors are $GABA_B$ receptors which activate G_0 proteins. The G_0 protein has two effects:

(1) By binding to voltage-dependent potassium channels in the presynaptic membrane it causes them to open. The resulting increase in K^+ efflux reduces the likelihood that N-type Ca^{2+} channels be activated by an action potential invading the terminal.
(2) It binds directly to the calcium channel, reducing Ca^{2+} entry into the terminal.

The combined effect of these actions reduces GABA release (*Fig. 1*).

Occasionally activation of presynaptic autoreceptors results in an increase in transmitter release. For example norepinephrine secretion is controlled by two populations of autoreceptor: an α2 receptor which decreases release and a β receptor which increases release.

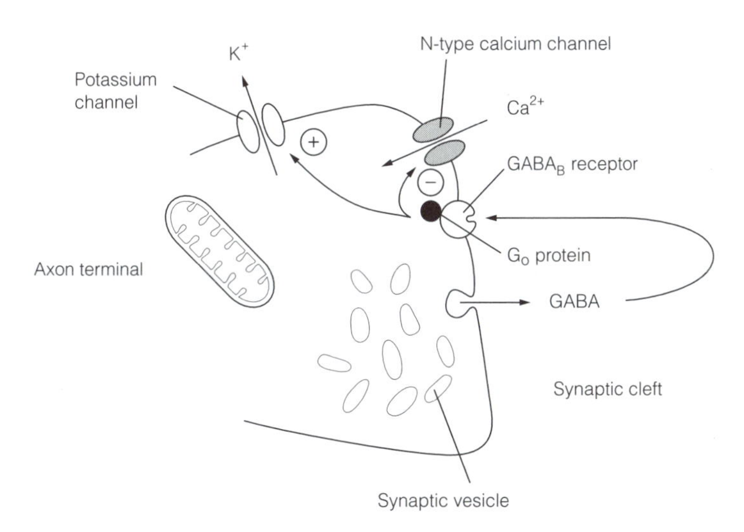

Fig. 1. $GABA_B$ autoreceptor actions.

Regulation of neurotransmitter synthesis

The synthesis of catecholamines and serotonin in their respective neurons is decreased by activation of the corresponding autoreceptor. Some dopaminergic neurons, for example, have autoreceptors that are members of the D2 family of dopamine receptors that couple to G_i proteins. One action of the G_i proteins is to lower calcium influx so reducing dopamine release. But G_i also inhibits adenylyl cyclase, lowering cAMP concentration. Tyrosine hydroxylase is the enzyme which catalyses a key step in the metabolic pathway leading to dopamine synthesis. The active form of this enzyme is phosphorylated by protein kinase A. Thus the activity of tyrosine hydroxylase is reduced by the autoreceptor-mediated decrease in cAMP, and dopamine synthesis is lowered.

Heteroceptors

Some presynaptic receptors are receptors for transmitters not secreted by the neuron in which they are situated. These are called **heteroceptors**. Such receptors regulate transmitter release. For example $GABA_B$ receptors exist presynaptically at glutamatergic synapses where they reduce glutamate release. It is assumed that they are activated by GABA that has diffused from neighboring synapses.

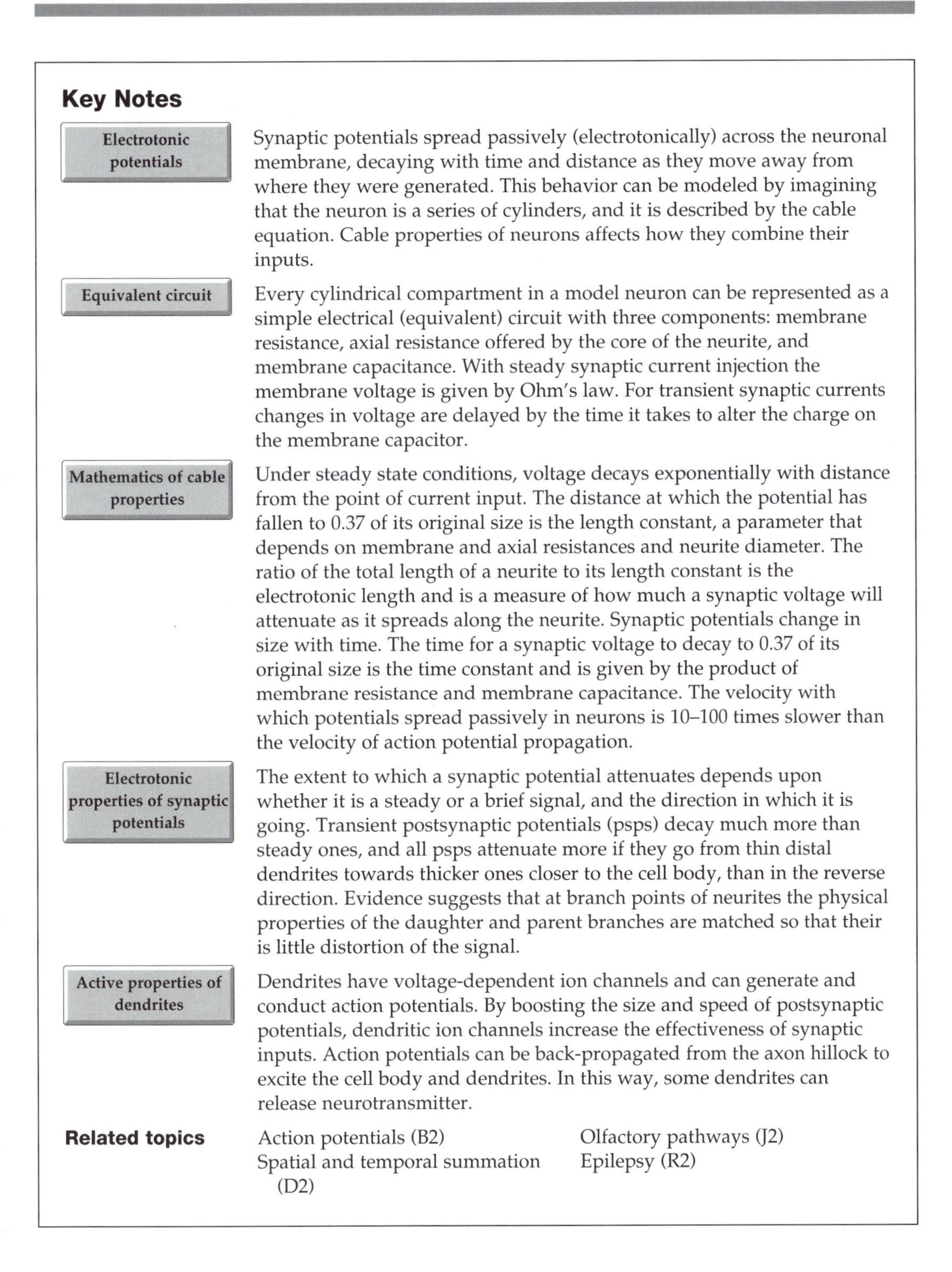

D1 Properties of neurites

Key Notes

Electrotonic potentials

Synaptic potentials spread passively (electrotonically) across the neuronal membrane, decaying with time and distance as they move away from where they were generated. This behavior can be modeled by imagining that the neuron is a series of cylinders, and it is described by the cable equation. Cable properties of neurons affects how they combine their inputs.

Equivalent circuit

Every cylindrical compartment in a model neuron can be represented as a simple electrical (equivalent) circuit with three components: membrane resistance, axial resistance offered by the core of the neurite, and membrane capacitance. With steady synaptic current injection the membrane voltage is given by Ohm's law. For transient synaptic currents changes in voltage are delayed by the time it takes to alter the charge on the membrane capacitor.

Mathematics of cable properties

Under steady state conditions, voltage decays exponentially with distance from the point of current input. The distance at which the potential has fallen to 0.37 of its original size is the length constant, a parameter that depends on membrane and axial resistances and neurite diameter. The ratio of the total length of a neurite to its length constant is the electrotonic length and is a measure of how much a synaptic voltage will attenuate as it spreads along the neurite. Synaptic potentials change in size with time. The time for a synaptic voltage to decay to 0.37 of its original size is the time constant and is given by the product of membrane resistance and membrane capacitance. The velocity with which potentials spread passively in neurons is 10–100 times slower than the velocity of action potential propagation.

Electrotonic properties of synaptic potentials

The extent to which a synaptic potential attenuates depends upon whether it is a steady or a brief signal, and the direction in which it is going. Transient postsynaptic potentials (psps) decay much more than steady ones, and all psps attenuate more if they go from thin distal dendrites towards thicker ones closer to the cell body, than in the reverse direction. Evidence suggests that at branch points of neurites the physical properties of the daughter and parent branches are matched so that their is little distortion of the signal.

Active properties of dendrites

Dendrites have voltage-dependent ion channels and can generate and conduct action potentials. By boosting the size and speed of postsynaptic potentials, dendritic ion channels increase the effectiveness of synaptic inputs. Action potentials can be back-propagated from the axon hillock to excite the cell body and dendrites. In this way, some dendrites can release neurotransmitter.

Related topics

Action potentials (B2)
Spatial and temporal summation (D2)
Olfactory pathways (J2)
Epilepsy (R2)

Electrotonic potentials

Action potentials are actively regenerated by opening of voltage-dependent ion channels at each point along an unmyelinated axon, or at the nodes of Ranvier in a myelinated one. Because of this, action potentials are propagated without loss of size along the axon. However, most of the synaptic potentials generated on the surface of a neuron are subthreshold. These potentials spread passively, **electrotonically**, in a manner determined only by the physics of the cell – decaying with time and distance. Physicists have derived an equation, the **cable equation**, that describes the spread of a current with time and distance along electrical cables. Because neurons can be regarded as a series of cylindrical compartments, this cable equation can be used to model the electrotonic spread of potentials along them. The **cable properties** of neurons are important because they govern how nerve cells integrate their synaptic inputs, and this specifies what computations they can do.

Equivalent circuit

Each cylindrical compartment of a neuron can be modeled as a simple electrical circuit, the **equivalent circuit** (*Fig. 1a*).

Current injected into a neuron will initially change the amount of charge stored on the membrane. The membrane acts as a **capacitor** because it is an insulator (a phospholipid bilayer) separating two conductors (the electrolyte solutions of cytoplasm and extracellular fluid). If the current injected is constant (i.e. at steady state) the voltage change caused by the current will be that predicted by Ohm's law, $V = IR$. The resistance, R, has two components: the **internal** or **axial resistance** offered by a cylinder of cytoplasm of unit length, r_i, and the **membrane resistance** of a cylinder of membrane of unit length, r_m. As the current spreads along a dendrite or axon, through r_i, some of it will leak across the membrane resistance, r_m. Consequently, as the distance away from the site of injection gets longer, the current – and thus the voltage change – gets smaller. This explains why synaptic potentials decay with distance. When the current is switched off, the voltage changes again. In a purely resistive circuit (i.e. one without a capacitance) the voltage would alter instantaneously with a step size given by Ohm's law. However, because real neurons have a capacitance, the voltage change is initially offset by alteration of the amount of charge stored on the membrane. The outcome is that the voltage changes gradually with time in an exponential fashion (*Fig. 1b*).

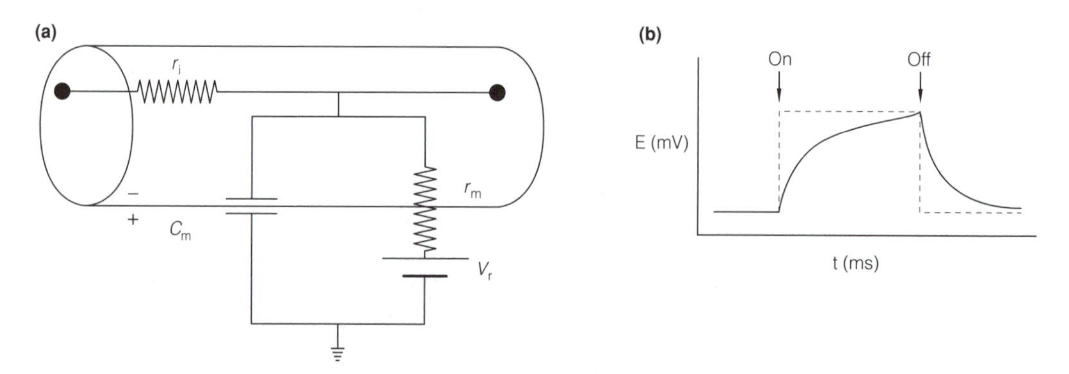

Fig. 1. (a) Equivalent circuit for neuron compartment of unit length: r_i is the internal or axial resistance (Ω cm^{-1}), r_m is membrane resistance (Ω cm), c_m is membrane capacitance (F cm^{-1}), V_r is the resting potential. The resistance of the external medium is assumed to be negligible. (b) Effect of transient current injection into a neurite. The membrane capacitance introduces a delay in the rise and fall of voltage. The dotted line depicts what would happen if the capacitance were zero.

Mathematics of cable properties

The cable equation can be solved under steady state conditions, in which capacitance can be ignored, to show that the exponential decay of a voltage, V_0, is given by:

$$V_x = V_0 e^{-x/\lambda}$$

where V_x is the voltage at some distance, x, along the neurite, V_0 is the voltage at $x = 0$ (i.e. where the synaptic potential is generated) and λ is the **length constant** (space constant and characteristic length are alternative names).

Making $x = \lambda$ gives

$$V_x = V_0 e^{-1}$$
$$= 0.37\, V_0$$

Hence λ is the length a voltage signal travels before it has decayed to 37% of its original size and so is a measure of how far a current can spread passively along a neurite. So, λ just shows how much the voltage decays with distance (*Fig. 2*).

The cable equation gives the value of the length constant as just:

$$= (aR_m/2R_i)^{\frac{1}{2}}$$

where a is the neuron radius, R_m and R_i are the specific membrane resistance and specific internal resistance, respectively. They are termed 'specific' because they apply to a particular neuron. Both R_m and R_i can be measured experimentally. R_m varies hugely in different neurons, in different parts of a neuron and even in the same region, over time, by between 10^3 and 10^5 Ω cm^2. This 100-fold difference will only change λ by 10-fold, however, because λ varies with the square root of the membrane resistance. Although R_i only ranges between 50 and 200 Ω cm, it has an appreciable effect on the value of λ. The length constant also depends on the radius of the neurite.

Neurites range from about 0.01 to 10 µm, giving an approximately 30-fold difference in λ; short for fine processes, long for thick ones. The **electrotonic length**, L, relates the actual length of a neurite to its length constant and so is a measure of how much a signal will attenuate over its whole length:

$$L = x/\lambda$$

For many projection neurons, electrotonically spread signals decay to virtually zero along their axons. This highlights the necessity for action potentials to

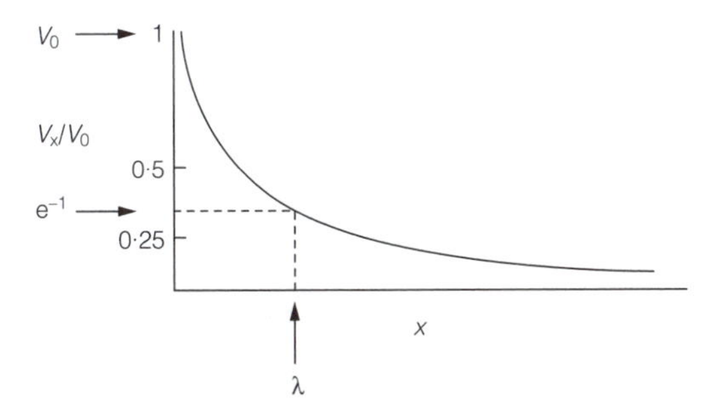

Fig. 2. Decay of voltage, V_x, with distance, x. The length constant, λ, is the distance at which the voltage has fallen to 1/e of its initial value.

transmit signals. By the same token, many interneurons have extremely short axons in which $\lambda \geq L$ and so action potentials are unimportant.

For transient signals, membrane capacitance cannot be ignored because it determines the rise time and decay time of the signal (*Fig. 2b*).

The cable equation shows that to a first approximation the fall of potential with time is exponential and described by:

$$V_t = V_0 e^{-t/\tau}$$

where V_t is the potential at time t, V_0 is the potential at $t = 0$ and τ is the **time constant**. The time constant is a measure of how fast currents rise or decay passively as they spread through neurons (*Fig. 3*). The value of τ is given by:

$$\tau = R_m C_m$$

C_m is the specific membrane capacitance. Thus, τ is independent of neuron radius. Moreover, as specific membrane capacitance is fairly constant at 0.75 μF cm^{-2}, the time constant essentially depends only on R_m, which in turn is determined by the number of open ion channels and their conductances.

It is important to note that the length constant and time constant are not static for a neuron, but change with time. As both depend on the membrane resistance, they alter whenever ion channels in the membrane open or close. Similarly the electrotonic lengths of dendrites (which depends on λ) change with time.

The cable equation shows that the velocity, θ, of electrotonic conduction of synaptic potentials is given by:

$$\theta = 2\lambda/\tau$$

Electrotonic conduction velocities are 10–100 times lower than action potential conduction velocities in a given neuron.

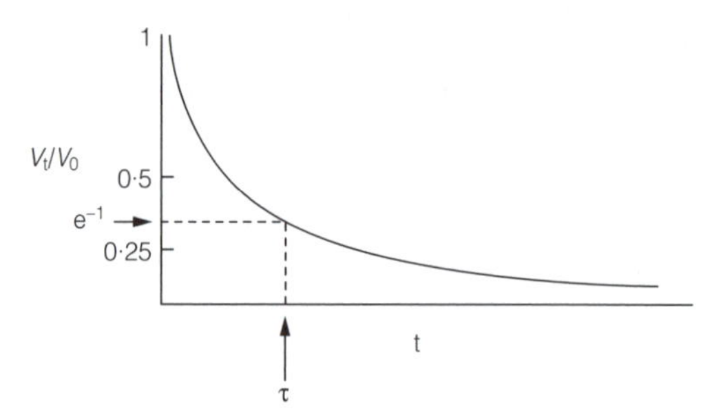

Fig. 3. Decay of voltage, V_t, with time, t. τ is the time constant for the decay.

Electrotonic properties of synaptic potentials

Under steady state conditions (approximated by long-lasting psps) the electrotonic length between synapse and axon hillock gives a good measure of the efficacy with which the synapse fires the cell. For a given size synaptic input, distal synapses are less effective than more proximal ones. Since for many neuron types L lies between 0.3 and 1.5, the attenuation of psps – even those at distal sites – is not great. Moreover, in some neurons, such as spinal motor neurons, signal decay is offset by their having larger synaptic conductances at distal than at proximal synapses.

For transient responses (fast epsps and ipsps), however, where the time constant is important as well as the length constant, the situation is different. Brief responses decay much more than steady ones, because a greater fraction of a brief current goes to charge the membrane capacitor.

The direction in which a postsynaptic potential spreads through a dendritic tree is important in determining its amplitude and time course. Decay is more pronounced for potentials travelling along dendrites towards the cell body (distal to proximal) than the reverse direction. This effect is more marked for fast epsps and ipsps than for slow ones. As a postsynaptic current flows from a thin distal dendrite with a high axial resistance into the larger proximal dendrites it faces a big fall in axial resistance, a **conductance load**. By Ohm's law, this causes the voltage change produced by the current to drop. Moreover, if the current is a transient one, as it moves from the distal dendrite into the rest of the dendritic tree much of it will go to charge its membrane capacitor, which takes time. So, as a psp on distal parts of the dendritic tree spreads passively towards the soma through increasingly larger dendrites it gets smaller and is progressively delayed.

Active properties of dendrites

Dendrites have voltage-dependent ion channels and under some circumstances can trigger and propagate action potentials. The fact that dendrites can be excitable greatly increases the complexity and sophistication of information processing by neurons.

Fast synaptic potentials originating at distal synapses on the dendritic tree suffer considerable weakening and slowing if they spread passively. Activation of voltage-dependent Na^+ and Ca^{2+} channels can boost the size and speed of electrotonic potentials. In this way, active dendrites enhance the efficacy of synaptic inputs.

In many types of neuron, action potentials triggered at the axon hillock are propagated not only down the axon, but also sweep over the cell body and invade the dendrites. Action potentials that go the 'wrong' way in a neuron are said to be **antidromic** or **back-propagated**. In pyramidal cell dendrites, back-propagated action potentials are large, long-lasting calcium spikes (the density of Na^+ channels in dendrites is too low to allow conduction of sodium spikes) that are needed to trigger the bursts of axonal sodium action potentials that are part of the normal repertoire of these cells. Back-propagated action potentials can also lead to release of neurotransmitter from dendrites. One site where this occurs is in the olfactory bulb (Topic J2). Here, reciprocal synapses, formed between mitral cells and granule cells, are in effect two synapses side by side. One is axodendritic, the other has vesicles in the dendrite and the axon is post-synaptic (*Fig. 4*). Back-propagation of action potentials into the granule cell dendrite causes it to secrete GABA which inhibits the mitral cell.

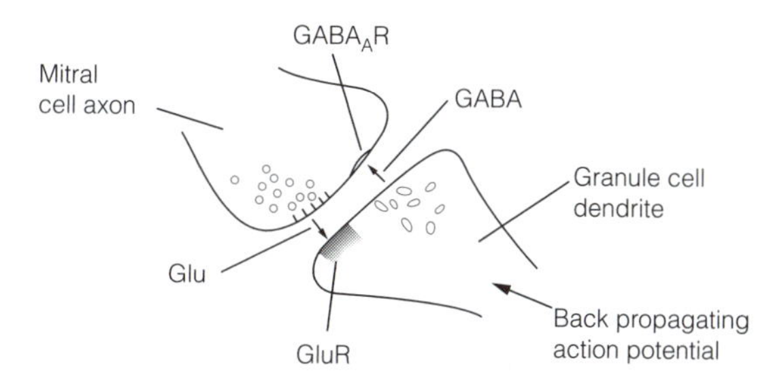

Fig. 4. Reciprocal synapses between mitral and granule cells in the olfactory bulb.

D2 Spatial and temporal summation

Key Notes

Neurons as integrators

Postsynaptic potentials (psps) generated on a neuron, both excitatory and inhibitory, add together (summate). If the result of this summation is that the axon hillock membrane potential is driven beyond threshold, the neuron will fire. So, whether or not a neuron will fire at any moment depends on how many excitatory and inhibitory synapses are active, and where they are located. It is by integrating synaptic inputs in this way that neurons act as computational devices.

Temporal summation

The summation of psps generated at slightly different times is temporal summation. The extent to which it occurs depends on the time constant: the longer this is the greater the summation. Slow potentials are more likely to be summed than fast ones. Temporal summation is nonlinear: successive psps have progressively smaller influence.

Spatial summation

The summation of potentials arriving on different parts of the neuron is spatial summation. It depends on the length constant and (for fast psps) the time constant. For distant sites summation is linear, but for sites that are close spatial summation is nonlinear; the size of the combined response is smaller than predicted from simple addition.

Related topic

Properties of neurites (D1)

Neurons as integrators

Many thousands of synapses are formed on a neuron, both excitatory and inhibitory. At any given time a subset of these will be activated to generate epsps and ipsps. A special property of these graded potentials is that they **summate**, or add together. If a sufficient number of epsps are produced, in summing they will drive the axon hillock membrane potential across the threshold for triggering action potentials and the neuron will fire. The axon hillock is crucial because, being the region of a neuron with the highest density of voltage-dependent sodium channels, it has the lowest threshold. If at any instant insufficient excitatory synapses are activated, or a high level of excitatory synaptic input is more than offset by the generation of ipsps from inhibitory input, then the axon hillock will not be driven across the threshold and the cell will not fire. So, neurons are decision-making devices. The decision to fire or not is actually taken by the axon hillock on the basis of whether the sum total of epsps and ipsps causes its membrane potential to become more positive than the firing threshold. It is this operation that constitutes information processing by individual neurons. In engineering terms, a synapse converts digital signals into analog signals. The neuron then integrates all of its analog signals over a short time and compares the result of that integration with a given threshold to decide whether to fire. When it does fire the signal (the action potential) is digital.

Experiments on pyramidal cells show that about 100 excitatory synapses, on average, must be activated at the same time to trigger an action potential. However the efficacy with which a synapse can influence firing depends on its position. Because postsynaptic potentials decay as they spread passively towards the axon hillock, a synapse far out on distal dendrite will be weaker than one closer to the cell body. In this context it is noteworthy that on pyramidal cells there are only about 250 inhibitory synapses on the cell body but 10 000 or so excitatory axodendritic synapses. The relative strength of a synapse in contributing to a neuron's output is its **weighting**. This need not be a fixed property but may change with time.

Temporal summation

If an afferent neuron fires a series of action potentials in quick succession (a **volley**), then the earliest psps generated in the postsynaptic cell will not have time to decay before the next psps arrive. Hence successive psps summate over time. This is referred to as **temporal summation**. Its properties are:

- The extent of the summation depends on the time constant of the postsynaptic cell. The shorter τ, the faster the psp decays, the higher the firing frequency of the volley needed to achieve a given degree of summation. An effect of slow psps is to enhance temporal summation.
- Temporal summation is nonlinear. Successive psps get slightly smaller because earlier psps reduce the ionic driving force for later ones.
- A sufficiently large temporal summation will cause the postsynaptic cell to reach firing threshold.

Spatial summation

The summing of postsynaptic potentials generated at separate points on the neuron surface is called **spatial summation** (*Fig. 1*).

The properties of spatial summation are:

- The extent of spatial summation is determined by the cable properties of the neuron. Although the spread of slow psps set up by transmitters acting at metabotropic receptors may approximate to steady state conditions, the spread of fast psps will depend on τ as well as λ.
- For synaptic inputs that are quite distant on a cell's surface summation is linear. The size of the potential change is just the algebraic sum of all the psps. However, for inputs that are adjacent, the resulting potential change is smaller than would be predicted from the sizes of the contributing psps, this is nonlinear spatial summation.
- The outcome of spatial summation can be cell firing.

Although temporal and spatial summation are described as separate processes, both occur together as a neuron is stimulated and it is their combined effect which dictates whether it will fire. The frequency with which a cell fires, and how long it fires, is determined by the amplitude and duration of the depolarization of the axon hillock membrane respectively.

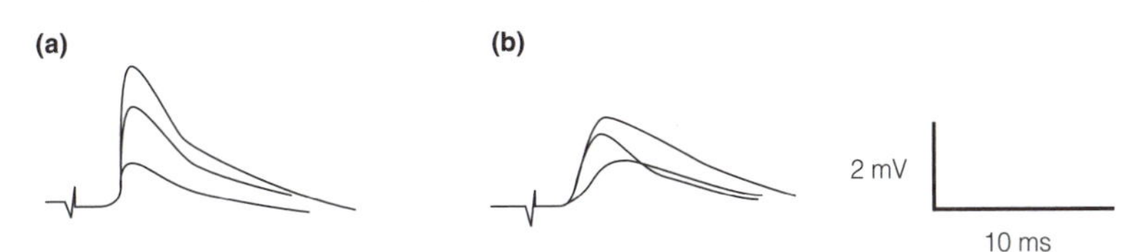

Fig. 1. Spatial summation. In each case the upper trace is the summed response of the two lower epsps generated at synapses: (a) a long way apart, (b) close together.

E1 ORGANIZATION OF THE PERIPHERAL NERVOUS SYSTEM

Key Notes

Principal divisions of the nervous system

The brain and spinal cord comprise the central nervous system, while the peripheral nervous system, divided into somatic, autonomic and enteric parts, is everything else.

Somatic nervous system

Thirty-one pairs of spinal nerves originating from the spinal cord and 12 pairs of cranial nerves arising from the brain make up the somatic nervous system. All spinal nerves are mixed nerves, containing both sensory and motor fibers. Of the cranial nerves only four are mixed: some are purely sensory and others purely motor. Every spinal segment gives rise to a pair of spinal nerves, each with a dorsal root containing sensory fibers and a ventral root with motor fibers. The cell bodies of the sensory fibers lie outside the spinal cord in the dorsal root ganglia.

Peripheral nerves consist of nerve fibers (axons surrounded by their associated Schwann cells) organized into bundles (fasciculi) and invested with connective tissue. Peripheral nerve fibers are classified by their diameters and conduction velocities.

Autonomic nervous system (ANS)

This visceral motor system originates with cell bodies in the brainstem and spinal cord that give rise to preganglionic myelinated axons which secrete acetylcholine (ACh). They synapse with postganglionic unmyelinated fibers in autonomic ganglia. The ANS has two divisions, sympathetic and parasympathetic. The sympathetic system arises from thoracic and lumbar spinal segments, has its autonomic ganglia close to the cord in the paravertebral chains or subsidiary ganglia, and its long postganglionic fibers secrete norepinephrine. The adrenal medulla secretes epinephrine into the bloodstream under the influence of preganglionic sympathetic innervation. The parasympathetic system originates from the brainstem and sacral spinal cord. Its autonomic ganglia are located on or near the innervated organ. The short postganglionic fibers secrete ACh.

Enteric nervous system (ENS)

The nervous system of the gut is organized into two highly interconnected cylindrical sheets of neurons embedded in the gut wall called the submucosal and myenteric plexuses. The enteric nervous system regulates gut function semi-autonomously although its activity is modified by the ANS.

Related topics

Organization of the central nervous system (E2)
Dorsal column pathways for touch sensations (G2)
Eye and visual pathways (H2)
Skeletal muscles and excitation–contraction coupling (K1)
Autonomic nervous system function (M6)
Early patterning of the nervous system (P1)

Principal divisions of the nervous system

The nervous system is comprised of the **central nervous system (CNS)** and **peripheral nervous system (PNS)**. These divisions are contiguous both anatomically and functionally. The CNS includes the brain and spinal cord. The PNS is everything else: namely nerve trunks going between the CNS and the periphery, and the networks of nerve cells with supporting glia in organs throughout the body. The PNS has three subdivisions, the somatic, autonomic and enteric nervous systems.

Somatic nervous system

The somatic nervous system structure reflects both the bilateral symmetry and segmented nature of the vertebrate body plan. It consists of 31 pairs of **spinal nerves**, each pair arising from a single segment of the spinal cord, and 12 pairs of **cranial nerves** which have their origin in specific regions in the brain. Both spinal and cranial nerves can carry axons entering the CNS, **afferent (centripetal) fibers** and axons leaving the CNS, **efferent (centrifugal) fibers**. Afferent fibers carry sensory information from skin, muscles, joints and viscera. The majority of these afferents are wired to **mechanoreceptors** which inform about mechanical forces impinging on the surface or produced within the body. Some act as **nociceptors** which respond to factors associated with tissue damage, and some (restricted to the skin) are connected to **thermoreceptors** which are temperature sensitive. Efferent fibers are the axons of motor neurons supplying skeletal muscles. The synapse between a motor neuron and a skeletal muscle fiber is called a **neuromuscular junction**.

All spinal nerves are **mixed**, that is, they contain both sensory and motor fibers. Of the cranial nerves only four are mixed (see *Table 1*). The olfactory, optic and vestibulocochlear are essentially pure sensory nerves, while the oculomotor, trochlear, abducens, accessory and hypoglossal are purely motor. The optic nerves, unique among the cranial nerves, develop as direct outgrowths of the brain so it could be argued that they and the retinae are part of the CNS. All other PNS components arise from the neural crest (see Topic P1).

Each spinal nerve is formed from a **dorsal root** which contains sensory fibers and a **ventral root** carrying motor fibers. The cell bodies of the primary afferent neurons lie within the **dorsal root ganglia (DRG)** just outside the spinal cord. There are a pair of DRGs for each spinal segment. Efferent neuron cell bodies lie within the spinal cord (*Fig. 1*).

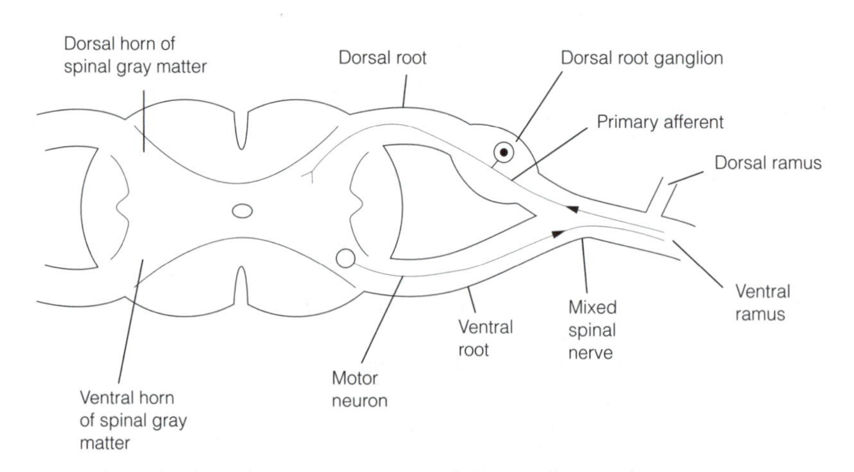

Fig. 1. Origin of a spinal nerve from a spinal cord segment.

Table 1. Peripheral nerves

Nerve	Type	Region of origin or destination in CNS	Function
Cranial nerves			
I Olfactory	Sensory	Olfactory bulb	Smell
II Optic	Sensory	Forebrain LGN (thalamus) Midbrain superior colliculus tectum	Vision Visual reflexes
III Oculomotor	Motor[a]	Midbrain	Motor to extrinsic eye muscles except superior oblique and lateral rectus, autonomic to intrinsic eye muscles
IV Trochlear	Motor	Midbrain	Motor to superior oblique extrinsic eye muscles
V Trigeminal	Mixed	Midbrain and hindbrain	Sensory from head and face, motor to jaw muscles
VI Abducens	Motor	Hindbrain	Motor to lateral rectus extrinsic eye muscles
VII Facial	Mixed[a]	Ventral lateral thalamus (sensory) Hindbrain (motor)	Sensory from tongue (taste) and palate Motor to face, parasympathetic secretomotor to submandibular, submaxillary salivary glands and lachrymal glands
VIII Vestibulocochlear	Sensory	MGN (auditory division) Hindbrain (vestibular division)	Sensory from inner ear (hearing and balance)
IX Glossopharyngeal	Mixed[a]	Hindbrain	Sensory from tongue (taste) Motor to pharyngeal muscles Parasympathetic secretomotor to parotid salivary glands
X Vagus	Mixed[b]	Hindbrain	Sensory from viscera Somatic motor to pharyngeal and laryngeal muscles Parasympathetic to viscera
XI Accessory	Motor	Medulla, spinal cord C1–C5	Motor to palate and some neck muscles
XII Hypoglossal	Motor	Medulla	Motor to tongue
Spinal nerves			
C1–8	Mixed		
T1–12	Mixed (including sympathetic autonomic T1–12)		
L1–5	Mixed (including sympathetic autonomic L1, 2)		
S1–5	Mixed (including parasympathetic autonomic S2, 3)		
Cx 1	Mixed		

[a]Including autonomic.

[b]Large autonomic component.

LGN, lateral geniculate nucleus; MGN, medial geniculate nucleus.

All peripheral nerves have a common basic structure. A **nerve fiber** consists of an axon together with accompanying Schwann cells. Several unmyelinated axons are invested by a single glial cell which comprises the **neurolemma**. In myelinated axons this term is reserved for the outer, nucleated, cytoplasm-rich portion of the Schwann cell. Nerve fibers are collected into bundles, **fasciculi,** surrounded by a connective tissue sheath, the **perineurium**. Within the fasciculus individual fibers are supported by a connective tissue network, the **endoneurium** which is continuous with the perineurium. A nerve may be one or several fasciculi all bound up by a connective tissue **epineurium**.

Two systems for the classification of PNS nerve fibers are in common use. They are based on fiber diameter and conduction velocity and are summarized in *Table 2*. The Erlanger and Gasser system is used to classify both afferents and efferents. The Lloyd and Hunt scheme is used exclusively to define sensory fibers.

Table 2. Classification of peripheral nerve fibers

Fiber (type/group)	Mean diameter (μm)	Mean θ (m s^{-1})	Functions (example)
Erlanger/Gasser classification (type)			
Aα	15	100	Motor neurons
Aβ	8	50	Skin touch afferents
Aγ	5	20	Motor to muscle spindles
Aδ	4	15	Skin temperature afferents
B	3	7	Unmyelinated pain afferents
C	1	1	Autonomic postganglionic neurons
Lloyd/Hunt classification (group)			
I	13	75	Primary muscle spindle afferents
II	9	55	Skin touch afferents
III	3	11	Muscle pressure afferents
IV	1	1	Unmyelinated pain afferents

Autonomic nervous system (ANS)

The ANS is the visceral motor nervous system. By definition it includes no sensory components. However the activities of the ANS are modified by sensory input that travels by way of the somatic nervous system and by the CNS. The target tissues of the ANS are smooth muscle, cardiac muscle, endocrine and exocrine glands, liver, the juxtaglomerular apparatus of the kidney and adipose tissue. The synapses of autonomic neurons with their target cells are called **neuroeffector junctions**.

The preganglionic neurons of the ANS have their cell bodies in motor nuclei of cranial nerves in the midbrain or medulla (see below), or the intermediolateral horn of the thoracic, lumbar and sacral spinal cord. Their axons are myelinated B fibers which secrete ACh. The preganglionic axons synapse with postganglionic neurons in **autonomic ganglia**. The axons of the postganglionic neurons are unmyelinated C fibers. The ANS has two divisions, the **sympathetic** and **parasympathetic**, the main distinguishing features of which are summarized in *Table 3*.

In general, the **preganglionic axons** of the sympathetic division are short and the **postganglionic axons** are long because the **sympathetic ganglia** lie close to the spinal cord in one of two locations:

(1) In paired **paravertebral chains** that lie just lateral of the vertebral column, running parallel to it in the neck and down the posterior wall of the thorax and abdomen.
(2) In **subsidary ganglia** of autonomic plexuses situated in the midline adjacent to major blood vessels.

The pathway taken by sympathetic axons is illustrated in *Fig. 2*.

Table 3. Divisions of the autonomic nervous system

Anatomy	Physiology	Postganglionic cell neurotransmitters
Craniosacral Preganglionic axons in cranial nerves III, VII, IX, X and spinal nerves S2, S3	Parasympathetic	Acetylcholine Vasoactive intestinal peptide
Thoracolumbar Preganglionic axons in spinal nerves T1–T12, L1, L2	Sympathetic	Norepinephrine (but acetylcholine at selected neuroeffective junctions) Neuropeptide Y Adenosine 5′-triphosphate

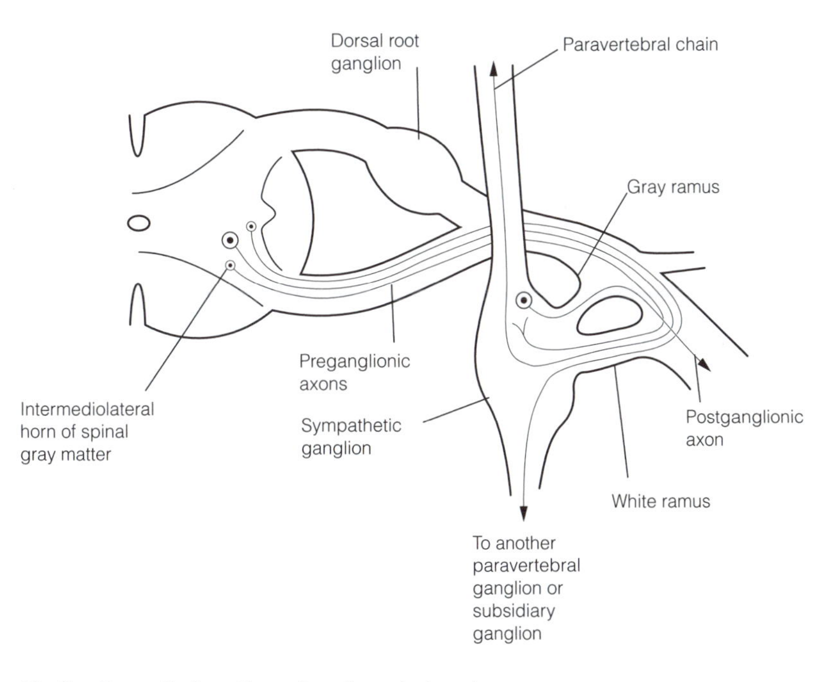

Fig. 2. Sympathetic pathway from the spinal cord.

Preganglionic axons may synapse in the nearest ganglion, traverse the paravertebral chain to synapse in subsidiary ganglia, or ascend or descend the chain to synapse in a ganglion at a different level. Preganglionic sympathetic axons can modify the actions of up to 100 postganglionic cells. This is an example of **divergence** and serves to disseminate and amplify neural activity. It is seen throughout the nervous system. In the sympathetic system, divergence may be achieved by direct synapses, via interneurons or via the local diffusion of transmitter. Probably all peripheral nerves contain postganglionic sympathetic axons because they supply the smooth muscle of blood vessels. The adrenal medulla is endocrine tissue that secretes epinephrine directly into the circulation in response to activity in the preganglionic sympathetic fibers that supply it. The adrenal medulla is therefore regarded as part of the sympathetic system.

Parasympathetic autonomic ganglia are all subsidary ganglia located close to the target organ. For this reason, in the parasympathetic division the preganglionic fibers are long and the postganglionic fibers are short. Although all the major organs, except the liver, have a parasympathetic supply, this division is far less extensive than the sympathetic. This is because only a few specialized blood vessels have a parasympathetic innervation, and divergence in this system is less. The functions of the ANS are discussed in Topic M6.

Enteric nervous system (ENS)

An interconnected network of about 10^8 neurons makes up the nervous system of the gut. It is organized into two thin cylindrical sheets that run along the length of the gut. The **myenteric (Auerbach's)** plexus lies between the longitudinal and circular smooth muscle layers and extends the whole length of the gut. The **submucosal (Meissner's)** plexus lies in the submucosa and extends from the pylorus of the stomach to the anus. There are extensive interconnections between these two plexuses. A number of amines (norepinephrine, ACh, serotonin), peptides and nitric oxide (NO) are employed as transmitters by this system. The ENS can act autonomously to coordinate gut motility and secretion. Its activity is modified by input from both divisions of the ANS.

E2 ORGANIZATION OF THE CENTRAL NERVOUS SYSTEM

Key Notes

Spinal cord

The human spinal cord contains some one hundred million neurons. Peripheral white matter and central gray matter can be seen by the naked eye. The gray matter of the spinal cord contains neuron cell bodies. Sensory neuron fibers enter the dorsal horn of the gray matter in an ordered fashion, larger diameter fibers entering more medially and extending deeper than smaller ones. Motor neuron cell bodies lie in the ventral horn of the gray matter. The spinal gray matter is divided on morphological grounds into 10 columns which on transverse section are Rexed laminae. Each lamina has a distinctive set of inputs and outputs. The white matter contains tracts of axons ascending or descending the cord. Neural tracts or pathways are named for their origin and destination.

The brain

The brain has three structural components. White matter consists of fiber tracts or pathways. Embedded in this are nuclei which are clusters of neuron cell bodies. Two large brain structures, the cerebrum and cerebellum are covered by cortex, a thin layer of gray matter densely packed with neurons.

Anatomically the brain has three principal divisions: hindbrain, midbrain and forebrain. The center of the brain is taken up with the cerebrospinal fluid (CSF)-filled ventricular system. The hindbrain consists of medulla, pons and cerebellum, and the midbrain is divided into a ventral tegmentum and a dorsal tectum. Together hindbrain and midbrain are the brainstem from which emerge most of the cranial nerves. With the exception of the cerebellum which organizes high level motor functions, the brainstem is concerned mainly with vital functions and functions requiring orchestrated activity of large parts of the brain (e.g. wakefulness).

The forebrain consists of diencephalon and cerebrum. The diencephalon contains the thalamus (a sensory structure) dorsally and the hypothalamus ventrally, implicated in temperature and endocrine regulation, and in appetitive behaviors. The cerebrum has two cerebral hemispheres heavily interconnected across the midline. Its surface is covered by cortex which has been subdivided into Brodmann areas and which has motor, perceptual and cognitive functions. The core of the cerebrum is occupied by the nuclei which form two neural systems. The basal ganglia form the extrapyramidal motor system and the limbic system (which includes cortical areas) is concerned with emotion, and learning.

Related topics

Organization of the peripheral nervous system (E1)
Meninges and cerebrospinal fluid (E5)
Early patterning of the nervous system (P1)

Spinal cord

The human spinal cord has about 10^8 neurons. It has 31 segments, each of which gives rise to a pair of spinal nerves. It ends at the level of the first lumbar vertebra. The lumbosacral nerve roots pass down the vertebral canal as the **cauda equina** so that they emerge from the vertebral column at their appropriate levels.

A transverse section through the spinal cord shows a butterfly shaped central **gray matter** which contains neuron cell bodies, **neuropil** (dendrites and short lengths of axon) and glia. The **white matter** surrounding the gray is largely axons in ascending and descending tracts and gets its color from the high content of myelin. In the middle is the **central canal**, which is continuous with the **ventricles** of the brain and contains **cerebrospinal fluid (CSF)**, though in adult humans it is often not patent.

Sensory fibers enter the spinal cord via the dorsal roots to synapse largely with cells in the **dorsal horns** of the spinal gray matter. Fiber sorting occurs as afferents enter the cord. Larger diameter fibers enter more medially and extend more deeply into the dorsal horn. Motor neurons cell bodies lie in the **ventral horns** of the spinal gray matter and their axons exit via the ventral roots. The distribution of afferents to dorsal roots and efferents from ventral roots is referred to as the **Bell–Magendie law**. Some visceral afferents, however, enter the spinal cord via the ventral roots.

Ten distinct regions can be distinguished on morphological grounds in the spinal gray matter. Each region occupies a long column that extends through the cord. On transverse section these columns appear as **Rexed laminae** (*Fig. 1*).

Each lamina has distinctive input–output relations which reflect a measure of functional specialization. Nociceptor afferents terminate on **dorsal horn cells (DHCs)** in lamina II. Cutaneous mechanoreceptor afferents terminate in deeper layers of the dorsal horn. Lamina VI is present only in spinal segments associated with limbs and receives sensory input from joints and muscles that provide information about the position and movement of the limb in space. Lamina VII includes the cell bodies of the preganglionic autonomic axons. Lamina IX houses both α and γ motor neurons which go to skeletal muscles.

The white matter is organized into columns or **tracts** specified by the origin (in the case of a descending pathway) or destination (for an ascending pathway) of the tract within the brain. For example, a tract that runs down the cord from

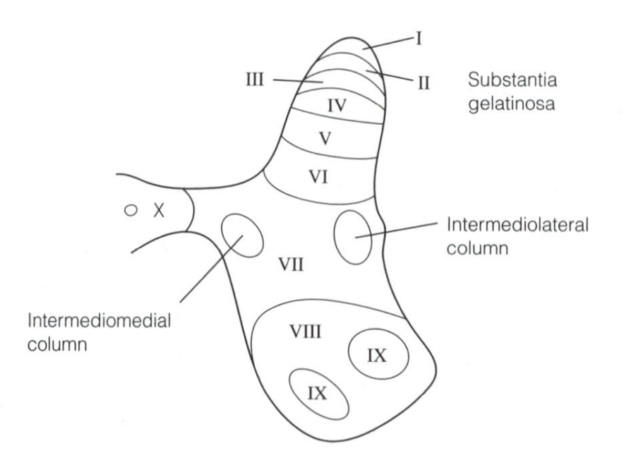

Fig. 1. Rexed laminae. Lamina VI is only present in spinal segments supplying the limbs.

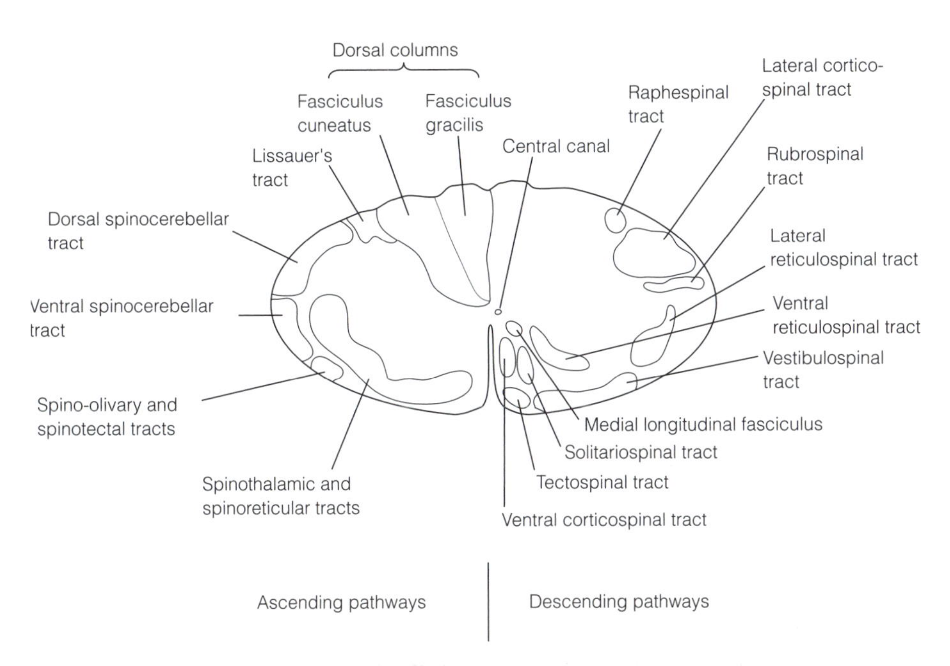

Fig. 2. Pathways in the spinal cord white matter.

the cerebral cortex is termed the corticospinal tract whereas an ascending pathway which terminates in the thalamus is the spinothalamic tract. *Fig. 2* shows the locations of the major tracts and may be consulted whenever a tract is referred to subsequently in the text.

The brain

There are three main structural components to the brain:

(1) **Tracts** or pathways enter the neuraxis at various levels, ascend or descend, and these together with internal tracts which go from one part of the brain to another constitute the white matter.
(2) **Nuclei** embedded in the white matter are clusters of neuron cell bodies and associated neuropil. Some neural structures are composed of groups of nuclei. The thalamus, for example, consists of about 30 nuclei, most of which have sensory functions.
(3) Two brain structures, the cerebrum and the cerebellum, are covered by **cortex**, a thin rind with a very high density of neuron cell bodies. In wiring terms, cortex appears to be a simple circuit between just five or so neuron types, repeated millions of times.

Together, the nuclei and cortex are the gray matter of the brain. **Neural systems** are comprised of interconnected nuclei and cortical regions that have a common function. The visual system, for example, consists of the retinae, thalamic nuclei, the visual cortex and the pathways between them.

In the human embryo at the end of the fourth week the CNS is a hollow tube, the **neural tube**, the **caudal** (back) end of which becomes the spinal cord (*Fig. 3*). At its **rostral** (front) end are three swellings, **primary vesicles**, which are the most fundamental anatomical divisions of the brain. These are the **hindbrain**, **midbrain** and **forebrain**. As embryogenesis proceeds, the forebrain differentiates into a caudal **diencephalon** and a rostral **telencephalon**, which in turn

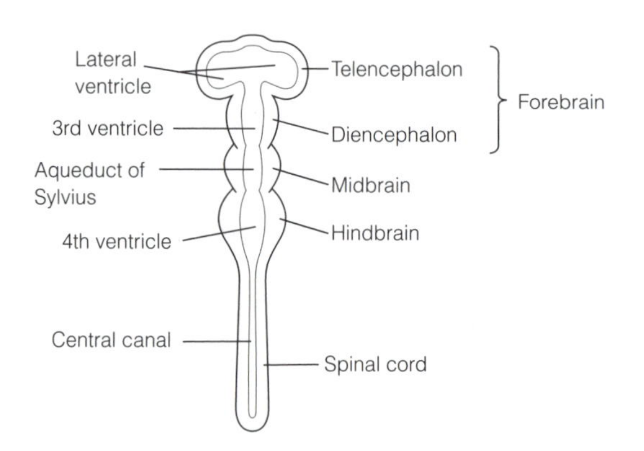

Fig. 3. The human embryo neural tube, at 28 days gestation.

acquires two lateral swellings, the **cerebral hemispheres**. Down the center of the neural tube is the CSF-filled ventricular system.

Much of the neural tube is divided into a dorsal **alar plate** in the midline of which runs the **roofplate**, and a ventral **baseplate** which has along its midline the **floorplate**. In the spinal cord and hindbrain these dorsal and ventral plates organize sensory and motor functions respectively. Such a clear distinction is not so obvious in midbrain or forebrain.

The hindbrain subsequently develops into a caudal **medulla** and a rostral **pons** and (from about 12 weeks) a dorsal outgrowth, the **cerebellum**. The midbrain, which in the adult is the smallest part, acquires a ventral **tegmentum**, in which are found cell bodies of dopamine-using neurons that are part of a **motivation** system, and a dorsal **tectum**, which organizes visual and auditory reflexes. The hindbrain and midbrain together are often referred to as the **brainstem**. Apart from the cerebellum which is concerned with quite high level motor functions, including motor learning, much of the brainstem is occupied with vital (life-support) functions; for example, autonomic regulation of the cardiovascular system, generating the rhythmic neural output required for breathing. In addition, the brainstem contains the nuclei of most cranial nerves. A core of highly interconnected nuclei extending through the brainstem constitutes the **reticular system**. Many of its neurons use amine transmitters. It is involved in orchestrating global brain functions such as attention, arousal, sleep and wakefulness and as such connects extensively with the forebrain.

The diencephalon in the adult is differentiated into a dorsal **thalamus**, which is predominantly a sensory structure, and a ventral **hypothalamus** concerned with thermoregulation, regulating endocrine systems, and goal-directed behaviors (eating, drinking and sexual behavior).

The dominant part of the telencephalon is the **cerebrum**, two cerebral hemispheres linked across the midline by about 10^6 axons that constitute the **corpus callosum.** The cerebrum is massively developed in humans. Each hemisphere is divided into four lobes named for the bones which overlie them (*Fig. 4*). The surface is covered by cortex and is highly convoluted giving it a high surface area in relation to its volume. The folds are called **gyri** (singular **gyrus**), and the creases between them **sulci** (singular **sulcus**). Most of the cerebral cortex is **neocortex** (new cortex) which has six layers. Cortical regions are mapped into

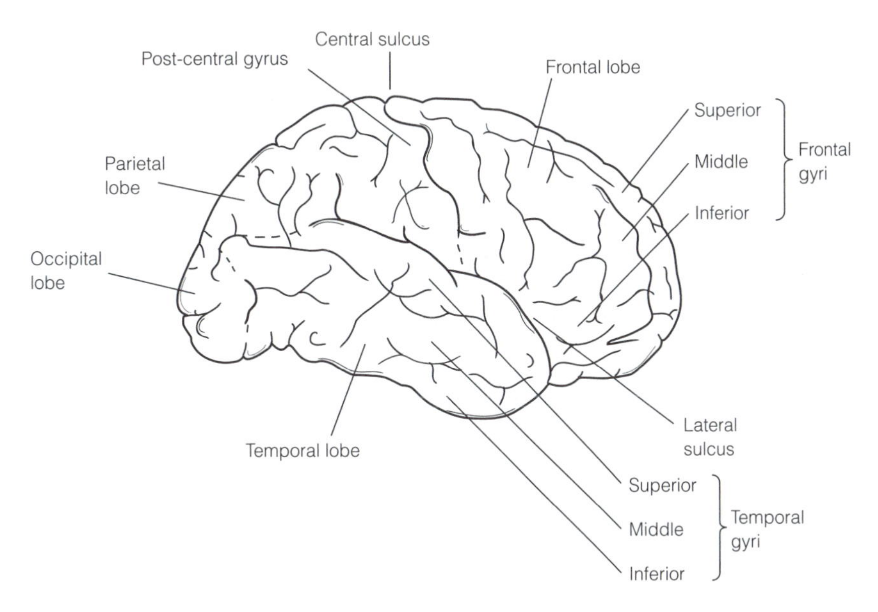

Fig. 4. Lateral surface of the human right cerebral hemisphere.

Brodmann areas on the basis of differences in **cytoarchitecture**, that is, their cellular composition and relative thickness of the layers. The significance of this is that the Brodmann map corresponds quite well to how functions are localized in the cortex though nowadays its main use is as a numerical guide.

The layers of the cerebral cortex are numbered from 1, nearest to the pial surface through to 6 which is the deepest (*Fig. 5*). Sometimes Roman rather than Arabic numerals are used to number the layers and the order is the same (i.e. I

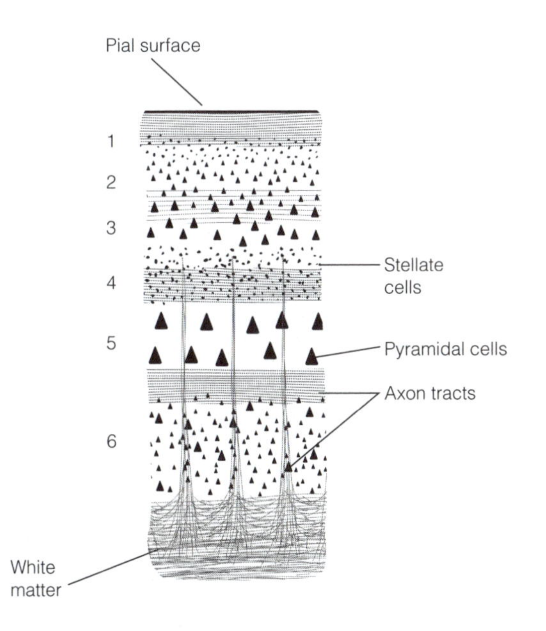

Fig. 5. Representative section through the neocortex (parietal lobe).

at the surface, VI the deepest). The layers contain different proportions of two types of neurons, pyramidal cells that are output cells, and stellate cells that are interneurons. Layer 1 consists mostly of axons that run parallel to the cortical surface. Layers 2 and 3 have small pyramidal cells that project to other cortical areas. Layer 4 is rich in interneurons and the site for the termination of most inputs to the cortex from the thalamus. Layer 5 has the largest pyramidal cells which project to subcortical nuclei, brainstem and spinal cord. Layer 6 pyramidal cells send their axons back to the same thalamic nucleus that supplied the inputs. Glial cells are found throughout the cortex. Tracts of axons run tangentially and radially through the cortex. The relative sizes of the layers differ with cortical function: for example, the sensory cortex has a thick layer 4 because of its large number of thalamic inputs, whereas in the motor cortex layer 5 is particularly extensive because these neurons project to the brainstem and spinal cord to mediate motor activity. The cerebral cortex is implicated in most brain activities, but is most often associated with the planning and execution of intentional movement, sensory perception and cognitive functions (those functions that use knowledge to solve problems).

Within the core of each hemisphere lie clusters of nuclei that form major components of two neural systems, the extrapyramidal motor system and the limbic system (*Fig. 6*). The **extrapyramidal motor system**, responsible for stereotyped patterns of movement, is a group of structures which together are referred to as the **basal ganglia**. These are the **striatum** which consists of **neostriatum**, itself composed of two nuclei, the **caudate** and **putamen**, and **paleostriatum** or **globus pallidus**. Anatomically the putamen and globus pallidus together form the **lentiform nucleus**. The striatum has extensive connections with two midbrain nuclei which are also included in the basal ganglia, the **subthalamus** and the **substantia nigra**.

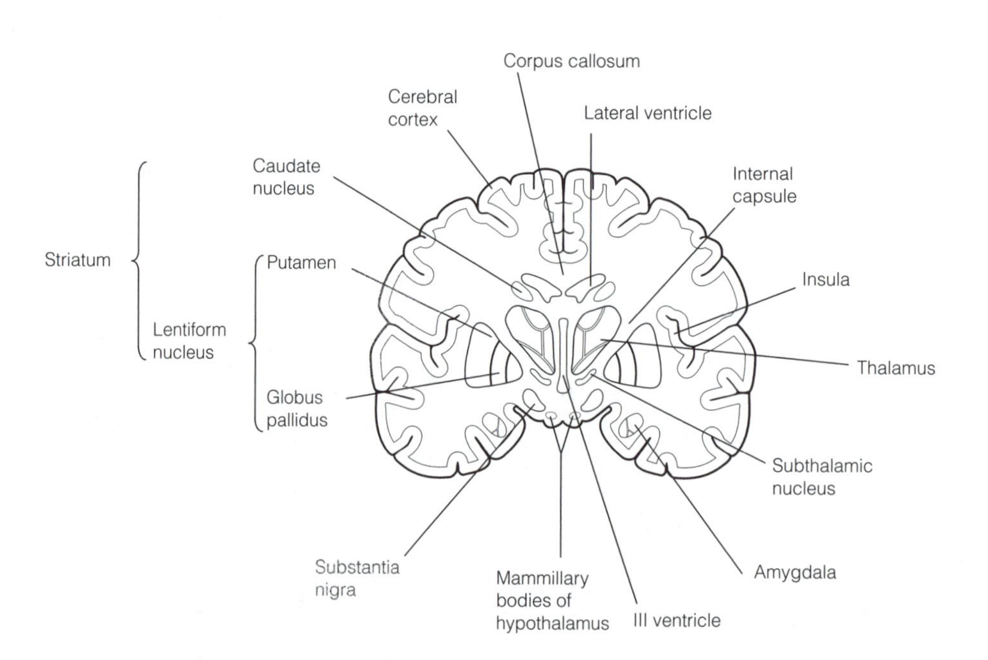

Fig. 6. Coronal section through the human cerebrum at the level of the posterior hypothalamus.

The **limbic system** is made from several heavily interconnected nuclei and several regions of cerebral cortex which form a ring around the diencephalon (*Fig. 7*). The cortical regions are the **cingulate gyrus**, which lies above the corpus callosum and has contributions from medial parietal and frontal lobes, and the **parahippocampal gyrus** and **uncus**, which are part of the medial surface of the temporal lobe.

The medial and underside of the temporal lobe is occupied by the **hippocampal formation**, most of which is the **hippocampus** and **subiculum**. The hippocampus is **archaecortex** (ancient cortex) and has only three layers. The subiculum is transitional cortex showing gradations from four through five to six layers where it merges with neocortex. All of the efferent fibers of the hippocampus and many of its afferents travel through the **fornix**. Limbic system nuclei include the **amygdala**, **septal nucleus**, and the **mammillary bodies** (which are part of the hypothalamus). The limbic system is implicated in emotion and its expression. The hippocampus is concerned with certain types of learning.

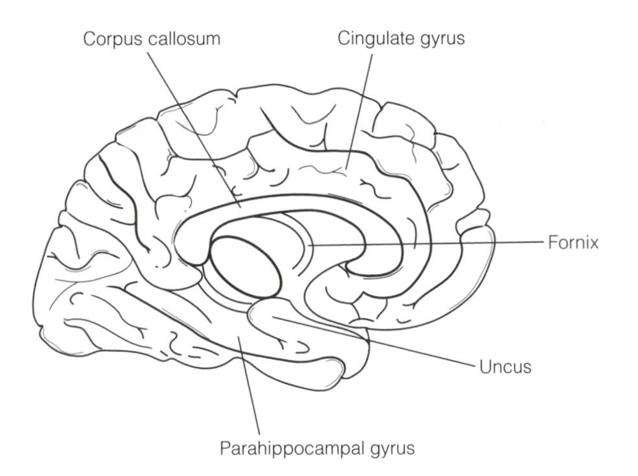

Fig. 7. Medial surface of the human left cerebral hemisphere.

E3 NEUROANATOMICAL TECHNIQUES

Key Notes

Histology

Staining with cationic dyes which bind nucleic acids reveals cell bodies. Silver stains are used to show neurites.

Histofluorescence

Neural pathways that use catecholamines or serotonin can be demonstrated by viewing brain slices that have been exposed to formaldehyde vapor under UV light.

Matching function and morphology

Injecting a neuron with a fluorescent dye at the conclusion of an intracellular recording experiment enables the morphology of a neuron to be matched to its electrophysiology.

Histochemistry

The location of a neuron or its terminals can be studied *in situ* by injecting a marker, which is taken up by the cell and can then be revealed by a suitable procedure, such as an enzyme assay. Some agents are taken up by the soma and transported towards the terminals, the location of which is thus demonstrated. Others are taken up by terminals and transported in the retrograde direction to reveal the location of the soma.

Immunostaining

Proteins can be localized in the brain by exposing tissue slices to antibodies raised against the protein. The location of this antibody is in turn revealed by reacting with a second anti-immunoglobulin (Ig)G antibody that has been tagged with a suitable marker such as an enzyme or a dye.

Autoradiography

Many studies, particularly those mapping specific neurotransmitter pathways, use radiolabeled markers. Autoradiography is a method of showing how the radiolabel is distributed in a brain slice. Slices can be prepared either for light or electron microscopy. The slice is coated with photographic emulsion which is developed later.

Related topics

Neuron structure (A1)
Morphology of chemical synapses (A3)
Early visual processing (H6)
Parallel processing in the visual system (H7)
Anatomy of the basal ganglia (L5)
Dopamine neurotransmission (N1)

Histology

Classical histology of the nervous system has used either **cationic dyes** or **silver stains**. Cationic dyes such as **cresyl violet** or **toluidine blue** bind to the negatively charged phosphate groups of nucleic acids in nucleus, nucleolus and Nissl bodies and so show cell bodies. The **Golgi** (and related) methods work by staining tissue with silver nitrate which is then reduced to silver. This produces uniform dark coloring of the whole neuron. The morphology of dendrites and axon can be studied in detail because this method stains only about 1% of neurons. A number of myelin-specific staining techniques are available for revealing myelinated projections.

Histofluorescence

Neural pathways using catecholamines or serotonin can be specifically identified by exposing freeze-dried brain slices to formic acid or formaldehyde vapor at 60°C. This converts the transmitters to isoquinolines which fluoresce at characteristic wavelength under UV light. Catecholamines appear as various hues of green, and serotonin as yellow-green.

Matching function and morphology

To relate the electrophysiology of a cell to its type at the end of an intracellular recording, a fluorescent dye such as Lucifer yellow can be injected through the microelectrode. The dye diffuses through much of the neurites so the morphology of the cell can subsequently be revealed by fluorescence microscopy of brain slices.

Histochemistry

The location and connections of neurons *in situ* can be revealed by a large number of compounds which are actively transported into neurons and then axonally transported. A few are taken up by the soma and transported in the **anterograde** direction to the terminals (e.g. leucoagglutinin, a lectin from the pea plant, *Phaseolus vulgaris*). Injecting leucoagglutinin into the region of the cell bodies thus shows where the terminals of the neurons are located. Some are taken up by terminals and transported in the **retrograde** direction (such as the fluorescent dye diamidino yellow) thus revealing the location of the cell bodies. Many are transported in both directions, for example, radiolabeled amino acids, **horseradish peroxidase** (**HRP**) or the **carbocyanine** lipophilic fluorescent dyes.

In a typical study the compound is injected into an appropriate site in the nervous system. After 1–3 days the animal is anesthetized and perfused with a fixative. The brain is removed and sectioned for analysis. The fluorescence dyes can be visualized directly by fluorescence microscopy, and radiolabeled compounds revealed by **autoradiography**. The distribution of HRP is shown by a histochemical reaction based on the ability of the enzyme to break down hydrogen peroxide. Retrograde HRP is used to identify the location of cell bodies associated with identified terminals, but HRP only reveals the morphology of a neuron fully when used in the anterograde direction

Immunostaining

Tissue slices incubated with specific antibodies will bind them avidly if they contain the corresponding antigen. The location of particular proteins such as enzymes or receptors can be determined using either polyclonal or monoclonal IgG antibodies. Antibodies can also be produced against small molecules, such as neurotransmitters, by first conjugating them to a protein. Antibody incubated with a brain slice will bind its antigen. The location of this first antibody is visualized by incubating with a second antibody raised against IgG in another species. The anti-IgG has previously been labeled in some way (e.g. radiolabeled, conjugated with an enzyme such as HRP or alkaline phosphatase or fluorescent dye, such as Texas red) so that it can be localized by an appropriate assay. In the latter case (illustrated in *Fig. 1*) the use of biotin/streptavidin serves to amplify the signal as many biotin molecules can be linked to a single IgG.

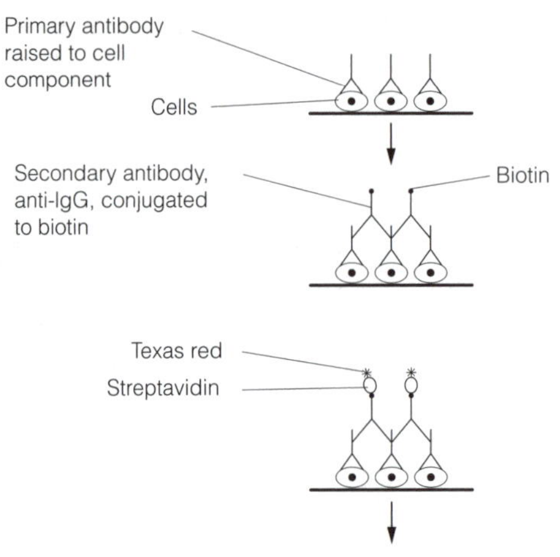

Fig. 1. Immunostaining. The marker illustrated here is the widely used biotin–streptavidin–Texas red system. Biotin is conjugated to a secondary antibody via covalent bonding. Streptavidin has a very high affinity for biotin. Texas red is a fluorescent dye bound to the streptavidin.

Autoradiography

To localize the distribution of radiolabel either by light or electron microscopy autoradiography is used. The method is of wide applicability but has proved useful in mapping neurotransmitter-specific pathways. Nerve terminals which release a classical transmitter will take up radiolabeled transmitter or an analog which thus acts as a marker for the terminals. In addition, high affinity radiolabeled ligands can act as markers for specific receptors. Once the tissue has been incubated with the appropriate labeled molecule it is fixed and cut into thin sections (3–5 μm in the case of light microscopy, 90 nm for electron microscopy). The sections are coated with a thin layer of photographic emulsion, kept in the dark in a freezer for a variable exposure time (days to months), then developed and stained.

E4 Brain imaging

Key Notes

Computer assisted tomography (CAT)

A CAT scan produces a series of X-ray images, each one a slice, of living brain. The technique can distinguish tissues that differ in their ability to transmit X-rays by as little as 1% and has a spatial resolution of 0.5 mm. Its main use is in the diagnosis of neurological disorders that can be revealed anatomically.

Positron emission tomography (PET)

PET scanning, by revealing the distribution in the brain (with a spatial resolution of between 4 and 8 mm) of a positron-emitting isotope can provide functional as well as anatomical information about the living brain. Using uptake of a radiolabeled glucose analog as a marker for neuron activity it can show which regions of the brain are involved in a variety of activities both in health and disease. Radiolabeled (positron emitting) neurotransmitters and receptor ligands are used similarly to study neurotransmitter pathways in the living brain.

Functional magnetic resonance imaging (fMRI)

Magnetic resonance imaging (MRI) takes advantage of the fact that some atomic nuclei (e.g. hydrogen) adopt different energy levels in response to a pulse of radio waves depending on their chemical environment. The nuclei return to their initial state in a characteristic relaxation time, emitting energy with a specific amplitude and frequency, and from this signal the concentration of the nuclei, how they are distributed across the brain and something of the chemical environment (e.g. water content) can be plotted. The technique has great diagnostic utility.

Related topics

Parallel processing in the visual system (H7)
Cortical control of voluntary movement (K6)
Strokes and excitotoxicity (R1)
Parkinson's disease (R3)

Computer assisted tomography (CAT)

The first imaging technique developed to allow visualization of the living brain, the CAT scan, interposes the head between a source which emits a narrow beam of X-rays and an X-ray detector (*Fig. 1*). A series of measurements is made of X-ray transmission. The source and detector are rotated as a pair through a small angle and a further series of measurements taken. This is repeated until the source and detector have rotated through 180º. The radiodensity of each region of the head is computed from the transmission data for all of the beams that have traversed that region, and the results are displayed visually. This provides a view through a single slice of brain lying at a known orientation. The key element here is the algorithm, and the computer software to implement it, which calculates the radiodensity for each point in the brain slice: this is the **computerized tomography**. By moving the head at right angles to the orientation plane for a short distance another section can be imaged. This is repeated until the whole brain has been scanned.

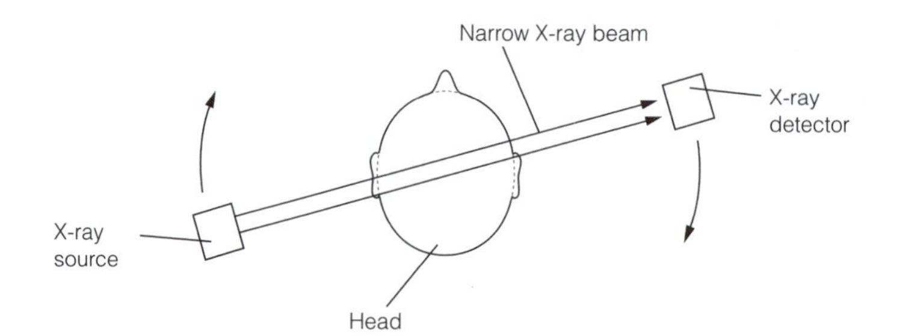

Fig. 1. Computer assisted tomography (CAT). Arrows depict the rotation of the scanner.

The method can distinguish tissues which differ in X-ray opacity by 1% (the lower the density the darker the image) with a spatial resolution of about 0.5 mm. Blood vessels can be seen by intravenous injection of radio-opaque dyes. This allows vascular disease, or tumors and abscesses with abnormal vascularities to be revealed.

Positron emission tomography (PET)

A remarkable feature of PET is that it provides insights into the function of the living brain as well as its anatomy. It uses the principles of computerized tomography in which a γ-ray detector is rotated around the head but the source is a positron (β particle)-emitting compound, either injected or inhaled, which enters the brain (*Fig. 2*). Compounds used include neurotransmitters, receptor ligands, and glucose analogs which are used for studying brain activity. Typically they are radiolabeled with ${}^{11}_{6}C$, ${}^{13}_{7}N$, ${}^{15}_{8}O$, or ${}^{18}_{9}F$ (which substitutes for hydrogen), isotopes with short half-lives which decay to the element with atomic number one less: a proton (p^+) within the nucleus decays to a neutron (n) emitting a positron (e^+, β) in the process. For example,

$${}^{13}_{7}N \rightarrow {}^{13}_{6}C + e^+$$

$$p^+ \rightarrow n + e^+$$

The positron travels a short distance before colliding with an electron (e^-). The two particles annihilate with the production of two γ-ray photons that shoot off in exactly opposite directions. These are detected simultaneously by a pair of detectors 180° apart. This coincidence detection permits localization of the site of the γ-ray emission, which is between 2 and 8 mm from the positron source, depending on the isotope used.

The spatial resolution of PET is about 4 to 8 mm, not as good as CAT, but it can be used to follow brain events over time.

The importance of PET in functional studies is illustrated by the use of the nonmetabolizable analog of glucose, **2-deoxyglucose (2-DG)**. This molecule crosses the blood–brain barrier, is transported into neurons and phosphorylated to 2-DG-6-phosphate, and so remains within the cell but not metabolized further. It thus acts as a marker for local glucose uptake and therefore of neuron activity. Imaging the distribution of $[{}^{18}_{9}F]$2-DG while subjects engage in sensory, motor or cognitive tasks reveals how these functions are localized in the brain. Related studies show that during transient increases in neuronal activity, the increase in local cerebral oxygen consumption (as measured by ${}^{15}_{8}O$ PET) does not match the increase in glucose utilization (as estimated from 2-DG PET). This implies that brief periods of brain activity can be supported by glycolysis.

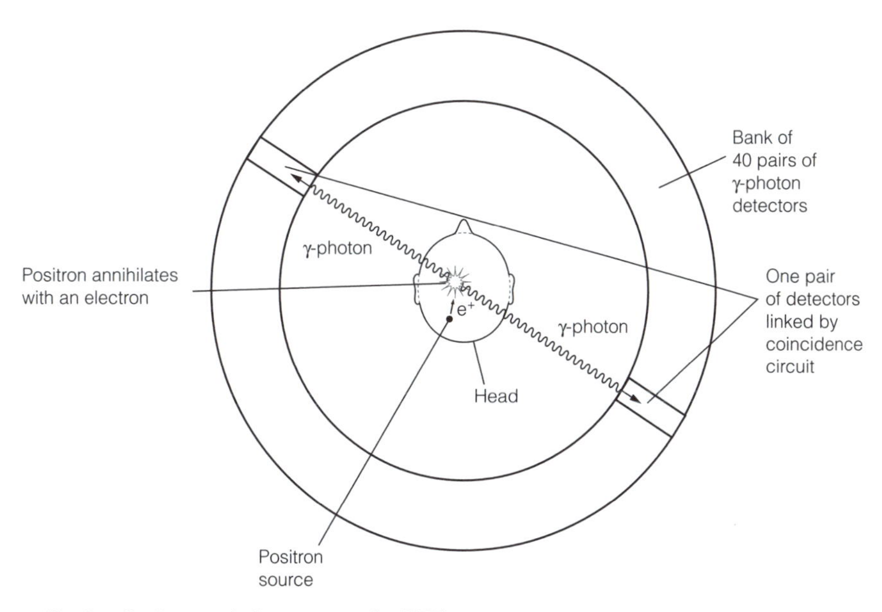

Fig. 2. Positron emission tomography (PET).

Functional magnetic resonance imaging (fMRI)

Like PET, MRI provides information about brain function as well as anatomy. It combines computerized tomography with **nuclear magnetic resonance (NMR)**. Nuclei with odd mass number, for example, ${}^{1}_{1}H$, or ${}^{13}_{6}C$ have a net angular momentum and generate a magnetic field along their spin axis. In a large external magnetic field, hydrogen nuclei can adopt one of two orientations: with their magnetic fields either parallel or anti-parallel to the external field. The parallel state has a slightly lower energy and normally a small excess of nuclei will be in this state (*Fig. 3*).

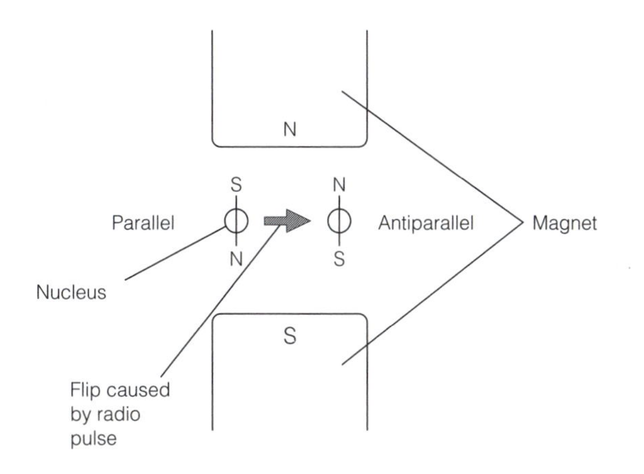

Fig. 3. The principle of nuclear magnetic resonance (NMR). A radiofrequency pulse will excite atomic nuclei, flipping them from the parallel state into the higher energy antiparallel state. Relaxation of the nuclei back into the low energy state generates the magnetic resonance imaging (MRI) signal.

Giving a brief pulse of electromagnetic radiation of the appropriate radio frequency causes some of the nuclei to absorb energy and flip into the high energy anti-parallel state. When this happens the nuclei are said to be in **resonance**. The amount of energy absorbed depends on the concentration of nuclei. The precise frequency needed depends not only on the nucleus (protons require 60 MHz whereas $^{13}_{6}C$ needs 24 MHz), but also on its chemical environment: the resonant frequency is changed by the magnetic fields of other nuclei in the vicinity. By changing the frequency it is possible to get information about the molecular neighborhood of the nucleus. In practise it is hard to change the radio frequency systematically. However the same information can be deduced by keeping the radio frequency constant and altering the external magnetic field.

Nuclei do not remain in the higher energy state but flip back to the lower energy state. It is the amplitude and frequency of the energy emitted when this happens that provides the MRI signal. This happens in a **relaxation time** (T) that is characteristic for the nucleus and its surroundings. For example, T for a proton in lipid is much shorter than for a proton in water. Brain images can be generated by an MRI scanner which either represents how concentration or relaxation time is distributed over a cross section of brain.

Imaging is achieved by having the magnetic field intensity vary in one direction across the brain. Signals emitted by nuclei have a higher frequency for greater magnetic field intensity. The signal frequency thus shows where across the brain it comes from. Changing the direction of the magnetic field allows brain slices in different orientations to be imaged.

Functional MRI has numerous clinical uses: mapping cerebral blood vessels, showing changes in extracellular space that accompany trauma or inflammation, diagnosis and following the progress of a variety of diseases (such as multiple sclerosis), and the precise localization of regions of stroke damage or tumors.

E5 MENINGES AND CEREBROSPINAL FLUID

Key Notes

Meninges

The brain and spinal cord are invested by three connective tissue layers, the meninges. Directly covering the brain is the pia mater, above which is the arachnoid mater. Between these layers lies the subarachnoid space which is filled with cerebrospinal fluid (CSF) and through which run blood vessels, branches of which enter the brain. Passive exchange of water and solutes across the pia mater keeps brain extracellular fluid and CSF in equilibrium. The tough outer layer is the dura mater which contains venous sinuses. Projections of the arachnoid mater herniate into the venous sinuses. At these arachnoid villi bulk flow of CSF from subarachnoid space into the venous circulation occurs. The potential space between the arachnoid mater and the dura mater is the subdural space. Traumatic rupture of the veins passing through this space from brain to venous sinuses causes subdural hemorrhage. Between the dura and the cranial bones is the extradural space through which run major arteries. Traumatic rupture of these results in extradural hemorrhage.

Cerebrospinal fluid (CSF) circulation and secretion

CSF is actively secreted by the choroid plexuses located in the ventricles. The direction of CSF flow is from lateral to third to fourth ventricles from where it enters the subarachnoid space. Finally it drains into the venous sinuses. Obstruction to the flow of CSF causes hydrocephalus.

About 500 cm^3 of CSF is secreted per day into a volume of between 100 and 150 cm^3. Choroid plexus epithelium contains a variety of active transport mechanisms. This results in secretion into the CSF of sodium, chloride, and bicarbonate but resorption of potassium, glucose, urea, and a number of neurotransmitter metabolites. The protein concentration of CSF is very much lower than that of blood plasma.

CSF and meningeal functions

The CSF acts as a sink for metabolites that eventually are dumped into the blood via arachnoid villi or choroid plexuses. Mechanical functions of CSF and meninges are to reduce the weight of the brain in the skull, to resist changes in intracranial pressure due to altered brain blood flow and to cushion the brain from impacts during violent movements of the head.

Related topics

Blood–brain barrier (A5)

Organization of the central nervous system (E2)

Meninges

The brain and spinal cord are surrounded by three connective tissue membranes, the meninges (*Fig. 1*).

The **pia mater** and **arachnoid mater** are together called the leptomeninges. In the **subarachnoid space** that lies between them run superficial cerebral blood vessels. These are invested by a leptomeningeal coat and suspended in the space

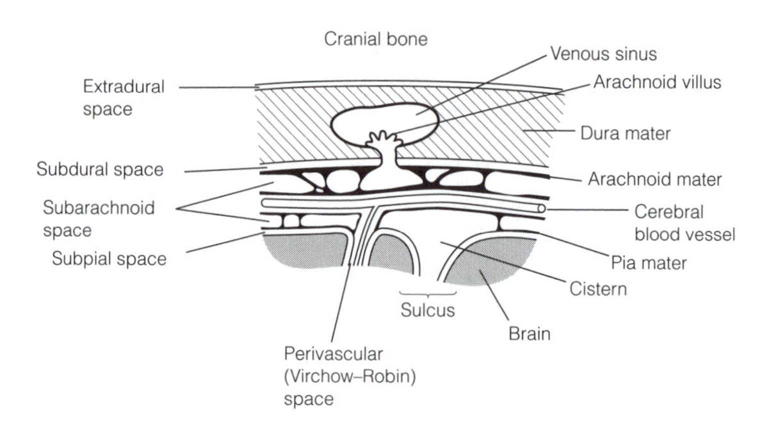

Fig. 1. The meninges.

by trabeculae. The subarachnoid space is filled with CSF. Branches of the subarachnoid vessels penetrate the brain, becoming surrounded by a cuff of pia mater that extends as far as the capillaries. The **perivascular (Virchow–Robin) space** between the vessel wall and the pia mater is continuous with the subarachnoid space. Here passive exchange of water and solutes across the pia mater keeps the CSF in equilibrium with brain extracellular fluid. At the cerebral capillaries the pia mater is lost and the single layer of capillary endothelial cells, with their basement membrane, are covered by glial cells (see Topic A5). Expanded regions of the subarachnoid space are **cisterns**. The lumbar cistern is the target for sampling CSF (**lumbar puncture**) because there is no risk of damage to the cord.

The **dura mater** is a thick tough outer layer with **venous sinuses** running through it. Small herniations of the arachnoid mater called **arachnoid villi (arachnoid granulations)** protrude through the dura into the venous sinuses. Here bulk flow of CSF into blood occurs via mesothelial tubes in the arachnoid villi that act as valves, closing when the pressure in the venous sinus exceeds that of subarachnoid space to prevent reflux of blood into the CSF.

The **subdural space** is a potential space between the dura mater and the arachnoid mater. It is traversed by cerebral veins entering the venous sinuses in the dura. Traumatic rupture of these vessels as they pass through the space causes **subdural hemorrhage**, which may present clinical problems at any time from the moment of injury to months later. Trauma which shears major vessels going from the dura mater into the cranial bone causes bleeding into the **extradural space** that opens up between meninges and skull. **Extradural hemorrhage** is a life-threatening surgical emergency. In the vertebral canal the dura mater forms a loose sheath leaving an **epidural space** between it and the canal wall. Injection of local anesthetics into this space produces an **epidural nerve block**.

Cerebrospinal fluid (CSF) circulation and secretion

CSF is actively secreted by choroid plexuses situated in the lateral, third, and fourth ventricles (*Fig. 2*). Flow of CSF is from the lateral ventricles through the **foramen of Munro** into the third ventricle, and then through the **aqueduct of Sylvius** into the fourth ventricle. From here it drains via three orifices, a medial **foramen of Magendie** and two lateral **foramina of Lushka**, to enter the

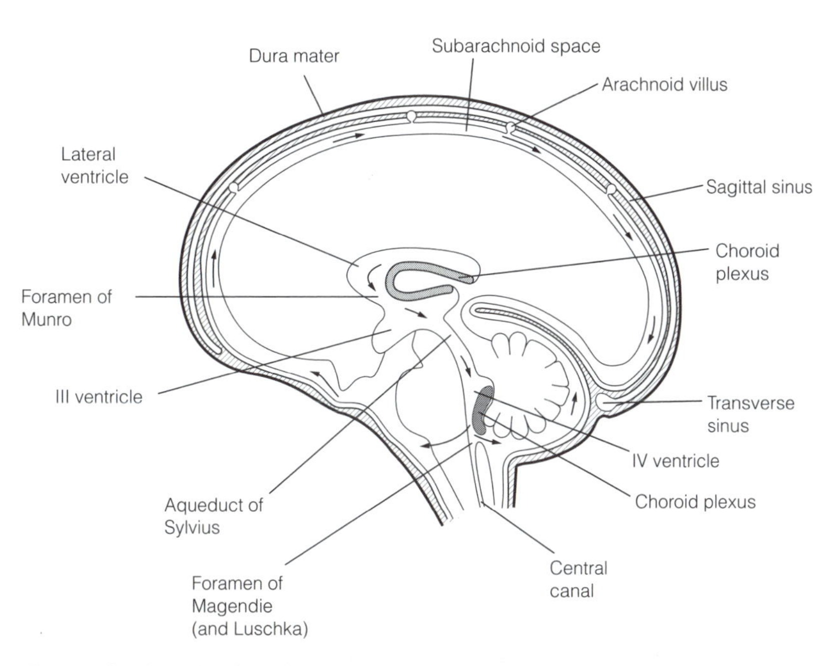

Fig. 2. Cerebrospinal fluid circulation (arrow shows the direction of bulk flow).

subarachnoid space. Here it equilibrates with extracellular fluid in the perivascular spaces. Finally it is dumped into the venous sinuses via the arachnoid villi.

Obstruction to the flow of CSF causes **hydrocephalus**, an accumulation of fluid in the cranium. This may increase CSF pressure, distending the ventricles and inflicting damage to the surrounding neural tissue. An obstruction of the ventricular system is **noncommunicating hydrocephalus**. It is the result of congenital malformation, scarring or tumors. In **communicating hydrocephalus** there is a failure of CSF flow from the arachnoid villi. This may happen if the concentration of protein in the CSF gets abnormally high, as with some spinal cord tumors, subarachnoid hemorrhage, meningitis, or acute peripheral neuropathy.

CSF secretion

Each **choroid plexus** consists of a cuboidal epithelium derived from the **ependyma**, the lining of the ventricles and spinal cord central canal, covering a core of highly vascular pia mater. In adult humans CSF is secreted at about 500 cm^3 day^{-1} into a steady state volume of 100–150 cm^3. Of this, about 30 cm^3 is in the ventricles and the rest in the subarachnoid space. CSF is turned over about every 5–7 h.

The choroid plexus secretes some substances and absorbs others specifically, most by active transport mechanisms. In this way it acts as a selective interface between blood and CSF, the **blood–CSF barrier**. The result is that by comparison with blood plasma CSF has somewhat higher Na^+, Cl^-, and HCO_3^- concentrations but lower K^+, urea, glucose and amino acid concentrations. Although the protein concentration of CSF is about 1000-fold lower than that of blood plasma, its higher ionic concentration gives the two fluids the same osmolality.

Some of the mechanisms involved in ion transport across the blood–CSF barrier are shown in *Fig. 3*. Na^+, K^+-ATPase on the apical border of the epithelial

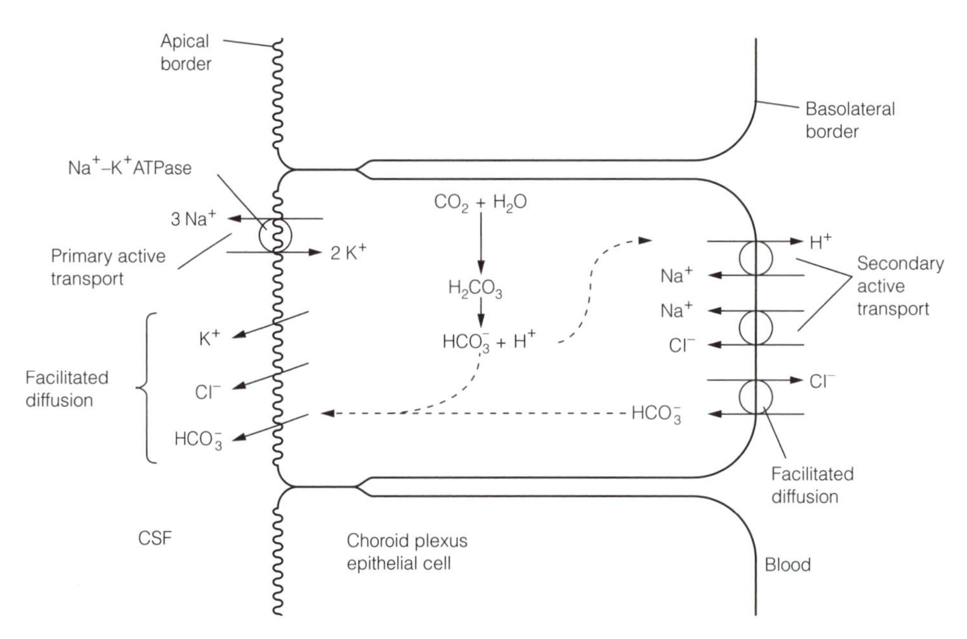

Fig. 3. Aspects of ion transport across the choroid plexus.

cell pumps sodium into the CSF. This generates a sodium gradient that drives two secondary active transport mechanisms bringing Na^+ across the basolateral border: Na^+-H^+ exchange and a Na^+-Cl^- symport. The Cl^- influx in turn drives a Cl^--HCO_3^- antiport. Bicarbonate brought into the cell in this way is added to that formed intracellularly by hydration of CO_2, a reaction greatly accelerated by the high levels of **carbonic anhydrase** present in the choroid plexus. The bicarbonate diffuses via an apical anion transporter into the CSF.

The ability of the choroid plexus to absorb materials from the CSF means that it acts as an excretory organ for the brain. It scavenges choline, dopamine and serotonin metabolites, urea, creatinine, and K^+, dumping them into the blood.

CSF and meningeal functions

The functions of the CSF are metabolic and mechanical. By equilibrating with brain extracellular fluid unwanted metabolites are removed to the blood, either via arachnoid villi or choroid plexuses. There are three mechanical effects:

(1) Because the subarachnoid space is a fluid-filled compartment in which the brain floats, the effective weight of the brain is reduced from about 1350 g to about 50 g.
(2) Adjustments to CSF and meninges prevent changes in intracranial pressure due to alterations in cerebral blood flow. When blood flow increases CSF is squeezed from ventricles into the subarachnoid space around the spinal cord. Here the dura mater is more elastic and stretches to accommodate the increase in volume. Longer-term increases in intracranial pressure can be offset by an increase in CSF flow into the venous sinuses through the arachnoid villi.
(3) The meninges support the brain and the CSF reduces the force with which the brain impacts the inside of the cranium when the head moves.

F1 Information representation by neurons

Key Notes

Information coding

The frequency with which a sensory neuron fires conveys information about the timing and intensity of a stimulus. How a sensory neuron is connected encodes the location of a stimulus and its qualitative nature (modality). Motor functions are similarly encoded. Often information is represented by the concerted activity of a number of cells – this is population coding.

Error protection

Errors in the transmission of information are minimized within the nervous system because action potentials are digital signals (which are intrinsically less error prone than other modes of signaling), information is encoded in mean frequency of firing, and population coding means that neural systems have redundancy.

Extracellular recording

A technique for recording from single cells or groups of cells in a variety of situations both *in vitro* and *in vivo*, extracellular recording works by amplifying the potentials that arise between a focal electrode close to the neuron(s) and a distant, indifferent electrode.

Related topics

Intensity and time coding (F2)
Cortical control of voluntary movement (K6)
Hippocampal learning (Q4)

Information coding

Neurons encode information by virtue of two properties. Firstly, the frequency with which a sensory neuron fires conveys information about the **duration** of a stimulus, its **intensity** and how the intensity changes over time. In the same way, motor neuron firing rate encodes the timing and force of contraction of a discrete population of muscle fibers. Secondly, the **address** of an afferent neuron, that is, how it is connected via its inputs and outputs, encodes the **spatial location** of a stimulus, and the qualitative nature of the stimulus or **modality**. The address of a motor neuron contributes to the type of movement executed and its direction. In both sensory and motor systems the accurate encoding of a given feature (e.g. skin temperature or the direction of a limb movement) often depends on activity in an array of cells. This is referred to as **population coding**.

Error protection

The accuracy of sensory coding or motor output depends on the fidelity with which action potentials transmit information. This is facilitated in three ways.

(1) The fact that action potentials are all-or-none makes them **binary digital** signals. At any given instant a length of axon is either transmitting an

action potential or it is not. Binary digital information coding is less prone to error (corruption of the signal by noise) than other modes of signaling because only two states need to be discriminated.

(2) With frequency modulation, the spurious absence or inclusion of the occasional action potential will not change the mean frequency of a train of action potentials much, unless the train is short.

(3) Finally, most stimuli are sensed, or motor output generated, by populations of neurons operating in concert. This makes for a degree of **redundancy**. Firing errors in a few neurons will be swamped by proper firing of the majority. Even if there are a sizeable number of rogue cells the system will not fail catastrophically. All that happens is that the information conveyed will be less precise. Systems that fail in this way are said to show **graceful degradation**.

Extracellular recording

The firing patterns of either single neurons or clusters of neurons in living animals in response to physiological stimuli are obtained by extracellular recording. This technique uses two fine electrodes usually of tungsten or stainless steel. One, the **exploring (focal) electrode** is placed as close as possible to the surface of the neuron of interest but does not impale it. The second, **indifferent electrode** is placed at a convenient distance. Neuron activity will cause currents to flow between the two electrodes. These currents are amplified and fed to a cathode ray oscilloscope (CRO) or to the analog-to-digital port of a computer running software to capture, store and analyse such data. By convention, if the exploring electrode is positive with respect to the indifferent electrode an upward deflection is recorded by the CRO. The polarity, shape, amplitude and timing of the recorded waveform generated by neural activity will depend on the position of the electrodes. The closer the exploring electrode is to the neuron the larger the measured signal. Changing the distance between the two electrodes or altering their relative positions will modify all of the above parameters. All of this can make extracellular recordings hard to interpret. The technique can be used in brain slices or other *in vitro* preparations, in anesthetized animals, or via **chronically implanted electrodes** (electrodes can be very precisely inserted into the brain under anesthetic, ahead of time, and attached to a connector cemented into the skull), which allow recordings to be made in conscious animals while they are behaving over long periods.

F2 INTENSITY AND TIME CODING

Key Notes

Static and dynamic coding

Stimulus intensity is encoded by the firing frequency of a neuron. The firing pattern of a sensory neuron is determined by the nature of its sensory receptor and the spatio-temporal characteristics of the stimulus. Slowly adapting receptors cause their afferents to fire at a rate that reflects the size of a constant stimulus so are said to show static responses. Rapidly adapting receptors result in their afferents showing reduced firing in response to application of a constant stimulus. These afferents respond to the rate of change of stimulus intensity and therefore show dynamic responses.

Stimulus intensity

The relationship between the intensity of a stimulus, and the firing frequency of a sensory neuron can be linear or more complicated. Many afferents fire at a rate that is proportional to the logarithm ($\log_{10}$) of the stimulus intensity, and this permits them to signal a wide range of intensities with the modest changes in firing rates that neurons are capable of.

Time coding

Temporal coding allows neurons to signal the precise timing of events in a way that frequency coding cannot. It requires that neurons fire only in response to imputs that occur at the same time. It is probably important in perception.

Related topics

Cutaneous sensory receptors (G1)
Sense of balance (G4)
Attributes of vision (H1)
Retina (H3)
Parallel processing in the visual system (H7)
Anatomy and physiology of the ear (I2)
Central auditory processing (I4)

Static and dynamic coding

Stimulus intensity is encoded by the mean frequency with which a sensory neuron fires. This is called **frequency modulated (FM) coding**. Broadly speaking the behavior of afferents falls into two categories depending on the nature of their sensory receptor. **Slowly adapting receptors** respond to a protracted stimulus for as long as the stimulus lasts, causing its sensory neuron to fire repetitively with a frequency that relates to the magnitude of the stimulus. These neurons exhibit **static (tonic)** responses to a constant stimulus. In contrast, **rapidly adapting receptors** respond only briefly to a constant stimulus because they soon become insensitive, or adapt, to it. These receptors respond best to changes in stimulus intensity. Their afferents show **dynamic (phasic)** responses. Many afferents display a mixture of dynamic and static responses.

Examples of static and dynamic responses are shown in *Fig. 1* which compares three classes of afferent in the skin that are wired to different types of mechanoreceptor. The **Ruffini organ** is slowly adapting so its afferent has a frequency of

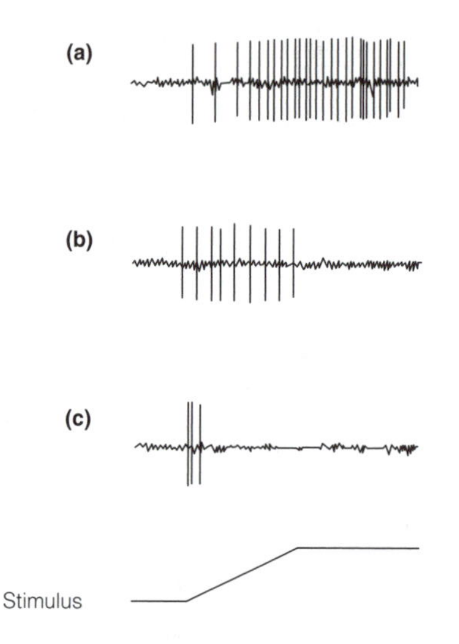

Fig. 1. Static and dynamic afferent neuron responses to skin displacement, p: (a) static response of Ruffini organ afferent; (b) dynamic response of Meissner's corpuscle afferent to velocity of displacement; (c) dynamic response of Pacinian corpuscle afferent to acceleration.

firing, *f*, that is directly proportional to the extent to which overlying skin is indented by a mechanical force. The **Meissner's corpuscle** is rapidly adapting and its afferent fires only when skin displacement is changing with time. It codes velocity of displacement. Finally, the **Pacinian corpuscle** adapts so rapidly that its afferents respond to acceleration of the skin. Hence the three afferents between them encode a wealth of dynamical information about the stimulus.

The beginning and end of a stimulus will be signaled by changes in the rates of firing of slowly adapting afferents, and by transient bursts of firing from rapidly adapting afferents. In this way stimulus duration is encoded.

Stimulus intensity

The relationship between stimulus intensity and response for static cells may be a simple linear one, as for example in skin thermoreceptor afferents. Often, however, the relationship is more complicated. Commonly, for example, the firing rate is linear with respect to the logarithm ($\log_{10}$) of the stimulus intensity. Many skin mechanoreceptors and photoreceptors fall into this category. This type of relationship allows a very wide range of stimulus intensities to be encoded by quite small differences in firing frequency. It has the disadvantage, however, that for high intensities the ability to discriminate between differences in intensity is compromised.

Time coding

With frequency modulation coding, obtaining accurate information about stimulus intensity requires that enough time elapses for a neuron to fire several action potentials. Hence, FM coding is not well suited to conveying information about the precise timing of events. To overcome this, neurons may act as coincidence detectors – that is, they fire only if inputs are simultaneous, thereby signaling the exact time at which the inputs arrive. This is called **temporal coding**. It requires that simultaneous inputs are subject to temporal summation (Topic D2), causing the postsynaptic cell to fire. For this to be an accurate time signal, these simultaneous inputs must not be summated with previous or successive inputs. This will only be the case if the time constant (Topic D1) is very short, about one millisecond. This is true of fine dendrites and of neurons being bombarded by inhibitory postsynaptic potentials.

Temporal coding is involved in the precise timing of events. For example, the source of a sound can be localized by measuring the time delay for sound entering both ears (Topic I4). Time coding is also thought to be crucial for perception. Different aspects of a stimulus (e.g., the color, shape and motion of an object) are processed in different neurons in widely dispersed brain regions (Topics H1 and H7). How does the brain produce a unified percept from all the disparate bits of information it has about the object? This question is known as the **binding problem**. One possible solution is that all the segregated bits of information pertaining to a single stimulus are 'bound' by simultaneous firing of the neurons involved. This would require precise timing. It is thought that oscillations in firing rates of neurons in the thalamus may be the signals required for binding.

F3 STIMULUS LOCALIZATION

Key Notes

Receptive fields

The region of a sensory surface which, when stimulated, causes a neuron to respond is the cell's receptive field. In sensory systems, proximal neurons have larger receptive fields than distal neurons because of convergence onto proximal cells of inputs from several distal neurons, and more complex receptive fields because proximal cells can receive inputs from many sources. Many receptive fields show lateral inhibition, in which the cell is excited if a stimulus is directed at the center of the field, but inhibited when it is directed onto the surround of the field, or vice versa. Lateral inhibition enhances contrast at sensory boundaries.

Topographic mapping

Sensory pathways are organized anatomically so that information about the location of a stimulus on a sensory space are preserved. In consequence many structures in the brain contain ordered maps of the sensory space. Three broad categories of map exist. Discrete maps are anatomically accurate representations of a sensory surface, though area is usually distorted, and reflect the presence of largely local interactions. Patchy maps have discontinuities which distort anatomical relations and represent interactions between distant parts of the body. Diffuse maps are not ordered by any property of the sensation.

Related topics

Dorsal column pathways for touch sensations (G2)
Retinal processing (H5)
Early visual processing (H6)
Central auditory processing (I4)
Olfactory pathways (J2)

Receptive fields

The spatial location of a stimulus on a sensory surface (skin, retina etc.) is given by which particular subset of neurons respond. The **receptive field (RF)** of a neuron is the region of a sensory surface which when stimulated causes a change in the firing rate of the neuron. Primary afferents generally have small RFs, the size of which is governed by the distribution of the cluster of sensory receptors which supply the afferent. Receptive fields of neighboring neurons responding to the same type of stimulus tend to overlap.

More proximal neurons in a sensory pathway have RFs that are composites of the RFs of more distal neurons. This gives rise to two features:

(1) In general, proximal neurons have larger RFs because of **convergence**: several afferents may synapse on a single, more proximal (i.e., downstream) neuron. Low convergence is seen where high **spatial resolution** (the ability to sense stimuli that are close together as independent) is important, such as between cones and bipolar cells in the retina. In contrast, high convergence is required where it is necessary to integrate weak signals from a number of receptors to achieve high sensitivity. This is the case between rods and bipolar cells in the retina, where it permits vision in dim light (Topic H5).

(2) The more proximal a neuron the more complex its RF. This is because proximal neurons receive inputs from a wider range of sources than distal neurons. This reflects the fact that extensive information processing occurs in sensory systems. Greater complexity of RFs also arises as a consequence of an extremely common characteristic of sensory pathways, **lateral (surround) inhibition**. At its simplest, this is where the RF of a neuron has two zones, a central area and a surround, from which opposite and antagonistic effects are produced in the cell when stimulated. It is seen in somatosensory, visual and auditory pathways. *Fig. 1* shows the receptive field of a somatosensory cell. Stimulation of the center causes an increase in firing so the RF is said to have an excitatory center. Stimulation of the surround reduces firing and is brought about by inhibitory interactions. A cell behaving in this way is described as an **on-center cell**. **Off-center cells** are also common. For the on-center cell, maximum firing rate would be seen with a stimulus that was just sufficient to stimulate the entire center. A larger stimulus that encroached upon the surround would be less effective, by causing some inhibition. In this way lateral inhibition sharpens spatial resolution and enhances contrast at boundaries between stimuli.

In skin mechanoreceptor afferents this improves **two-point discrimination**. By similar means, light–dark contrast at edges is enhanced in the retina, and tone discrimination sharpened by central auditory neurons. In general, lateral inhibition happens between neurons coding the same type of sensation. However color vision depends on lateral inhibition between cells that respond to different wavelengths (see Topic H6).

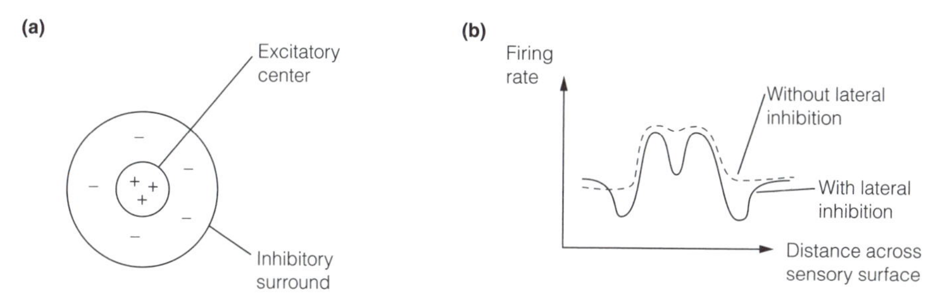

Fig. 1 Lateral inhibition. (a) Receptive field of an on-center sensory neuron showing lateral inhibition; an off-center neuron would have an inhibitory center and an excitatory surround. (b) Contrast enhancement in the presence and absence of lateral inhibition.

Topographic mapping

In most sensory pathways primary afferents are wired to specific subsets of more central neurons in a strictly ordered fashion so that nearest neighborhood relations are conserved. This means that information about stimulus location is not lost in more proximal parts of a pathway. This arrangement is called **topographic mapping**. RFs are aligned to produce an ordered map across brain structures such as the thalamus or the cerebral cortex. These maps are neural representations of a sensory surface or some feature of a sensation. Key examples are: somatotopic maps which represent skin surface, retinotopic maps that reflect the visual fields and tonotopic maps which represent the pitch of a sound. In addition are numerous motor maps, particularly in the cerebral and cerebellar cortices in which movements are represented in a systematic way. The motor mapping is preserved in descending pathways so that connections with motor neurons are precisely those needed to execute the mapped movement.

Three broad types of map are recognized, thought to be determined by the extent of the connections between the neurons involved in the mapping:

(1) **Discrete maps** like somatotopic or retinotopic maps are anatomically accurate and complete representations of a sensory surface, though they are usually distorted in that the area of the surface is not faithfully proportioned. Fingers and lips get far more than their fair share of the cortex in somatotopic maps. Discrete maps arise because neurons are connected mostly to their neighbors, allowing **local interactions** between cells. In other words, most of the comparisons the central nervous system needs to make of, for example an image, are between adjacent pixels of retina.

(2) **Patchy maps** consist of several domains within each of which the body is represented accurately. However adjacent domains map regions that are not anatomically close or which are disoriented. Cerebellar motor maps are of this kind and said to exhibit **fractured somatotopy**. Patchy maps arise because whereas some groups of neurons are locally connected others are wired to distant neurons allowing **global interactions** to take place. Serving a tennis ball requires the coordination of movements in distant parts of the body.

(3) **Diffuse maps** are those which have no underlying topography. Distinct smells are mapped to particular sites in the olfactory bulb but not in any orderly fashion. Smells are not arranged within the brain in any systematic way by property.

F4 Stimulus quality

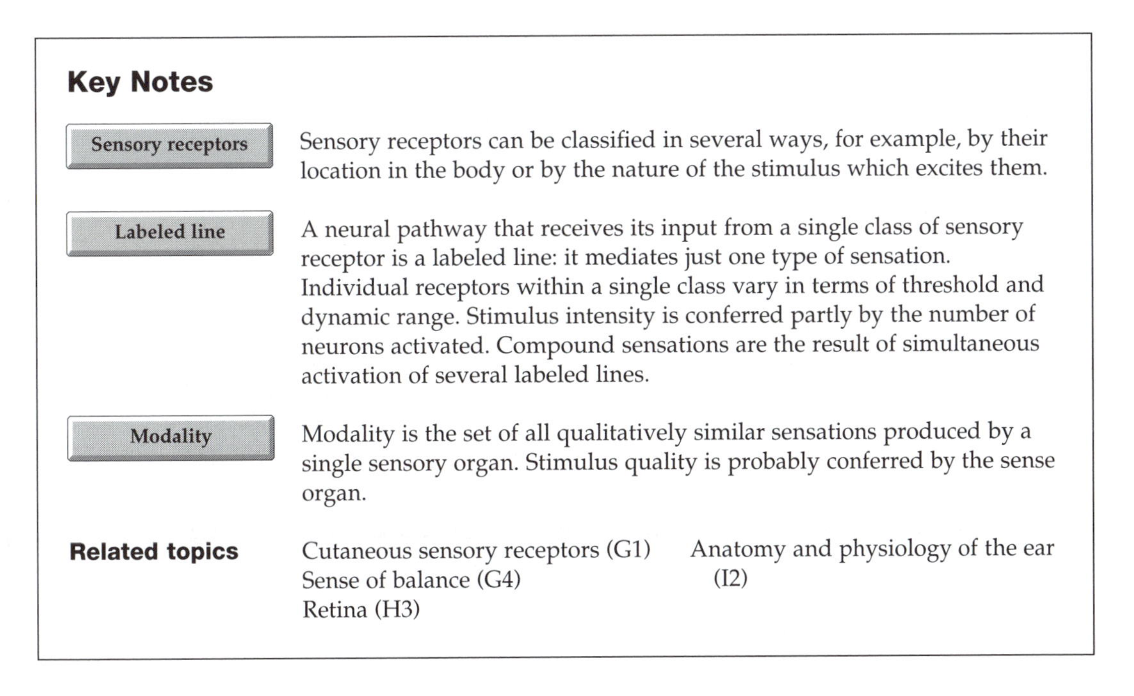

Key Notes

Sensory receptors — Sensory receptors can be classified in several ways, for example, by their location in the body or by the nature of the stimulus which excites them.

Labeled line — A neural pathway that receives its input from a single class of sensory receptor is a labeled line: it mediates just one type of sensation. Individual receptors within a single class vary in terms of threshold and dynamic range. Stimulus intensity is conferred partly by the number of neurons activated. Compound sensations are the result of simultaneous activation of several labeled lines.

Modality — Modality is the set of all qualitatively similar sensations produced by a single sensory organ. Stimulus quality is probably conferred by the sense organ.

Related topics — Cutaneous sensory receptors (G1); Sense of balance (G4); Retina (H3); Anatomy and physiology of the ear (I2)

Sensory receptors

Sensory receptors can be classified in a variety of ways. These are summarized in *Tables 1, 2* and *3*.

Labeled line

The neural pathway that is connected to a single class of sensory receptor and which, when stimulated, gives rise to a readily identifiable type of sensation is called a **labeled line**. The correspondence between receptor class and the nature of the sensation occurs because a sensory receptor responds only to a specific type of stimulus (e.g., photons, pressure). For example, a labeled line exists for

Table 1. Sensory receptors classified by location

Location	Organ/receptor	Sense
Exteroreceptors		
Special	Retina	Vision
	Cochlea	Hearing
	Olfactory epithelium	Smell
	Gustatory epithelium	Taste
	Vestibular inner ear	Balance
Superficial[a]	Cutaneous mechano-, thermo- and nociceptors	Touch, temperature and pain
Proprioceptors		
Deep[a]	Muscle/joint mechanoreceptors	Body position and movement
Interoceptors		
Visceral	Visceral mechanoreceptors	Visceral senses

[a] These are classified as somatosensory receptors.

Table 2. Sensory receptors classified by nature of the stimulus

Receptor	Stimulus	Sense
Photoreceptors	Light	Vision
Mechanoreceptors	Mechanical forces	Hearing, balance, touch Proprioception, visceral stretch
Thermoreceptors	Heat	Temperature
Chemoreceptors	A diverse variety of molecules	Olfaction, taste

Note: Nociceptors are mixed and may be mechano-, thermo or chemoreceptors or **polymodal**, that is responding to several stimuli.

Table 3. A classification of sensation quality

Organ	Modality	Submodality
Retina	Vision	Gray scale brightness, color
Cochlea	Hearing	Tone
Olfactory epithelium	Smell	No agreed primary qualities
Tongue epithelium	Taste	Salt, sweet, sour, bitter
Vestibular inner ear	Balance	Direction of gravitational field, angular acceleration of the head
Muscle and joint mechanoreceptors	Proprioception	
Visceral mechanoreceptors	Visceral stretch	
Skin mechanoreceptors	Touch	Light touch, pressure, vibration/flutter
Skin warm thermoreceptors	Warmth	
Skin cold thermoreceptors	Cold	
Skin/visceral nociceptors	Pain	
Skin itch receptors	Itch	

skin warming. This is because warm thermoreceptors respond optimally to increases in skin temperature. Even receptors in a single class may differ from each other in their properties. Individual receptors may vary in the strength of the stimulus that will make their afferent fire on 50% of the occasions it is delivered, the **threshold stimulus**. Individual warm thermoreceptors respond over different ranges of temperature, that is, they differ in the **dynamic range** over which they operate. The sensation mediated by a labeled line involves activating a population of afferents. This population coding contributes to the intensity of a sensory experience. Many topographic maps are of single or a few closely related labeled lines. Some topographic maps are really a series of embedded submaps, each representing a labeled line.

Many perceived sensations do not correspond to what can be produced by activating a single labeled line. These **compound sensations** arise from the activation of several receptor types by a single stimulus. By this means a rich variety of higher order sensory experience is made possible (e.g. texture, wetness).

Modality

The concept of **modality** is ill defined in the neuroscience literature. Some authors state that it is the sensation derived from stimulating a single class of receptor. By this definition there are as many modalities as there are types of sensory receptor and a labeled line is then the pathway devoted to a given modality. An alternative definition is that modality is the group of qualitatively similar sensations detected by a particular sensory organ and recognizes **submodalities** based on perception. In this scheme, some submodalities correspond to a receptor class, others encompass a diversity of receptors. A version of this approach is given in *Table 3*. Recent experiments suggest that stimulus quality is determined by the sense organ. Surgically rerouting visual pathways to auditory cortex resulted in animals which behaved as if they interpreted input into the redirected pathway as light, not sound. This further suggests that sensory cortex may be a rather general purpose machine.

G1 Cutaneous sensory receptors

Key Notes

Receptor potentials

Sensory receptors respond to a stimulus with a change in membrane potential, a receptor potential. In vertebrates this is depolarizing at all receptors, except photoreceptors which hyperpolarize, and inner ear hair cells which show responses of either polarity. A cutaneous sensory receptor is part of a primary afferent. In other sensory systems the sensory receptor is a separate receptor cell. Receptor potentials are small amplitude, analog, passively conducted potentials that decay with time and distance (c.f. synaptic potentials). Receptors adapt to a constant stimulus in that the response declines with time. Receptor potentials that are sufficiently large will trigger action potentials in sensory pathways, if they do so directly they are called generator potentials.

Cutaneous mechanoreceptors

Skin mechanoreceptors respond to mechanical forces. They are classified as slowly or rapidly adapting and within each of these they fall into two types. Type I have small receptive fields (RFs) with clear boundaries and are concerned with shape and texture sensation. Type II have large RFs with fuzzy edges. The density of receptors is variable, being highest in fingertips and lips. Skin regions with a high density of receptors have a bigger area on somatotopic maps than regions with low density.

Cutaneous thermoreceptors

Thermoreceptors are slowly adapting. Warm receptors increase firing in response to a rise in skin temperature, whereas cold receptors respond to decreased temperature. Thermoreceptors are poor at signaling absolute temperature or slow changes.

Nociceptors

Mechanical nociceptor afferents are Aδ fibers responsible for the sensation of sharp pricking pain. They sensitize (show increased response with time) to prolonged heat. Polymodal nociceptor afferents are C fibers and respond to intense mechanical forces, heat and a number of chemicals released during tissue damage. They cause the sensation of poorly localized burning pain.

Related topics

Dorsal column pathways for touch sensations (G2)
Anterolateral systems and descending control of pain (G3)
Stimulus localization (F3)
Stimulus quality (F4)

Receptor potentials

Sensory receptors convert the stimulus to which they are sensitive to a change in membrane voltage by making the membrane more permeable to one or more ions. This process is called **transduction** and is different in different receptors. For somatosensory systems the sensory receptor is the modified ending of the primary afferent neuron, and is depolarized directly by the stimulus. In all other sensory systems the sensory receptor is a specialized cell type which forms synaptic connections with the first afferent neuron. On these, alterations on membrane potential translate into modulations in sensory cell neurotransmitter release with corresponding effects on the primary afferent. In vertebrates, all sensory receptors, except photoreceptors, depolarize in response to stimulation. In contrast, photoreceptors hyperpolarize when exposed to light. The hair cells of the inner ear responsible for balance and hearing depolarize or hyperpolarize depending on the stimulus.

The stimulus-evoked change in membrane potential is called a **receptor potential**. In some sensory systems, for example somatosensory systems, the effect is to trigger actions potentials if the stimulus is sufficiently strong. Receptor potentials which generate action potentials directly are often referred to as **generator potentials**.

Receptor potentials share many of the properties of synaptic potentials (*Fig. 1*). They are small amplitude, graded in size depending on the stimulus strength, electrotonically (passively) conducted over the receptor cell surface or along neurites, and consequently, decay with time and distance as dictated by the cable properties of the cell. A generator potential will trigger action potentials for as long as it remains beyond the firing threshold, the frequency of firing will be higher the greater its amplitude. All receptor potentials eventually result in altered sensory neuron firing. In engineering terms, all stimuli are encoded as **analog** signals, and sensory systems act as **analog–digital converters**. The significance of this is that although analog signals can be summated, and such integration amplifies the effects of weak stimuli, they are error prone in a way action potentials are not (see Topic F1).

Receptors demonstrate **adaptation**, a decline in response over time to a constant stimulus, and are classified as **slowly** or **rapidly adapting**.

Cutaneous receptors are classified as mechanoreceptors, thermoreceptors and nociceptors. Their properties are summarized in *Table 1*.

Cutaneous mechanoreceptors

Mechanoreceptors are classified as slowly or rapidly adapting and, separately, as being of two types, type I and II, distinguished by their location and RFs. Type I are superficial, lying at the boundary of epidermis and dermis and have small RFs with well defined boundaries. These include **Meissner's corpuscles** (*Fig. 2a*) and **Merkel's discs** (*Fig. 2b*). Type II are deep in the dermis and have

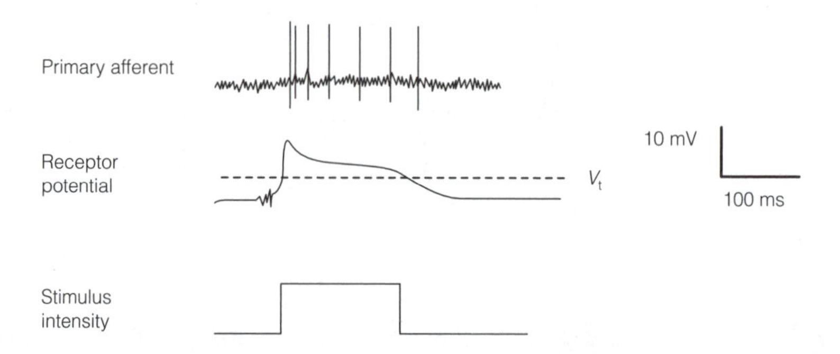

Fig. 1. Receptor (generator) potential (middle trace) and discharge (upper trace) of a slowly adapting cutaneous mechanoreceptor afferent in response to 150 ms indention of skin (lower trace). V_t, *threshold voltage.*

Table 1. Cutaneous receptors

Receptor	Adaptation		Fiber type	Sensation
Mechanoreceptors				
Meissner's corpuscle	RA1	velocity	Aβ	Touch, flutter, stretch
Pacinian corpuscle	RA2	acceleration	Aβ	Vibration
Merkel's disc	SA1	velocity and displacement	Aβ	Touch, pressure
Ruffini corpuscle	SA2	displacement	Aβ	Stretch
Lanceolate ending[a]	RA1	velocity	Aα	Hair movement
Pilo-Ruffini ending[a]	SA2	displacement.	Aβ	Hair movement
Hair follicle receptor[a]	RA1	displacement.	Aβ	Hair movement
Thermoreceptors				
Warm, bare nerve ending	SA		C	↑Skin temperature
Cold, bare nerve ending	SA		Aδ	↓Skin temperature
Nociceptors				
Mechano- bare nerve ending	Nonadapting		Aδ	Sharp pain
Polymodal bare nerve ending	Nonadapting		C	Burning pain

[a] Hairy skin only.

large RFs with poorly defined edges, and include **Ruffini corpuscles** (*Fig. 2c*) and **Pacinian corpuscles** (*Fig. 2d*).

Type I receptors are more directly concerned with form and texture perception than type II receptors. The density of type I receptors varies across the body surface, being highest in the fingertips, lips and tongue and lowest in the trunk. Areas with higher density have proportionally greater representations in somatotopic maps. Receptor convergence varies with receptor; whereas each Merkel's disc afferent receives input from two to seven receptors, a one-to-one ratio is the case for Pacinian corpuscles and their afferents.

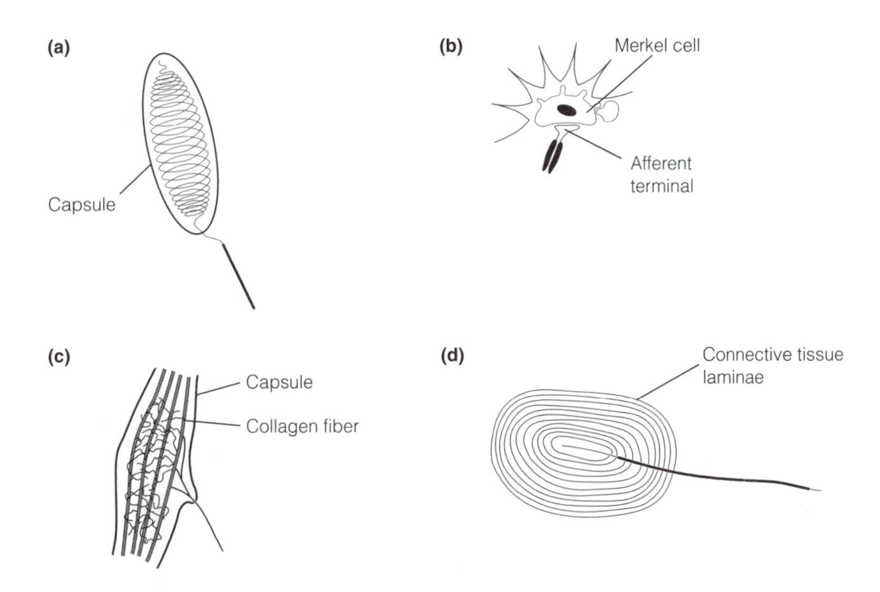

Fig. 2. Morphology of glabrous (non-hairy) skin mechanoreceptors: (a) Meissner's corpuscle; (b) Merkel's disc; (c) Ruffini corpuscle; (d) Pacinian corpuscle.

Meissner's corpuscle signals are important in adjusting grip force because they are very sensitive to small movements of an object over the skin of a grasping hand. Human skin is sensitive to vibration over a wide range of frequencies (5–500 Hz). For frequencies < 40 Hz the term flutter is used and this sensation is largely attributable to Meissner's corpuscles. Higher frequency vibration is detected by Pacinian corpuscles, which follow a sinusoidal vibration stimulus by triggering a single action potential per period. Optimal sensitivity is about 200 Hz. Frequencies in this range can be perceived even at skin indentations less than 1 µm.

Transduction has been most extensively studied for Pacinian corpuscles. Skin indentation force is transmitted through the corpuscles to deform the neurite within. This causes the opening of tetradotoxin insensitive Na^+ channels in the membrane and so a brief depolarization. Membrane potential returns to resting extremely fast because the receptor adapts. Adaptation is caused by the corpuscle, which consists of concentric layers of connective tissue, so that an applied force is transmitted only transiently before being dissipated as shear force causing the layers to slide over each other.

Human skin is either **hairy** or **glabrous** (non-hairy). Innervation of hairy skin differs in having a lower density of Merkel's disks and in possessing additional types of mechanoreceptor closely associated with hairs.

Cutaneous thermoreceptors

Thermoreceptors are slowly adapting, signaling the rate of change of temperature. Warm receptors increase their discharge rate in response to increasing skin temperature, whilst cold receptors respond to falling temperature. They are much more sensitive to rapid than slow temperature changes, and are poor indicators of absolute temperature. The maximum firing frequencies are at higher temperatures for warm receptors (*Fig. 3*). Thermoreceptor afferents get input from three to four receptors, and have very small RFs (1 mm diameter in glabrous skin), yet infrared radiation is very poorly localized.

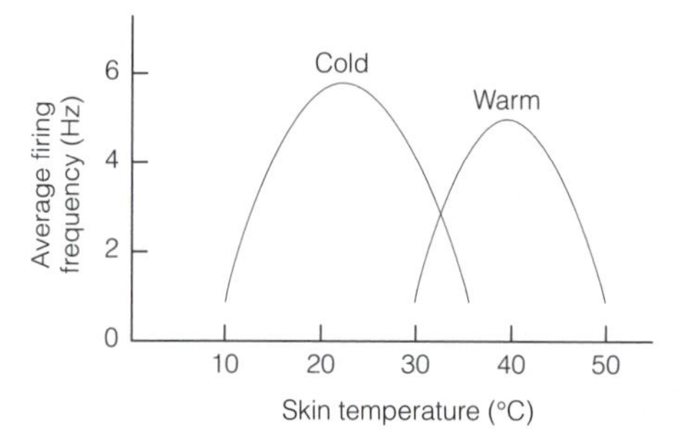

Fig. 3. Frequency response of populations of cutaneous cold and warm thermoreceptors.

Nociceptors

There are two quite distinct populations of nociceptors. A mechanical nociceptor is a bare nerve ending, one of five to 20 branches of an Aδ myelinated axon. Nociceptor afferents with a wide range of conduction velocity (7–30 m s^{-1}) have RFs composed of small spots, each corresponding to a receptor, over an area 2–3 mm across. Mechanical nociceptors have a high threshold to brief skin heating, but show **sensitization**, a lowered threshold, to prolonged heat. These receptors are responsible for the perception of sharp pricking pain which is well localized. Following mechanical trauma it is the first pain to be felt.

Polymodal nociceptors respond to skin puncture, temperature in excess of 46°C and a number of chemicals released by tissue injury including K^+, H^+, bradykinin and histamine. The relationship between stimulus strength and response of their unmyelinated C fiber afferents is linear. Because C fiber conduction is so slow, the pain they produce arrives last after a mechanical injury. Polymodal nociceptors give rise to ill localized burning pain and is poorly tolerated. Visceral pain has a similar quality. Release of bradykinin and prostaglandin E_2 from damaged tissue reduces the threshold of nociceptors to mechanical and thermal stimuli so that the site of an injury becomes more sensitive to painful stimuli. Even non-noxious stimuli may elicit pain. This phenomenon is called **primary hyperalgesia**.

G2 DORSAL COLUMN PATHWAYS FOR TOUCH SENSATIONS

Key Notes

Dorsal column–medial lemniscal (DCML) pathway

Each dorsal root receives input from a skin dermatome. Mechanoreceptor and proprioceptor afferents enter the dorsal roots to synapse with interneurons involved in spinal reflexes in the dorsal horn. A branch from each afferent ascends the spinal cord in the dorsal columns to synapse with neurons in the dorsal column nuclei (DCN) in the medulla. Lateral inhibition occurs in the DCN. Axons of the DCN cross the midline and ascend in the medial lemniscus to terminate in the ventro-posterolateral thalamus. From here neurons project to the primary somatosensory cortex (SI). Somatotopic mapping at each stage preserves stimulus location (the somatosensory cortex has several somatotopic maps over its surface, each representing a different class of receptor), and the dynamic features of stimuli captured by the receptor are transmitted faithfully right up to the cortex. The somatosensory cortex is organized into radially arranged columns, each of which gets input from a single type of receptor in a particular place on the skin. Adjacent regions of skin have adjacent columns. SI is concerned with tactile discrimination and stereognosis, the ability to perceive shape by touch. The secondary somatosensory area (SII) gets input from both sides of the body and is involved in guiding movement in the light of somatosensory input.

Descending connections

Reciprocal connections between the somatosensory cortex and DCML system nuclei are formed which have the same somatotopic mapping as the ascending pathway. These descending connections probably act to filter somatosensory inputs.

Related topics

Stimulus localization (F3)
Stimulus quality (F4)
Cutaneous sensory receptors (G1)
Anterolateral systems and descending control of pain (G3)

Dorsal column–medial lemniscal (DCML) pathway

The region of skin innervated by a dorsal root is a dermatome. Cutaneous mechanoreceptor primary afferent axons relaying cutaneous mechanoreceptor and proprioceptor signals, enter the spinal cord via the dorsal roots to synapse with interneurons, **dorsal horn cells (DHCs)**, in deep Rexed laminae. These neurons mediate or modify spinal reflexes. Each afferent sends a collateral up the dorsal columns to synapse with neurons in the **dorsal column nuclei (DCN)** in the medulla. The **cuneate nucleus** receives input from C1–8 and T1–6, whereas the **gracile nucleus** gets its inputs from T7–12 and L1–5. Dorsal column nuclei are a site for lateral inhibition (*Fig. 1*).

Axons of the dorsal columns nuclei neurons cross the midline to ascend on the opposite side of the spinal cord as the **medial lemniscus**, terminating in the **ventroposterolateral (VPL)** division of the ventrobasal thalamus (*Fig. 2*). VPL

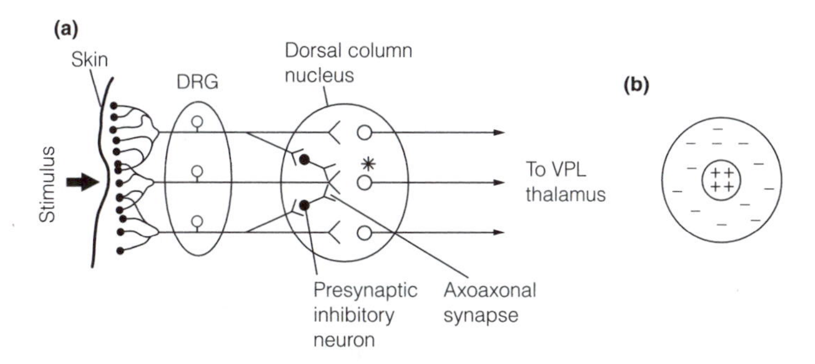

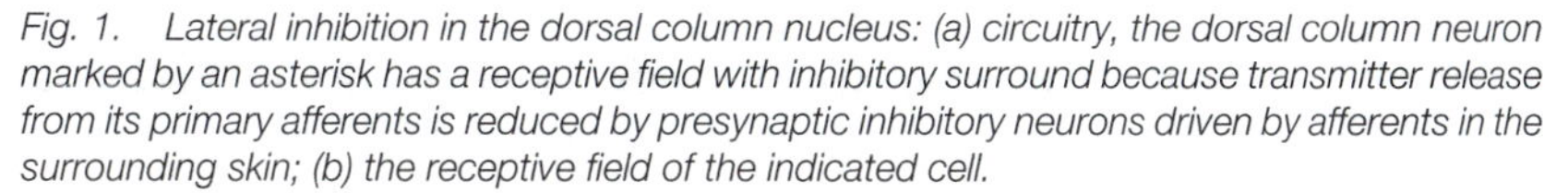

Fig. 1. Lateral inhibition in the dorsal column nucleus: (a) circuitry, the dorsal column neuron marked by an asterisk has a receptive field with inhibitory surround because transmitter release from its primary afferents is reduced by presynaptic inhibitory neurons driven by afferents in the surrounding skin; (b) the receptive field of the indicated cell.

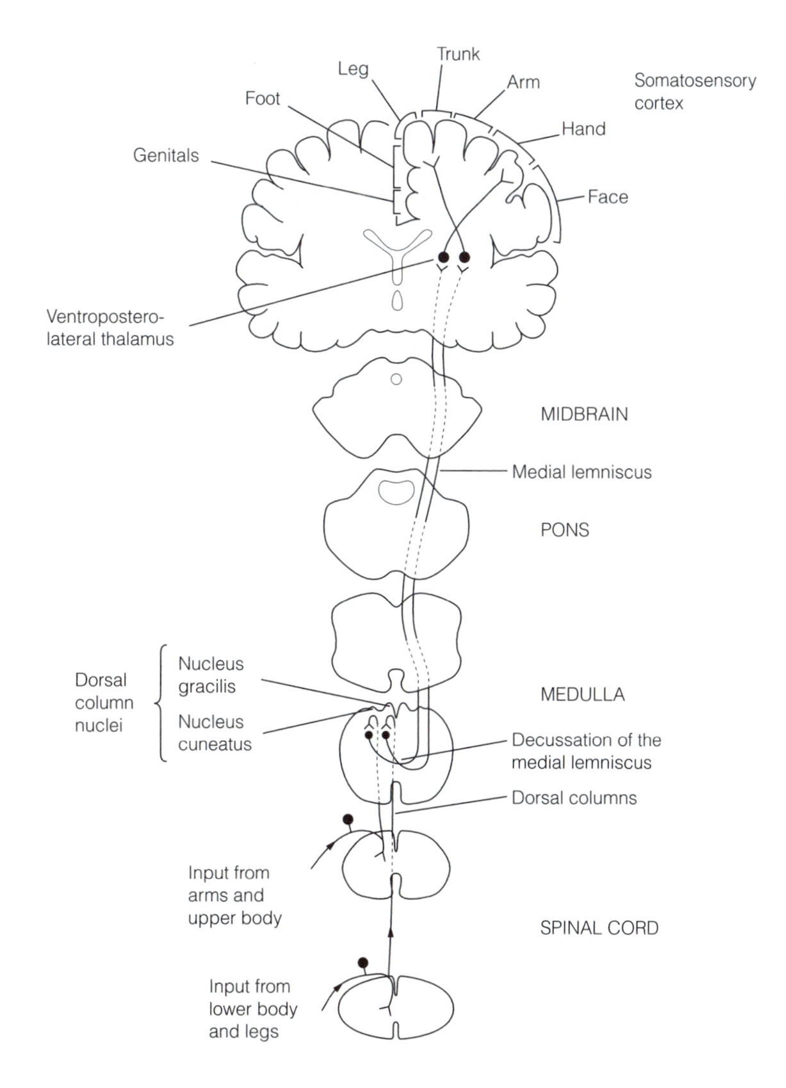

Fig. 2. The dorsal column–medial lemniscal system. All neurons shown are excitatory.

neurons give rise to thalamo-cortical axons, which project to the **primary somatosensory cortex** SI (Brodmann's areas 1, 2, 3a and 3b) situated over the postcentral gyrus. SI neurons in turn project to SII (*Fig. 3a*).

The general properties of the DCML system are:

- The great strength of its excitatory synaptic connections.
- The properties of its neurons are matched to those of the sensory receptors supplying them, so the dynamic features of stimuli that are captured by the receptor are transmitted with high fidelity through the whole system.
- Somatotopic mapping preserves localization at every stage. Body maps are found in dorsal column nuclei, VPL and somatosensory cortex. Each of the four regions in SI has a distinct map. Cutaneous input maps to the core of the VPL and then to areas 1 and 3b whereas proprioceptor input maps more peripherally in the VPL thalamus and then to areas 2 and 3a (*Fig. 3b*).

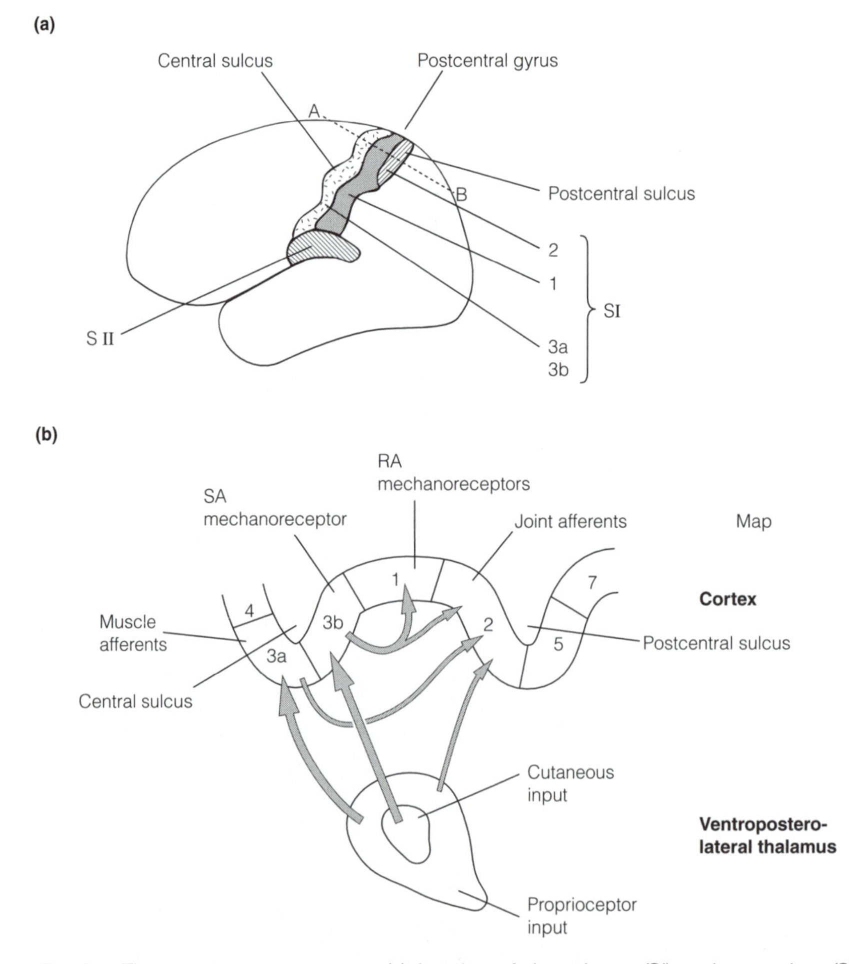

Fig. 3. The somatosensory cortex: (a) location of the primary (SI) and secondary (SII) somatosensory cortex of the left cerebral hemisphere, lateral aspect. Numbers refer to Brodmann areas; (b) interconnections of the thalamus and somatosensory cortex viewed across line A–B in (a).

Neurons in SI are organized into radially arranged **columns**. Each column gets input from just a single type of receptor, and from a specific location. Adjacent locations are represented in adjacent columns in a somatotopic manner. Extensive neural connections exist within a column, connections between columns are sparse.

The first cortical stage in somatosensory processing is in area 3b to which most VPL neurons project. Cells in 3b then project to layer IV of areas 1 or 2. The RFs of neurons in area 3b are relatively simple and those in areas 1 and 2 are more complex. Lesion studies show that area 3b is important for all tactile discrimination, area 1 is concerned with analysis of texture and area 2 with **stereognosis**, the ability to perceive the three dimensional shape of an object by touch. In addition to cutaneous input, area 2 gets proprioceptor input (directly and from 3a) and has reciprocal connections with the motor cortex. These are not involved in modifying ongoing movements but may inform the motor system of the sensory consequences of moving.

The secondary somatosensory cortex (SII) gets inputs directly from the ventrobasal thalamus and from SI. Many neurons in SII have bilateral receptive fields, that is, stimuli in the corresponding region on both sides of the body will evoke a response. Inputs from the contralateral body surface arise as a direct consequence of the **decussation** (crossing over) of the medial lemniscus. Inputs from the ipsilateral body surface enter SII from the contralateral side via the corpus callosum. By integrating information from both sides of the body, SII is the first stage in the formation of a unified percept of the whole body. It enables tactile discriminations learned using one hand to be easily performed with the other – the interhemispheric transfer of tactile discrimination.

SII is important in controlling movement in the light of somatosensory input via its connections with the motor cortex. In addition SII has inputs to the limbic cortex and so to the hippocampus and amygdala. By this route tactile learning is brought about.

Descending connections

The somatosensory cortex has reciprocal connections with all of the subcortical structures which relay sensory input to it, the VPL thalamus, dorsal column nuclei and DHCs. The descending pathway is made by the corticospinal (pyramidal) tract either directly or via its connections with the brainstem reticular nuclei. These back projections have a somatotopic mapping precisely in register with the ascending DCML pathway. They are probably the vehicle by which somatosensory input can be selectively filtered as an attention mechanism.

G3 Anterolateral systems and descending control of pain

Key Notes

Anterolateral pathways

Small diameter afferents carrying temperature and nociceptor input together with a few mechanoreceptor afferents enter the dorsal horn, terminating on neurons of the anterolateral pathway which mediates temperature, pain and poorly localized (crude) touch sensation. Most axons of these neurons cross the midline within one or two spinal segments to ascend in the anterolateral columns. There are three anterolateral pathways. The biggest is the spinothalamic tract, the axons of which terminate in the ventroposterolateral (VPL) thalamus, and is responsible for conscious pain perception. The spinoreticular pathway is partly ipsilateral, and via its connections with the central laminar thalamic nucleus is concerned with arousal in response to pain. The spinomesencephalic pathway terminates in the midbrain structures that regulate nociceptor input into the central nervous system (CNS).

Dorsal pathways

Some nociceptor input is by way of pathways outside the anterolateral columns. This input goes via dorsal horn cells that project through dorsal columns to the lateral cervical nucleus or dorsal column nuclei. This input ends up in the VPL thalamus.

Cortical involvement in pain

Although thalamic neurons transmitting nociceptor input project to the somatosensory cortex, and cortical cells show nociceptor responses, ablation of the cortex has no effect on pain perception. The cingulate cortex is concerned with the emotional response to pain.

Descending control of nociception

Control of nociceptor input into the CNS is exerted at two levels. At the spinal cord the input into the spinothalamic tract from small diameter primary afferents is inhibited by concurrent activity in large diameter mechanoreceptor afferents, via enkephalinergic interneurons in the substantia gelatinosa. Descending brain pathways, the transmitters of which include enkephalins, serotonin and norepinephrine, inhibit transmission in spinothalamic neurons. Opioid drugs probably exert some of their analgesia by acting as agonists of opioid receptors in the brainstem and spinal cord.

Pain syndromes

Pain arising from internal organs is often referred to the body surface because nociceptor input from different sources converges in the spinal cord. Rewiring of central connections may be responsible for phantom pain that follows amputation of limbs. Pain can be caused by activity of sympathetic neurons, or damage to central pain pathways.

Related topics

Stimulus localization (F3)
Stimulus quality (F4)
Cutaneous sensory receptors (G1)
Dorsal column pathways for touch sensations (G2)

Anterolateral pathways

The primary afferents of the anterolateral pathways are small diameter dorsal root ganglion (DRG) cells driven by thermoreceptors or nociceptors and DRG cells with large receptive fields (RFs) stimulated by mechanoreceptor input. The anterolateral system thus mediates temperature and pain sensations and poorly localized (crude) touch sensation (*Fig. 1*).

The primary afferent axons are situated laterally in the dorsal roots; they enter the **dorsolateral tract (Lissauer's tract)** where they divide into ascending and descending branches giving off collaterals that enter the dorsal horn usually within one to two segments; they terminate in laminae I and II and V–VIII of the spinal gray matter, to synapse with spinal neurons of the anterolateral pathway in the same segment. The destinations of nociceptor afferents is shown in *Fig. 1*. Most axons of these neurons cross the midline within one or two segments to ascend in the anterolateral columns. Three distinct pathways exist.

(1) Most anterolateral axons comprise the **spinothalamic tract** (STT) which arise from laminae I and V–VII and terminates in the thalamus. About 10%

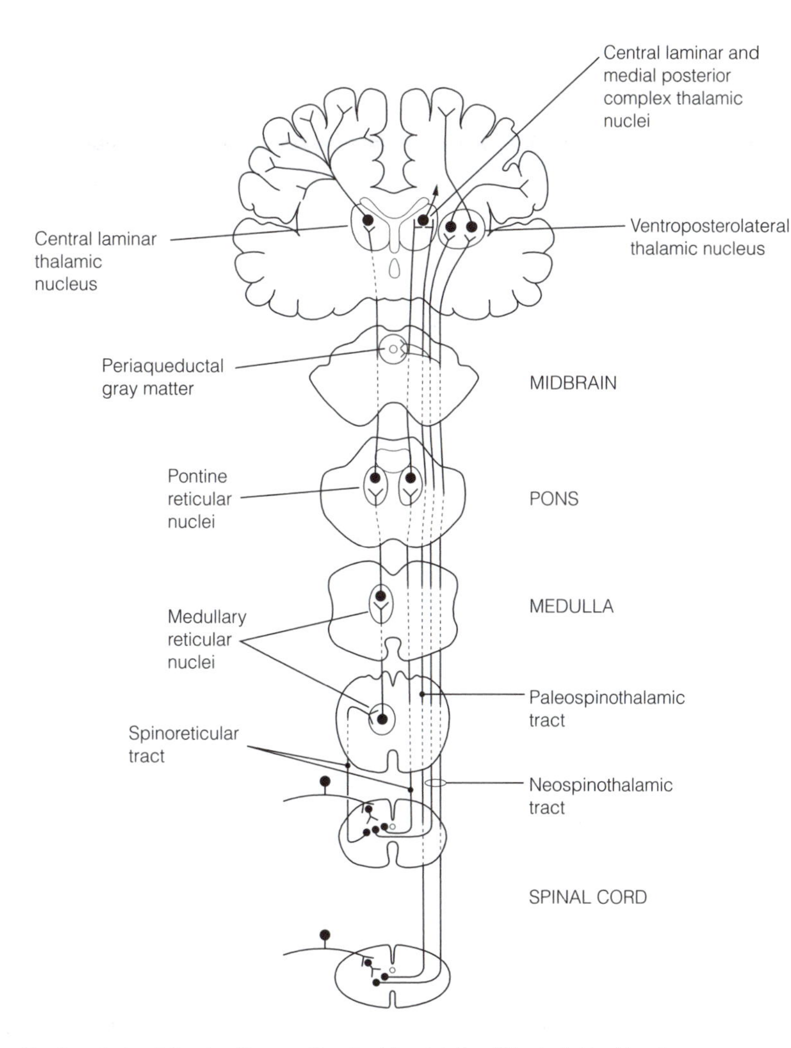

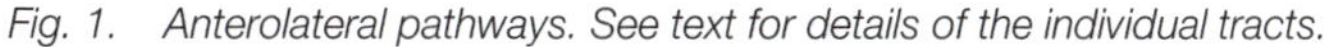
Fig. 1. Anterolateral pathways. See text for details of the individual tracts.

of STT axons from laminae I and V go to the VPL thalamus in a precise somatotopic fashion. These **neospinothalamic** neurons have small RFs and specifically convey either nociceptor, thermoreceptor or wide dynamic range mechanoreceptor input. By 'wide dynamic range' it is meant that a neuron will respond to a large range of stimulus intensities and so must receive inputs from receptors with both low and high thresholds. STT neurons terminate in the VPL alongside medial lemniscus neurons from the corresponding region. Although the two systems converge on separate cells within the VPL, dorsal column–medial lemniscal (DCML) input into the VPL may be needed for the STT to perform its task of localizing sharp pain.

Most STT axons terminate in the medial nucleus of the posterior complex (POM) or central laminar (CL) nucleus of the thalamus. These **paleospinothalamic** neurons arise from laminae VI–VII, have large RFs and are not well organized somatotopically. Although many neurons in POM and CL are optimally driven by nociceptor input and are involved in awareness of poorly localized burning pain, these thalamic nuclei are responsible for producing arousal in response to a wide range of sensory stimuli.

(2) Some nociceptive anterolateral neurons, particularly from laminae VII and VIII of cervical spinal segments, synapse with neurons in the reticular system of the medulla and pons. These constitute the spinoreticular pathway. Unlike the STT pathway many spinoreticular axons do not cross the midline but ascend ipsilaterally. Since the reticular system has extensive connections with the CL nucleus of the thalamus, the spinoreticular pathway is a route for arousal in response to pain.

(3) A **spinomesencephalic** pathway arises from laminae I and V (largely as collaterals of exclusively nociceptive neospinothalamic fibers) and terminates in the superior colliculus (**spinotectal** fibers) or periaqueductal gray matter (PAG) of the midbrain. This pathway regulates descending pathways which inhibit the degree of nociceptor input at the spinal level (see below).

Dorsal pathways

Two pathways for transmitting nociceptor input exist outside the anterolateral columns. They account for the fact that **anterolateral cordotomy** (a surgical procedure to cut the anterolateral columns at a specific spinal level to treat intractable pain) is often followed by the recovery of some pain sensation. These other pathways are derived from dorsal horn cells (DHCs), which project via dorsal columns either to the lateral cervical nucleus in the upper cervical cord or to dorsal column nuclei. Axons of these nuclei ascend in the medial lemniscus to the VPL thalamus.

Cortical involvement in pain

The role of the somatosensory cortex in pain perception is difficult to assess. In monkeys, nociceptive neurons in the thalamus project to neurons in the somatosensory cortex. These cortical cells respond to nociceptor input, however there is no clear nociceptor somatotopic mapping. While positron emission tomography shows activity in SI and SII in response to painful thermal stimulation, clinical ablation of large areas of somatosensory cortex has no obvious effects on pain perception. Human brain imaging shows that cingulate cortex activity increases in pain. The central laminar nucleus of the thalamus receives inputs from anterolateral neurons and has connections with the cingulate cortex. This cortical region is part of the limbic system that is concerned with emotion, so is probably involved in the affective consequences of pain.

Descending control of nociception

Nociceptor driven activity of the spinothalamic tract can be reduced by concurrent activity in large diameter (Aα and Aβ) mechanoreceptor afferents. The proposed mechanism, the **gate control theory**, is illustrated in *Fig. 2*.

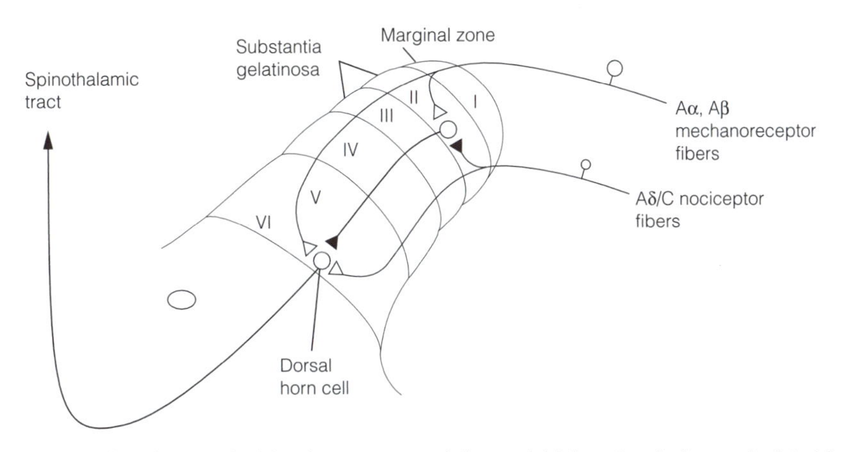

Fig. 2. Circuitry required by the gate control theory. Inhibitory terminals are depicted by closed triangles.

Nociceptor C-fiber stimulation causes prolonged firing of DHCs in lamina V by releasing the excitatory peptide transmitter, **substance P**. Concurrent large diameter stimulation, after initial excitation, produces inhibition via interneurons in the **substantia gelatinosa** (laminae II and III). Hence the extent to which nociceptor input is transmitted to the STT depends on the amount of activity in large diameter fibers. The gate control mechanism has been proposed as an explanation for reduced pain sensation reported by rubbing a wounded area, by **transcutaneous electrical nerve stimulation** and by **acupuncture**.

Nociceptor input is inhibited by brainstem descending pathways (*Fig. 3*). A major component is the **periaqueductal grey (PAG)**, which is a small region of gray matter surrounding the aqueduct of Sylvius in the midbrain. Electrical stimulation of this region in conscious animals and humans produces profound analgesia, without loss of other sensations, a phenomenon called **stimulus-produced analgesia**. The PAG is rich in enkephalin-containing neurons. These excite the descending **antinociceptor** pathway indirectly by inhibiting an inhibitory γ-aminobutyrate (GABA)ergic neuron. This is an example of **disinhibition**. PAG antinociceptor neurons stimulate serotonin- and enkephalin-containing cells in the **nucleus raphé magnus (NRM)**, located in the midline of the medulla, and noradrenergic cells of the **lateral tegmental nucleus**.

The inhibition of DHCs by these transmitters is by several mechanisms:

(1) Direct inhibition occurs via axodendritic synapses on the spinothalamic tract neurons in the dorsal horn. Serotonin, enkephalin and norepinephrine act through G protein-linked receptors [serotonin (5-HT_1), μ-opioid and α2 adrenoceptors, respectively] to hyperpolarize the spinothalamic tract neurons by opening potassium channels.

(2) **Presynaptic inhibition** of transmitter release from nociceptor afferent terminals is produced by the serotonergic and noradrenergic descending axons, which make axoaxonal synapses with the afferent terminals. Secreted serotonin and norepinephrine act on G protein-linked receptors to

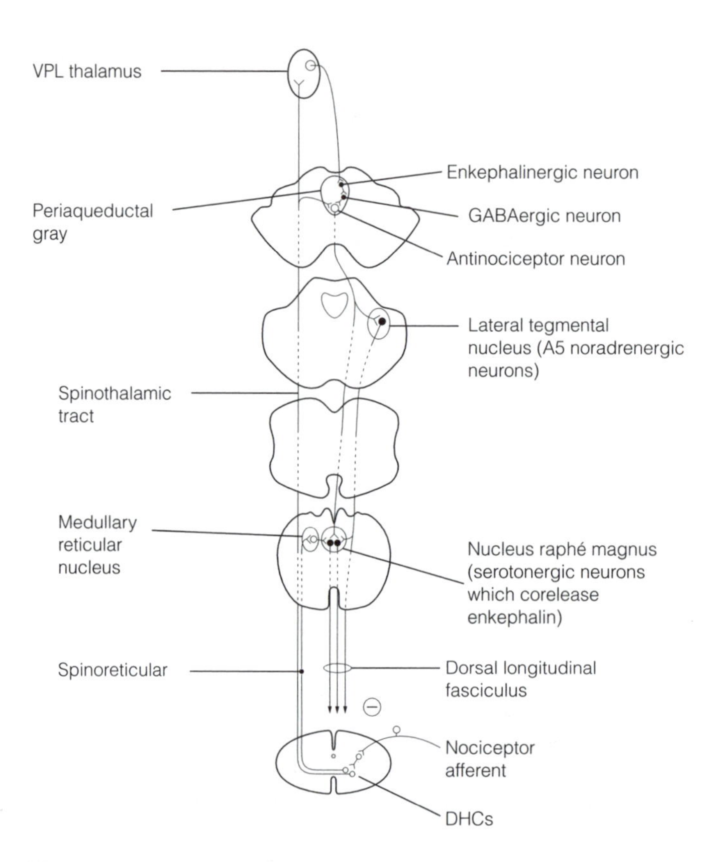

Fig. 3. Descending pathways inhibiting nociceptor input showing neurotransmitter involvement. NRPG, nucleus reticularis pars gigantocellularis; VPL, ventroposterolateral thalamus.

close Ca^{2+} channels in the nociceptor terminal. This shortens the time of any action potential invading the nociceptor terminal, reducing its transmitter release. Presynaptic inhibition is common in the spinal cord.

(3) Indirect inhibition is brought about by enkephalin-using interneurons in the substantia gelatinosa. These are activated by the descending serotonergic and noradrenergic axons. The enkephalinergic neurons act both postsynaptically on the STT neurons by opening K^+ channels and presynaptically on the nociceptor primary afferent terminals by closing Ca^{2+} channels. At both of these sites, enkephalin acts predominantly through μ-opioid receptors. **Opioid** drugs such as morphine, heroin and pethidine probably exert some of their analgesia by agonist actions at opioid receptors in the brainstem and spinal cord. However, a large component of opioid analgesia results from a change in the emotional response to pain, presumably brought about via opioid receptors in the frontal cortex

How the antinociceptor pathway is activated is not known. However:

- The PAG and NRM both receive excitatory inputs from the spinothalamic tract, the latter via a medullary reticular nucleus (see *Fig. 3*), hence the system may be subject to negative feedback; intense nociceptor input activates the antinociceptor pathway.

- **Emergency analgesia** is well documented in humans subjected to severe injury in a highly arousing context, for example, sports or battlefield injuries. It has a rapid onset, lasts only a few hours, is specifically localized to the site of the injury (so a descending somatotopic mapping is at work) and does not impair spinal and autonomic reflexes. This contrasts with **stress analgesia** in which the analgesia is generalized.

Pain syndromes

Commonly, activation of nociceptors in the viscera results in pain perceived at the body surface. This **referred pain** arises when cutaneous and visceral nociceptors synapse on a common DHC (*Fig. 4*). The brain cannot distinguish the source of the signal but is wired in such a way that firing of the DHC is interpreted as coming from the surface. Referred pain can be a useful aid to diagnosis.

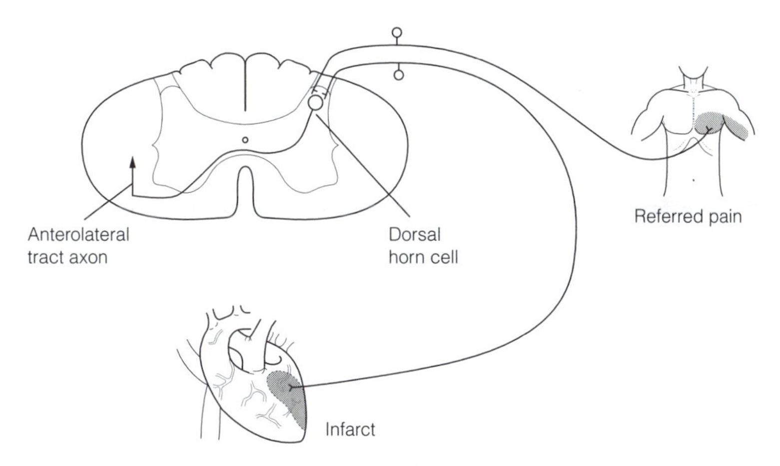

Fig. 4. Referred pain. Typically the pain of ischemic heart disease is referred to the chest and left arm.

The loss of afferent input to the spinal cord can result in abnormal sensations, frequently pain, so called **deafferentiation pain**. It occurs in traumatic avulsion, in which dorsal roots are torn from the cord, or with amputation of limbs or organs (uterus, breast) when it is called **phantom pain** (so called because it is associated with the feeling that the amputated structure still remains). Phantom pain has been attributed to spurious firing of DHCs resulting from the lack of proprioceptor feedback. However, there is evidence that rewiring of central connections of the somatosensory cortex occurs in deafferentiation which contributes to the phantom limb phenomenon. Following the loss of a hand, for example, the region of SI which once received sensory input from the hand is thought to be rewired by projections from adjacent areas of SI that subserve facial sensation. Tactile stimulation of the face now evokes the phantom limb sensation.

Sympathetic activity can result in pain, **causalgia**, possibly caused by cross-talk between postganglionic sympathetic fibers and C-type nociceptor afferents, both of which are unmyelinated. **Central pain syndrome** can occur as a consequence of damage (e.g. vascular lesions) to regions in nociceptive pathways (e.g. ventrobasal thalamus).

G4 SENSE OF BALANCE

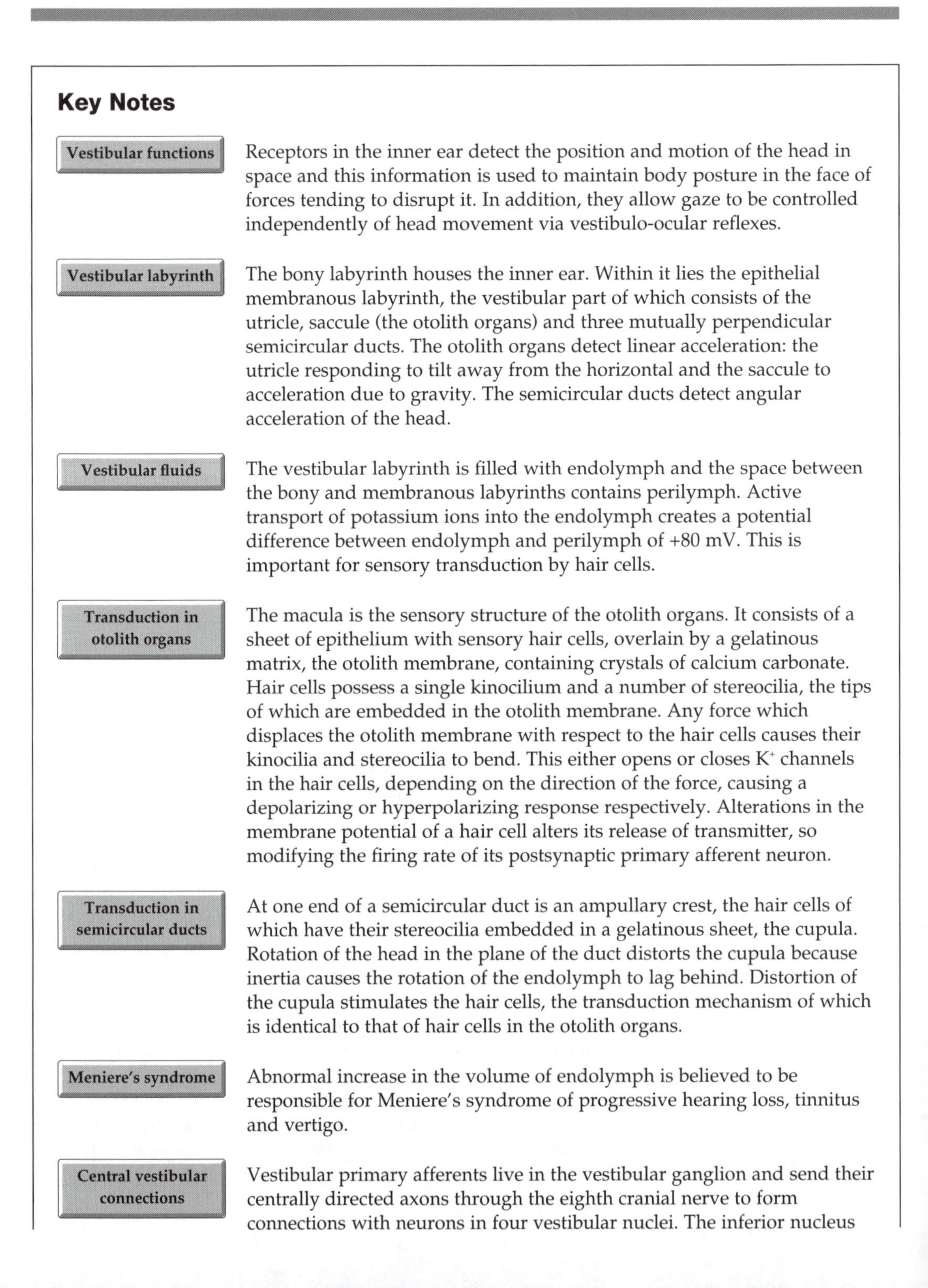

Key Notes

Vestibular functions

Receptors in the inner ear detect the position and motion of the head in space and this information is used to maintain body posture in the face of forces tending to disrupt it. In addition, they allow gaze to be controlled independently of head movement via vestibulo-ocular reflexes.

Vestibular labyrinth

The bony labyrinth houses the inner ear. Within it lies the epithelial membranous labyrinth, the vestibular part of which consists of the utricle, saccule (the otolith organs) and three mutually perpendicular semicircular ducts. The otolith organs detect linear acceleration: the utricle responding to tilt away from the horizontal and the saccule to acceleration due to gravity. The semicircular ducts detect angular acceleration of the head.

Vestibular fluids

The vestibular labyrinth is filled with endolymph and the space between the bony and membranous labyrinths contains perilymph. Active transport of potassium ions into the endolymph creates a potential difference between endolymph and perilymph of +80 mV. This is important for sensory transduction by hair cells.

Transduction in otolith organs

The macula is the sensory structure of the otolith organs. It consists of a sheet of epithelium with sensory hair cells, overlain by a gelatinous matrix, the otolith membrane, containing crystals of calcium carbonate. Hair cells possess a single kinocilium and a number of stereocilia, the tips of which are embedded in the otolith membrane. Any force which displaces the otolith membrane with respect to the hair cells causes their kinocilia and stereocilia to bend. This either opens or closes K^+ channels in the hair cells, depending on the direction of the force, causing a depolarizing or hyperpolarizing response respectively. Alterations in the membrane potential of a hair cell alters its release of transmitter, so modifying the firing rate of its postsynaptic primary afferent neuron.

Transduction in semicircular ducts

At one end of a semicircular duct is an ampullary crest, the hair cells of which have their stereocilia embedded in a gelatinous sheet, the cupula. Rotation of the head in the plane of the duct distorts the cupula because inertia causes the rotation of the endolymph to lag behind. Distortion of the cupula stimulates the hair cells, the transduction mechanism of which is identical to that of hair cells in the otolith organs.

Meniere's syndrome

Abnormal increase in the volume of endolymph is believed to be responsible for Meniere's syndrome of progressive hearing loss, tinnitus and vertigo.

Central vestibular connections

Vestibular primary afferents live in the vestibular ganglion and send their centrally directed axons through the eighth cranial nerve to form connections with neurons in four vestibular nuclei. The inferior nucleus projects to the contralateral ventral posterior thalamus and so to a region of parietal cortex close to SI. This pathway mediates the conscious perception of balance.

Related topics

Anatomy and physiology of the ear (I2)
Brainstem postural reflexes (K5)
Oculomotor control (L7)

Vestibular functions

The sense of balance is conferred by receptors which detect the position and motion of the head in space. The receptors are located in organs which are part of the **inner ear (labyrinth)** located in a hollow **vestibule** and three **semicircular canals** within the petrous portion of the temporal bone. Conscious perception of balance is normally overshadowed, except when head acceleration is high, by visual and proprioceptive cues to head position and motion. Vestibular input is used to maintain body posture in the face of forces which shift the center of mass to cause pitch (rocking backwards and forwards), yaw (rocking from side-to-side) or roll (rotation of the body around the long axis), by adjusting output to anti-gravity muscles.

Vestibular input is also used to execute eye movements which are independent of head movement. These **vestibulo-ocular reflexes** constitute one of several mechanisms for maintaining a fixed gaze (see Topic L7).

Vestibular labyrinth

Within the **bony labyrinth** which contains all the inner ear structures lies the **membranous labyrinth**, a sensory epithelium subserving hearing and balance (*Fig. 1*). The vestibular labyrinth, concerned with balance, consists of two **otolith organs**, the **utricle** and **saccule** and three **semicircular ducts**. The sensory structure of the otolith organs, the **macula**, which detects linear acceleration, is horizontal in the utricle and vertical in the saccule for a person standing upright. The effect of this is that the utricle is sensitive to tilting of the head (pitch and yaw) whilst the saccule is sensitive to vertically acting forces such as the acceleration due to gravity. The three semicircular ducts are mutually approximately orthogonal. Each contains a sensory structure, the **ampullary crest**, which detects angular acceleration in the plane in which the duct lies. Using the signals coming from all six semicircular ducts the brain computes the magnitude and direction of the angular acceleration of the head.

Vestibular fluids

The vestibular labyrinth is filled with **endolymph** which has a potassium concentration of about 160 mM and a sodium concentration of about 2 mM, and has a composition similar to that of intracellular fluid. It is secreted by a specialized epithelium, the **stria vascularis**, lining the outer wall of the **cochlear duct**, and drains into a venous sinus of the dura via the **endolymphatic sac**. The space between the bony and membranous labyrinths is filled with a cerebrospinal-like fluid, **perilymph** that is secreted by arterioles of the periosteum (the connective tissue layer covering the bone) and which drains into the subarachnoid space via the **perilymphatic duct**. The high potassium concentration in the endolymph arises because the marginal cells of the stria vascularis have Na, K-ATPase on their basolateral border allowing them to concentrate potassium ions for secretion into the endolymph (*Fig. 2*). The K^+ transport results in the endolymph having a potential difference of + 80 mV. Since the resting potential of the hair cell is about – 60 mV, the effective potential across its apical border is 140 mV. Hence there is both a sizeable electrical and chemical gradient favoring

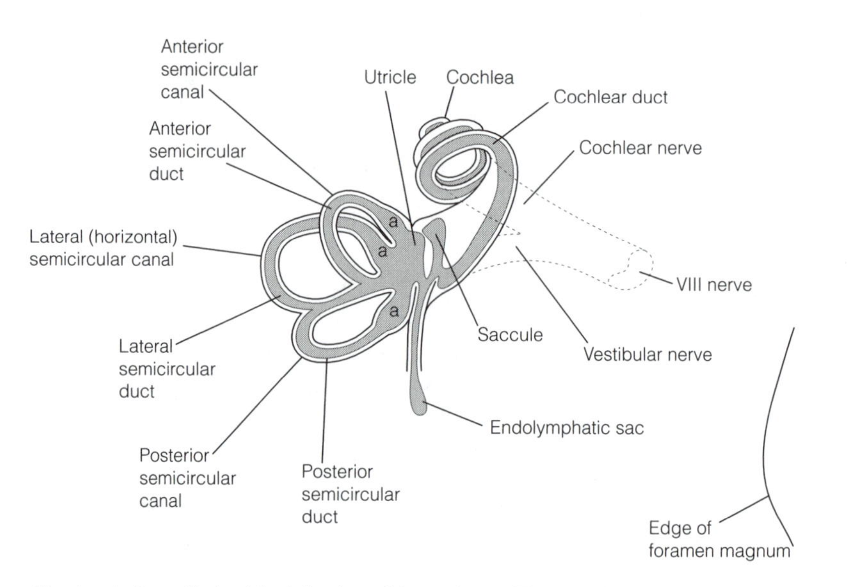

Fig. 1. Left vestibular labyrinth viewed from above. The membranous labyrinth is shaded. a, ampullae of the semicircular ducts.

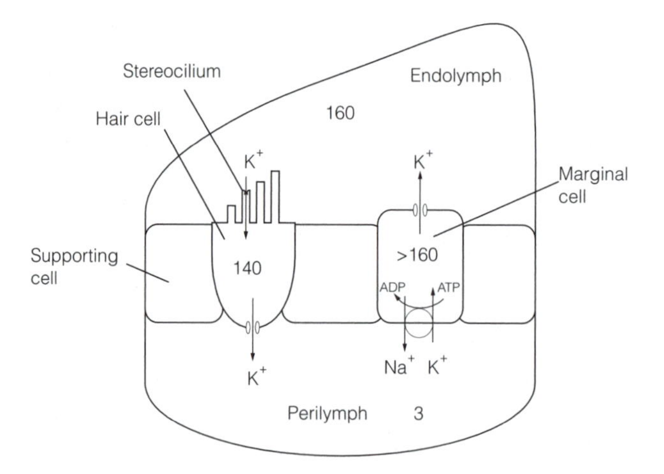

Fig. 2. Simplified model of K^+ transport in the inner ear. Figures are approximate concentrations of K^+ (mM).

facilitated passive diffusion of K^+ across the hair cell. This ensures that hair cells are extremely sensitive.

Transduction in otolith organs

The maculae consist of an epithelial sheet of **supporting cells**, embedded in which are an array of **hair cells**, which are sensory epithelial cells. Each hair cell is innervated at its base by a single vestibular afferent and an efferent fiber. The apical border of a hair cell has a single motile kinocilium, resembling a cilium, and 40–100 stereocilia, microvilli which are progressively shorter the further they are from the kinocilium (*Fig. 3*). This defines an **axis of polarity** for a hair cell, with a direction going from the smallest stereocilium to the kinocilium. Stereocilia lying along this axis are connected at their tips, those perpendicular to the axis are free.

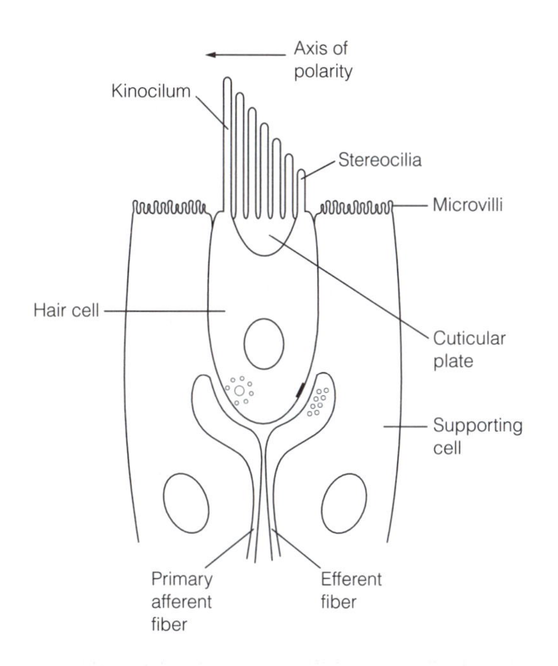

Fig. 3. The hair cell of an otolith organ surrounded by supporting cells of the sensory epithelium.

The kinocilium and stereocilia are embedded in a gelatinous matrix called the otolith membrane, which contains tiny crystals of calcium carbonate, **otoliths** (**otoconia, statoconia**). The hair cell is at rest if no force acts on the otolith membrane to cause the stereocilia to pivot. In this state the tension in the links (*Fig. 4*) connecting adjacent stereocilia is slight so only about 10% of the potassium ion channels gated by these links are open, causing a small depolarization. This is sufficient to sustain tonic release of an excitatory transmitter, probably glutamate, which maintains baseline firing of the primary afferent. Head tilt in the direction of the axis of polarity causes the otolith membrane to pull on the stereocilia, making them pivot, so increasing the tension in the connecting links. This opens the K^+ channels in the tips of the stereocilia, allowing K^+ influx to

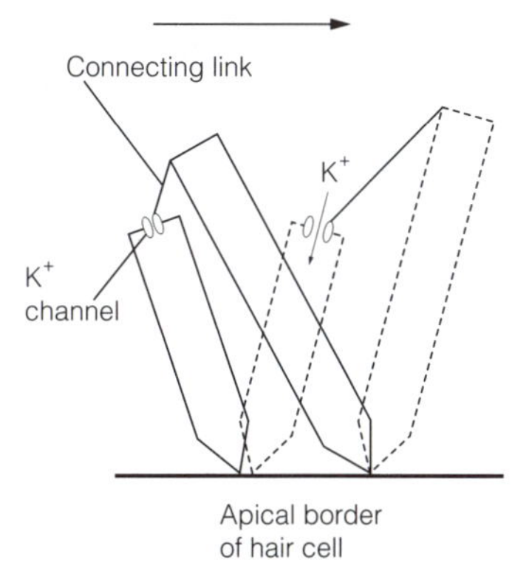

Fig. 4. Transduction by otolith hair cells. Two stereocilia are shown with their connecting link. When pivoted to the right (dotted lines) the connecting link is stretched, opening K^+ channels to cause an influx of K^+.

depolarize the hair cell, increasing its transmitter release and raising the afferent firing rate. Tilt in exactly the opposite direction, by reducing tension in the links, causes hair cell hyperpolarization which reduces primary afferent firing. Tilt which is perpendicular to the axis of polarity of a hair cell has no effect because stereocilia are not linked in this direction. Tilts in intermediate directions cause graded receptor potentials. The responses of individual otolith afferents are proportional to tilt angle and adapt only with prolonged stimulation.

Transduction in semicircular ducts

Velocity is a vector quantity, it has a magnitude (speed) and direction. For circular motion, such as head rotation, even if the angular speed is kept constant, the direction in which the velocity vector is acting is changing continuously. Hence head rotation is an angular acceleration.

Both ends of each semicircular canal insert into the utricle. Within the canal is the endolymph-filled semicircular duct. At one end of each duct is a dilation, the **ampulla**, in which sits the ampullary crest. Vestibular hair cells in the crest have their stereocilia embedded in a gelatinous sheet, the **cupula**, which stretches from the crest to the foot of the ampulla. Rotation of the head maximally stimulates hair cells in the canals lying in the same plane as the rotation. Rotation of the endolymph lags behind head rotation because of its inertia, so the endolymph exerts a pressure distorting the cupula, bending the stereocilia. The transduction mechanism is identical to that of hair cells in otolith organs. Because the cupula is not an ideal pressure transducer the signals transmitted by the duct afferents measure angular acceleration for slow and fast rotations, but encode velocity for mid-range rotation speeds. Semicircular ducts on each side lying in the same plane operate in pairs. Head rotation that causes depolarization of hair cells in the horizontal duct of the left ear will hyperpolarize hair cells in the horizontal duct of the right ear. Since the anterior duct on one side lies on approximately the same plane as the posterior canal on the other side these make up the two further pairs.

Meniere's syndrome

Typically this disorder involves progressive hearing loss, **tinnitus** (high or low pitched sounds generated within the head), in one ear, and episodes of **vertigo** (loss of balance sensation often accompanied by nausea and vomiting). It is associated with an increase in the volume of the endolymph causing herniations and ruptures of the membranous labyrinth. It may be caused by a viral infection of the ear which impairs reabsorption of endolymph. Cutting the vestibular nerve or destroying the labyrinth on the affected side can abolish the vertigo but has no effect on the other symptoms.

Central vestibular connections

The vestibular primary afferents (about 20 000 on each side) are pseudobipolar cells with their cell bodies in the **vestibular (Scarpa's) ganglion**. Their axons run in the vestibulocochlear (eighth cranial) nerve to enter the vestibular nuclei which lie laterally in the medulla and pons. There are four vestibular nuclei. Their roles in postural reflexes and vestibulo-ocular reflexes are dealt with in Topics K5 and L7 respectively.

The pathway for conscious perception of balance are axons of the inferior vestibular nucleus which cross to the contralateral side, ascend close to the medial lemniscus to terminate in the ventral posterior thalamus. Cortical representation is in the lateral parietal cortex, adjacent to the SI map of the head. Other representations may exist in the superior temporal cortex adjacent to the auditory area.

H1 Attributes of Vision

Key Notes

Visual perception

From two-dimensional retinal images the brain constructs a three-dimensional percept by which it can identify what is present in the world and where it is. Different aspects of the visual image (color, form, movement and depth perception) are processed in parallel by distinct neural pathways. Visual perception arises because the brain has internal representations which it compares with retinal images to make hypotheses about what they are. This allows objects to be recognized even if the image is poor quality or from any viewpoint. Object recognition is facilitated by perceptual constancy in which parameters such as size and color are preserved over big differences in viewing conditions. Some internal representations are developmentally programed but most are learned.

Sensitivity

The human eye responds to light between 400 and 700 nm over a 10^{11}-fold intensity range, though discrimination declines for higher light levels.

Acuity

The ability to see fine detail is highest at the central retina, the fovea. In ideal conditions two points are separable if they subtend an angle of one arc minute at the retina. Acuity falls with decreasing light levels.

Depth perception

The perception of distance comes from monocular cues for distant objects and binocular cues for nearer objects. Monocular cues include parallax, perspective and shadows. Binocular vision, stereopsis, arises because each eye gets a slightly different view of the world so the image of an object may lie on different points in each retina. For small degrees of retinal disparity the brain reconstructs a single percept and computes the object distance from the disparity.

Color vision

Color vision allows boundaries to be discerned simply on the basis of differences in the wavelength composition of reflected light. It requires a minimum of two types of receptor that respond over different wavelengths so that two values for brightness can be assigned for each part of an image. From this some color perception is derived. This dichromatic vision is the case for most mammals. Many primates, including humans, have trichromatic vision mediated by three types of receptor, which allows three brightness values to be ascribed to an object. The brain compares these values to give the perception of color.

Related topics

Retina (H3)
Early visual processing (H6)
Parallel processing in the visual system (H7)

Visual perception

Vision has been defined as the process of discovering from images what is present in the world and where it is. This requires the brain to use a two-dimensional shifting pattern of light intensity values on the two retinae (the light sensitive layers of the eyes) to form a representation of the form of an object, its color, movement, and position in three-dimensional space. There is good evidence that each of the visual subroutines (color, form, movement and depth perception) are handled simultaneously by distinct (but interdependent) neural pathways. This is called **parallel processing**. It contrasts with serial processing in which a task is segmented into several subroutines which must be executed, in sequence, one after the other. Parallel processing has the advantage of speed. The final visual representation is a unified percept in which all the independently processed subroutines must somehow be combined together. How this might be achieved by the brain is termed the binding problem, which applies not only to vision but to other sensory modalities as well.

Visual perception involves processing of the retinal image so that its key features can be abstracted. The visual system is more concerned with regions of the visual world that are changing in time (movement) and space (contrast) than regions that are static. It is widely accepted that perception requires the existence of internal representations of the visual world which allow the brain to make hypotheses about what the retinal image is. Internal representations account for the fact that the visual system shows **pattern completion**, in that it can generate a complete percept even when the raw sensory data is incomplete or corrupted by noise, and **generalization**, the ability to recognize objects from a wide variety of vantage points. Internal representations are probably encoded in the firing patterns of assemblies of neurons. Some internal representations are specified during development and are immutable, but most probably depend on early learning. Mental images of objects are thought to be manifestations of the internal representations of the objects and can be manipulated by most people in predictable ways. Whenever an unresolvable mismatch occurs between the sensory input and the internal representation the result is a **visual illusion**.

Perceptual constancy is a key property of vision. Visual perception can be invariant over wide differences in the properties of the retinal image. For example, with **size constancy**, familiar objects do not diminish in size in proportion to the reduction of the retinal image, but appear larger than they should. **Color constancy** preserves the colors of objects in the face of alterations in the wavelength composition of the light source. Perceptual constancy permits successful object recognition under a wide variety of ambient conditions.

Sensitivity

The human eye is sensitive to the electromagnetic spectrum between the wavelengths 400 nm (violet) and 700 nm (red). The range of light intensity to which we are exposed is huge, about 10^{11}-fold. Although the human eye can respond to a single photon of light, 5–8 photons arriving within a short time are required to give the experience of a flash of light in the dark adapted state. Because intensity is encoded by the visual system logarithmically, it is difficult to distinguish differences in intensity at high light levels.

Acuity

Visual acuity is the ability to see fine detail. It is highest at the central region of the retina, the **fovea**, and depends on the ambient illumination. Under ideal conditions two points of light can be separately resolved if they subtend an angle at the retina of one arc minute. For ruled gratings acuity is much better: lines can be resolved if separated by only a few arc seconds.

Loss of acuity in dim light occurs because insufficient photons fall on the retina to build up a complete image within the time span over which the light sensitive cells can integrate the energy.

Depth perception

The retinal image is two-dimensional but from it the visual system can infer the three-dimensional structure of the world. There are both monocular and binocular clues to depth perception. Monocular clues are most important for distant objects, where binocular cues cannot be used, and include:

- **Parallax**. Movement of the head causes an apparent movement of near objects with respect to distant ones. The closer the object the bigger this apparent movement.
- **Perspective**. Parallel lines appear to converge with distance. Artists from the early 1400s onwards used perspective as a major depth cue in painting.
- The relative sizes of objects of known dimensions.
- **Occultation**, in which a more distant object is partly hidden by a nearer one.
- Shadows.

The binocular cue to depth perception is called **stereopsis**. It is possible only for the field of view for which the two monocular visual fields overlap. Because the eyes are about 6.3 cm apart each has a slightly different view of the world and the image of a nearby object falls onto different horizontal positions on the left and right retina, a phenomenon called **retinal (binocular) disparity**. This can be visualized by viewing a scene first through one eye, then the other, when nearby objects appear to jump sideways; **binocular parallax**. When the eyes are made to converge so as to fixate on a nearby point the images are formed at the fovea on each retina, and the two images are perceived as fused into a single point. All other points that are fused appear so because their images lie at corresponding positions on the left and right retinas. Points in space that lie closer to or beyond those that form images at corresponding positions will generate binocular disparity (*Fig. 1*). Images of these points will also fuse. The brain is able to compute depth from disparity simply by comparing where the same pattern lies on left and right retinae. Stereopsis does not require form, movement or color. For binocular disparity that is too great (binocular disparity > 0.6 mm or 2° arc) fusion cannot occur and two images are seen; **double vision (diplopia)**.

Color vision

Color vision permits boundaries to be seen between regions that have equal brightness, provided that the spectrum of wavelengths they reflect is different. The spectrum of light reflected from an object depends on the wavelength composition of the illuminating light and the reflectance of the surface, but color vision is not just a matter of measuring all the wavelengths in the reflected light.

Color vision requires a minimum of two types of receptor that respond over different wavelength ranges. This is dichromatic color vision and is the case for all mammals except for old world monkeys, apes and humans. With two receptors the visual system can assign two brightness values for each object. By comparing these values, colors may be perceived. For example, if an object reflects more short wavelength light it will appear brighter to a short wavelength receptor than a long wavelength receptor, and will be seen as blue. If an object reflects more long wavelength light it will be seen as red. In the case that an object reflects equal amounts of short and long wavelength light it will appear to be monochrome, either white or shades of gray depending on the intensity of the light.

Human color vision is trichromatic because the eye has three populations of receptors that can function in daylight (cones), each sensitive to a different (but

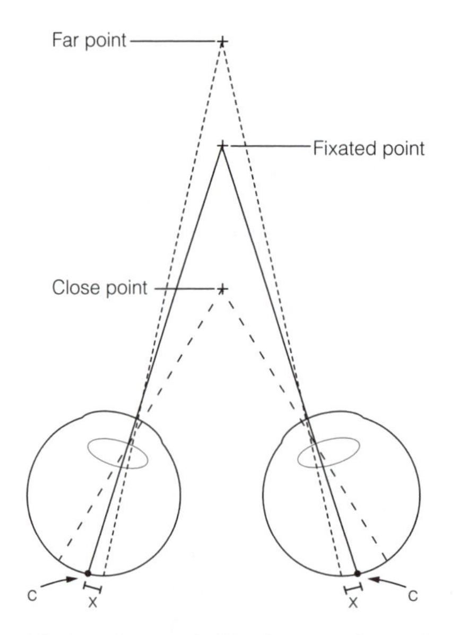

Fig. 1. Stereopsis. The fixated point produces images on corresponding positions, c, on the retinae so that the images are fused. The images of the far point are displaced from the corresponding points by a distance x giving a binocular disparity = 2x. A similar argument applies to the close point images.

wide and overlapping) range of wavelengths. The three types of cones have their maximum absorption corresponding approximately to violet, green and yellow light. The wavelength of the light does not affect the character of the response of the cone: a given cone simply has a higher probability of absorbing a photon which is close to its peak wavelength. This means that the visual system has no way of detecting the absolute wavelength composition of any light. The trichromatic visual system abstracts three brightness values for an object and comparisons of these values determines the color.

Color vision has several remarkable properties, including:

- **Color constancy**. An object can be viewed under a variety of light sources with different spectral compositions (e.g., neon lighting, sunlight or tungsten light) and appear to be the same color even though the wavelengths of light it reflects in each case will be quite different.
- **Perceptual cancellation**. While some colors in the same pixel of visual space perceptually mix to produce other color categories (e.g., blue and green mix to give cyan) complementary colors (e.g., red and green) do not perceptually mix; reddish-green colors are never seen.
- **Simultaneous color contrast**. This is the perceptual facilitation of complementary colors that occurs across boundaries. For example, a gray disc within a red background looks slightly green while a gray disc in a green background appears to be slightly red. Each of these features can be accounted for in terms of visual system physiology (Topic H7).

H2 Eye and visual pathways

Key Notes

Structure of the eye

The eye has three layers. The tough outer sclera maintains its shape and serves as an attachment for extraocular muscles. The choroid is pigmented to prevent light being reflected within the eye. The inner layer is the light sensitive retina. At the front the sclera becomes the transparent cornea responsible for most of the refraction of light rays entering the eye. The front of the choroid forms the ciliary body and iris. The biconvex lens is attached to the ciliary body by a suspensory ligament. The iris is a diaphragm surrounding the pupil and contains smooth muscles which act as pupillary sphincter or dilator. The anterior chamber of the eye lies in front of the lens and contains aqueous humor which determines the pressure of the eyeball. Behind the lens is the vitreous humor, a refractile medium.

Anatomy of visual pathways

The optic nerves meet at the optic chiasm where nerve fibers from the nasal half of each retina cross to the other side. Beyond this point nerve fibers enter the optic tract. A small number go to the pretectum which controls pupil and accommodation reflexes, others go to the superior colliculus, which mediates many visual reflexes, but most go to the lateral geniculate nucleus of the thalamus. From here the optic radiation goes to the primary visual cortex located in the occipital cortex. This last pathway is responsible for visual perception.

Visual reflexes

The amount of light entering the pupil can be altered 30-fold by changing the size of the pupil. The pupil light reflex causes pupil constriction in bright light. Light shone in one eye causes a light reflex in both eyes. The reflex pathway goes by way of optic nerve axons to the pretectum, which sends output to preganglionic parasympathetic neurons in the accessory oculomotor nucleus. The postganglionic cells lie in the ciliary ganglion and their axons supply the pupillary sphincter. Light stimulation elicits contraction of the muscle. For close objects greater refraction is needed to focus the image. This is achieved by the accommodation reflex; contraction of the ciliary body eases tension in the suspensory ligament, allowing the lens to become more spherical. The accommodation reflex is brought about by parasympathetic action on ciliary muscles. When observing a close object the eyes converge so that both eyes can fixate on it (the vergence reflex) and the pupils constrict, which increases depth of field and acuity.

Related topics

Attributes of vision (H1)
Oculomotor control (L7)
Autonomic nervous system function (M6)

Structure of the eye

The eye consists of three layers enclosing its contents, the sclera, the choroid and the retina (*Fig. 1*). The **sclera** is a thick, stiff outer layer of connective tissue. At its anterior end it is continuous with the cornea. At its posterior end it becomes the dura mater covering the optic nerve. The function of the sclera is to maintain the shape of the eyeball and to serve as an attachment for the extraocular muscles. The **cornea** is a curved circular transparent layer at the front of the eye. Its outer layer is continuous with the **conjunctiva**, an epithelial sheet covering the front of the eyeball. Most of the focusing power of the eye is due to refraction of light by the cornea. The **choroid** is a thin, highly vascular layer, dark brown in color because of the presence of **choroidal pigment cells**. By absorbing light it prevents total internal reflection within the eye.

Anteriorly the choroid becomes the **ciliary body** and the **iris**. The ciliary body gives rise to numerous, thin **zonular fibers** which attach to the capsule of the lens and constitute the **suspensory ligament**. Inside the ciliary body lies the ciliary muscle composed of smooth muscle fibers arranged in both radial and circular directions.

The iris is essentially a diaphragm surrounding a central hole, the **pupil**. The iris contains two intraocular muscles which act in concert to control the size of the pupil. Innermost is a flat ring of circularly arranged smooth muscle fibers, the **pupillary sphincter**. Surrounding the sphincter is a thin layer of radially organized myoepithelial cells which form the **pupillary dilator**.

The innermost layer of the eye is the light-sensitive **retina**. The structure and function of the retina are dealt with in Topics H3, H4 and H5.

The eye is divided into two by the lens. In front, the anterior and posterior chambers contain aqueous humor and behind the lens lies the vitreous humor. **Aqueous humor** is actively secreted by the epithelium of the ciliary body into the posterior chamber. It percolates through the pupil to the anterior chamber from where it drains into the venous system via the **canals of Schlemm** located in the **irideocorneal angle**. The pressure of the aqueous humor determines the

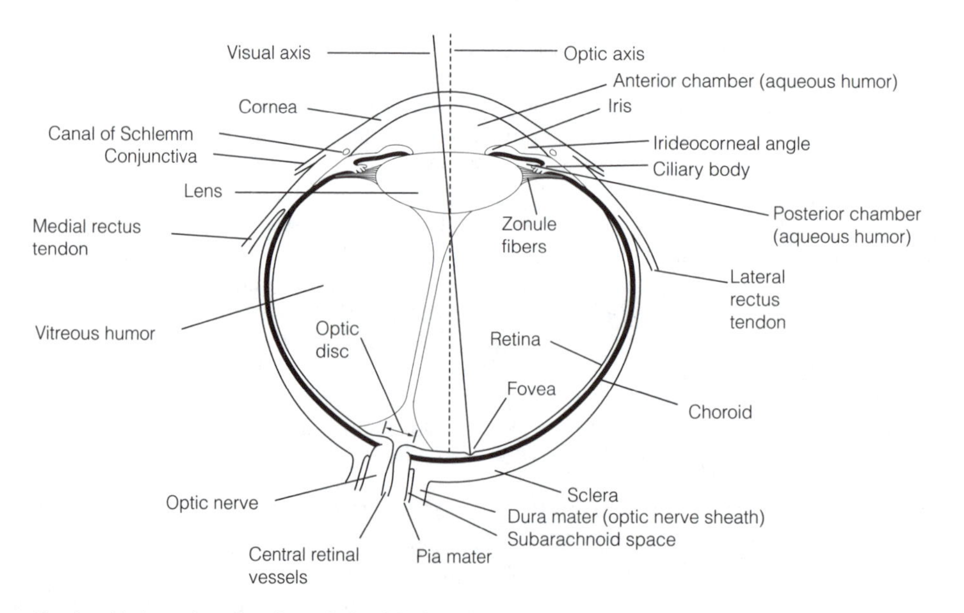

Fig. 1. Horizontal section through the right human eye.

pressure of the eyeball. This is normally less than 3 kPa. Obstruction to the proper drainage of the aqueous humor causes raised intraocular pressure, a condition called **glaucoma** which can result in blindness due to the reduced perfusion of blood through the retina. Aqueous humor is a vehicle for metabolic requirements (e.g. glucose, amino acids and ascorbate) for the lens and cornea which have no blood supply. **Vitreous humor** is a gel of extracellular fluid which refracts light rays appropriately so that they come to focus on the retina.

The biconvex **lens** of the human eye has a diameter of 9 mm. It is encapsulated within an elastic connective tissue membrane which is attached to the suspensory ligament.

Anatomy of visual pathways

The optic nerves meet in the midline at the **optic chiasm** (*Fig. 2*). Here, 53% of optic nerve fibers, those from the nasal halves of the retina, cross to the contralateral side in the **optic decussation**. Axons from the temporal halves of

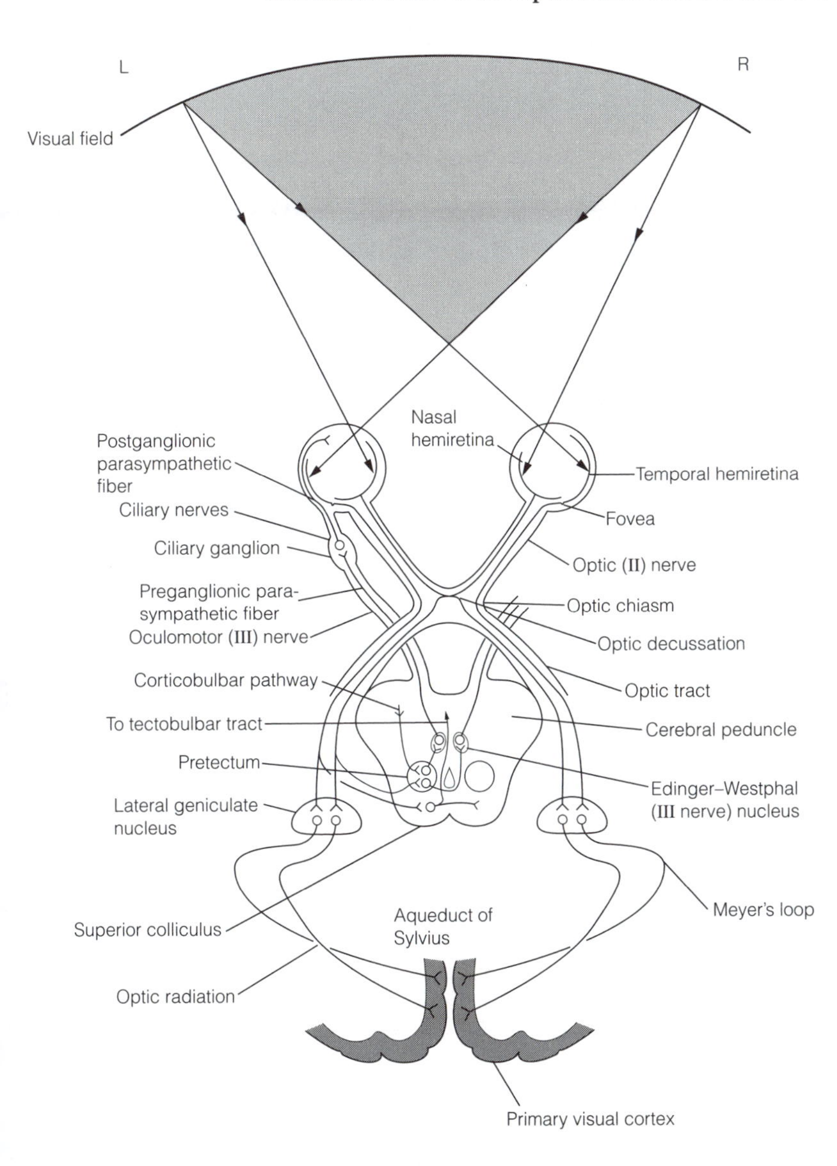

Fig. 2. Visual pathways. Reflex pathways are shown complete only on the left side. The direction of light rays from left and right halves of the visual field into the eyes is depicted. Note that light from the left visual field falls on the right halves of each retina (nasal hemiretina of the left eye, temporal hemiretina of the right eye) and light from the right visual field goes to the left hemiretinae. Binocular vision is possible only in the shaded region.

the retina remain on the ipsilateral side. Retinal axons leave the optic chiasm to enter the **optic tracts** from where they go to three targets. A small proportion go to the **pretectum** of the midbrain which controls pupil and accommodation reflexes (see below). Other axons go to the **superior colliculus** in the tectum of the midbrain which organizes several visual reflexes. The great majority of axons go to the **lateral geniculate nucleus** (LGN), part of the thalamus. From here, the **optic radiation** sweeps to the medial aspect of the pole of the occipital cortex, most axons terminating in layer IV of Brodmann area 17, the **striate** or **primary visual cortex (V1)**. The retina–LGN–visual cortex pathway is responsible for visual perception. Visual defects characterized clinically can provide clues to the site of lesions within the visual system (*Fig. 3*).

Visual reflexes

The **pupil light reflex** controls the amount of light entering the eye by altering pupil size. This ranges between 1.5 and 8 mm in diameter, being maximal in complete darkness. Although this allows only a 30-fold change in light entry

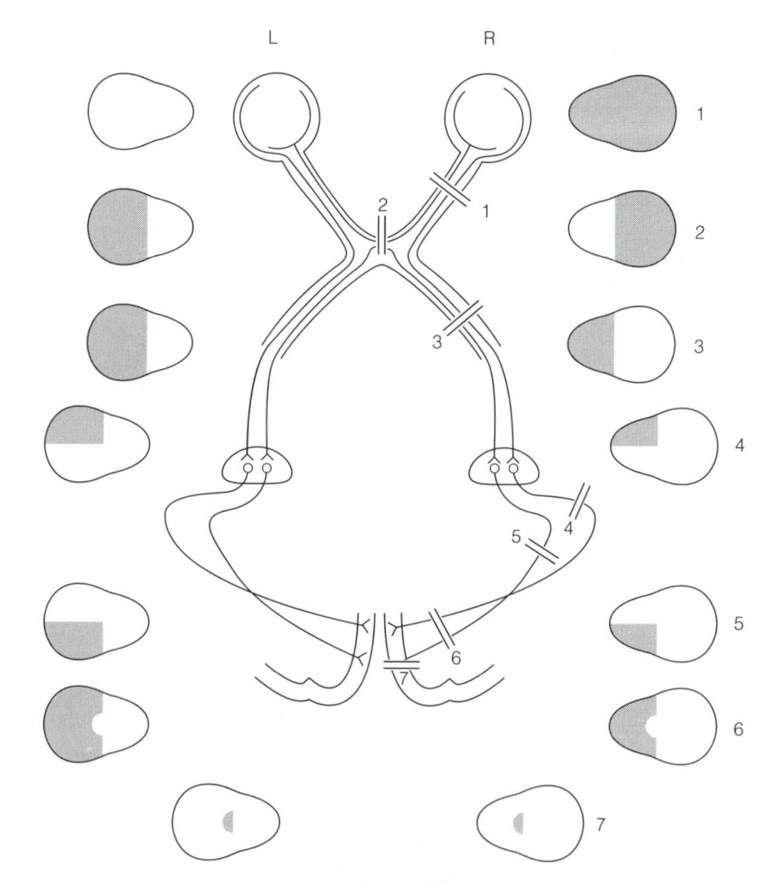

Fig. 3. Visual defects arising from damage to visual pathways. Lesion 2 commonly arises from compression of the central part of the optic chiasm by a pituitary tumor. Optic tract lesions (3) are rare. Optic radiation damage is usually due to an infarct or a tumor in the temporal lobe (4) or parietal lobe (5). Lesion 6 is usually caused by occlusion of the posterior cerebral artery. The fovea area is spared because it is supplied by the middle cerebral artery. Damage to one occipital lobe pole is usually caused by trauma and, since the fovea has by far the largest representation, selective loss of foveal vision is typical (lesion 7).

(which is small compared with the range of light intensity the visual system can experience) the reflex is useful because it operates over the light levels typically encountered during daylight. Light shone in one eye produces pupil constriction of the same eye (the **direct** reflex) and of the contralateral eye (the **consensual** reflex) because of reciprocal crossed connections in the midbrain. The reflex pathway is shown in *Fig. 2*. Optic nerve axons synapse in the pretectum which sends output to preganglionic parasympathetic fibers in the **Edinger–Westphal (accessory) oculomotor** nucleus. These autonomic fibers travel in the oculomotor nerve to the ciliary ganglion which lies in the orbit. Postganglionic fibers from there go to the pupillary sphincter. Light stimulation of optic nerve fibers excites the parasympathetic terminals to release acetylcholine, which contracts the sphincter. The latency of the reflex is about 200 msec. Lesions of the optic and oculomotor nerves, or of the midbrain can be diagnosed by examining defects in the pupil light reflex (*Fig. 4*).

For close objects, light rays are diverging as they enter the eye and so greater refraction is needed to bring them to focus at the fovea. This is achieved by the **accommodation reflex**. Contraction of the ciliary muscles pulls the ciliary body forwards and inwards, easing the tension in the suspensory ligament and lens capsule, allowing the lens to become more spherical and reducing its focal length. The stimulus for the accommodation reflex is blurring of the retinal image. This is monitored by the visual cortex which projects to the pretectum via the corticobulbar pathway. Via connections between the pretectum and the Edinger–Westphal nucleus, parasympathetic fibers are activated which contract the ciliary muscles. Accommodation occurs in both eyes equally and takes nearly 1 second to execute.

Observing a close object, in addition to accommodation, causes convergence of the visual axes of both eyes, the **vergence reflex**. This enables both eyes to fix their gaze on an object. In addition, the degree of convergence provides a cue for stereopsis, since the closer an object is, the greater the convergence must be. Vergence can be triggered by a blurred retinal image or by consciously altering

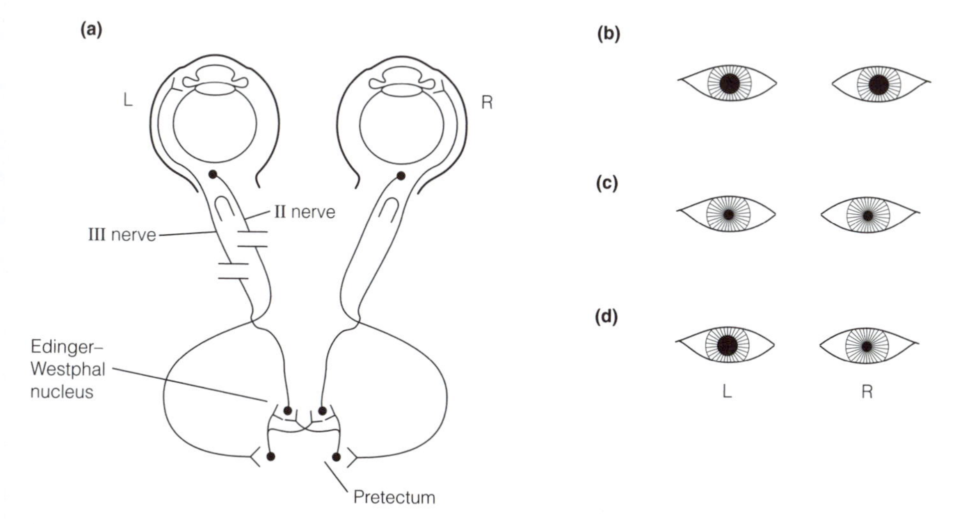

Fig. 4. Altered pupil reflexes following damage to either optic (II) or oculomotor (III) nerves on the left side: (a) reflex pathway; (b) optic nerve damage, left eye stimulated; (c) optic nerve damage, right eye stimulated; (d) oculomotor nerve damage, left or right eye stimulated.

gaze to a point at a different distance. The circuitry is from the visual cortex to a region of the frontal cortex concerned with the planning and execution of eye movements (see Topic L7).

Accommodation and convergence are both accompanied by pupil constriction which has two effects. Firstly, it increases the depth of field, the distance range in which objects are in focus. Secondly, it reduces spherical aberration (a defect of lenses in which parallel rays of light are not brought to focus at the same point); this improves acuity when looking at near objects. Pupil constriction in this instance is mediated by a pathway from the primary visual cortex to the pretectum. The **Argyll–Robertson pupil** fails to constrict in response to light, but will constrict during accommodation and convergence. It results from damage to the light reflex pathway in the region of the tectum or aqueduct of Sylvius.

H3 Retina

Key Notes

Structure of the retina

The light sensitive retina has five basic neuron types arranged in several layers. The outer nuclear layer consists of the cell bodies of the photoreceptors and the inner nuclear layer contains the cell bodies of the retinal interneurons, bipolar cells, horizontal cells and amacrine cells. The innermost ganglion cell layer harbors the output cells of the retina which send their axons into the optic nerve. The number of ganglion cell axons is 100-fold less than the number of photoreceptors, which indicates that a great deal of visual processing must go on in the retina. Only ganglion cells are excitable, all other retinal neurons signal via electrotonic potentials. The central 1.5 mm of retina is the fovea and has the highest visual acuity. The optic disc, the site where optic nerve and blood vessels pierce the retina, accounts for the blind spot.

Rod and cone cells

Rod photoreceptors, situated throughout the retina except the fovea and optic disc, are extremely sensitive to light and are used in dim light vision. In daylight rod cells become saturated and are unresponsive. Rod cell vision has a low acuity because many rod cell signals converge, which, while maximizing light sensitivity causes loss of information about location of an image. Cone cells are found at the fovea. They are 1000-fold less sensitive to light than the rods, fail in dim light, but do not saturate except in very bright light and so operate in daylight vision. Daylight vision has high acuity because there is little convergence of cone signals. There are three populations of cones, distinguished by the range of wavelengths to which they are sensitive. Short wavelength (blue) cones constitute only a few percent of all cones and are absent from the centre of the fovea. Medium (green) and long (red) wavelength cones are randomly but patchily distributed, so color vision cannot be used to see fine detail. The maximum sensitivity of the human eye is to yellow light but this shifts towards the green in low light conditions when rod cells become active. Going from bright to dim light causes a massive increase in sensitivity of the retina to occur – dark adaptation – which takes about 30 min.

Color blindness

Virtually all color blindness is genetic and caused by loss or abnormality of cones. Trichromats retain all their cones, but have a defect in one type. Dichromats lack one population of cone cell, while monochromats, missing two or all three types of cone, have no color vision. Since the genes for the medium and long wavelength cone pigments lie on the X chromosome a defect in either, resulting in red-green color blindness, is an X-linked recessive trait predominantly afflicting males.

Related topics

Attributes of vision (H1)
Retinal processing (H5)

Structure of the retina

The retina is the light sensitive innermost layer of the eye. It contains five distinct neuron types, interconnected in circuits that are repeated millions of times. Under light microscopy, the retina is seen to be composed of several layers (*Fig. 1*).

Closest to the choroid is a single layer of pigmented epithelial cells. These contain melanin and absorb light not absorbed by the retina so that it is not reflected back to degrade the image. The outer nuclear layer contains the cell bodies of the photoreceptors. The inner nuclear layer consists of the cell bodies of retinal interneurons, bipolar cells, horizontal cells and amacrine cells. The ganglion cell layer contains ganglion cell bodies, the axons of which provide the output of the retina via the optic nerve. Ganglion cell axons only become myelinated at the optic disc (see below). The two plexiform layers are the locations for the connections between the retinal cells. Note that although light must pass through the full thickness of the retina before striking the photoreceptors the retina is highly transparent. Each human retina has around 10^8 photoreceptors, but an output via only about 10^6 optic nerve axons. This is a massive convergence and shows that considerable processing of visual input is carried out by the retina.

Although all the cells in the retina (except pigment cells) are neurons, only the ganglion cells are able to fire action potentials. Photoreceptors and retinal interneurons signal by way of electrotonic potentials.

The gaze of the eye is usually adjusted so that the images are brought to focus at the **fovea**. This region of the retina with a diameter of 1.5 mm has the greatest visual acuity. The high visual acuity at the fovea is in part due to:

- The great density of photoreceptors.
- The displacement of overlying layers of the retina to the side so that light hits the photoreceptors directly.
- The lack of blood vessels.
- The fovea lying on the optical axis of the eye so image distortion by the optics (e.g. spherical or chromatic aberration) is minimal.

About 4 mm from the fovea towards the nose lies the **optic disc**, where optic nerve fibers and retinal blood vessels pierce the retina. This region lacks photoreceptors and accounts for the **blind spot** that occurs in the visual fields.

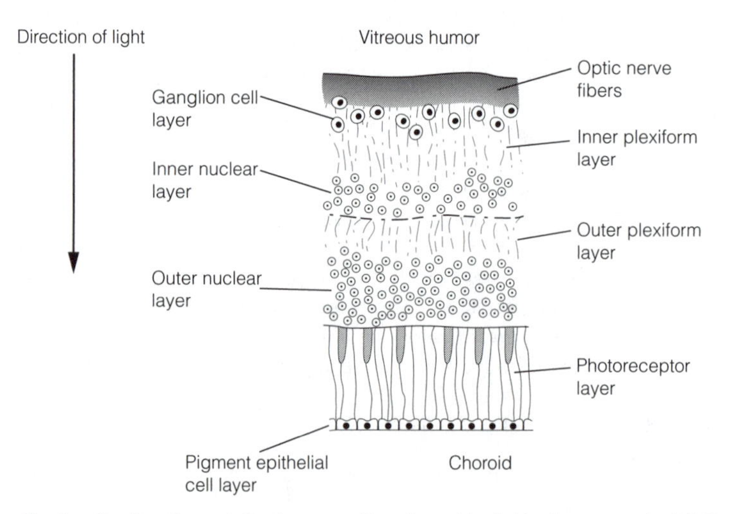

Fig. 1. Section through the human retina viewed by light microscopy (× 1500).

Retinal detachment can occur as a result of head trauma, the separation occurring at the interface between the pigment epithelial cells and the photoreceptors.

Rod and cone cells

There are two populations of photoreceptors, rods and cones. Only about 10% of light entering the eye succeeds in exciting photoreceptors, the rest is scattered or absorbed.

Rod cells are 20-fold more numerous than cones and are distributed across the entire retina except for the fovea and the optic disc (*Fig. 2*). They are about 1000 times more sensitive to light than cone cells and are used for **scotopic** (dim light) vision. Indeed, rod cells saturate in daylight in which state they become unresponsive. The high sensitivity of rod cells comes about partly because they integrate responses to incoming photons over a long period (approximately 100 ms). The disadvantage of this is that rods are unable to discern flickering light if the flicker rate is faster than about 12 Hz. A second feature contributing to rod cell sensitivity is that they are able to greatly amplify the effect of the photons impinging upon them. Scotopic vision has a low **acuity** for two reasons. Firstly, the image formed on the peripheral retina is quite distorted. Secondly, many rods converge onto a single bipolar cell. Although this maximizes the chance of a bipolar cell responding to a dim light signal, because it can capture light signals from a large area of retina, by the same token information about localization is less precise.

Cone photoreceptors are most dense at the fovea and their numbers fall off sharply beyond 5 degrees of it. They have a low sensitivity to light and do not saturate except in very intense light, so are the receptors used for **photopic** (daylight) vision. Photopic vision has high acuity because there is little or no convergence between cone cells and bipolar cells. Cone cells integrate photon responses over a short time and so are able to resolve a flicker frequency of less than about 55 Hz. There are three populations of cone cells which differ in their spectral sensitivity (*Fig. 3*).

Although more properly called short (S), medium (M) and long (L) wavelength cones they are often referred to as blue, green and red cones respectively although their peak sensitivities are not best described by these colors. The S cones are sensitive to wavelengths down to 315 nm, however the normal eye does not see wavelengths shorter than 400 nm because they are absorbed by the lens. Absorption of UV light is an important cause of **cataract**, in which the lens becomes progressively more opaque, which is especially prevalent in countries at lower latitudes where sunlight is most intense.

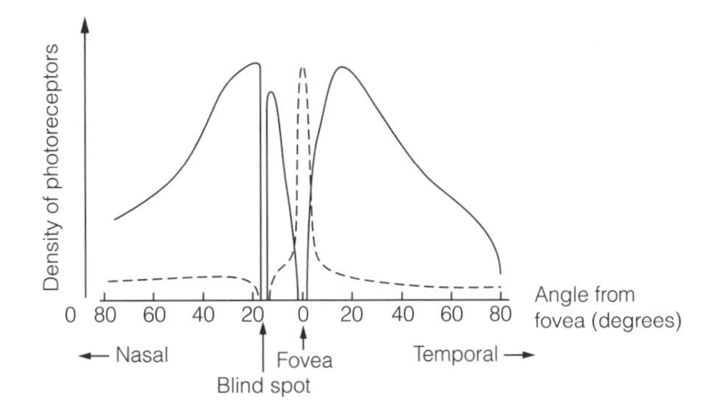

Fig. 2. Distribution of cone (----) and rod (———) photoreceptors in the human retina.

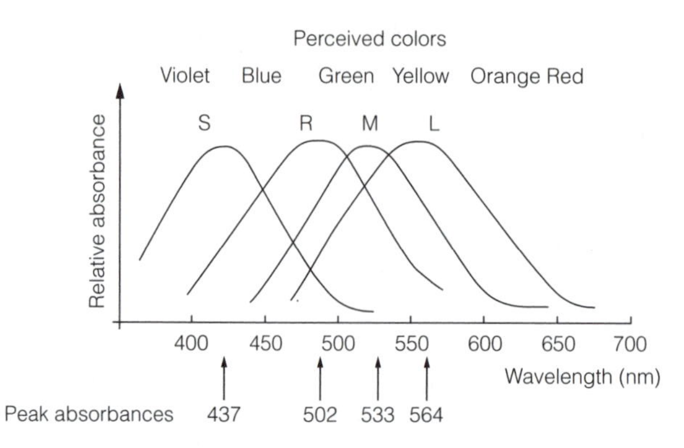

Fig. 3. Spectral sensitivities of photoreceptors: S, short wavelength cones; R, rods; M, medium wavelength cones; L, long wavelength cones.

Color vision requires comparisons of the relative strengths in the outputs of the S, M and L cones. S cones constitute only about 5–10% of the total number of cones and are absent from the center of the fovea. This is because the eye suffers from **chromatic aberration** in which short wavelength light is not brought to focus at the same point as longer wavelengths and so causes slight blurring of the image. This would compromise high acuity vision. In consequence, color vision at the central fovea is dichromatic and furthermore, M and L cones are distributed randomly leaving patches in which there is only one population of cone. These features mean that color vision is coarse grained and cannot resolve fine detail. Scotopic vision is achromatic because all rod cells have the same spectral sensitivity curve. They are unable to distinguish between wavelengths on the rising and falling limbs of the spectral sensitivity curve that excite the cell to the same extent.

Under scotopic vision the wavelength sensitivity of the eye is determined by the rod cells and peaks at around 500 nm. Under photopic conditions the wavelength sensitivity is governed by cones and is maximal at 555 nm. This shift in wavelength sensitivity between scotopic and photopic vision is called the **Purkinje shift**. It accounts for the fact that as dusk falls the last color sensations to be lost are blues and greens.

When moving from bright to very dim light the sensitivity of the retina to light increases a million-fold over a period of 30 min or longer. This is called dark adaptation and is a property of photoreceptors. **Dark adaptation** has two phases (*Fig. 4*). The first is due to cone cells which increase sensitivity about 100-fold, the second longer phase is due to rod cells.

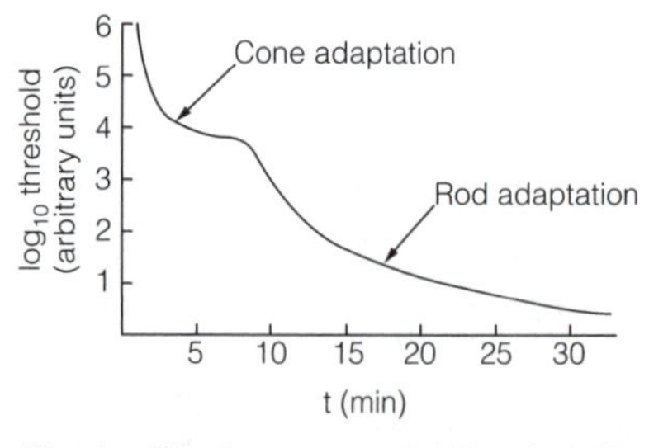

Fig. 4. The timecourse of dark adaptation.

Rod cells are 50 to 100-fold less sensitive to red light than L cones in the dark adapted state (note the big gap between the spectral curves in *Fig. 3*). Hence dim red light can be used for photopic vision tasks (e.g. reading) by individuals, such as astronomers, who need to work with their scotopic (rod-cell mediated) vision dark adapted. **Light adaptation** occurs when going from dim to brightly lit conditions. It is very much faster than dark adaptation.

Color blindness

Almost all **color blindness** is genetic in origin and results from loss or abnormality of cone cells. **Trichromats** retain all three cone populations but have abnormalities in the opsins (visual pigments) of S, M or L cones, most commonly the M opsin. **Dichromats** have only two cone populations and their visual defect is more severe than that of the trichromats. Dichromats are classified as **protanopes** (without L cones), **deuteranopes** (lacking M cones) and **tritanopes** (lacking S cones). Most affected, but rare, are **monochromats** who, lacking two or all three types of cone cell, have no color vision. Those without any cones have no photopic vision and are effectively blind in daylight.

Abnormality of S opsin or loss of S cones is quite rare and afflicted individuals may be unable to distinguish colors having a short wavelength component (violet) from those without (yellow); both appear gray. The gene for S opsin is on chromosome 7 and defects in S cone vision are inherited as an autosomal dominant trait. In defects of M or L cone vision it is not possible to discriminate red from green or either from gray. M and L opsin genes lie on the X chromosome so red–green color blindness is an X-linked recessive trait. Not surprisingly it is much more common in males (4–8% in Europe, depending on ethnicity) than females (about 0.4%).

Loss of rod cell function has also been described. Afflicted individuals have only narrow, central visual fields and, lacking scotopic vision, are blind whenever the light level falls below that required to excite cones.

H4 PHOTOTRANSDUCTION

Key Notes

Photoreceptor structure

Photoreceptors have an inner segment (containing the nucleus) which has a synaptic terminal and an outer segment, the plasma membrane of which is invaginated into deep folds to form discs. The visual pigments, rhodopsin of rods or cone opsins, are situated in the disc membrane. Photoreceptors cannot divide but their outer segments are continuously turned over.

Photoreceptor transduction

Photoreceptors in the dark have quite a low (depolarized) membrane potential because of the influx of sodium and calcium ions through cyclic nucleotide-gated ion channels in the outer segment plasma membrane. Light causes a hyperpolarizing receptor potential by closing the channels. Transduction in rod cells starts when photons are captured by the prosthetic group in rhodopsin, retinal, which undergoes photo-isomerization. This activates the rhodopsin which in consequence couples with a G protein, transducin. Transducin stimulates a phosphodiesterase that hydrolyzes 3′,5′-cyclic guanosine monophosphate (cGMP), reducing its concentration, so closing the cyclic nucleotide-gated channels. Subsequently the photo-isomerized retinal dissociates from the rhodopsin leaving the pigment bleached. Dark adaptation is the regeneration of rhodopsin.

Related topics

Slow neurotransmission (C3)
Receptor molecular biology (C4)
Retina (H3)

Photoreceptor structure

Rod and cone photoreceptors have similar structures (*Fig. 1*). Photoreceptors have diameters ranging from 1 to 4 μm, being smaller at the fovea, a factor contributing to the higher acuity achieved by this area. The inner segment contains the nucleus, is rich in mitochondria and has an axon-like process connected to a synaptic terminal called a **spherule** in rods and a **pedicle** in cones. The inner segment is connected to the outer segment via a thin waist through which runs a single cilium. The outer segment in the cone cell has its plasma membrane invaginated into numerous closely packed parallel folds, forming discs. In rod cells the discs are pinched off the plasma membrane to become completely intracellular. The disc membrane is densely packed with **visual pigment**. In rod cells this is **rhodopsin** ('visual purple', so-called because it absorbs most blue and green light). Each population of cone cells has its characteristic **cone opsin**. The outer segment is continually regenerated from the base, whilst its apical tip is phagocytosed by pigment epithelial cells at the rate of three to four discs per hour. Failure of this phagocytotic mechanism may underlie some forms of the X-linked disorder **retinitis pigmentosa**. Photoreceptors are neuroepithelial cells and are incapable of mitotic division.

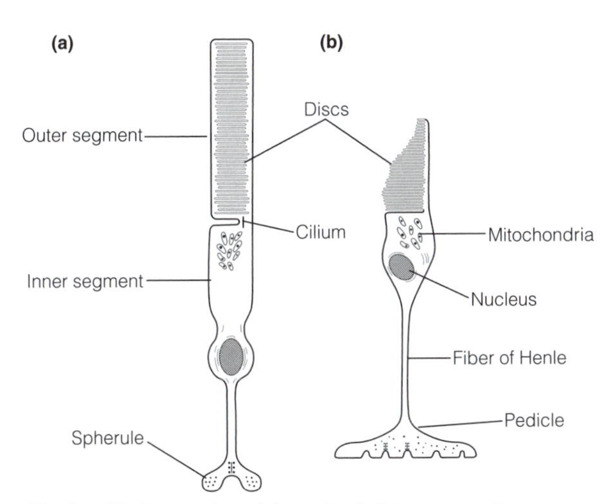

Fig. 1. Photoreceptors: (a) a rod cell; (b) a cone cell.

Photoreceptor transduction

The resting potential of the photoreceptor plasma membrane in the dark is quite low, about –40 mV. Light produces a hyperpolarizing receptor potential, the amplitude of which is related to the light intensity (*Fig. 2*).

The hyperpolarization is produced by the light-evoked closure of cyclic nucleotide-gated cation channels that are open in the dark. The normal, relatively depolarized, state of the photoreceptor is caused by the flow of a **dark current**, as shown in *Fig. 3*.

The cyclic nucleotide-gated cation channel allows Na^+ and Ca^{2+} ions to flow into the outer segment in darkness. Na^+ ions are actively extruded by the Na^+-K^+ ATPase in the inner segment. Ca^{2+} leaves the photoreceptor via a Na^+-K^+-Ca^{2+} transporter. When light photons strike the outer segments a cascade of biochemical events is initiated which results in the closure of the cation channels, reducing the dark current, and hyperpolarizing the photoreceptor. The transduction process in rod cells is well understood. Rhodopsin is a member of the G protein receptor superfamily with seven transmembrane segments. It consists of a protein, **opsin**, and a prosthetic group, **retinal**. Retinal is the aldehyde of **retinol (vitamin A)** from which it is synthesized by retinol dehydrogenase. Retinol cannot be synthesized *de novo* in mammals and hence must be supplied in the diet. Lack of vitamin A in the diet causes night blindness and if prolonged results in irreversible damage to rod cells.

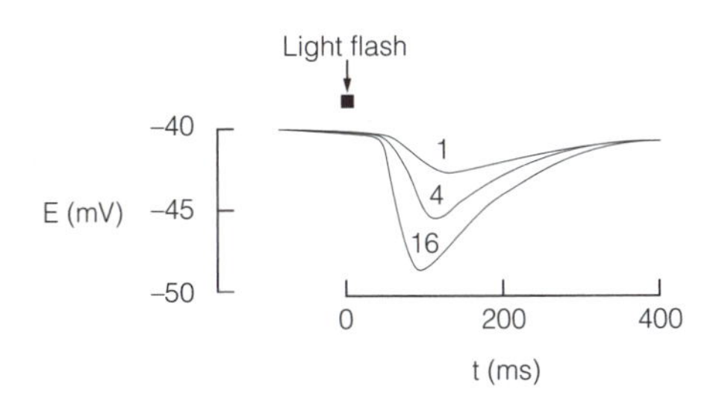

Fig. 2. Cone receptor potentials in response to light flashes of increasing relative intensity (1, 4 and 16).

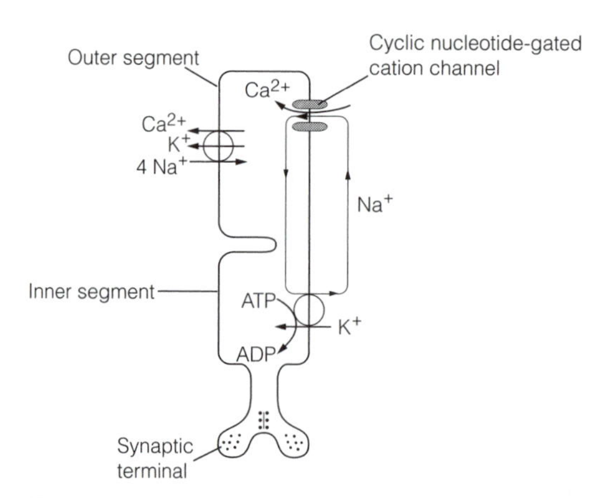

Fig. 3. The ionic basis of the rod photoreceptor dark current.

In the dark, retinal is present as the 11-*cis*-isomer. Light causes photo-isomerization to the all-*trans* isomer. The isomerization occurs within a few picoseconds of the photon being absorbed and triggers a series of conformational changes in the rhodopsin to form photoexcited rhodopsin (R*). Photoexcited rhodopsin couples with a G protein, **transducin** (abbreviated either to **T**, or, because it is a G protein, to $\mathbf{G_t}$), and (in a manner exactly analogous to G protein coupling to metabotropic receptors outlined in Topic C3) exchange of guanosine 5′-diphosphate (GDP) for guanosine 5′-triphosphate (GTP) occurs. The GTP-bound form of the transducin alpha subunit activates a phosphodiesterase (PDE) which catalyzes the hydrolysis of cGMP to 5′-GMP. This reduces the concentration of cGMP in the photoreceptor and so the cation channels, normally kept open by the cyclic nucleotide, close. Each cation channel allows only a small current to pass (~ 3 fA) so the total dark current in a rod is only about 20 pA. The opening of each channel requires three molecules of cGMP with high positive cooperativity. This means that small changes in cGMP concentration effect big changes in the number of open channels. This sequence of events is summarized in *Fig. 4*.

This second messenger cascade has a large amplification. A single photon activates about 500 transducin molecules, closes hundreds of cation channels, blocking the influx of 10^6 Na^+ ions to cause a hyperpolarization of about 1 mV.

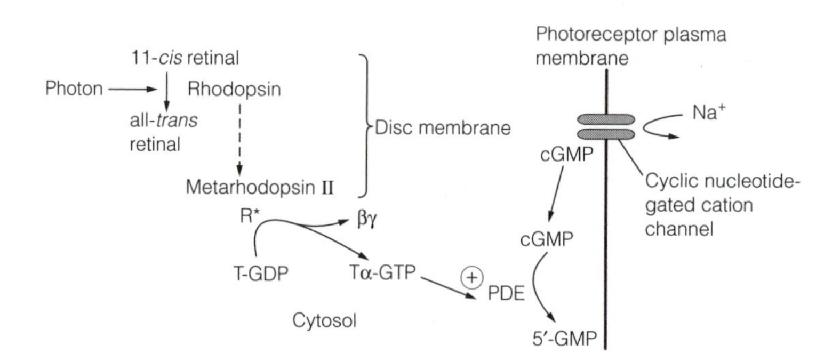

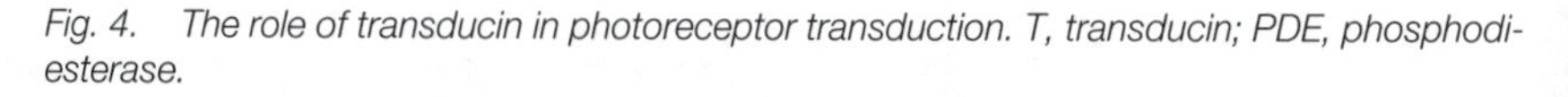
Fig. 4. The role of transducin in photoreceptor transduction. T, transducin; PDE, phosphodiesterase.

Several mechanisms act sequentially to terminate the cascade:

- Like other G proteins, transducin has an intrinsic GTPase which hydrolyses the bound GTP to GDP, stopping the activation of PDE.
- Photoexcited rhodopsin is phosphorylated by **rhodopsin kinase**, and then binds **arrestin** which blocks the binding of transducin.
- Within a few seconds the bond between retinal and opsin in photoexcited rhodopsin spontaneously hydrolyzes and the all-*trans* retinal diffuses away from the opsin. In high light levels most of the rhodopsin exists in this dissociated state in which it is described as being **bleached** and the rod said to be **saturated**. Regeneration of rhodopsin occurs in the dark: retinal isomerase catalyses the isomerization of the all-*trans* isomer to the 11-*cis* isomer which then reassociates with the opsin. This process underlies **dark adaptation**.

Restoration of the dark state in addition requires the synthesis of cGMP. This is catalyzed by guanylyl cyclase.

Light adaptation, in which photoreceptors become less sensitive in the face of steady state light exposure, allows them to respond to levels of illumination that vary by as much as four orders of magnitude (*Fig. 5*). Light-evoked closure of the cation channels reduces Ca^{2+} influx, so the Ca^{2+} concentration in the rod outer segment falls. Since Ca^{2+} normally inhibits the guanylyl cyclase needed for cGMP synthesis, this drop in Ca^{2+} concentration increases the production of cGMP, offsetting its destruction by the light.

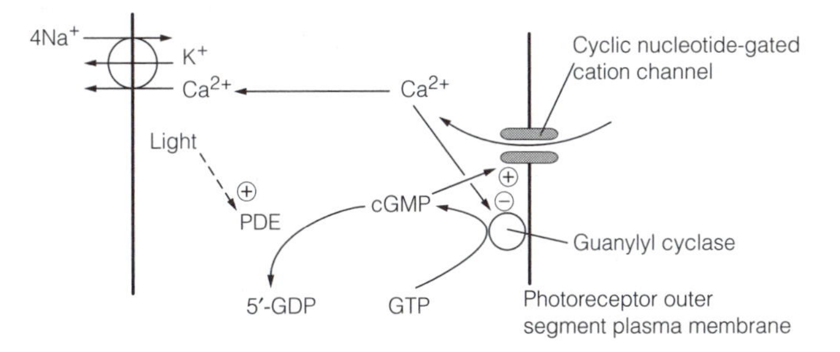

Fig. 5. The role of Ca^{2+} in photoreceptor light adaptation.

H5 Retinal processing

Key Notes

Bipolar cells and on and off channels

Photoreceptors synapse with either invaginating bipolar cells which depolarize in response to light, or flat bipolar cells which hyperpolarize when light stimulated. Midget bipolar cells, which get their input from cones, are of either type and synapse directly with ganglion cells, which respond in the same way to light as their bipolar cells; on ganglion cells increase firing, while off ganglion cells are silenced, by light. Hence the retina has on channels, formed from cone–depolarizing bipolar cell–on ganglion cell, and off channels, routed through cone–hyperpolarizing bipolar cell–off ganglion cell. The transmitter of all photoreceptors is glutamate and the opposite responses of the two types of bipolar cell are because they have different glutamate receptors. On channels signal the presence of local bright patches and off channels local dark regions of an image.

Horizontal cells and lateral inhibition

Bipolar and ganglion cell receptive fields are circular and divided into centre and surround. Light stimulation has the opposite effect on the cell depending on whether it falls on the center or the surround. This is lateral inhibition and its effect is to enhance contrast at boundaries. It is mediated by γ-aminobutyrate (GABA)ergic horizontal cells that form reciprocal connections with adjacent photoreceptors. Because horizontal cells are heavily interconnected, the surround signal they generate represents average light intensity over a given area of retina.

Ganglion cells

There are two populations of ganglion cell. P ganglion cells are small, slowly conducting, get their input from single cone types, are wavelength selective and show sustained responses. They are implicated in form and color vision. P ganglion cells exhibit color opponency, a sort of lateral inhibition in which they are excited by one type of cone but inhibited by one or both of the others. M ganglion cells in contrast are large, rapidly conducting, get input from middle and long wavelength cones together and show transient responses. M ganglion cells detect brightness (but not color) contrast, and movement.

Rod signaling

In daylight only the cone channels are functional. At dusk rod cells become sensitive but transmit their signals via gap junctions to cone cells, boosting their function to maintain high acuity and color vision. When it is very dark cone cells fail and the rods cells signal exclusively through their own pathway via depolarizing (rod) bipolar cells and amacrine cells.

Amacrine cells

These interneurons, with neurites that have properties of both axons and dendrites, are extremely diverse. Different types are involved in rod cell pathways, surround inhibition and in signaling the direction of movement of a stimulus.

Related topics

Overview of synaptic transmission (C1)
Slow neurotransmission (C3)
Neurotransmitter inactivation (C7)
Properties of dendrites (D1)
Stimulus localization (F3)
Retina (H3)

Bipolar cells and on and off channels

Photoreceptors synapse with bipolar cells. Two types of bipolar cell can be distinguished on the basis of both morphology and physiological responses. Those with processes that form **triad ribbon** synapses (*Fig. 1*) deep in the photoreceptor terminal are **invaginating bipolar** cells and they depolarize in response to light striking the photoreceptor. Triad ribbon synapses are so-called because they have three postsynaptic components, the bipolar cell dendrite and dendrites of two horizontal cells. By contrast, **flat bipolar** cells form superficial **basal** synapses with photoreceptors and hyperpolarize in response to light shone on the photoreceptor.

Cone cells form synapses with **midget bipolar** cells (so-called because of their size) of either one or the other type. Midget bipolar cells synapse directly with **ganglion cells** which respond to light in the same sense as their bipolar cells. This arrangement gives rise to two labeled lines: **on channels** are formed by cone–depolarizing bipolar cell–on ganglion cell, whereas **off channels** are cone–hyperpolarizing bipolar cell–off ganglion cell. On ganglion cells are depolarized and increase their firing rate as a function of light intensity. Off ganglion cells are silenced by hyperpolarization (*Fig. 2*).

All photoreceptors use glutamate as a transmitter. The opposite responses of invaginating and flat bipolar cells come about because they have different glutamate receptors. On invaginating bipolar cells, tonic release of glutamate in the dark is inhibitory. When light hyperpolarizes the photoreceptor glutamate release is suppressed, and inhibition lifted, so the bipolar cell depolarizes. For flat bipolar cells the response to tonic glutamate release is excitatory and light, by reducing that excitation, causes the bipolar cell to hyperpolarize.

On channels respond with increased firing to light levels that are greater than the local average. Off channels show increased firing in response to dark regions (i.e. where light levels are lower than the local average). In this way the existence of separate on and off channels is a mechanism to enhance the boundaries between regions that reflect different amounts of light. It is one of several processes by which the visual system is adapted to respond preferentially to stimulus change as opposed to steady state stimulation.

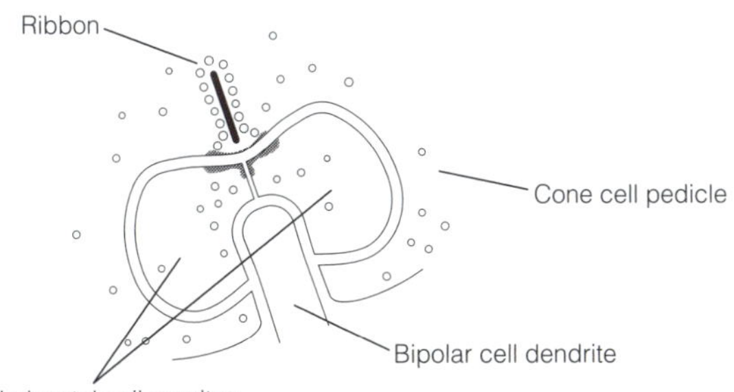

Fig. 1. A triad ribbon (invaginating) synapse.

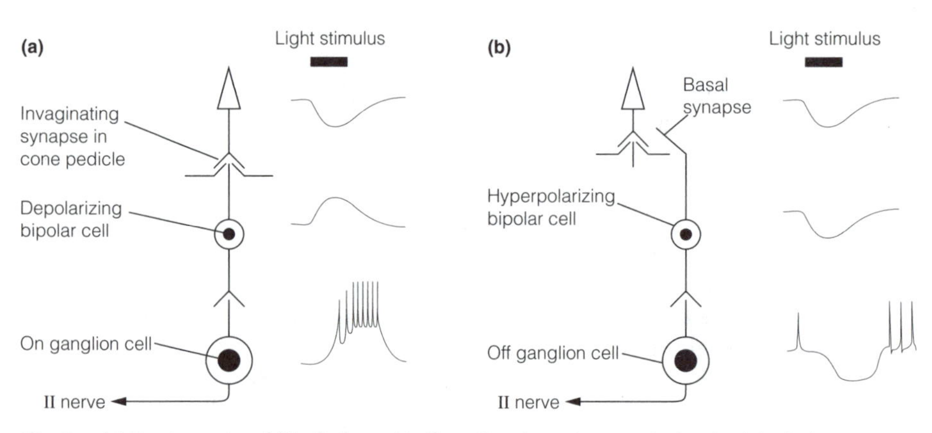

Fig. 2. (a) On channel and (b) off channel in the retina. In each case electrophysiological responses of the cells to light stimulation recorded intracellularly is shown on the right. All cells depicted use glutamate as a neurotransmitter.

Horizontal cells and lateral inhibition

An important mechanism for enhancing contrast in the retina is lateral inhibition brought about by **horizontal cells**. Lateral inhibition can be seen in the receptive fields (RFs) of both bipolar and ganglion cells. These are circular and divided into an inner center and an outer surround. Stimulation of these two regions separately produces opposite effects on the cells. In the case of an on ganglion cell, for example (see *Fig. 3*), background firing rate is dramatically increased by illumination of the center but silenced by light on the surround. When light fills the whole of the RF there is little change in the background firing rate. Off ganglion cell RFs have the converse response, with center illumination producing inhibition and surround illumination excitation.

Lateral inhibition arises because horizontal cells form reciprocal connections at triad synapses between neighboring photoreceptors. In the dark, horizontal cells are excited by glutamate release from photoreceptors but themselves

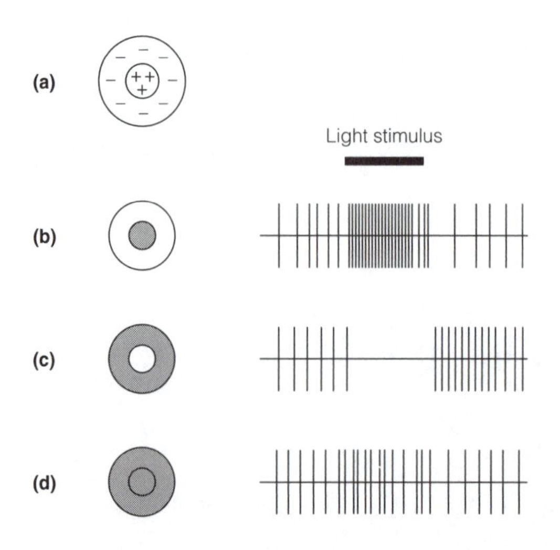

Fig. 3. Extracellular recording from on ganglion cells: (a) receptive field; (b) central illumination; (c) surround illumination; (d) overall illumination.

release GABA which tends to inhibit the photoreceptors. Light which hyperpolarizes surrounding photoreceptors causes them to secrete less glutamate so reducing horizontal cell excitation. This means that GABA release from the horizontal cells is lowered which in turn allows the central cone to depolarize somewhat, so that it releases more glutamate. The final step in the sequence depends precisely on the type of bipolar cell the central cone synapses with. If it is a depolarizing (on) bipolar cell the increased glutamate will cause it to hyperpolarize, as glutamate is inhibitory at invaginating synapses. If it is a hyperpolarizing (off) bipolar cell the increased glutamate will make the bipolar cell depolarize. Note that in each case the bipolar cell response (and hence that of the ganglion cell with which it synapses) is the opposite for surround compared with central illumination (*Fig. 4*).

Horizontal cells are extensively interconnected via gap junctions forming a network which spans an area of retina termed the **S space**. The S space horizontal cells provide the signal for surround inhibition and it is thought that this signal is a measure of the mean luminance over quite a wide area of retina.

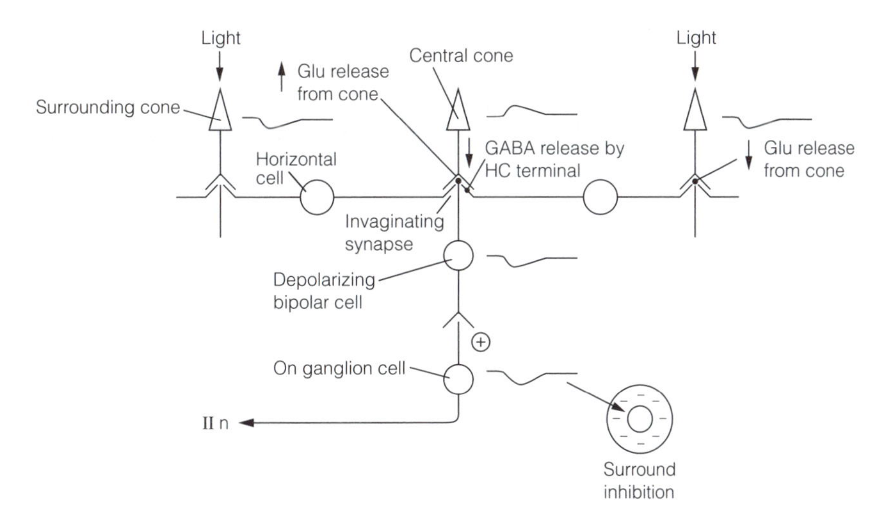

Fig. 4. The mechanism of lateral inhibition by horizontal cells (HC).

Ganglion cells

Ganglion cells are the output cells of the retina. Their axons become myelinated at the optic disc and form the optic (second cranial) nerve. Ganglion cells are the only retinal cells capable of firing action potentials. In the macaque old world monkey (which has vision similar to that of humans) there are two major populations of ganglion cells.

Parvocellular (P, Pβ) ganglion cells are small and by far the most numerous, there are about a million in each retina. **Magnocellular (M, Pα) ganglion** cells are large and number about 100 000 per retina.

These two types differ in several important respects:

- P cells have smaller RFs than M cells.
- Because of their smaller size, P cells have a lower conduction velocity than M cells.
- Whereas P cells often show sustained responses, M cells respond transiently to a prolonged visual stimulus.

- P cells are usually wavelength selective whereas M cells are not.
- M cells are much more sensitive than P cells to low contrast stimuli.

From these differences it can be inferred that P cells must get their input from single cones, or from several cones with the same wavelength sensitivity (S, M or L). By contrast, M cells get input from M and L cones together (but not S cones), and from rods. Hence P, but not M, cells must be involved in color vision. The low contrast sensitivity of M cells shows them to be important in scotopic vision, and their rapid, transient responses makes them adapted for motion detection. The small RFs of P cells, and their sustained responses are suitable for fine form discrimination. The distinct functional properties of P and M ganglion cells is the starting point for parallel processing in the visual system.

Most ganglion cells have RFs that exhibit lateral inhibition (see *Fig. 3*). Because the M cells get combined input from two types of cones they are called **broad band** cells and their RFs measure only brightness contrast. P cells, however, are **color opponent cells**. They have RFs that are excited by one type of cone cell but inhibited by another. Two types of P cell can be distinguished by the nature of their RFs (*Fig. 5*). Most common are the concentric single opponent (red–green) cells in which the RF compares input from M and L cones. Coextensive single opponent cells do not have a center surround pattern but are either excited by S cones and inhibited by a combined M plus L cone signal (or vice versa). These are also called blue–yellow opponent cells because combining the inputs of M and L cones gives the sensation of yellow.

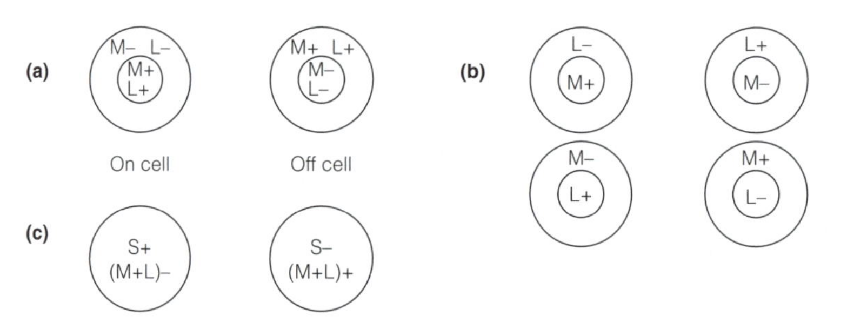

Fig. 5. Receptive fields of retinal ganglion cells: (a) M ganglion cells; (b) concentric single opponent (red–green) P cells; (c) coextensive single opponent (blue–yellow) P cells.

The concentric single opponent cells respond slightly differently to small or large spots of light. For example, a green on–red off cell will be equally stimulated by a small green or white spot that covers the center of the RF because the white light contains the wavelengths that excite the green cones. The same cell will be excited by a large green spot but not by a large white spot because the white light contains the wavelengths which stimulates the red cone surround inhibition. A large red spot will silence the cell. In general, concentric single opponent cells are more wavelength selective for large stimuli than small ones. For small spots they cannot distinguish red or green from white light, but they can signal brightness.

Rod signaling

Signaling by rod cells depends on the light intensity. At high light levels rods are saturated, and only cone on and off channels operate. In the partially dark adapted eye (e.g. at dusk) rod cells come on stream but signal through gap junctions to neighboring cones. This effectively augments cone cell function so maintaining acuity and color vision. However, when it is very dark (e.g. moonless night sky) cone cells fail even with signal boosting from rods. Rod cells in the dark will have a greater influx of Ca^{2+} via the dark current. One effect of this increase in Ca^{2+} is to close the gap junctions between the rod and cone cells. Rod signaling is now relayed via depolarizing (rod) bipolar cells which synapse with a population of amacrine cells. The effect of this is to increase contrast sensitivity.

Amacrine cells

Amacrine cells have no axons but their extensive neurites share properties of both axons and dendrites. They are a very diverse group morphologically, and most of the neurotransmitters identified in the nervous system are used by one or other of the 30 or so types of amacrine cell.

Amacrine cells are implicated in rod signaling, surround inhibition and detecting the direction of motion of an object across the visual field. Dopaminergic amacrine cells are only about 1% of all amacrine cells but their long dendrites interconnect, possibly via gap junctions, to form a network. These cells get input from cone bipolar cells so the network is able to signal average illumination which is used to produce surround inhibition. In the dark adapted eye, ganglion cells become much more sensitive to light by virtue of the fact that this dopamine surround inhibition is turned off. Some ganglion cells are sensitive to the direction of motion of a stimulus. Direction sensitivity is conferred by amacrine cell circuits.

H6 Early visual processing

Key Notes

Lateral geniculate nucleus

The sorting of fibers in the optic chiasm means that the left lateral geniculate nucleus (LGN) maps the right side of the visual field. The primate LGN has six layers, two magnocellular layers get input from magnocellular (M) ganglion cells and contain movement sensitive cells, four parvocellular (P) layers are innervated by P ganglion cells and cells in these layers are wavelength selective. Each layer gets a precise retinotopic input from just one eye. The properties of LGN cells are similar to those of the ganglion cells which supply them and have circular receptive fields (RFs) with surround inhibition. The LGN projects to the primary visual cortex and receives extensive connections from it which may be involved in visual attention.

Primary visual cortex

The striate cortex (V1) in the occipital lobe is the primary visual cortex and gets a retinotopic projection from the LGN, with the fovea occupying a disproportionately large area. The M and P LGN cells project to different sublayers in layer 4C of the cortex, giving rise to distinct streams of information flow through the cortex. Most cells in V1 have elongated RFs and respond to linear features rather than to spots. Simple cells are position sensitive. Complex cells are less sensitive to position than simple cells, many having a preference for linear stimuli moving at right angles to the long axis of their receptive fields.

Orientation columns

V1 is divided into radial columns which extend the full thickness of cortex. All cells lying within a column respond to linear features having approximately the same orientation. All orientations are represented for each point on the retina. These orientation columns are ordered so that a smooth gradient for orientation exists; columns with the same orientation are aligned in stripes across the cortex.

Binocular cells

Many cells in V1 get input from both eyes, but most show ocular dominance in that they are preferentially driven by one eye. These cells occur in discrete ocular dominance columns that are aligned in stripes across the cortex in which ipsilateral and contralateral eye dominance alternates. Binocular cells get input from corresponding positions on the two retinae and measure retinal disparity from which the visual system computes the depth of an object in three-dimensional space.

Hypercolumns

The volume of cortex in which every orientation is mapped for corresponding positions on both retinae is called a hypercolumn. It consists of a complete set of orientation columns and ocular dominance columns for a single pixel of the visual field so it can be regarded as the basic unit of V1.

Related topics

Organization of the central nervous system (E2)
Attributes of vision (H1)
Eye and visual pathways (H2)
Parallel processing in the visual system (H7)

Lateral geniculate nucleus

The pathway for visual perception commences with the retinogeniculate fibers, axons of ganglion cells that end in the LGN. Because of the manner in which fibers are sorted in the optic chiasm, the left optic tract and left LGN carry axons from the left side of both retinae. Thus the left LGN represents the right side of the visual field (see *Fig. 1*, Topic H2).

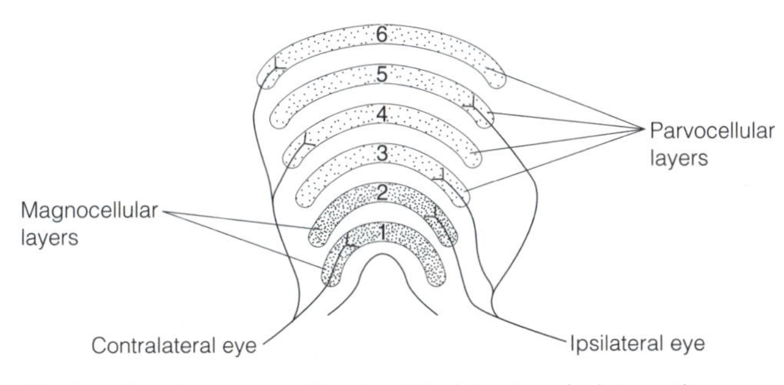

Fig. 1. The structure and inputs of the lateral geniculate nucleus.

The primate LGN has six layers (*Fig. 1*). The two most ventral are the **magnocellular** (large-cell) **layers** which receive input from M ganglion cells. Dorsal to these are the four **parvocellular** (small-cell) **layers** that are innervated by P ganglion cells. Interleaved between these major layers are **koniocellular layers** containing very small cells. These are thought to receive input from small slowly conducting retinal ganglion cells with large dendritic trees, and large RFs, which signal average illumination. LGN cells have circular RFs with surround antagonism. They show little or no response to diffuse light covering the whole receptive field. Each layer in the LGN gets input from only one eye and no cells show binocular responses (responses to both eyes). It is not until the visual cortex is reached that input from both eyes is integrated. The responses of the LGN cells match those of the ganglion cells which supply them, so on and off channels remain independent and P cells display precisely the same color opponency properties as retinal ganglion cells. There is a very precise topographic (retinotopic) mapping from the retina onto each of the layers of the LGN, with the representation of the fovea taking up about half of the nucleus. The maps in each layer are in precise register with each other so that any given vertical axis through the LGN passes through cells with RFs representing the same place in visual space.

The LGN contains two populations of neurons. Those that project to the visual cortex are **geniculostriate** neurons. In addition there is a substantial population of smaller interneurons the exact function of which is not known. Furthermore only about 20% of the synaptic connections on geniculostriate neurons are from retinal ganglion cells. Other synapses are made by back projections from the visual cortex and by the reticular formation. They may play a role in visual attention, modifying geniculostriate neuron responses so that only a selected proportion of retinal input is transmitted through to the visual cortex.

Primary visual cortex

The fibers of the optic tract terminate in the **striate** cortex (area 17) on the medial surface of the tip of the occipital lobe. This region is the primary visual cortex (V1). Precise retinotopic mapping is maintained up to V1 with the fovea having a disproportionate representation.

There are at least three parallel streams of information into the primary visual cortex. The movement sensitive M LGN cells input into layer $4C_{\alpha}$, the P LGN cells which are wavelength sensitive go to layer $4C_{\beta}$, whereas the koniocellular layers of the LGN project to layers 2 and 3. These streams remain quasi-independent throughout the visual system. The connections of the primary visual cortex are illustrated for the primate in *Fig. 2*.

The great majority of cells in V1 have elongated RFs with both inhibitory and excitatory regions, and respond to bars, slits, edges and corners rather than to spots of light. Most fall into two categories based on their RF properties, simple or complex cells. Both are orientation selective, in that they respond to linear features in only a narrow range of orientations.

(1) **Simple cells** are pyramidal cells found mostly in layers 4 and 6. They are highly sensitive to the position of a stimulus on the retina. They have small oval RFs with center-surround antagonism (*Fig. 3*). A simple cell gets its input from a linear array of LGN cells having the same RF properties, so the RF of the simple cell emerges as a consequence of the RFs of the LGN inputs.
(2) **Complex cells** are most abundant in layers 2 + 3 and 5. They have larger RFs than simple cells and because they lack distinct inhibitory or excitatory regions a stimulus of the appropriate orientation anywhere in the RF evokes a response. Hence complex cells are much less fussy about position than simple cells. Many complex cells show a preference for movement orthogonal to the long axis of the stimulus. There is ongoing controversy about the inputs to complex cells. In the original **hierarchical model** of the

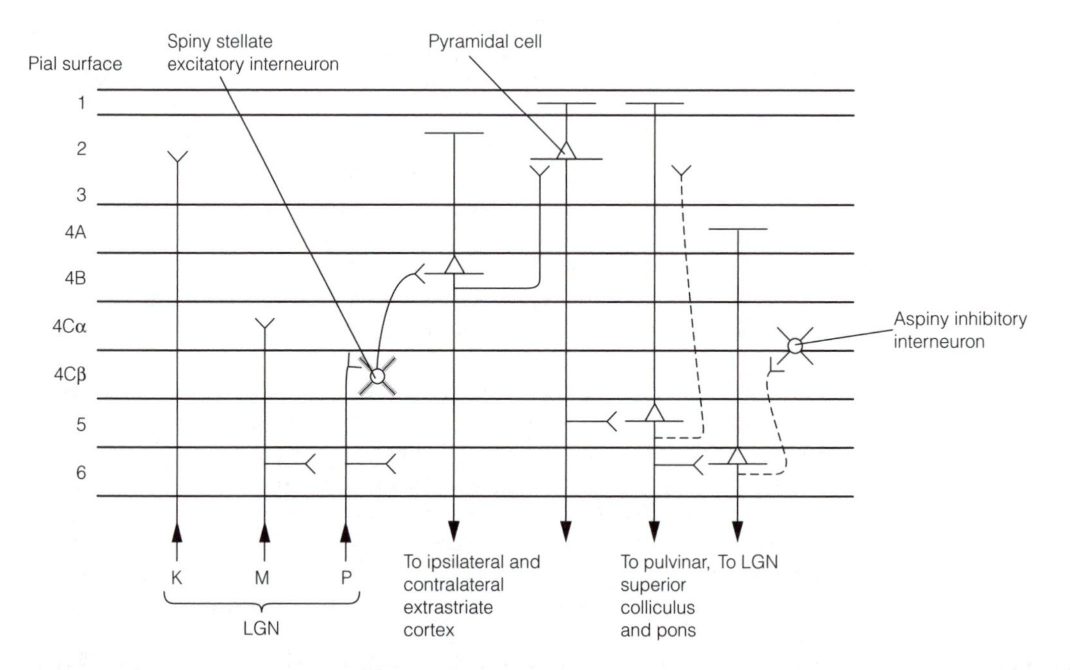

Fig. 2. Canonical circuitry in the primary visual cortex illustrated for parvocellular (P) LGN input. Magnocellular (M) circuitry (not illustrated) has its input to 4Cα. Spiny stellate cells here send axons to pyramidal cells in 4B and these send collaterals to pyramidal cells in layers 5 and 6 directly rather than via more superficial layers. Koniocellular input is directly to pyramidal cells in blobs of 2 + 3. Feedback collaterals are dotted.

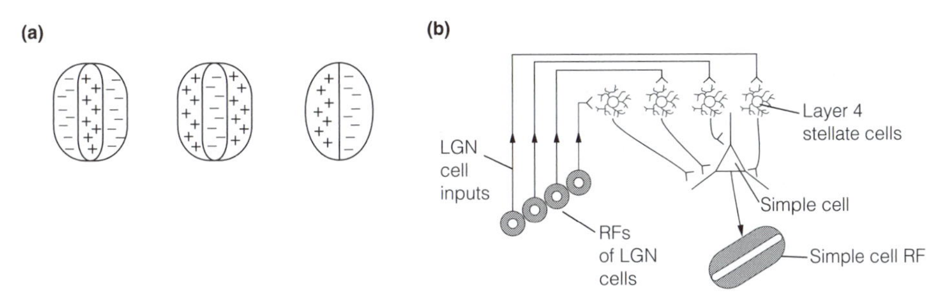

Fig. 3. (a) Receptive fields of three simple cells; (b) a diagram depicting how lateral geniculate nucleus (LGN) cells contribute to the simple cell receptive field (RF), four on-center LGN cell RFs generate an on-centre simple cell RF.

visual cortex, complex cells received their inputs from simple cells. More recent evidence suggests that some complex cells get their input directly from the LGN.

Orientation columns

In common with other sensory cortex, the primary visual cortex is divided into radial columns 30–100 µm across, in each of which all cells respond preferentially to linear features with a given orientation. These are called **orientation columns**. The cortex is organized so that adjacent columns have an orientation preference that differs by only about 15° – in other words, orientation is represented in a systematic way across the cortex. Columns which have the same orientation are arranged in stripes across the cortex. The obvious inference, that orientation selectivity is how the visual system represents straight line segments which can be built up to give the form of an image, need not be true. Computer modeling shows that orientation selectivity is a property of neural networks that learn the curvature of curved surfaces from their shading. Hence orientation selectivity might, counter intuitively, be concerned with representations of curves rather than linear features in the visual world.

Binocular cells

V1 is the first region in which input from both eyes is combined. Many cells, particularly in layers 4B and 2 + 3 show binocular responses in that they can be driven by either eye. This is a necessary condition for stereopsis. Most **binocular cells** show a preference for one eye, a phenomenon referred to as **ocular dominance**.

The RFs of binocularly driven cells resemble those of simple or complex cells, lie in corresponding positions in the two retinae, have identical orientation properties and finally have similar arrangements of excitatory and inhibitory regions.

Similar input from both eyes into arrays of binocular cells is needed for perception of a fused image. To the extent that inputs into these cells is unequal they measure retinal disparity and so the depth of an object in three-dimensional visual space. Cells that respond to visual disparity have been discovered in the primate visual cortex (including V1). These are responsible for stereopsis (see Topic H1).

Cells which have the same ocular dominance (e.g. those that are driven preferentially by the ipsilateral eye) occupy **ocular dominance columns** that are situated in long stripes about 500 µm across. Columns representing ipsilateral and contralateral input alternate regularly over the cortex which, when visualized at the level of layer 4C, looks like the pattern of stripes on a zebra.

Hypercolumns

A higher order modularity to the structure of the primary visual cortex is thought by many authors to be the basic structural unit (*Fig. 4*); called a **hypercolumn**, it represents a given corresponding position for both retinae, and maps every orientation for that position. It consists of a full thickness slab of cortex with an area of about 1 mm^2 containing a complete set of orientation columns for both ipsilateral and contralateral ocular dominance. The retinotopic map in V1 occurs because adjacent pixels of the retina map to adjacent hypercolumns in an orderly fashion.

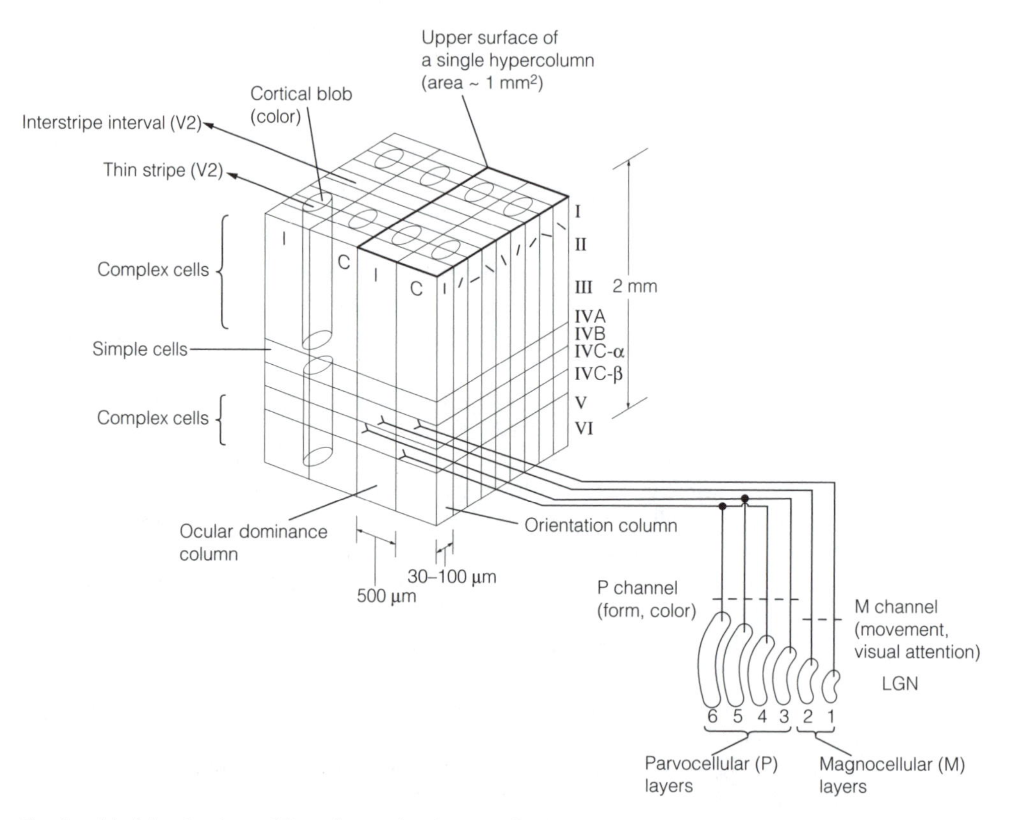

Fig. 4. Modular structure of the primary visual cortex. Cortical layers are designated by Roman numerals. I, ipsilateral; C, contralateral (blobs are described in Topic H7).

H7 PARALLEL PROCESSING IN THE VISUAL SYSTEM

Key Notes

Parallel processing in V1

There are three relatively independent pathways for processing information in the visual system. The magnocellular (M) pathway gets its input from M lateral geniculate nucleus (LGN) cells which synapse with neurons in 4Cα. The M pathway is color blind and involved in analysis of moving stimuli, control of gaze and stereopsis. The parvocellular (P) systems arise from P LGN cells which synapse with cells in 4Cβ. There are two P systems. The parvocellular–interblob pathway is concerned with form, its cells are orientation selective and binocular. The parvocellular-blob pathway mediates color vision; wavelength selective cells in this pathway have double opponent receptive fields (RFs) in which the center is excited by some cones and inhibited by others, whilst the converse situation occurs in the surround. Three features of color perception – color constancy, perceptual cancellation and simultaneous color contrast – can be explained by the manner in which double opponency is organized.

Extrastriate visual cortex

All cortical areas involved in vision other than V1 are together referred to as extrastriate visual cortex. It includes much of the occipital cortex and parts of the parietal and temporal cortex. The secondary visual cortex (V2) receives input from V1 which then projects to other extrastriate cortex. The three streams of visual information remain segregated in V2, as revealed by stripes in cytochrome oxidase staining, and throughout the extrastriate visual cortex. The M pathway goes via thick stripes in V2 to V3 and then V5. Destruction of human V5 causes a loss of ability to see objects in motion. The parvocellular–interblob pathway goes via V2 interstripes to V3 and V4, whilst the parvocellular–blob pathway goes from V2 thin stripes to V4, cells of which show color constancy. Destruction of human V4 causes loss of color vision.

Where and what streams

Beyond V5 and V4 information is divided into two streams. From V5 a dorsal stream to the medial superior temporal cortex and the posterior parietal cortex is concerned with object location. The ventral stream from V4 to the inferotemporal cortex is concerned with recognition of objects. The two streams are called the 'where' and the 'what' streams respectively.

Related topics

Attributes of vision (H1)
Retinal processing (H5)
Early visual processing (H6)
Oculomotor control (L7)

Parallel processing in V1

Three relatively independent pathways, each of which processes different aspects of vision in parallel, in a quasi-autonomous fashion, can be delineated in the primary visual cortex.

The **magnocellular pathway** from M (P_α) ganglion cells to M LGN cells has its input to spiny stellate cells in layer $4C_\alpha$. These excitatory interneurons synapse with pyramidal cells in layer 4B which show orientation and direction selectivity. These cells send axon collaterals to pyramidal cells in layers 5 and 6. Layer 5 cells project to subcortical regions, the pulvinar (a thalamic nucleus involved in visual attention), the superior colliculus and pons. Layer 6 pyramidal cells go to the extrastriate cortex. The M pathway is specialized for analysis of motion. Its outputs via layer 5 are important in visual attention and gaze reflexes. Some cells in the M pathway are binocular so it contributes to stereopsis. Because it originates with ganglion cells which combine input from two classes of cone cell it is not wavelength selective, the M system is color blind.

There are two **parvocellular pathways**. They arise from P (P_β) ganglion cells via P LGN cells which synapse with spiny stellate cells in $4C_\beta$. Like the M pathway the interneurons connect with pyramidal cells in 4B. However in the parvocellular paths, 4B cells (which are orientation selective simple cells) synapse with pyramidal cells in layers 2 + 3 which then relay with deep pyramidal cells in layer 5 (see *Fig. 2*, Topic H6). Segregation of the two parvocellular pathways occurs in layers 2 + 3. When stained for the mitochondrial enzyme **cytochrome oxidase** layers 2 + 3 show pillars of high activity, **blobs** (see *Fig. 4*, Topic H6). Each blob is centered on an ocular dominance column. Between the blobs lies the **interblob** region. Cells in the interblob region are orientation selective, binocularly driven, complex cells. They are neither wavelength selective nor motion sensitive. They are part of the **parvocellular–interblob** (PI) pathway which processes high resolution analysis of form in the visual world. By contrast cells in the blobs are wavelength selective, show poor orientation and are monocular. The **parvocellular–blob** (PB) pathway mediates color vision. Blob pyramidal cells get direct input from the koniocellular LGN layers but the function of this input is not yet understood.

Wavelength selective blob cells are **double opponent** cells with RF properties derived from their inputs, the single opponent parvocellular LGN cells. Double opponent cells have center-surround antagonist RF configuration, they signal color contrast and come in four classes categorized by their preferred stimuli. The left cell in *Fig. 1* is excited by L cones in the center and inhibited by L cones in the surround. In addition it is inhibited by M cones centrally but excited by M cones in the surround. The preferred stimulus for this cell is a red spot on a green background. However the cell gives **off responses** if exposed to a green spot on a red background (*Fig. 2*).

Unlike single opponent cells which are excited by small spots of white light, double opponent cells are unaffected by white light stimuli of any size, so they are more selective detectors of color contrast. The organization of double opponent cells RFs explains some of the properties of color vision (see Topic H1).

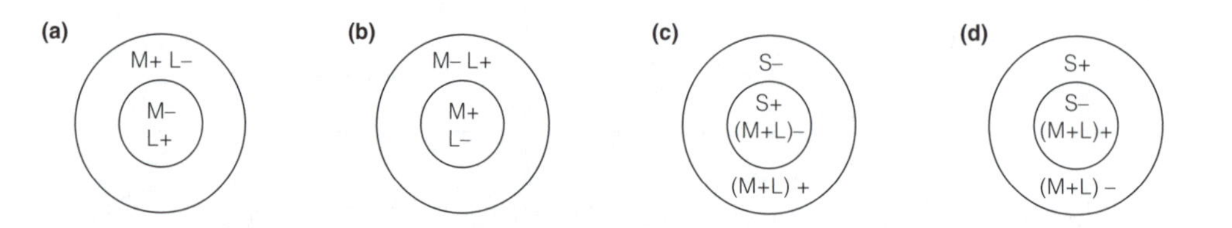

Fig. 1. Double opponent cells in V1 blobs. Preferred stimuli: (a) red spot, green surround; (b) green spot, red surround; (c) blue spot, yellow surround; (d) yellow spot, blue surround.

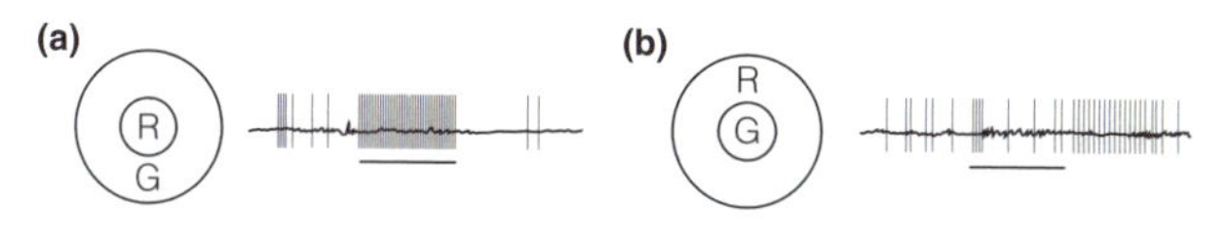

Fig. 2. Responses of the double opponent cell in Fig. 1(a): (a), preferred stimulus; (b) off response which might account for successive color contrast (see text for details).

Color opponancy is the way in which the brain computes color constancy; it is not understood in detail but is partly accounted for by the behavior of double opponent cells. A shift in the wavelength composition of light will produce equal but opposite effects on the responses of the center and surround of double opponent cells. There will be little effect overall on the RF of the cell which will continue to signal the same color. On the scale of the entire visual field, color constancy is thought to involve comparing red–green brightness, blue–yellow brightness (from color opponent cells) and total brightness (added outputs of S, M and L cones) over large areas of retina.

Perceptual cancellation is explained by the way in which color opponency happens to be organized as red (R) versus green (G) and yellow (R + G) versus blue channels. Since mutual antagonism occurs between red and green or between yellow and blue only one color in each pair can be seen at a single pixel of the retina at any time.

Simultaneous color contrast can also be accounted for by the properties of double opponent cells. For example, the cell in *Fig. 2* cannot discriminate between a green stimulus to its surround or a red stimulus to the center; the response is the same for both. So a grey disc viewed in a green background is interpreted as red. A similar mechanism underlies successive color contrast in which complementary after-images appear after staring at a uniform patch of color (*Fig. 2b*).

Extrastriate visual cortex

The segregation of visual information for motion, form and color in V1 is maintained in the **extrastriate visual cortex**, which is a term applied to all of the visual cortex except V1. The extrastriate cortex of primates contains about 30 regions that can be differentiated on the grounds of cytoarchitecture, connections and physiological properties. Most have a retinotopic map of some aspect of the visual world. It includes not only occipital cortex areas 18 and 19 but also areas of parietal and temporal cortex. In humans it is estimated that almost one-half of the cerebral cortex is implicated in vision, more than is devoted to any other single function. This implies that vision is the most complex task that the brain performs. The terminology usually adopted for extrastriate cortex is based on studies of the macaque monkey. It is thought that most regions in this primate have counterparts in humans. The location and connections of the major visual cortical areas are depicted in *Fig. 3a* and *3b* respectively.

Most outputs from V1 go to **V2**, the **secondary visual cortex**, which occupies part of area 18. V2 shows a characteristic cytochrome oxidase staining pattern, alternating thick and thin stripes running at right angles to the V1/V2 border. Pathway tracing techniques and electrophysiological studies reveal how the magnocellular and parvocellular pathways continue into V2 and beyond.

Cells in the V2 thick stripes are motion sensitive and binocular, being driven by a preferential retinal disparity. The thick stripe receives inputs from layer 4 of the interblob region of V1 and sends much of its output, via **V3,** to the **medial temporal (MT)** visual cortex, **V5**. Lesions of human V5 result in loss of the

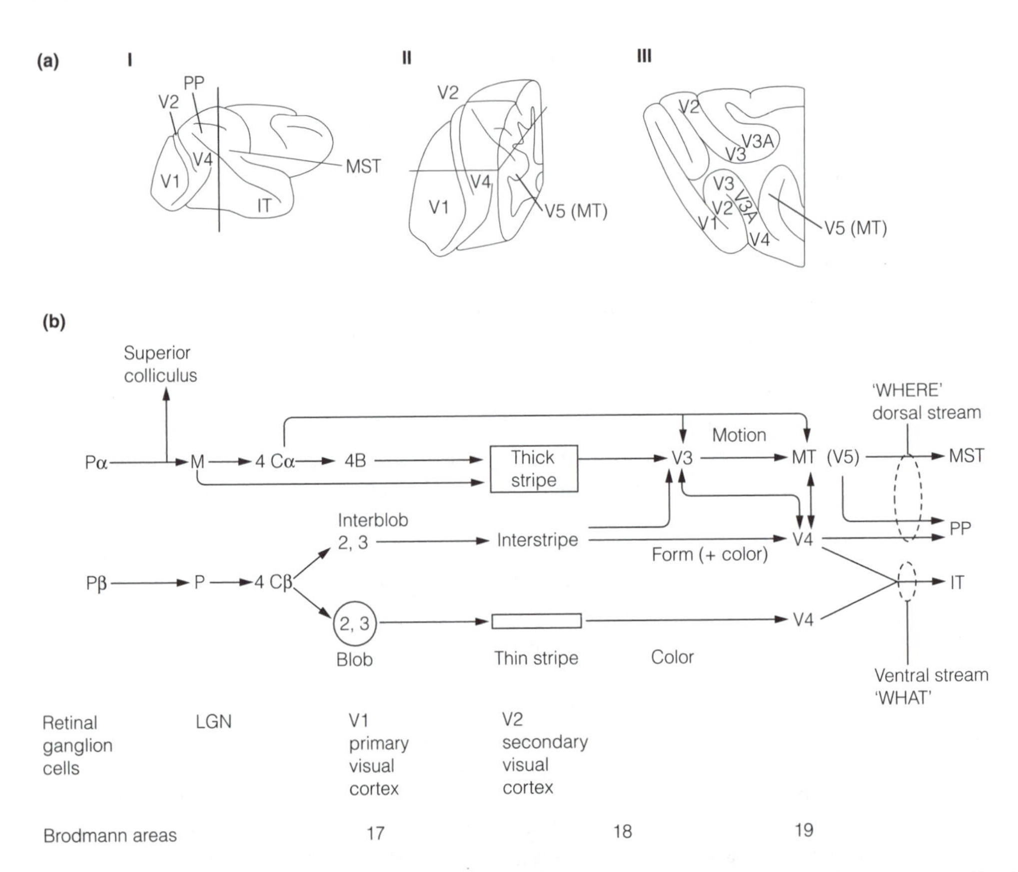

Fig. 3. Parallel processing in the visual system. (a) Anatomy of the visual areas in the macaque: (i) left cerebral hemisphere; (ii) coronal section through the posterior third of the hemisphere; (iii) horizontal section. Modified from Kandel, Schwartz, Jessell (eds) (1991) Principles of Neural Science, *3rd edn. (b) Flow diagram of M and P channels in the primate visual system. IT, Inferotemporal cortex; MST, medial superior temporal cortex; MT, medial temporal cortex; PP, posterior parietal cortex.*

ability to perceive motion (**akinetopsia**). Hence the V2 thick stripe–V3–V5 (MT) connection is the extension of the magnocellular pathway (*Fig. 3b*), and concerned with motion and depth perception.

The interstripe region of V2 gets its inputs from the V1 interblob regions (layers 2 + 3) and sends outputs to V3 and then to **V4**. Many cells in V3 and a proportion of those in V4 are orientation selective and these represent a continuation of the PI parvocellular pathway. It is primarily concerned with form perception.

The blobs of V1 project to the thin stripe of V2 which in turn sends outputs to visual area V4. Thin stripe V2 cells, and some V4 cells are both wavelength selective and show color constancy, so the blob – V2 thin stripe – V4 route is the extension of the PB parvocellular pathway for color vision. This is supported by the loss of color vision (**achromatopsia**) and inability to recall colors that occurs in patients with damage to V4.

Although the M, PI and PB pathways operate in parallel they are not completely independent. Reciprocal pathways exist between V3–V4 and V5–V4 which presumably allow interactions between M and PI systems, both of which contribute to stereopsis. Interaction of motion and form analysis is probably required for the identification of moving objects. There seems to be no cross talk, however, between M and PB pathways. The M system is color blind and for **equiluminant** stimuli (those varying in color but not in brightness) which can only be perceived by the PB system, the perception of motion vanishes. The PI system must receive input from wavelength selective cells in V4 since it is able to use color contrast to localize borders as part of its role in analyzing form. However form information seems not to be available to the PB pathway. When viewing equiluminant blocks of color they appear to 'jump around' because the PB system is unable to localize boundaries.

Saccades are rapid stereotyped movements of the eyes which serve to bring an object at the periphery of the visual field to the fovea (Topic L7). During saccades black and white gratings detectable only to the M system vanish while equiluminant stimuli detectable by the parvocellular pathway remain visible; thus the M, but not the P system, is shut down during saccades. This means that the M (motion) system is not confused by the rapid eye movement. The response times of the cells in the P system are sufficiently slow that they are unaffected by the shifting image.

Where and what streams

Parallel processing beyond V5 and V4 results from the segregation of information into two streams. The dorsal stream, largely from MT, goes to the **medial superior temporal (MST)** and **posterior parietal (PP)** cortex. Cells in the PP cortex have large receptive fields (RFs), show selectivity for size and orientation of objects and fire as a monkey makes hand movements to grasp an object. Many cells show gaze-dependent responses, i.e. their firing depends on where an animal is looking. Lesions to MST and PP in primates results in **optic ataxia**, in which visuospatial tasks are profoundly affected, but are without any affect on the ability of animals to recognize objects.

By contrast the ventral stream from V4 to **the inferotemporal cortex (IT)** is crucial for object recognition. Cells of the IT cortex have extremely large RFs, usually bilateral, are sensitive to form and color but are relatively unfussy about object size, retinal position or orientation. Many of these cells respond selectively to specific objects such as hands or faces. Unusually for visual cortex, the IT area has no retinotopic map. Lesions of the IT cortex cause **visual agnosia** in which animals fail to perform or learn tasks that require the recognition of objects. Visuospatial tasks are unaffected.

The very different functions of the dorsal and ventral streams are epitomized by their being referred to as **where** and **what** streams respectively. Clinical data suggests that a similar dichotomy exists in humans. Optic ataxia occurs with posterior parietal damage. These patients have no difficulty in recognizing objects but cannot seem to reach for, or grasp them. By contrast patients with damage to the occipito-temporal cortex fail to recognize common objects, including once familiar faces, a disorder termed **prosopagnosia**. These people experience no difficulty in understanding where objects are located in space or how to reach for, or avoid them.

Bilateral loss of V1 causes total loss of visual perception. However there are examples of primates and humans with this damage who are able to avoid obstacles whilst moving through space at a frequency much higher than chance. This phenomenon is called **blindsight**. Humans possessing it report that they are completely unaware of the visual world and do not understand how they are able to navigate through space. Blindsight is mediated by a pathway that goes directly from the magnocellular LGN to the thick stripe of V2. This provides input to the where system.

I1 Acoustics and Audition

Key Notes

Sound waves

Sound is longitudinal pressure waves in a medium. For a sinusoidal wave $f = c/\lambda$, where f is frequency, c is velocity (331 m s^{-1} for sound in air), and λ is wavelength. The frequency of the wave is the pitch of the sound.

Sound pressure amplitude

The change in pressure produced by a sound wave is the sound pressure amplitude (P). By comparing P with a reference amplitude at the threshold of human hearing a sound pressure level (SPL) can be calculated. The unit of SPL is the decibel. Differences in SPL are perceived as differences in the loudness of a sound. The phon is the unit of loudness and is the SPL of a tone with frequency of 1000 Hz.

Sensitivity of human hearing

In the young the detectable frequency range is between 20 Hz and 20 kHz. Pure tones that differ by only 2–6 Hz can be discriminated if played in succession. If pure tones are played simultaneously, however, the resolution plummets to about one-third of an octave.

Related topics

Anatomy and physiology of the ear (I2)
Peripheral auditory processing (I3)
Central auditory processing (I4)

Sound waves

Sound is the oscillation of molecules in a medium. The energy of the oscillations is transmitted as a longitudinal wave in which the medium is alternately compressed and rarefied. This can be measured as periodic oscillations in the pressure of the medium (*Fig. 1*).

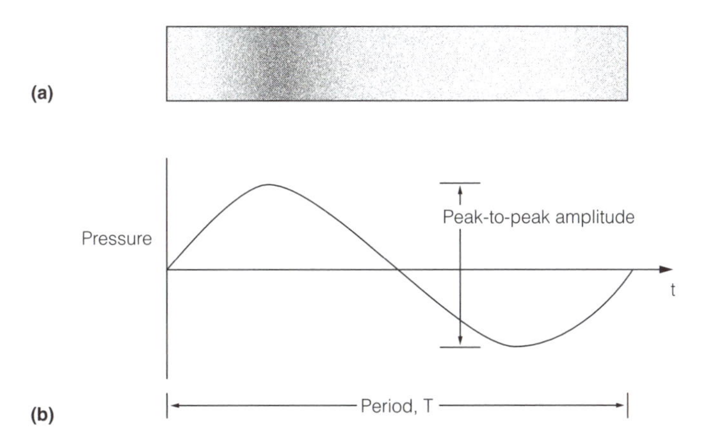

Fig. 1. Sound waves: (a) density of air molecules during propagation of a longitudinal pressure wave; (b) sine wave representation of a pressure wave.

For a sine wave the **period**, T, is the time taken for one complete cycle. The **frequency** of the wave, the perceived **pitch** of the sound, is the reciprocal of the period:

$$f = 1/T$$

The **wavelength** is given by:

$$\lambda = c/f$$

where c is the speed at which the wave is transmitted through the medium. For sound in air $c = 331 \text{ m s}^{-1}$.

The frequency range of human speech is between 250 and 4000 Hz which closely matches the peak sensitivity of the human ear.

Sound pressure amplitude

The amplitude of a sound wave is the total change in pressure that occurs during a single cycle. Because of the huge range in sound wave amplitudes, P, it is expressed in a logarithmic scale as a ratio of a reference pressure, P_{ref}:

$$\textbf{Sound pressure level (SPL)} = 20 \log_{10} P/P_{ref}$$

P_{ref} is 2×10^{-5} Pa, a sound pressure which is at the threshold of human hearing, since 50% of individuals can hear such a sound at a frequency (3000 Hz) to which the ear is at maximum sensitivity. The unit of SPL is the **decibel** (dB). Each 10-fold increase in SPL is equivalent to 20 dB.

For example if $P = 2 \times 10^{-4}$ Pa i.e., 10-fold greater than P_{ref}, then:

$$\begin{aligned} \text{SPL} &= 20 \log_{10} 10 \\ &= 20 \text{ dB} \end{aligned}$$

Sound pressure levels in excess of 100 dB can result in damage to hearing, and at 120 dB auditory pain results.

Differences in SPL are perceived as differences in the **loudness** of a sound. Loudness is the perceived magnitude of a sound but it varies with frequency for tones that have the same SPL. It is defined as the SPL of a **tone** (a single frequency sound) with a frequency, f, of 1000 Hz which matches the loudness of a sound. The unit of loudness is the **phon**. By definition a tone with $f = 1000$ Hz will have identical dB and phon values. For equivalent SPLs, a tone of $f = 4000$ Hz sounds the loudest; loudness falls off dramatically above this value and below 250 Hz. At a frequency of 4000 Hz (for SPL > 30 dB) the relationship between SPL and loudness is that a 10-fold increase in SPL is perceived as a four-fold increase in loudness.

Sensitivity of human hearing

The frequency response of the human ear is from 20 Hz to 20 kHz optimally, but rapidly narrows with age, with most of the loss occurring at the higher frequencies. By 50 years the upper limit averages about 12 kHz. Sensitivity varies with frequency. Greatest sensitivity or **auditory acuity** corresponds to a frequency range of 1000–4000 Hz. Over this bandwidth pure tones differing in frequency by only 2–6 Hz can be discriminated if they are heard in succession.

Over the same bandwidth the smallest interval in conventional Western music, the semitone, corresponds to frequency differences of 62–234 Hz. The same resolution does not, however, apply to pure tones played simultaneously, which must differ by about one third of an octave, the **critical bandwidth**, before they can be discriminated. However, virtually no naturally occurring sounds or musical notes are pure tones, but have harmonics that span many octaves and permit them to be resolved even when simultaneous.

I2 Anatomy and physiology of the ear

Key Notes

Middle ear

The middle ear converts pressure waves in the air to vibrations of perilymph in the inner ear. Sound waves striking the ear drum cause it to vibrate. This vibration is transmitted by three articulated middle ear bones – the malleus, incus and stapes – to the oval window and so to the perilymph. Because perilymph is incompressible it is set in motion *en masse*, with the pressure being transmitted through to the round window. The surface area of the oval window is 20 times less than that of the ear drum so that the pressure at the oval window is correspondingly greater. This results in a four-fold amplification of the sound across the middle ear. Two middle ear muscles, upon contraction, act on the middle ear bones to reduce sound transmission. They are activated by tympanic reflexes which may afford some protection against loud sounds.

Inner ear

The auditory inner ear is the cochlea, a bony canal arranged in a coil. Within it lies the cochlea duct, a part of the membranous labyrinth, which divides the cochlea cross section into three compartments. Within the cochlea duct is the scala media which contains endolymph. On either side lie the perilymph-containing scala vestibuli and scala tympani. These are continuous with each other at the apex of the cochlea so that vibrations of the oval window are propagated through the scala vestibuli to the scala tympani and then to the round window.

Pressure waves moving through the perilymph cause the basilar membrane on the floor of the scala media to oscillate. The organ of Corti, a sheet of epithelium running the length of the cochlea duct rests on the basilar membrane. Hair cells in the organ of Corti have their stereocilia embedded in a gelatinous matrix, the tectorial membrane, above. Vibrations of the basilar membrane cause it to shear with respect to the tectorial membrane and so bend the stereocilia to and fro. This results in periodic hair cell depolarization and hyperpolarization by the same transduction mechanism that operates in vestibular hair cells. The periodic changes in the release of transmitter from the hair cells alters the firing of the auditory primary afferents with which they synapse.

Because the basilar membrane differs in width, mass and stiffness systematically along its length, different frequencies of sound make the membrane vibrate maximally at different distances along it. This is the basis of pitch discrimination.

Related topics

Sense of balance (G4)
Acoustics and audition (I1)

Middle ear

The function of the middle ear is to convert pressure waves in the air to vibrations of the perilymph in the inner ear. Sound waves pass along the **external**

auditory meatus striking the **tympanic membrane** (ear drum) which resonates faithfully in response. The ear drum is critically damped in that it stops vibrating the instant the sound ceases. A sound at hearing threshold causes the ear drum to vibrate with an amplitude of about 0.01 nm, one-tenth the diameter of a hydrogen atom! The movement of the ear drum is transferred with an overall efficiency of about 30% to the fluid in the inner ear by a lever system, composed of three **ear ossicles**, lying in the **tympanic cavity** (middle ear) (*Fig. 1*).

The **malleus** (hammer) is fixed at its thin end (the handle) to the tympanic membrane. Its thick end (the head) articulates with the head of the **incus** (anvil) via a saddle-shaped joint. The long process of the incus makes a ball and socket joint with the head of the **stapes** (stirrup). The base of the stapes is attached by an annular ligament to the **oval window** (fenestra vestibuli). The malleus vibrates with the tympanic membrane. Inward movement locks the joint between the malleus and the incus driving the long process of the incus inward, pushing the stapes in the same direction to exert a pressure on the perilymph beyond the oval window. This pressure wave is transmitted through the perilymph to cause a compensatory bulge of the **round window** (fenestra cochleae). Outward movement reverses these motions. Since the area of the oval window is 20 times smaller than the tympanic membrane, the pressure (force per unit area) at the oval window is proportionally greater. This is important because perilymph is incompressible and so must be driven to vibrate *en masse*. This needs more force than it takes to transmit sound waves through air. In addition it results in an amplification of the sound by about 20 dB, corresponding to a four-fold increase in loudness, by the middle ear.

There are two middle ear muscles, the **tensor tympani** and **stapedius**. When they contract together the handle of the malleus and the tympanic membrane are pulled inwards and the base of the stapes is pulled away from the oval window. This reduces sound transmission by 20 dB, especially for low frequen-

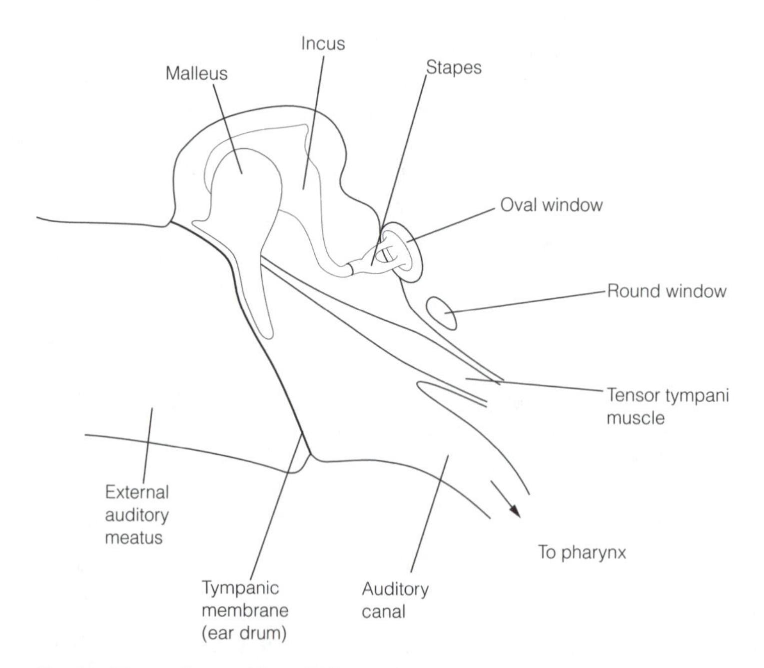

Fig. 1. The anatomy of the middle ear.

cies. Reflex contraction of these muscles in response to loud noise may prevent damage to the inner ear but since the reaction time is 40–60 ms this **tympanic reflex** affords no protection against brief loud sounds. The **auditory canal** connects the middle ear to the pharynx which allows the air pressure to be equalized with ambient pressure; this is important when going to high altitude. Impairment to middle ear function causes **conduction deafness**. The most common cause is **otosclerosis**, a bone disease in which the stapes becomes fused to the oval window. It is amenable to surgical correction.

Inner ear

The auditory part of the inner ear is the **cochlea**, a bony canal 3.5 cm long, which spirals two and three-quarter turns around a central pillar, the **modiolus**. Within this lies a tubular extension of the membranous labyrinth, the **cochlear duct**, attached to the modiolus and the outer wall of the cochlea. This divides the cochlea into three compartments, the inside of the **cochlear duct**, the **scala media**, which contains endolymph, the **scala vestibuli** and the **scala tympani**.

Both of these latter compartments contain perilymph and are continuous with each other via a small gap known as the **helicotrema** situated at the apex end of the cochlea, where the cochlear duct ends blindly (*Fig. 2*). Pressure waves generated at the oval window are propagated through the scala vestibuli into the scala tympani and so to the round window where the energy dissipates. During their passage the pressure waves cause oscillations of the **basilar membrane**, the floor of the scala media on which rests the sensory apparatus, the **spiral organ of Corti** (*Fig. 3*).

The spiral organ is a narrow sheet of columnar epithelium running the length of the cochlear duct. The epithelium consists of supporting pillar cells and Hensen's cells, and sensory hair cells resembling those in the vestibular apparatus (see Topic G5). A single row of 3500 **inner hair cells** form **ribbon synapses** with myelinated axons of large bipolar cells (type I) in the **spiral ganglion** of the cochlear nerve. Each inner hair cell is innervated by about 10 such axons, a large degree of divergence. There are about 12 000 **outer hair cells** arranged in three rows. These are innervated by an unmyelinated axon from small bipolar cells (type II) in the spiral ganglion, each of which synapses with 10 hair cells, representing considerable convergence.

Cochlea hair cells lose their kinocilia during development. The tips of the tallest stereocilia are embedded in the overlying **tectorial membrane**, a matrix of mucopolysaccharides and proteins. Oscillations of the basilar membrane in

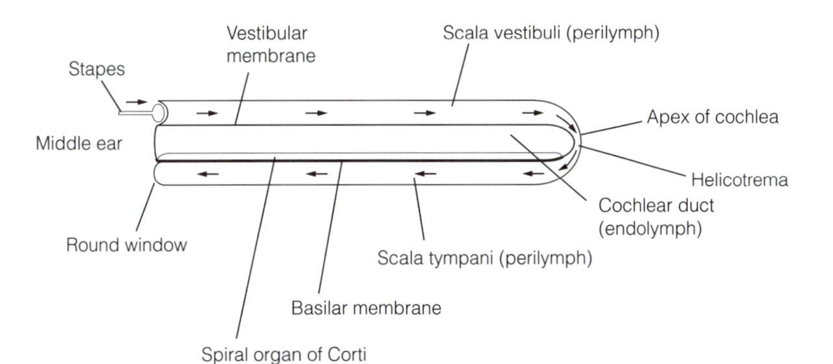

Fig. 2. The cochlea, depicted unfurled. Arrows show the direction of propagation of sound waves through the perilymph.

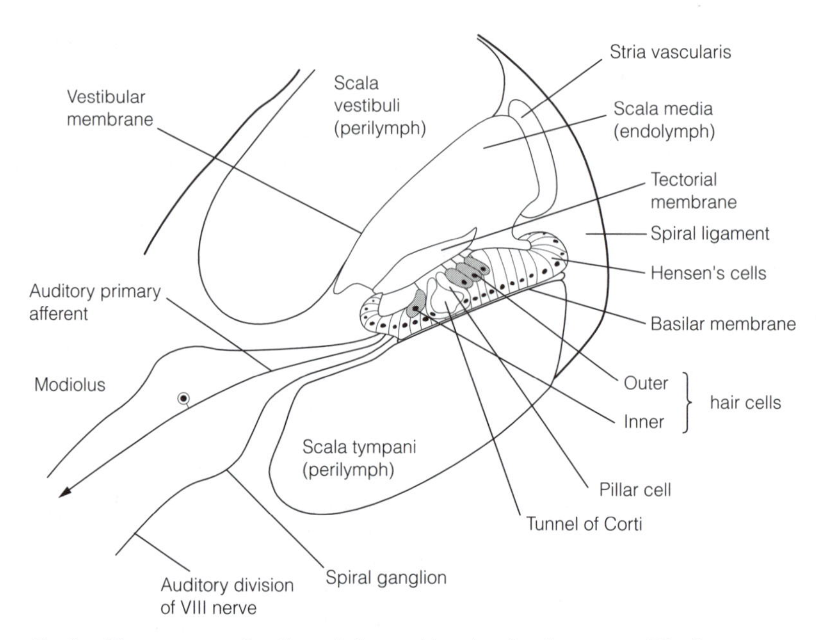

Fig. 3. Transverse section through the cochlea showing the organ of Corti.

response to a sound stimulus causes it to shear with respect to the tectorial membrane, bending the stereocilia first one way and then the other. This results in periodic depolarization and hyperpolarization of the hair cell, producing cyclical alterations in the tonic secretion of glutamate. The transduction mechanism for hair cells is like that of vestibular hair cells (see Topic G5).

A sound stimulus causes a **travelling wave** (like that generated by twitching the free end of a rope fixed at its other end) to spread from the base to the apex of the basilar membrane. High frequencies cause vibration at the basal end whereas low frequencies cause vibration towards the apex. This frequency sorting is a result of the continuous variation in the width, mass and **stiffness** of the basilar membrane along its length. The basilar membrane is narrow (50 μm) and stiff at the base, wider (500 μm) and less stiff at the apex. The relationship between frequency and length is logarithmic. At a given frequency, as the sound pressure level (SPL) increases so does the amplitude of the displacement and the length of the basilar membrane vibrating.

Outer hair cells (OHCs) contract in a voltage-dependent manner. Depolarization causes them to shorten. The speed with which they change length is so fast that they are able to follow the high frequency voltage changes produced by sound stimuli. By this means OHCs augment the vibrations of the basilar membrane, a process called **cochlear amplification**. It probably contributes to the high sensitivity and fine tuning to frequency exhibited by the basilar membrane, since these features are lost when OHCs are selectively damaged by **aminoglycoside antibiotics** such as streptomycin. Cochlear amplification causes vibrations of perilymph that are transmitted to the oval window across the middle ear in the 'wrong' direction to the tympanic membrane which now acts as a loudspeaker producing **otoacoustic emissions**. They may occur spontaneously or be evoked by sound. They are usually pure tones, inaudible and are not the cause of tinnitus, the origin of which is unknown. Otoacoustic emissions are not necessary for normal audition but provide insights into ear function and as a basis for clinical tests.

I3 Peripheral auditory processing

Key Notes

Primary auditory afferents

Primary auditory afferents have their cell bodies in the spiral ganglion and send their central axons to the pons via the eighth cranial nerve. Auditory afferents fire spontaneously and increase their firing in response to a tone. Most are sharply tuned in that, for low sound pressures, they show highest sensitivity for a narrow range of frequencies.

Coding of sound frequency

Frequency is coded in two ways. Generally for frequencies above 3000 Hz the frequency response of an afferent depends on where along the basilar membrane it is from. This is a place coding: specifically the mapping of frequency to position is referred to as tonotopic mapping. For lower frequencies, afferents fire during a particular phase of a sound wave. This is called phase locking. Since a population of afferents is involved, any given afferent need only fire occasionally. This is an example of temporal coding.

Coding of sound level

An auditory afferent responds only to a limited range of sound pressure levels (SPLs). The full range is encoded by afferents with different dynamic ranges. Afferents that display the highest rate of spontaneous firing are the most sensitive. Efferents from the superior olivary complex synapse with hair cells, with the effect that auditory afferent sensitivity is reduced. This allows them to respond to high sound levels

Related topics

Intensity and time coding (F2)
Acoustics and audition (I1)
Central auditory processing (I4)

Primary auditory afferents

The primary afferents have their cell bodies in the **spiral ganglion** located in the modiolus. Their centrally-directed axons project through the vestibulocochlear (VIII) nerve to synapse in the cochlear nuclei of the lower pons. In humans, about 30 000 type I afferents from inner hair cells provide the bulk of the output from the cochlea. Three-quarters of the hair cells [the outer hair cells (OHCs)] send their output to only about 3000 type II afferents. The nature of type II bipolar cell signalling is unknown. Auditory afferents fire spontaneously.

In response to a tone, type I afferents show an increase in firing which adapts. When the sound stops, firing ceases for a brief period. Hence they exhibit both dynamic and static responses (*Fig. 1*). Responses of type I afferents plotted as **tuning curves** (*Fig. 2*) show that they are sharply tuned at low sound pressures. The frequency to which the unit is most sensitive is the **characteristic frequency (CF)**. At high SPLs the primary afferents respond to a much wider frequency range.

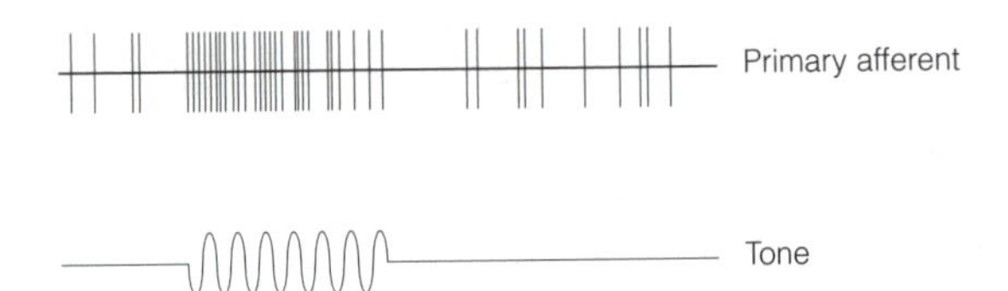

Fig. 1. Firing of auditory primary afferent (upper trace) in response to a tone (lower trace).

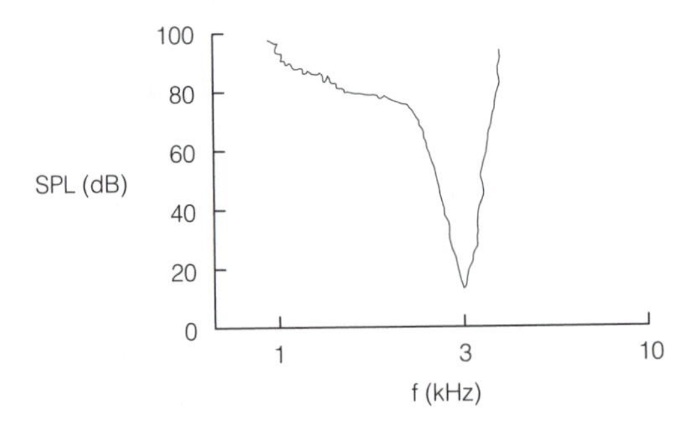

Fig. 2. Tuning curve for a type I cochlear afferent. The plot shows the minimum SPL required to evoke a response over a range of frequencies.

Coding of sound frequency

The nervous system has two ways of encoding the frequencies of a sound, **place coding** and **temporal coding**. Place coding is possible because the CF of an afferent is determined by its position of origin along the cochlea. Fibers with successively lower CFs are found closer to the apex of the cochlea. This means that the tuning of the basilar membrane and its afferents are matched. This mapping of frequency to position is known as **tonotopic mapping** and is retained in the central auditory pathway. Place coding is most important for frequencies above 1–3 kHz. For lower frequencies temporal coding is important and uses the property that afferents fire with greatest probability during a particular phase of a sound wave, **phase-locking**. It is only necessary that an individual afferent fires during some cycles if a group of cells is involved. Moreover if different groups phase-lock onto different parts of the cycle than a whole population of cells acting in concert can encode frequency.

Coding of sound level

Auditory afferents have a **dynamic range** of about 30 dB beyond which further increase in SPL has no additional effect. The full range of SPL (0–100 dB) is signaled by afferents with different sensitivities. Cells with the same CF may differ in threshold SPL by 70 dB.

The sensitivities of afferents can be modified by efferents which have their cell bodies in the **superior olivary complex**. Medial olivocochlear (OC) neurons synapse with OHCs, releasing acetylcholine which hyperpolarizes them, thereby reducing their cochlear amplification effect on the basilar membrane. Thus, in response to medial OC neuron activity the sensitivity of the basilar membrane, and so of the type I afferents, will be reduced. This allows the afferents to respond to higher sound levels. Indeed, medial OC neurons fire at greatest rates when sound levels are high. At low sound levels, medial OC neurons fire at low rates so cochlear amplification is high, and type I afferent sensitivity approaches maximum. In summary, the function of the efferents is to alter the gain of the cochlear amplifier.

14 Central auditory processing

Key Notes

Central auditory pathways

Primary auditory afferents terminate in the cochlear nuclei in the pons. Ventral cochlear axons go to the superior olivary complex on both sides. This structure projects to the nuclei of the lateral lemniscus, and is primarily concerned with localizing the direction of a sound source. The dorsal cochlear nucleus projects directly to the contralateral nucleus of the lateral lemniscus. The nuclei of the lateral lemniscus sends axons to the inferior colliculus of the tectum, which in turn projects to the medial geniculate nucleus (MGN). The auditory radiation which originates in the MGN goes to the primary auditory cortex, and is responsible for conscious sound perception. Though the largest auditory pathway is contralateral, extensive connections across the midline ensures interactions between sides.

Cochlear nuclei

Distinct cell types within the cochlear nuclei are able to process different features of a sound stimulus. Bushy cells signal exact timing information to the medial superior olivary nucleus which by comparing input from both ears is able to localize sound. Stellate cells are adapted to signal sound level. Many cells are finely tuned to particular frequencies and show lateral inhibition which sharpens this tuning.

Tonotopic mapping

Maps in which frequency is represented in a systematic way occur in all auditory structures. In humans all frequencies have about equal neural representation. In the primary auditory cortex isofrequency columns are found perpendicular to the cortical surface. These are arranged in bands which form an ordered tonotopic map.

Sound level

Cells responding to differences in sound level are found throughout the auditory system. Some are finely tuned to a characteristic sound level. In humans there are no maps of sound level.

Localization of sound

The source of a sound can be localized in the vertical plane (elevation) and in the horizontal plane (azimuth). Elevation is signaled by the delay caused by sound waves being reflected from the pinna (external ear). The azimuth of a sound source is measured in two ways. For higher frequencies, the difference in sound level between the ear nearest and that furthest from the sound source is computed by neurons in the lateral superior olivary nucleus. This nucleus projects to the tectum which controls eye and head reflexes in response to sound. For lower frequencies, cells in the medial superior olivary nucleus compute the phase difference that occurs because sound entering the ear furthest from the source is slightly delayed. These time differences are mapped topographically in the medial superior olivary nucleus (MSO). In the auditory cortex most cells respond preferentially to input in the contralateral ear and are either excited or inhibited by ipsilateral input.

Related topics

Early visual processing (H6)
Acoustics and audition (I1)
Peripheral auditory processing (I3)

Central auditory pathways

Primary auditory afferents bifurcate to form terminals in both the ventral and dorsal **cochlear nuclei**. From the ventral cochlear nucleus axons run to the superior olivary complex (SOC) on both sides and the contralateral **inferior colliculus** (IC). The auditory fibers that cross the pons constitute the **trapezoid body** (TB). Some axons from the TB enter the trigeminal and facial nerves forming the motor side of the tympanic reflex. The SOC compares input from the two ears to compute the whereabouts of a sound source. The SOC projects to the **nuclei of the lateral lemniscus**. The dorsal cochlear nucleus sends axons directly to the contralateral nucleus of the lateral lemniscus (*Fig. 1*).

The nucleus of the lateral lemniscus projects to the IC. The IC relays with the **medial geniculate nucleus (MGN)** of the thalamus. The MGN sends its output via the **auditory radiation** to the **primary auditory cortex**, A1 (Brodmann's areas 41 and 42) located in the superior temporal gyrus. The pathway via the MGN and auditory cortex mediates conscious auditory perception. The biggest auditory pathway is contralateral, however reciprocal connections between the nuclei of the lateral lemniscus (via **Probst's commissure**) and between the IC (via the **commissure of the inferior colliculus**) ensures extensive interactions between the input from both ears. These ascending sensory pathways are matched by descending projections. The auditory cortex sends axons back to both the MGN and the IC and the latter in turn projects to the SOC and the

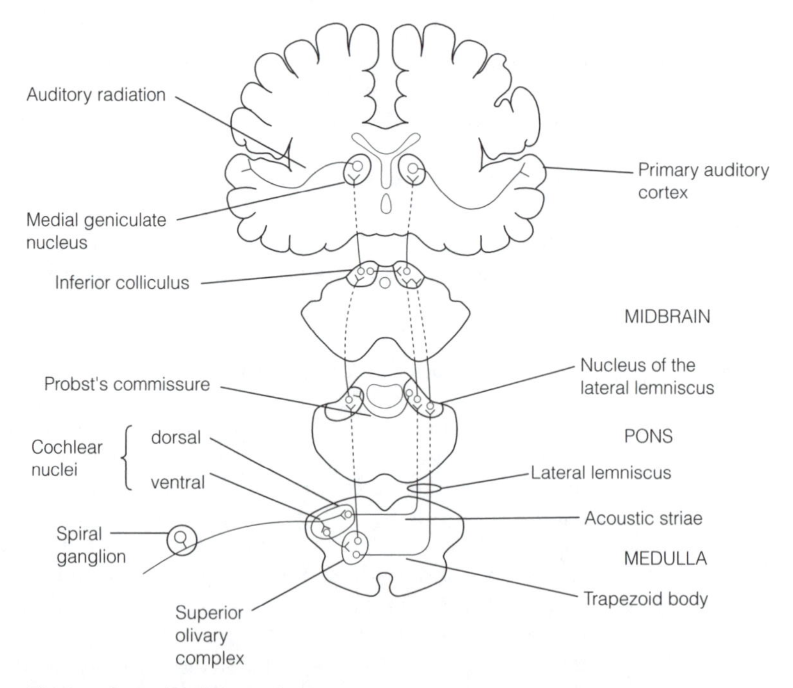

Fig. 1. Central auditory pathways.

cochlear nuclei. The SOC sends efferents to the spiral organ which modify the output of cochlear hair cells (see Topic I3).

Cochlear nuclei

Central auditory pathways process three features of sound input in parallel: tone, loudness and timing. From the last two the brain calculates the location of the sound in space. Parallel processing begins in the cochlear nuclei.

Several neuron types are present in the cochlear nuclei that can be distinguished both by their morphology and responses. Common are the **bushy cells** of the ventral nucleus which reproduce with high fidelity the firing pattern of the primary afferents, including phase-locking. Bushy cell output goes to the MSO. Because this output precisely signals the timing of sound, the nucleus is able to compare the input from both ears to compute the location of a sound source. By contrast, the ventral nucleus **stellate cells** are not well phase-locked but have a much greater dynamic range than bushy cells which makes them good at signaling sound level. This suggests parallel processing of timing and sound level in the auditory system.

Receptive fields (RFs) of auditory neurons are called **response maps** and are plotted in the same way as the primary afferent tuning curve. Five classes of cell can be distinguished in the cochlear nuclei on the basis of their RFs. Type I cells have a purely excitable RF that precisely matches primary afferent tuning curves, but all the other types have inhibitory responses which arise by lateral inhibition and which fine tune their frequency response. Type IV cell axons are the main output of the dorsal nucleus and *Fig. 2* shows a typical receptive field for a type IV cell.

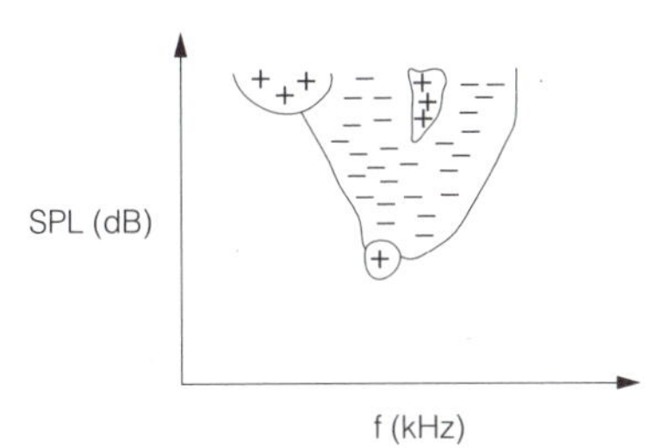

Fig. 2. Response map of a type IV cell in the dorsal cochlear nucleus. Excitatory region, +; inhibitory region, –.

Tonotopic mapping

Tonotopic maps are found in the cochlear nuclei, SOC, inferior colliculus and auditory cortex. Some structures have multiple maps. The cochlear nucleus is divided into **isofrequency strips**, each containing cells with similar characteristic frequencies. Strips representing increasingly higher frequencies are found progressively more posteriorly. In A1 **isofrequency columns** are aligned perpendicularly to the cortical surface and pass through all six layers of the cortex. These are arranged in isofrequency bands running mediolaterally over A1, with low frequencies represented rostrally and high frequencies caudally (*Fig. 3*). There are at least three other tonotopic maps in the auditory cortex. Adjacent maps are always mirror images of each other. In humans there is no great over-representation of particular frequencies.

Some regions of auditory cortex (e.g. **secondary auditory cortex**) are less well tonotopically organized and contain cells that respond to a wider range of frequencies.

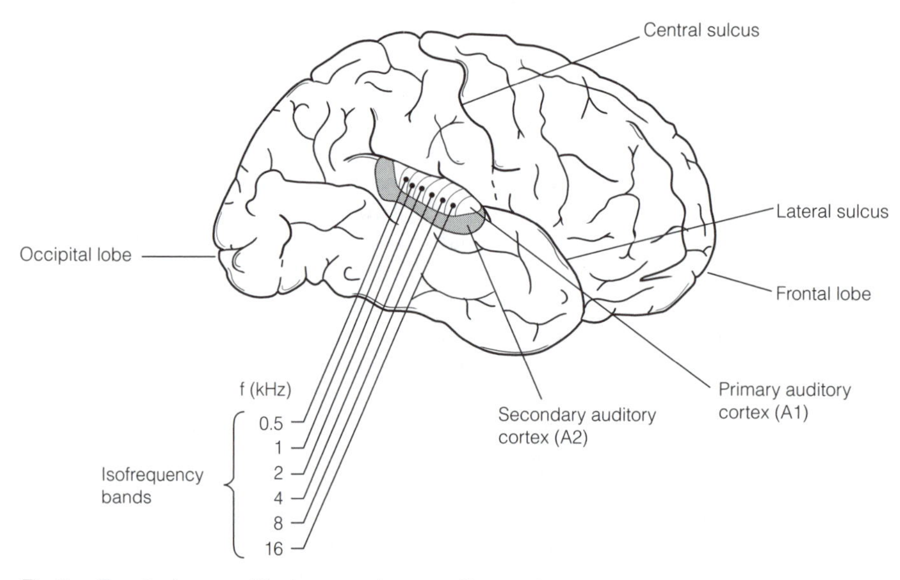

Fig. 3. Tonotopic map of the human primary auditory cortex.

Sound level

Cells throughout the auditory system respond to differences in sound level and fall into two broad classes. **Monotonic** cells have sigmoid plots of sound level against firing rates. Nonmonotonic cells are more finely tuned with a maximum firing rate at a characteristic sound level. Mapping of sound level occurs in bats which **echolocate** but have not been found in other species, including humans.

Localization of sound

The ability to localize the source of a sound in space is very important in avoiding danger. The coordinates of a sound source in vertical and horizontal planes are **elevation** and **azimuth** respectively. Different mechanisms are involved in determining these two coordinates.

For finding elevation the **pinna** of the outer ear is crucial. Sound waves entering the ear do so by two routes: one is direct and the other, reflected from the pinna, will be slightly delayed in arriving at the ear drum. Sound coming from different directions in the vertical plane will be reflected differently, because of the peculiarities of the shape of the pinna, and so have different delay times (*Fig. 4*). The auditory system uses the delay times to compute the sounds' position in the vertical plane. Although the human pinna is small and immobile it is important in this respect.

The superior olivary complex uses two methods to localize sound in the horizontal plane. Both compare input into the two ears and so constitute **binaural sound localization**; using them azimuth can be pinpointed with a precision of about one degree of arc.

Interaural level differences (ILDs)

If the head is orientated so that one ear is closer to the sound source, then the head forms a shadow which reduces the sound level entering the other ear. Using ILD to find azimuth is most accurate for high frequencies. The brain can use ILDs as low as 1 dB to compute azimuth.

Neurons of the **lateral superior olivary nucleus** (LSO) have a tonotopic map restricted to largely high frequency input. LSO neurons receive inputs from both

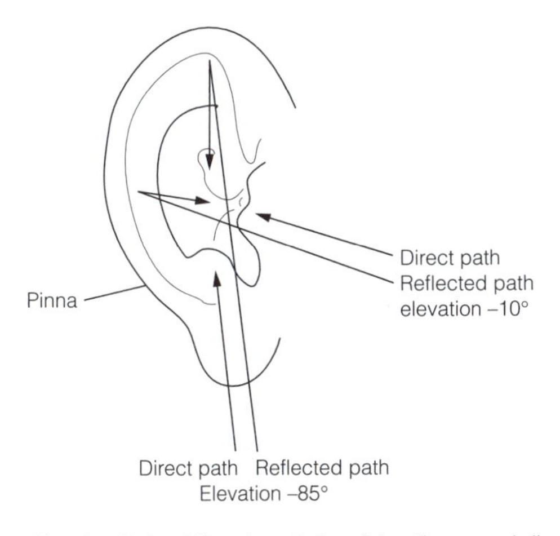

Fig. 4. Role of the pinna in localizing the sound direction in elevation.

ipsilateral and contralateral cochlear nuclei. However the contralateral route is by way of a glycinergic inhibitory neuron. Equal sound level in both ears causes overall inhibition of the LSO neuron and increasing the sound level in the contralateral only serves to augment the inhibition. However, increased sound level to the ipsilateral ear causes LSO firing. Maximum firing rate is seen when the ILD is 2 dB or more. Corresponding cells in the opposite LSO will show reverse responses to the same sound. The LSO projects to part of the inferior colliculus (IC). The IC connects extensively with the deep layers of the **superior colliculus** to form an **auditory space map** in register with the retinotopic map (Topic H2). Hence the superior colliculus is implicated in the auditory reflexes organizing gaze and head rotation towards the sound source.

Interaural time differences (ITDs)

A sound wave enters the closer ear slightly earlier than it enters the further one. For low frequencies (< 3 kHz) this results in a phase difference, in which the time delay is less than one period, which can be analyzed by neurons capable of phase-locking. At higher frequencies, input into the furthest ear is delayed by more than a single period, and this makes phase-locking unreliable, so ITDs cannot provide an unambiguous cue to location. ITDs as short as 20 μs can be detected.

The neural system for measuring ITDs depends on cells in the **medial superior olivary nucleus** (MSO) acting as coincidence detectors. The MSO has inputs from bushy cells in both cochlear nuclei that phase-lock in response to low frequency stimuli. If a phase difference exists between the two ears then the bushy cells corresponding to the furthest ear will fire slightly later. MSO cells fire maximally when ipsilateral and contralateral signals arrive at precisely the same time, and MSO circuitry is so arranged that this occurs for just one population of cells for any particular time delay.

In *Fig. 5*, if both ipsilateral and contralateral input arrives at the same time, cell C will produce the greatest response. If the ipsilateral signal were delayed (this corresponds to a sound coming from the contralateral side) cell D or E would have the highest firing rate. Conversely, contralateral signal delay (ipsilateral sound) will cause the biggest responses from A or B. In fact in the real

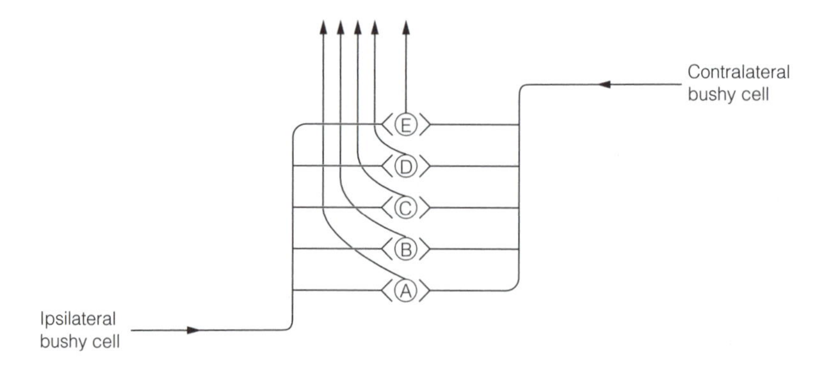

Fig. 5. *Medial superior olivary (MSO) nucleus circuit for measuring interaural time differences. Each MSO cell (A–E) acts as a coincidence detector and fires when it gets simultaneous ipsilateral and contralateral input. This will happen for a given MSO cell only when interaural time differences are negated by the differences in the length of the neural pathways from ipsilateral and contralateral ears to the MSO cell. See text for further details.*

auditory system the contralateral pathway is intrinsically longer so the cell giving the greatest response to zero ITD would not be C but a cell in the E-ward direction. A contralaterally placed sound will excite cells in the A-ward direction. The importance of this asymmetry is that without it both MSO nuclei would generate identical signals for all sound directions, with it each MSO nucleus signals best a sound source on the contralateral side.

Most cells in A1 are binaural in that they respond to input from either ear, but most are preferentially responsive to the contralateral ear. Such cells fall into two categories, and are localized in separate populations of cortical columns, depending on whether ipsilateral input excites (summation columns) or inhibits (suppression columns). Summation and suppression columns are arrayed alternately at right angles to isofrequency strips (c.f. ocular dominance columns, Topic H6). Neurons in summation columns have large location RFs, whereas, those in suppression columns are more finely tuned, responding only to a narrow range of azimuth values (*Fig. 6*).

No orderly map of sound location has been discovered in the auditory cortex, but lesions of A1 (in cats) do impair localization of sound on the contralateral side.

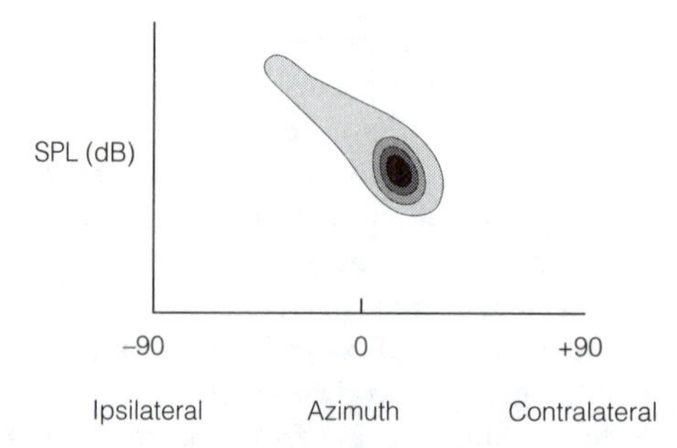

Fig. 6. *Location tuned neuron in a suppression column of the auditory cortex. Higher density of shading corresponds to a greater firing rate.*

J1 OLFACTORY RECEPTOR NEURONS

Key Notes

Olfactory epithelium

Bipolar olfactory receptor neurons lie in the olfactory epithelium. Their dendrites extend to the surface of the epithelium to form a swelling with a cluster of olfactory cilia. Their axons form the first cranial nerve and synapse with neurons in the olfactory bulb.

Olfactory transduction

Odor molecules are detected by odorant receptors in the olfactory cilia. The receptors are a family of about 1000 G protein-coupled receptors in which the associated G protein (G_{olf}) is related to G_s, the stimulatory G protein that is positively coupled to adenylyl cyclase. Most odorant receptors are positively coupled to the cyclic adenosine monophosphate (cAMP) second messenger system. The rise in cAMP on binding of an odorant molecule opens a cation channel, resulting in a depolarizing generator potential that is graded according to the concentration of odorant. Each olfactory sensory neuron expresses just a single species of odorant receptor and each odorant receptor binds a range of related molecules with differing affinities.

Related topics

Slow neurotransmission (C3) Taste (J3)

Olfactory epithelium

A sense of smell in humans is important in feeding and probably also in sexual behavior. Olfactory epithelium lies in the dorsal nasal cavity. It consists of bipolar **olfactory receptor neurons (ORNs)** and supporting cells. A dendrite emerges from one pole of the ORN and extends to the surface of the epithelium, forming a knob which gives rise to a cluster of six to 12 immobile **olfactory cilia** which lie within a mucus layer which is secreted by the supporting cells. The centrally-directed unmyelinated axon passes, in the olfactory (first cranial) nerve, through the **cribriform plate** of the **ethmoid** bone to synapse with cells in the **olfactory bulb**. Human olfactory epithelium contains about 10^8 ORNs. The mucus provides a medium to absorb airborne odor molecules which then reach the high surface area presented by the densely packed layer of olfactory cilia.

Olfactory transduction

Odor molecules are usually small (M_r < 200 Da), lipid soluble and volatile. Initially they bind to **odor-binding proteins** in the mucus which probably act to concentrate the odor molecules in the vicinity of the cilia. Odor molecules are recognized by **odorant receptors** in the cilia plasma membrane. These are G protein-coupled receptors and about 1000 members of this receptor superfamily have been identified in mammals. Each odorant receptor, unlike the G protein-coupled receptors for neurotransmitters, binds a range of related odor molecules with various affinities. Each odor molecule can interact with two to six odorant receptor subtypes. Each ORN probably expresses just a single subtype of odorant receptor. Because odorant receptors are relatively nonspecific, individual ORNs respond to a number of odors, collectively referred to as the **molecular receptive range**. The mammalian nervous system is able to discriminate some 10 000 distinct odors, presumable on the basis of precisely which array of odorant receptors (and so which sensory neurons) are stimulated, and with what relative intensities.

Odorant receptors are coupled to G proteins which are closely related to the G_s proteins that stimulate adenylyl cyclase, and termed $\mathbf{G_{olf}}$. Most odorant receptors are linked to the cAMP second messenger system (see Topic C3). Binding of an odor molecule causes a rise in cAMP within about 50 msec. This activates a cyclic-nucleotide-gated (CNG) channel, a nonspecific cation conductance allowing the flow of Na^+, K^+ and Ca^{2+} ions (*Fig. 1*). This results in a depolarization of the dendritic knob of the ORN which spreads electrotonically across the cell to trigger action potentials at the axon hillock. The generator potential is graded with an amplitude that signals the concentration of the odor molecule. However, a maximal response is produced by the opening of only a small fraction (3–4%) of the CNG channels available. The concentration range that can be signaled by firing of the ORN is narrow, about a 10-fold difference.

High odor concentration or prolonged exposure permit a high Ca^{2+} influx through the CNG channels. This ion has a number of modulatory effects in olfactory receptor neurons. Ca^{2+} activates heme oxygenase 2, an enzyme that synthesizes carbon monoxide (CO) which can activate **guanylyl cyclase** (GC) as shown in *Fig. 1*. Because Ca^{2+} also inhibits GC, there is no overall activation of the cyclase in the target ORN. However, CO is freely diffusible so it can activate GC in adjacent unstimulated ORNs, producing cyclic guanosine monophosphate which binds to and opens the CNG channels. In this way odorant excitation spreads to a cluster of ORNs. Since neighboring ORNs respond to the same odors this does not produce a loss of specificity. ORNs show **adaptation** to protracted stimulation. Ca^{2+} binds to calmodulin which can then bind to CNG channels, reducing the efficacy with which the cyclic nucleotides can open them. Hence Ca^{2+} attenuates the size of the generator potential.

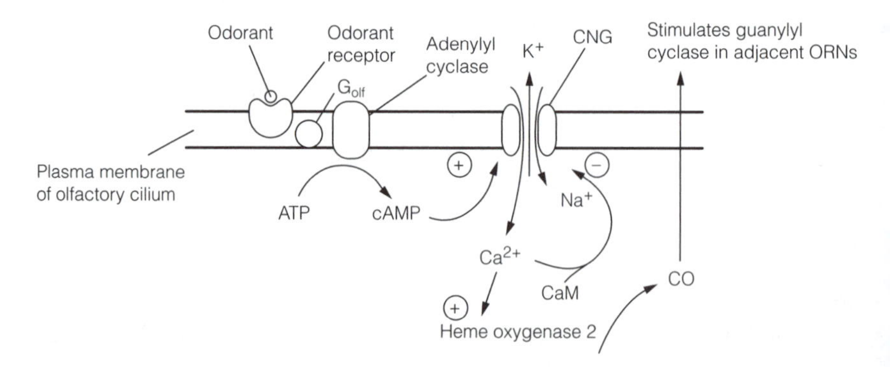

Fig. 1. Olfactory transduction in olfactory receptor neurons mediated by receptors coupled to the cAMP second messenger system. CaM, calmodulin; CNG, cyclic-nucleotide-gated channel.

J2 Olfactory pathways

Key Notes

Olfactory bulb

Olfactory receptor neurons synapse with mitral cells or tufted cells (M/T) and periglomerular cells within glomeruli in the olfactory bulb. Glomeruli are odor specific: each one receives terminals from receptor neurons that respond to the same set of odors. Lateral inhibition increases contrast between glomeruli that respond to similar odorants, and positive feedback circuitry enhances signals within glomeruli. Connections from the brainstem alter the responsiveness of M/T cells in the context of the behavioral state (e.g. hungry or sated).

Central olfactory connections

M/T neurons project to the olfactory cortex via the olfactory tract. Olfactory cortex is three layered palaeocortex (old cortex), the only cortex to receive sensory input directly rather than by way of the thalamus. Olfactory cortex projections to hypothalamus and amygdala are important for emotional and motivational aspects of odors, a pathway to the hippocampus is concerned with smell memory, and output via the thalamus to orbitofrontal cortex mediates the conscious perception of smell.

Related topic

Properties of neurites (D1)

Olfactory bulb

The axons of the ORNs run in the olfactory nerve to make excitatory synapses on the dendrites of **mitral cells** or **tufted cells** (M/T cells) and short axon **periglomerular cells** in the olfactory bulb. M/T cells send their axons into the olfactory tract. Periglomerular cells, together with granule cells, are inhibitory interneurons of the olfactory bulb. **Olfactory glomeruli** are spherical zones some 150 μm across in which extensive synaptic connections, many reciprocal, are made. The olfactory bulb contains about 2000 glomeruli. Each glomerulus receives the terminals of 25 000 ORNs which respond to the same odors. Hence, glomeruli are odor-specific functional units (*Fig. 1*). Low concentrations of a given odor molecule activate cells in the single glomerulus which gets input from the ORNs bearing odorant receptors with the highest affinity for the molecule. At higher concentrations, cells in other glomeruli are activated as their ORN odorant receptors' low affinity binding sites for the molecule are occupied. Each glomerulus has dendrites from about 75 M/T cells. Presumably, the M/T cells integrate weak inputs from a large number of ORNs to generate a sufficiently strong signal.

Olfactory bulb circuitry mediates at least two kinds of activity:

(1) Interglomerular processing enhances contrast between neighboring glomeruli. By dampening responses from glomeruli with slightly different odor specificities, odor discrimination is heightened. This occurs by lateral inhibition brought about via the reciprocal dendrodendritic synapses between M/T cells and granule cells. Via these synapses M/T cells excite

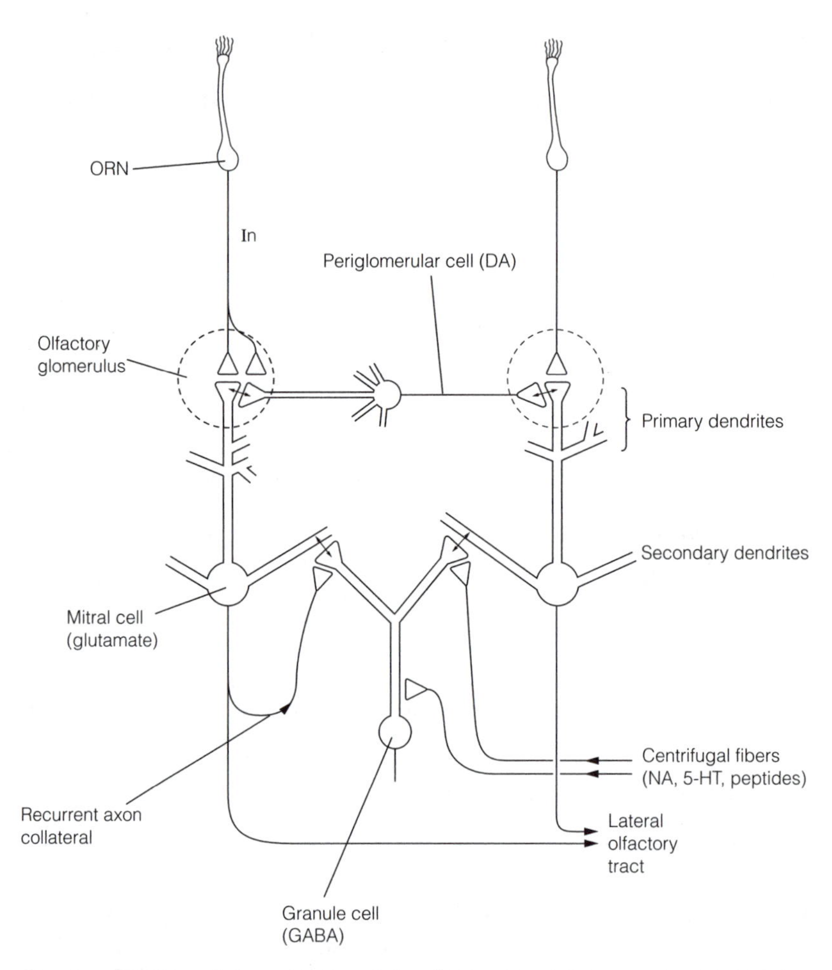

Fig. 1. Circuitry of the olfactory bulb. Reciprocal synapses are denoted by ↔. Neurotransmitters used by specific cell types are shown in parentheses: DA, dopamine; GABA, γ-aminobutyric acid; 5-HT, serotonin; NA, norepinephrine.

granule cells, which then inhibit the same, and adjacent M/T cells. The M/T cells excite the granule cells by means of back propagated action potentials (see Topic D3) sweeping into the M/T secondary dendrites which synapse with the granule cells.

(2) Intraglomerular processing involves positive feedback to enhance signals in M/T cells within a glomerulus.

There is a topographical organization of the fibers of the olfactory nerve and their projections to the olfactory bulb. Thin strips of olfactory epithelium running in an anteroposterior direction run to neighboring glomeruli. A given odor excites a particular array of glomeruli across the olfactory bulbs, an **odor image**. The higher the concentration of the odor molecule the bigger the area activated.

Central olfactory connections

M/T cell axons project via the **olfactory tract** to the **olfactory cortex**. This cortex is unusual in two respects. Firstly, it is **palaeocortex** (old cortex), structurally resembling the forebrain cortex of nonmammalian vertebrates, in only having three layers (c.f. neocortex, see Topic E2). Secondly, it is the only cortex to receive sensory input directly, rather than via the thalamus.

There are five regions of the olfactory cortex with distinct connections and functions but all receive input from the olfactory tract. The **anterior olfactory nucleus** gives rise to axons that cross the midline in the **anterior commissure** to go to the contralateral olfactory bulb (*Fig. 2*). The **anterior perforated substance** (called the **olfactory tubercle** in nonprimates) sends output to the posterior hypothalamus. This pathway, together with that to the **corticomedial amygdala**, which then projects to the medial hypothalamus, is concerned with the affective and motivational aspects of odors and directly influences feeding and mating. A pathway from the olfactory tract to the entorhinal cortex, which sends its entire output to the hippocampus, presumably encodes olfactory components of episodic memories (see Topic Q1).

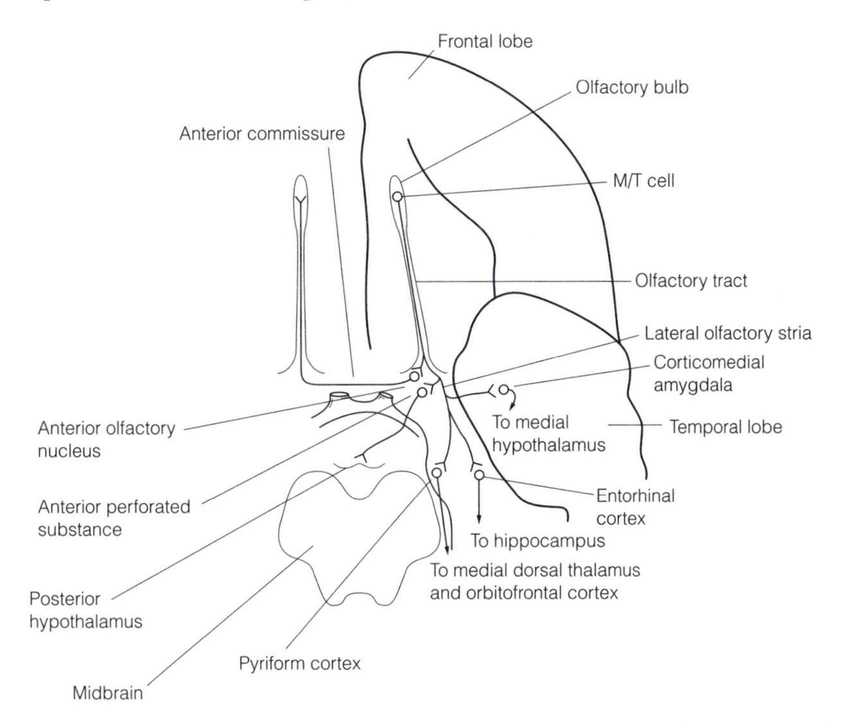

Fig. 2. Connections of the left olfactory cortex viewed from below. M/T, mitral and tufted cells.

A large part of the olfactory cortex is the **pyriform cortex**. This is concerned with olfactory discrimination. It send axons which terminate in the **medial dorsal thalamus**, which in turn projects to the **orbitofrontal cortex**. This cortex mediates the conscious perception of smell.

Olfactory processing is subject to considerable modulation. The olfactory bulb receives inputs from noradrenergic and serotonergic neurons in the brainstem and cholinergic neurons in the forebrain. In addition the anterior perforated substance (olfactory tubercle) receives a projection from the brainstem dopaminergic system. These various inputs are implicated in modifying olfaction on the basis of the behavioral state of the animal, and in olfactory learning. This is thought to be particularly important in feeding and mating behavior. In rats, for example, mitral cell responses to food odors depend on whether the animal is hungry or sated.

J3 TASTE

Key Notes

Gustation

The sense of taste provides a way in which harmful foods may be avoided and nutritious foods selected. Classically four tastes are defined, salty, sweet, sour and bitter, but a fifth due to glutamate is now recognised. It is one of several senses (others include olfaction) involved in the sensory experience evoked by having food in the mouth. Taste sensations help to regulate autonomic responses to feeding.

Taste buds

Taste buds are clusters of neuron-like epithelial cells, the gustatory receptor cells. Microvilli on the apical border of gustatory receptor cells are in contact with the contents of the mouth via taste pores. Receptor cells make synaptic connections with gustatory primary afferents, the axons of which travel through the VII, IX, or X cranial nerves. Taste buds are found not only in the tongue but also in the pharynx and upper esophagus.

Taste transduction

Salt taste is mediated by receptor cell depolarization brought about by the opening of amiloride-sensitive sodium channels. Hydrogen ions (sour taste) produce depolarization by blocking voltage-dependent K^+ channels. The sensation of sweetness involves the taste molecule binding to a G protein-linked receptor coupled (usually) to the cyclic adenosine monophosphate (cAMP) second messenger system. This causes depolarization by closing a potassium channel. There are multiple transduction pathways for bitter molecules but all result in receptor cell depolarization. Glutamate taste is mediated by metabotropic glutamate receptors (mGluR4).

Related topics

Slow neurotransmission (C3)   Retina (H3)

Gustation

The sense of taste, **gustation**, provides a means of avoiding potentially noxious foodstuffs or selecting for foods which have a high energy content. Classically, four tastes are described, salty, sour, sweet and bitter, on the basis that no cross-adaptation occurs between them. A fifth taste, umami, produced by monosodium glutamate, is now recognised. Plant alkaloids, some of which are toxic in high concentrations, are extremely bitter. A sour taste may signify a food degraded by microbiological action. By contrast a sweet food has a high content of sugars and so a readily available supply of metabolic energy. The sensory experience produced by having food in the mouth is called **flavor perception**, and relies on several sensory modalities. Apart from olfaction and gustation , information about food texture is provided by mechanoreceptors and proprioceptors in the mouth and jaw innervated by trigeminal sensory neurons. Taste sensations are important in triggering or modifying autonomic responses to feeding (e.g., salivation, gastric secretion and changes to gastrointestinal motility).

Taste buds

Gustatory receptor cells are epithelial cells but have many neuron-like features. They are organized into small clusters called **taste buds**, containing 50–150 members, together with supporting cells (*Fig. 1*). As with other epithelial cells, gustatory receptor cells are continually replaced from basal cells about every 10 days. Microvilli on the apical border of each receptor cell project through **taste pores** in the gustatory epithelium, bringing them into contact with the contents of the mouth. The microvilli carry out taste transduction.

Receptor cells form synaptic connections with gustatory primary afferent neurons. Each afferent branches to synapse with receptor cells in more than one taste bud. The axons of the primary afferents run through the facial (VII), glossopharyngeal (IX), and vagus (X) cranial nerves.

Taste buds are located in the epithelium of the tongue, palate, pharynx, epiglottis and the upper part of the esophagus. In the tongue they are present in small projections, **papillae**, of which there are three kinds. Circumvallate and foliate papillae located on the posterior tongue are supplied by the IX nerve and each contain thousands of taste buds. Fungiform papillae are scattered around the edge of the anterior tongue, are supplied by the VII nerve, and have only a few taste buds each. A few taste buds are present in the epiglottis and esophagus and are innervated by the X nerve.

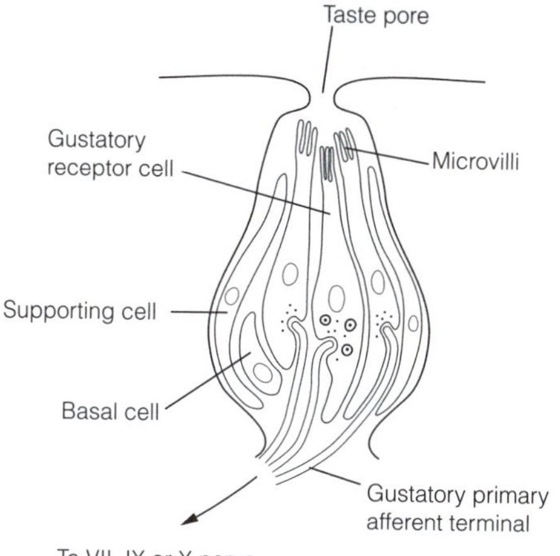

Fig. 1. A taste bud.

Taste transduction

Many of the ions or molecules responsible for taste sensation are hydrophilic and freely diffusible. Those which are hydrophobic include plant alkaloids which may bind to proteins in the saliva, equivalent to odorant binding proteins, for presentation to gustatory receptor cells. Transduction involves changes in membrane conductance which causes a depolarizing generator potential, triggering action potentials, calcium influx and so neurotransmitter release. Gustatory receptor cells have voltage-dependent Na^+, K^+ and Ca^{2+} and are excitable.

Salt taste is caused by Na^+ ions. Salt transduction (*Fig. 2a*) occurs by the influx of Na^+ through an amiloride-sensitive Na^+ channel which depolarizes the receptor cell (i.e. producing a generator potential) so causing it to fire.

H^+ ions responsible for sour (acid) sensation causes a generator potential by blocking voltage-dependent K^+ channels in the apical membrane which at rest

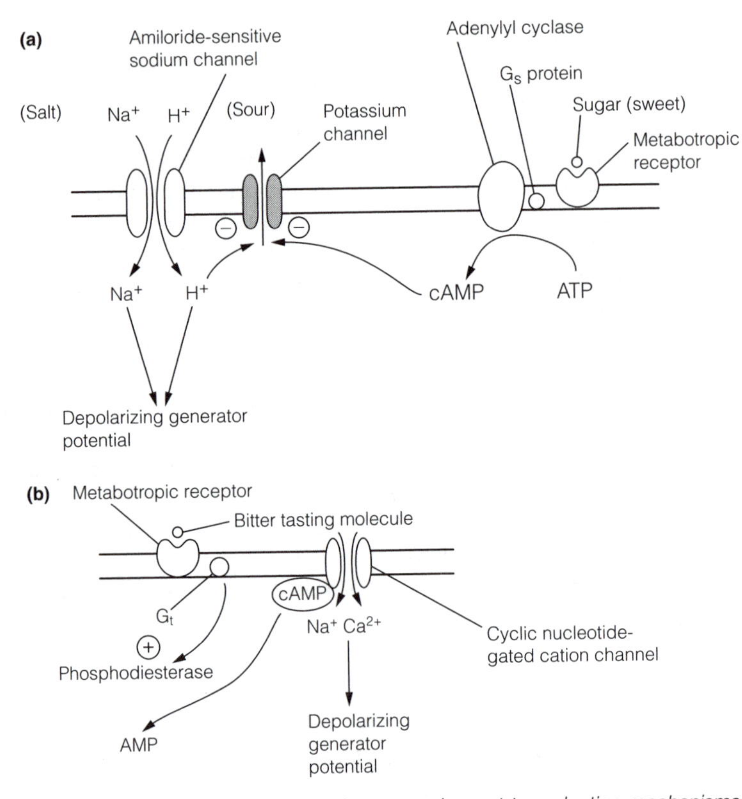

Fig. 2. Taste transduction: (a) salt, sour and sweet transduction mechanisms, note that an amiloride-sensitive Na^+ channel is implicated in both salt and sour taste; (b) one of several mechanisms involved in bitter transduction.

carry an outward, hyperpolarizing current. Blockade of other channels by protons may also contribute. The amplitude of the generator potential is proportional to the H^+ concentration.

Sugars, some amino acids and some proteins produce sweet sensations by interacting with metabotropic G protein-linked receptors coupled to second messengers. Sugars activate adenylyl cyclase and the consequent rise in cAMP produces depolarization by closing a K^+ channel.

Multiple pathways mediate bitter taste transduction. This reflects the wide diversity of molecules that are bitter flavored: divalent salts, alkaloids, some amino acids and some proteins. Divalent salts and quinine block K^+ channels and so produce depolarization by reducing an outward potassium current. In a mechanism having striking parallels with phototransduction, some bitter tasting agents bind metabotropic receptors coupled to transducin (G_t) which activates a phosphodiesterase (*Fig. 2b*). This enzyme breaks down cAMP, lowering its concentration within the cytoplasm of the gustatory receptor cell. The final consequence of this is dissociation of cAMP from a cyclic-nucleotide-gated cation conductance, allowing influx of Na^+ and Ca^{2+} and so depolarization. In addition, some bitter stimuli activate G protein-coupled receptors that stimulate phospholipase C.

The umami taste sensation produced by L-glutamate involves metabotropic glutamate receptors of the mGluR4 subtype.

J4 Gustatory pathways

Key Notes

Anatomy of gustatory pathways

Gustatory primary afferent cell bodies lie in the ganglia of either the VII, IX, or X cranial nerves. Their axons terminate in the nucleus of the solitary tract (NST) in the medulla. While some NST cells project to the lateral hypothalamus for regulating autonomic responses to feeding, others project to the ipsilateral ventral posterior medial thalamic nucleus. This nucleus sends axons to the ipsilateral cortex taste area I, located adjacent to the somatosensory area for the tongue in the postcentral gyrus, which mediates conscious taste perception. Taste area II in the insula is thought to be involved in emotional responses to taste.

Gustatory coding

Gustatory afferents are broadly tuned. Those in the facial nerve respond best to salt or sweet stimuli, those in the glossopharyngeal respond preferentially to sour and bitter stimuli, while vagal afferents measure the extent to which the ionic concentration differs from extracellular fluid. The classic taste sensations do not correspond to separate labeled lines, neither are they topographically represented in the brain.

Related topic

Taste (J3)

Anatomy of gustatory pathways

The gustatory primary afferents of cranial nerves VII, IX and X have their cell bodies in the **geniculate**, **petrosal** and **nodose** ganglia respectively. Their centrally directed axons end in the rostral portion of the **nucleus** of the **solitary tract** (**NST**) which lies in the dorsal medulla (*Fig. 1*). Gustatory primary afferents secrete glutamate and substance P.

Some NST cells project to the lateral hypothalamus which organizes autonomic responses to feeding. Gustatory neurons in the NST project via the central tegmental tract to the ipsilateral **ventral posterior medial nucleus** of the thalamus, terminating on a population of small cells distinct from those receiving somatosensory input from the tongue or pharynx. These cells send their axons to the ipsilateral cortex. So, unlike most sensory pathways, that for taste is uncrossed. Taste area I (Brodmann's area 43) is located on the dorsal wall of the lateral sulcus (see *Figs 4* and *6*, Topic E2) at the junction with the insula and adjacent to the somatotopic mapping of the tongue. It is thought to be concerned with the conscious perception of taste. Taste area II is in the insula, a region of cortex buried deeply in the lateral sulcus, which may be concerned with the affective aspects of taste.

Gustatory coding

Afferents in the VII nerve commonly exhibit preferences for either salty or sweet stimuli, whereas most of those in the IX nerve, supplied by the posterior tongue, are tuned to acids (sour) or bitter stimuli. These findings support the observation, in humans, that the anterior edge of the tongue is most sensitive to salt and sweet tastes whereas the back of the tongue is more sensitive to acid and bitter flavours. Vagal (X) afferents are broadly tuned but on average respond best to

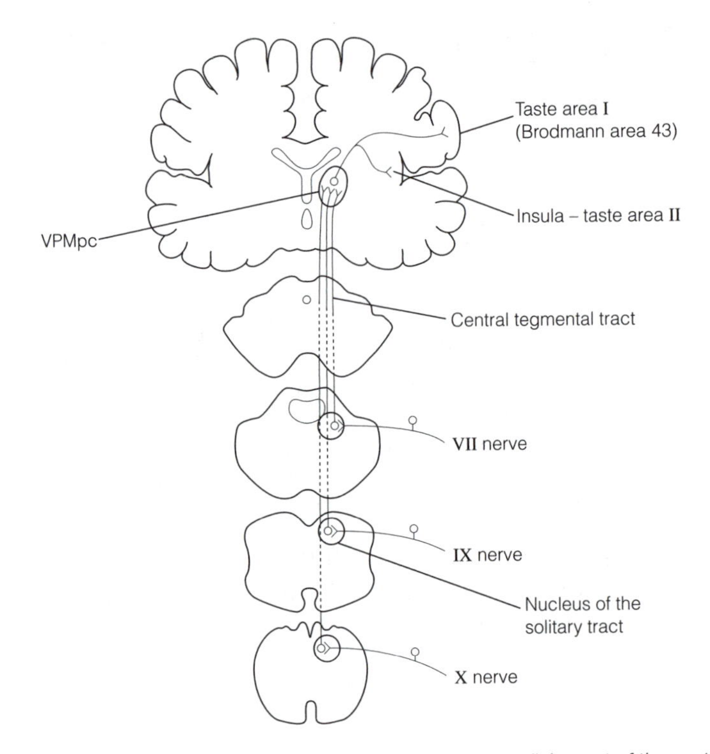

Fig. 1. Central gustatory pathways. VPMpc, parvocellular part of the ventral posterior medial nucleus of the thalamus.

Na^+ and H^+. In addition, many units respond to distilled water. These neurons exhibit their lowest firing in 154 mM NaCl, firing with increasing rates either as salt concentration increases or decreases from this value. Vagal afferents thus appear to measure to what extent the pharyngeal contents differ in ionic concentration from extracellular fluid.

The fact that gustatory neurons are generally quite nonspecific argues against the existence of labeled lines corresponding to the classical taste sensations. Furthermore, there is no topographical organization apparent in gustatory pathways. Hence distinctive taste sensations arise from neurons with opponent receptive fields that compare the outputs of differently tuned populations of afferents. This is analogous to how color vision arises from opponent processing that compares output from just three populations of cone photoreceptors (see Topic H5).

K1 SKELETAL MUSCLES AND EXCITATION–CONTRACTION COUPLING

Key Notes

Muscle structure

Skeletal muscles are composed of bundles of striated muscle fibers, multinucleate cells derived by the fusion of many myoblasts. A muscle fiber contains several parallel myofibrils, each divided into a series of sarcomeres, the basic contractile element. Sarcomeres contain thin filaments and thick filaments. The actin-containing thin filaments are attached to the ends of the sarcomere and extend towards its middle where they interdigitate with the myosin thick filaments. In response to an increase in Ca^{2+} within the muscle fiber the thin filaments slide over the thick filaments, shortening the sarcomere.

Physiology of muscle contraction

The length of a muscle determines the force of contraction by dictating the degree of overlap of the thin and thick filaments, and the tension exerted by its elastic components. Over the normal working range for a muscle, the relationship between force and length is not linear but is more complex and the neural control of muscle contraction must compensate for this to ensure that the force of contraction is appropriate for any given load. Muscle contraction can be isometric, in which case length remains constant and the force increases until it matches a load to be supported, or isotonic in which, by exerting a constant force the muscle shortens to move a load.

Neuromuscular junction

The neuromuscular junction (nmj) is the synapse between a motor neuron and a muscle fiber. Acetylcholine (ACh) released from the nerve terminal activates nicotinic cholinergic receptors (nAChR) in the postjunctional membrane, the endplate, causing it to depolarize. Secretion of ACh from a single vesicle causes a miniature endplate potential (mepp) of 0.4 mV. Release of ACh from many vesicles in response to the arrival of an action potential in the terminal causes summation of many mepps to give an endplate potential, a large depolarization sufficient to trigger muscle fiber action potentials.

Excitation–contraction coupling

Action potentials are propagated into invaginations of the muscle fiber plasma membrane. This depolarization causes the release of Ca^{2+} from the sarcoplasmic reticulum (SR), a subcellular organelle that acts as an internal calcium store. The rapid rise in intracellular Ca^{2+} triggers muscle contraction. This link between membrane events and contraction mediated by calcium is excitation–contraction coupling. Active transport drives Ca^{2+} back into the SR so that the muscle relaxes.

Neuromuscular blocking agents

Muscle relaxants, which block neuromuscular transmission, are used to paralyze skeletal muscles during surgery. They fall into two categories. Nondepolarizing drugs are competitive antagonists of nAChR, and their effects can be reversed by acetylcholinesterase (AChE) inhibitors which increase ACh concentrations in the cleft. Depolarizing drugs are nAChR agonists that produce blockade by preventing excitation–contraction coupling.

Related topics

Fast neurotransmission (C2) Elementary motor reflexes (K3)

Muscle structure

Intentional movements are executed by **skeletal muscles**, muscle attached by tendons to the skeleton. About 40% by mass of a human is skeletal muscle, composed of **striated** muscle fibers, which are named for their striped appearance under light microscopy caused by the regular arrangement of proteins within them.

Skeletal muscle fibers are formed by the fusion of numerous **myoblasts**, so each fiber is a **syncytium**, a multinucleated structure that functions as a single unit. Fiber diameters range from 10 to 100 μm, and they may be many centimeters in length in a large muscle, but do not usually extend along the whole muscle. Many muscle fibers are grouped into bundles called **fasciculi** (or **fascicles**) by a connective tissue sheath, the **perimysium**. Within a fasciculus the fibers are supported by areolar connective tissue packing called **endomysium**. Fasciculi bundled together and enclosed by the **epimysium** form a muscle.

Within each muscle fiber the contractile apparatus is organized into a number of parallel **myofibrils**, each about 1 μm in diameter. These extend the length of the fiber and are composed of repeating units called **sarcomeres** (*Fig. 1*) which give the appearance of alternating dark (**anisotropic**, A) and light (**isotropic**, I) bands along its length. Each I band is intersected by a thin **Z line**.

Within each sarcomere are two types of protein filaments. **Thin filaments** are attached to the **Z disc** (corresponding with a Z line) at each end of the sarcomere and extend towards the center, and **thick filaments** interdigitate between them. Thin filaments consist of **actin**, **tropomyosin** and a trimeric protein, **troponin**, one subunit of which is a calcium binding protein with a similar structure to calmodulin (Topic C3). On binding of Ca^{2+} to troponin the conformation of the thin filaments changes so that it interacts with the thick filament protein, **myosin**. The thin filaments are pulled inwards by this interaction which is powered by the hydrolysis of adenosine 5′-triphosphate (ATP). The result is a

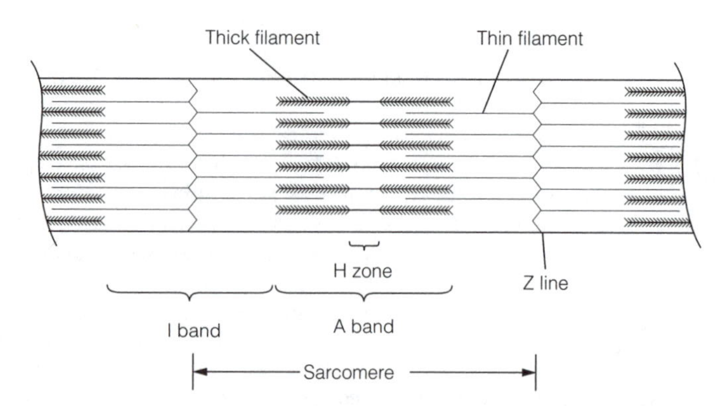

Fig. 1. Organization of protein filaments in a myofibril sarcomere.

shortening of the sarcomere, the basis of muscle contraction (details of the biochemistry of which can be found in *Instant Notes in Biochemistry*, 2nd edn).

Physiology of muscle contraction

The force of contraction of a muscle depends on its length. This is related to two factors, firstly the extent of overlap between thick and thin filaments in sarcomeres and secondly the degree to which elastic elements in the muscle are stretched. The relationship between length and force can be modeled by regarding the muscle as a spring. A spring has a resting length, L_0, at which it has no tendency to shorten. Once stretched beyond its resting length, a tension (a force due to the stretching) is exerted by the spring. In the case of an ideal spring this tension force, F, is proportional to its change in length; that is:

$$F \propto \Delta L$$
$$\text{or } F = k\,\Delta L$$

Since the change in length, ΔL, is just the stretched length, L, minus the resting length:

$$F = k\,(L - L_0)$$

The constant of proportionality, $k = F/(L - L_0)$, is the **stiffness** of the spring; it is a measure of how much load must be applied to stretch it by a given amount.

Like springs, skeletal muscles have a resting length, L_0, where elastic recoil force is zero, which coincides roughly with the length at which maximum contractile force is possible because there is optimal overlap of thin and thick filament. Moreover, *in vivo* muscles at rest are held close to this value, so under physiological conditions muscles are operating optimally. However close to its resting length muscle does not behave like an ideal spring; its stiffness is not constant but changes with length in a complicated manner.

Precise control of muscle contraction requires neural feedback mechanisms that compensate for this nonlinear behavior of muscle stiffness over the normal working range of muscles. This is achieved by muscle spindle and Golgi reflexes (Topic K2). Acting together they ensure that the extent of muscle contraction matches the loads imposed on the muscle and the required movement

Muscle contraction can serve two purposes. It can support a load without moving it: in this case muscle length remains constant and the force developed by the muscle must equal the load. If the load is increased the muscle must generate a greater tension if it is to stay the same length. This is called **isometric** contraction, and is typically the case for postural muscles. Alternatively, the contraction can do external work to move a load through some distance: in this **isotonic** contraction a constant tension is exerted and the muscle shortens by the amount necessary to move the load.

Neuromuscular junction

The axon of a motor neuron divides into a number of branches at the surface of the muscle fiber. Each branch ends in a bouton which forms a synapse with the muscle fiber, called a **neuromuscular junction** (**nmj**). The cleft of the nmj (*Fig. 2*) is about 50 nm across. The postjunctional membrane, the **endplate**, which is thrown into junctional folds, has an extraordinarily high density of nAChR concentrated under the active zones where ACh is released.

Overlying the endplate is a collagenous basement membrane (**basal lamina**) to which is bound AChE. Soluble forms of the same enzyme are also secreted into the cleft.

nAChR are members of the ligand-gated ion channel superfamily of receptors

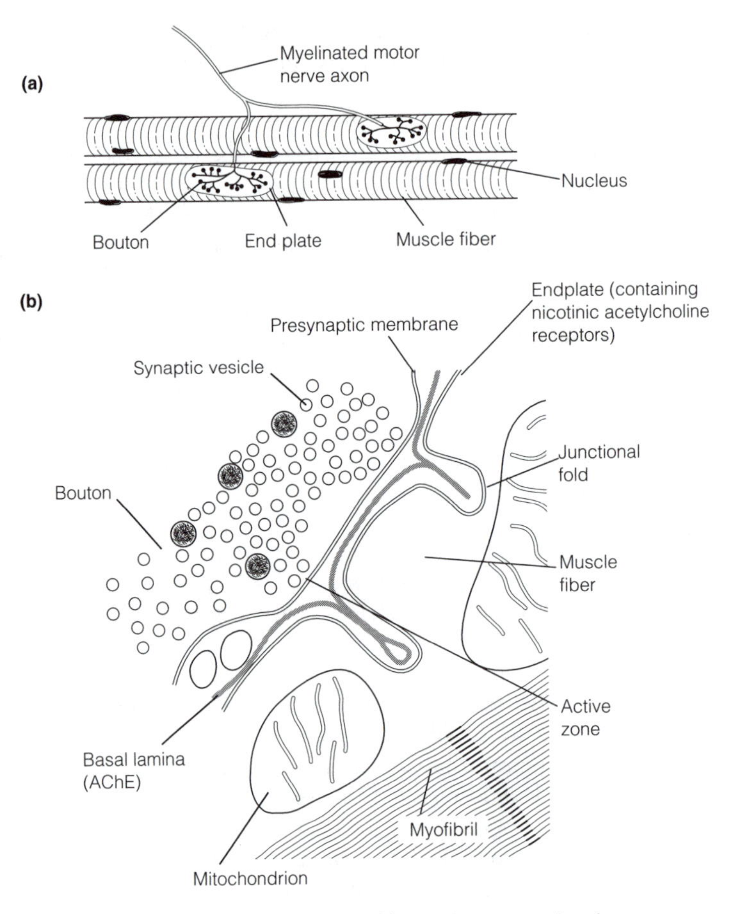

Fig. 2. The neuromuscular junction: (a) a motor neuron forming synapses on two muscle fibers (× 150); (b) a drawing of an electron micrograph of a neuromuscular junction.

and mediate fast transmission by ACh (Topic C4). Each of the two α subunits binds a molecule of ACh causing the opening of a cation channel which allows influx of Na^+ and efflux of K^+. Since the reversal potential (see Topic C2) for the current flowing through the nAChR is close to 0 mV, activating them causes depolarization.

Spontaneous release of a single quantum of ACh at the nmj causes a 0.4 mV depolarization at the endplate called a **miniature endplate potential (mepp)**. The arrival of an action potential at the motor nerve terminal triggers the release of 200–300 quanta which results in a massive depolarization, the summed effect of all the individual mepps, an **endplate potential (epp)** to about – 20 mV. This greatly exceeds the threshold for activating voltage-dependent sodium channels in the muscle membrane, so the effect of the epp is to set up an action potential which is propagated over the muscle fiber membrane. The motor endplate is unique among vertebrate synapses in that firing of the motor neuron almost invariably results in the triggering of muscle action potentials (compare this with typical synapse behavior, Topic D2).

The concentration of ACh reaches 1 mM in the nmj within about 200 µsec of the arrival of an action potential at the motor nerve terminal, but within a millisecond or so the ACh concentration has fallen back to baseline levels because of the high activity of AChE in the cleft (see Topic C7). The enzyme hydrolyses ACh to choline and acetate. Choline is taken back into the nerve terminal via a Na^+-dependent transporter.

Excitation–contraction coupling

Calcium is an absolute requirement for skeletal muscle contraction. By binding to troponin it triggers a conformational change in the thin filament allowing the interaction between actin and myosin which powers contraction. The sequence of events by which a muscle action potential mobilizes Ca^{2+} to bring about contraction is referred to as **excitation–contraction coupling**.

The plasma membrane of a muscle fiber, the sarcolemma, has deep invaginations called **transverse (T) tubules** (*Fig. 3*) located in register with the Z discs of the myofibrils. Closely associated with each T tubule is a pair of terminal cisternae. These are part of the **sarcoplasmic reticulum (SR)**, a specialized subcellular organelle of muscle fibers derived from smooth endoplasmic reticulum. The cisternae and T tubule complex is known as a **triad**. The SR contains a high concentration of Ca^{2+}. Under the electron microscope the cisternal membrane and T tubule membrane are shown to be linked by a number of **endfeet**.

Action potentials propagated over the muscle plasma membrane sweep into the T tubules which carry the depolarization within a millisecond to all the myofibrils in the muscle. This activates a population of modified L-type Ca^{2+} channels in the tubules. These channels are atypical in that while they undergo a conformational change in response to depolarization they have a low probability of opening, so the influx of external Ca^{2+} through them is small. However, they trigger the opening of Ca^{2+} channels called **ryanodine receptors** located in the terminal cisternae, allowing the rapid diffusion of Ca^{2+} from the SR to the cytoplasm where the free Ca^{2+} concentration at rest is very low (about 10^{-8} M). So, the source of Ca^{2+} for skeletal muscle contraction is the internal store in the SR. The ryanodine receptor is a large transmembrane protein of the SR membrane which forms the endfeet. It consists of four identical subunits arranged around a central Ca^{2+}-selective pore with a conductance about 10-fold greater than that of the L-type Ca^{2+} channels. It is so-called because it binds the alkaloid, **ryanodine**, from the plant *Ryania speciosa*. The ryanodine receptor

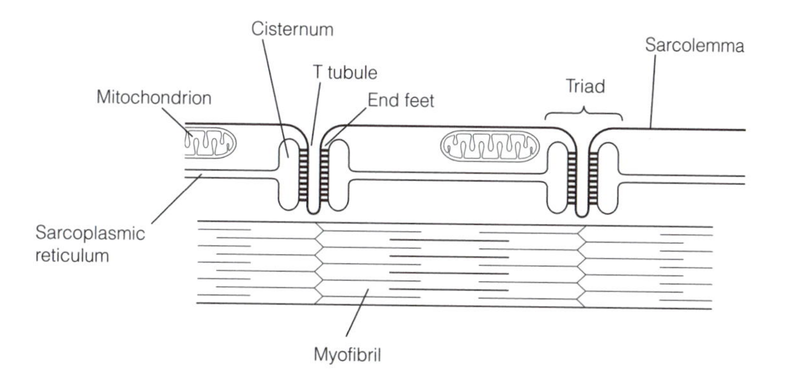

Fig. 3. The functional relationships between the triad components and myofibrils in a muscle fiber. Note the endfeet between the cisternae and the T tubules.

primary structure has a high homology with the inositol triphosphate receptor (see Topic C3) which serves a similar function.

The Ca^{2+} concentration in the muscle fiber cytoplasm reaches about 10 μM very rapidly, triggering contraction. The Ca^{2+} concentration falls rapidly as Ca^{2+} is actively pumped back into the SR via a Ca^{2+}-ATPase transporter and this permits relaxation.

Neuromuscular blocking agents

Blockade of transmission at the nmj is used during surgery to produce relaxation of skeletal muscle. They are effective within 1 min of injection. Muscle relaxant drugs fall into two categories depending on their mode of action and all have a structural resemblance to ACh.

Nondepolarizing drugs, such as **tubocurarine**, are competitive antagonists of nAChR. The duration of action of these drugs ranges from 15 to 60 min. Their action can be reversed rapidly by AChE inhibitors, such as **neostigmine**, that cause an increase in the concentration of ACh which can then compete with the drug for the nicotinic receptor.

Depolarizing drugs, of which **succinylcholine** is the only agent of clinical importance, are nAChR agonists. Initially the binding of the agonists opens the nicotinic receptor channel causing persistent depolarization of the endplate. This first causes generalized disorganized contractions of muscles called **fasciculations**, and is followed by **flaccid paralysis** as the T tubule Ca^{2+} channels inactivate and excitation–contraction coupling fails. This early stage in the action of depolarizing drugs (called **phase I** block) arise as a result of an ACh-like depolarization and so is augmented rather than reversed by AChE inhibitors. With continuing exposure **phase II** block occurs in which the nAChR either desensitizes, or suffers open channel blockade by the drug. Phase II block can be reversed by AChE inhibitors. Succinylcholine is rapidly hydrolyzed by circulating esterases, so its duration of action is only about 5 min.

K2 Motor units and motor pools

Key Notes

Motor units

A motor unit consists of a motor neuron together with all the muscle fibers it innervates, which ranges from six to a few thousand. In mammals, each muscle fiber gets input from just one motor neuron. An action potential in a motor neuron causes a twitch, a single contraction, in all the fibers it supplies. Individual twitches summate to produce tetanus, a prolonged maximal contraction, if a series of action potentials arrive in a short time. There are three types of motor unit. Slow twitch (S) units drive type 1 muscle fibers that are adapted for aerobic metabolism and capable of sustaining low forces for very long periods. These dominate in postural muscles. Fast twitch fibers [divided into fatigue resistant (FR) and fast fatigue (FF)] innervate type 2 muscle fibers and can produce large forces rapidly, but only for short periods.

Motor pools

A motor pool is the set of motor neurons that innervates a single muscle. The force of contraction of a muscle is determined by the firing frequencies of individual motor neurons, and by the number of motor neurons in the pool that are firing. Larger forces are generated by recruiting an increasing number of motor units. Recruitment generally (but not always) follow the size principle in which the smaller motor neurons come on line before the larger ones. This gives the order S–FR–FF.

Motor unit disorders

Myasthenic diseases are those in which transmission at the neuromuscular junction (nmj) is compromised. Most common is myasthenia gravis in which autoantibodies are made against nicotinic cholinergic receptors (nAChR). The result is that the endplate becomes less sensitive to acetylcholine (ACh). Muscular dystrophies are disorders in which muscle fibers die and are replaced at abnormally high rates. The X-linked recessive Duchenne dystrophy is the commonest. Trauma that severs α motor neuron axons causes flaccid paralysis of the disconnected muscle.

Related topics

Skeletal muscles and excitation–contraction coupling (K1)
Elementary motor reflexes (K3)
Synaptogenesis and developmental plasticity (P5)

Motor units

The final functional component of motor pathways is the **motor unit**. It consists of a motor neuron and the muscle fibers it innervates. In mammals each muscle fiber is supplied by only one motor neuron. However each motor neuron synapses with anything from six to a few thousand muscle fibers within a single muscle, the **innervation ratio**. The size of a motor unit is related to the precision of motor control required of a given muscle. Finely regulated muscles (e.g. extraocular eye muscles) consist of small motor units, less finely regulated muscles have larger ones. The fibers of a single unit are scattered widely throughout a muscle so no part of a muscle is controlled by just one motor unit.

A single action potential in the motor neuron causes a **twitch**, a single contraction, in all of the muscle fibers to which it is attached (*Fig. 1a*). The contraction and relaxation of muscle fibers is very much longer than the muscle action potential of about 3 msec. This is because of the time taken for the Ca^{2+} to activate the contractile machinery and for it to be pumped back into the sarcoplasmic reticulum. If a volley of action potentials is fired and there is insufficient time for the muscle to relax between successive impulses the twitches summate to increase the force which oscillates about a plateau value. This is called **unfused tetanus** (*Fig. 1b*). If the firing frequency is greater and reaches fusion frequency, the plateau is at maximum force, is smooth, and this is known as **fused tetanus** (*Fig. 1c*).

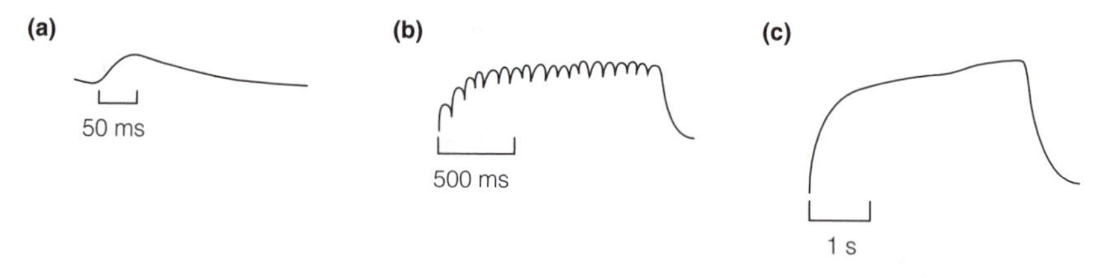

Fig. 1. Muscle fiber contraction: (a) single twitch; (b) unfused tetanus (firing frequency 12 Hz); (c) fused tetanus (30 Hz). Note the increase in force of contraction in going from (a) to (c).

Three types of motor unit can be distinguished by the firing behavior of their motor neurons and the properties of their muscle fibers.

The most numerous are the **slow twitch (S)** motor units which take about 50 msec to develop peak force and show little decline in force after even 1 h of repetitive stimulation. The motor neurons of S units are small, have a low conduction velocity and quite long refractory periods because they contain a high density of Ca^{2+}-activated K^+ channels which cause a long after-hyperpolarization. This limits maximum firing frequencies to quite low rates, but fused tetanus is achieved at low frequencies (15–20 Hz). The **type 1** muscle fibers of S motor units are rich in mitochondria, have high activities of Krebs cycle enzymes (which permits them to be selectively stained histologically) which fits them for high rates of aerobic metabolism. The type 1 muscle fibers are found particularly in the middle of muscles, which are the most vascular parts. S motor units are capable of exerting low force for very long times. They form the bulk of the antigravity or postural muscles of the trunk and legs, which are described as **red** muscles because of their high myoglobin content.

In contrast, FR and FF are fast twitch units which maximally contract in 5–10 msec. With repetitive stimuli, **fatigue resistant (FR)** units can sustain moderate force for 5 min or so before a steady decline sets in that takes many minutes. **Fast fatigue (FF)** motor units can achieve the greatest force of the three types, but with repetitive stimuli the force falls precipitously after 30 sec or so. The motor neurons of both FR and FF units are large with high conduction velocities. For brief periods they fire at high rates but action potential volleys are of short duration, particularly for FF units. Fast twitch units contain **type 2** muscle fibers which require firing frequencies of 40–60 Hz to produce fused tetanus.

Type 2 fibers come in two varieties that differ in their metabolism. The **type 2b** fibers of FF motor units are anerobic which explains why these units fatigue so quickly. **Type 2a** fibers, found in FR units, are intermediate between types 1 and 2b in terms of metabolism. Both FR and FF are adapted for producing rapid, large forces and so are found particularly in muscles involved in executing fast movements. Muscle in which fast twitch units predominate is called **white muscle** because its low myoglobin content makes it much paler than red muscle.

In motor units the properties of the muscle fibers and motor neurons are matched for optimal performance. This is brought about because muscle fiber properties are determined by the motor neurons which innervate them. If type 1 muscle fibers are denervated and the axon of an FF unit sprouts to establish new connections with the denervated fibers, they acquire the characteristics of type 2b muscle fibers. Athletic training may influence motor unit plasticity to produce a motor unit profile adapted to the nature of the sport.

Motor pools

Motor neurons that innervate the same muscle form a common **motor pool**. This pool is localized to motor nuclei of the brainstem and spinal cord. Spinal motor nuclei extend over several spinal segments. Axons of motor neurons leave the ventral horn of the spinal cord to run in the spinal nerve of the same spinal segment. Sorting of fibers destined for the same muscle but originating from different spinal segments occurs in the **nerve plexuses**. Axon collaterals of motor neurons ascend and descend a few segments to influence the behavior of other motor neurons in the same pool.

The force of contraction of a muscle is determined by the motor pool in two ways, the rate at which individual motor neurons fire and the number of motor neurons in the pool that are firing. Small increases in force are met mostly by increased firing rate, but larger contractions involve increasing the number of active motor units, a process called **recruitment**. This is done in an orderly manner. In general, the earliest units to be recruited are S, followed by FR and finally FF, an order determined by the **size principle**. Two effects are at work here. Firstly, the intrinsic cable properties of the motor neurons, secondly the organization of synaptic inputs onto them.

Cable properties are important because small cells offer a bigger resistance (known as the **input resistance**, R_i) to the flow of current than large ones. Ohm's law dictates that the relationship between the membrane voltage, V, and the current, I, flowing into a cell is given by;

$$V = IR_i$$

from which,

$$R_i = V/I$$

Hence a given current will produce a greater change in membrane voltage in a small cell (with large R_i) than it will in a large cell (with small R_i). Neurons within a motor pool are excited by common inputs. For a given sized synaptic current input into cells in the pool, the small cell body of an S motor neuron will have a larger excitatory postsynaptic potential than the larger cell body of a fast twitch unit, because the S cell has the greater input resistance (*Fig. 2*). This means that the weakest inputs recruit the S units, because they have the lowest threshold for synaptic activation. As the inputs to the pool get progressively stronger the other motor neurons are excited in turn.

The second effect determining the size principle is that the synaptic inputs to

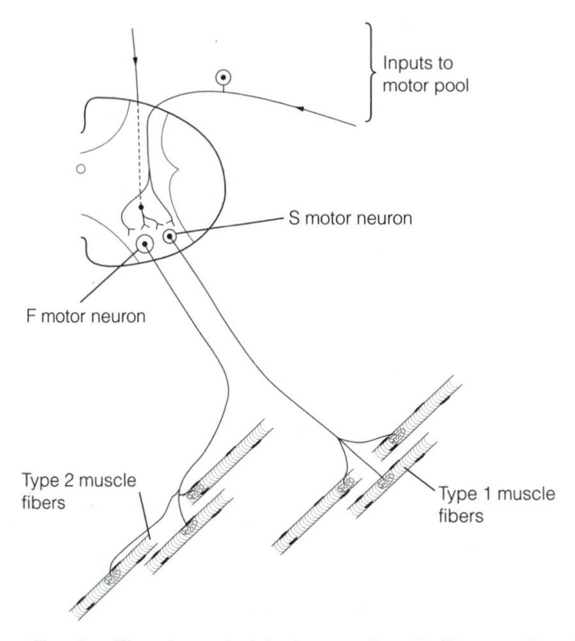

Fig. 2. The size principle in recruitment. The smaller slow twitch (S) motor neurons are recruited before fast twitch (F) motor neurons because they have a bigger excitatory postsynaptic potential in response to a given input.

the three classes of motor unit are partitioned in such a way that as input strength increases so motor units are recruited in the sequence S–FR–FF. Recruitment does not always obey the size principle. In some instances synaptic inputs are arranged so that large motor neurons get more excitation than small ones. Cutaneous afferents in humans, for example, preferentially excite high threshold fast twitch motor units and, again in humans, fast twitch units in intrinsic muscles of the hand are recruited first.

Motor unit disorders

Normally the endplate potential (epp) generated by motor neuron firing considerably exceeds the threshold for firing muscle fiber action potentials. The difference between the epp amplitude and the muscle firing threshold is the **safety margin** for nmj transmission. This is compromised in **myasthenic** diseases which arise either presynaptically, as a result of reduced ACh release, or postsynaptically due to defects in nAChRs or AChE. In **Lambert–Eaton myasthenic syndrome** autoantibodies are produced against voltage-dependent Ca^{2+} channels, greatly decreasing their numbers in the active zone and so reducing the probability of ACh release. **Myasthenia gravis** is an example of a postsynaptic disorder in which autoantibodies are directed against the nAChR. This causes enhanced internalization and degradation of the receptor so the muscle fiber becomes less responsive to ACh.

Muscular dystrophies are a group of diseases characterized by an increased turnover of muscle fibers. The most common is **Duchenne dystrophy**, an X-linked recessive disorder in which a large cytoskeletal protein, **dystrophin**, is abnormal or absent. Afflicted muscle fibers are weak and fatally damaged by normal mechanical forces. High rates of formation of new fibers from myocytes occur but they resemble fetal muscle fibers. They are small and do not propagate action potentials effectively. The number of motor units and their recruitment is normal in muscular dystrophies.

Peripheral nerve injury which severs motor neuron axons causes permanent **flaccid paralysis** and loss of stretch reflexes (see Topic K3) in the affected muscle, which consequently suffers disuse atrophy. The denervated muscle fibers synthesize large numbers of nAChRs which become inserted throughout the plasma membrane rather than being restricted to the endplate. These **extra-junctional receptors** cause muscle fibers to become exquisitely sensitive to ACh (**denervation supersensitivity**) and they exhibit minute contractions, fibrillations, to the low concentration of circulating ACh. Surgical nerve repair may effect recovery, if done with sufficient precision, since peripheral axons will regrow at a few millimeters per day to reinnervate muscle fibers.

K3 ELEMENTARY MOTOR REFLEXES

Key Notes

Properties of reflexes

Reflexes are stereotyped responses to sensory input mediated by reflex arcs, which generally include a sensory neuron, a motor neuron and one or more interneurons. Reflexes may be monosynaptic, disynaptic or polysynaptic depending on whether their circuits have one, two or more than two central synapses. The elapsed time between a stimulus and a response, the reflex latency, is determined by the time for conduction in the circuit and the number of synapses. Reflexes show facilitation, a response to multiple inputs greater than the sum of individual inputs, because connections of sensory neurons overlap on interneurons. Reflexes are modified by experience in a variety of ways such as habituation, sensitization and conditioning.

Muscle spindle reflexes

Muscle spindle (myotatic) reflexes are monosynaptic reflexes which cause a muscle to contract when it is stretched. They control muscle length by negative feedback. The sensory component of the reflex is the muscle spindle, which contains small intrafusal fibers that lie in parallel with the ordinary extrafusal fibers. There are two types of intrafusal fiber. Nuclear bag fibers signal muscle length, and rate of change of length (velocity) when the muscle is stretched. Nuclear chain fibers signal muscle length. Sensory neurons (types Ia and II) from the intrafusal fibers synapse with motor neurons supplying the same and synergistic muscles. A stretch reflex has a rapid component, which can be elicited by tapping the tendon of a muscle, followed by a slow component. In clinical practice the rapid component is called a tendon reflex. Intrafusal fibers receive input from γ fusimotor neurons. Contraction of the intrafusal fiber by γ fusimotor activity keeps it taut over all muscle lengths, so that it is kept sensitive to stretch whatever the muscle length. During a movement muscles shorten. To allow this, muscle spindle reflexes must be overridden. This is achieved by coactivation, the simultaneous firing of both the α motor neurons and γ fusimotor neurons so that both extrafusal and intrafusal fibers shorten together.

Inverse myotatic reflex

Inverse myotatic reflexes control muscle tension by negative feedback. Golgi tendon organs located in tendons measure muscle tension. Their Ib sensory neurons synapse, via dedicated Ib inhibitory neurons, with α motor neurons which go to the same and synergistic muscles. An increase in muscle tension causes inhibition of the α motor neurons, reducing the force of contraction of the muscle, and hence the tension.

Control of muscle stiffness

In many normal situations it is not possible to maintain a constant muscle length and constant muscle tension at the same time. Hence rather than controling muscle length and tension independently, central nervous system (CNS) motor systems probably control muscle stiffness.

Related topics	Intensity and time coding (F2) Skeletal muscles and excitation–contraction coupling (K1)	Motor units and motor pools (K2)

Properties of reflexes

The simplest operation the nervous system can execute is the **reflex** which couples sensory input to motor output. A reflex is a stereotyped response to a particular stimulus: when it involves the autonomic nervous system it is an **autonomic reflex**; when it occurs in the somatic nervous system it is a **motor reflex**. Reflexes are mediated by specified neural circuits, sometimes called **reflex arcs** (*Fig. 1*), which consist of a sensory neuron, a motor neuron and usually interneurons interposed between the two, which may be excitatory or inhibitory.

In humans, all but one reflex arc includes interneurons so they have several central synapses (three in *Fig. 1*) and are said to be **polysynaptic**. In the case that the reflex arc has only a single interneuron the reflex is **disynaptic**. The only example of a **monosynaptic** reflex, in which there is no interneuron, is the stretch reflex (see below).

A sensory neuron will form synapses with several interneurons (or motor neurons in the case of a monosynaptic reflex). Usually the effect of afferent firing is to produce quite large excitatory postsynaptic potentials (epsps) on a few neurons and more modest epsps in a larger group, depending on the number of boutons made. The connections made by several sensory neurons on interneurons overlap and this provides for the possibility of integration. **Facilitation** results if the number of action potentials generated by several inputs arriving at the same time exceeds the combined effect of the inputs applied separately. **Spatial facilitation** occurs if activation of several afferents causes spatial summation in the postsynaptic cells. Sometimes firing several afferents at the same time gives a reflex response smaller than the sum of individual afferent responses. This is **occlusion** and occurs when one input is sufficient to drive almost an entire group of neurons beyond threshold, so additional inputs to the same group produce little extra effect. **Temporal facilitation**,

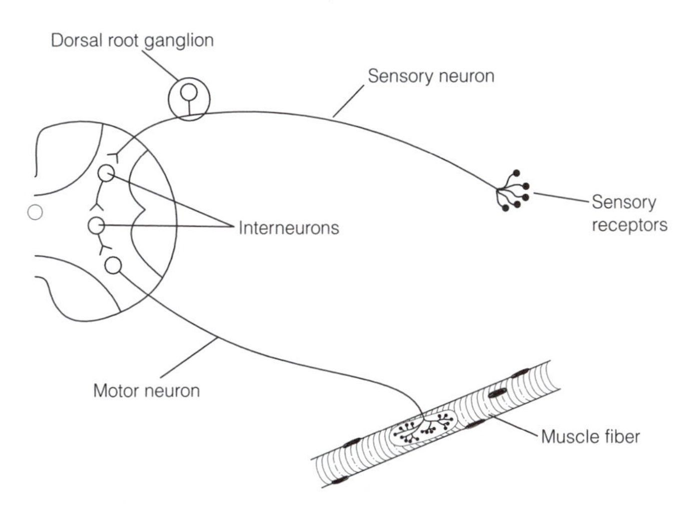

Fig. 1. A polysynaptic spinal motor reflex arc.

results when firing of afferents occurs with a high enough frequency to cause temporal summation.

The time between the stimulus and response is called the **reflex latency** or **reflex time**. It results chiefly from the conduction time along afferent and efferent fibers, but also includes the time taken for sensory transduction and for activation of the effector (excitation–contraction coupling or excitation–secretion coupling). A small interval is taken up by the **synaptic delay**, usually between 0.5 and 1 msec. All other things being equal, reflex latencies will reflect the number of central synapses.

Reflex time gets shorter as stimulus intensity increases, due to summation. An increase in afferent firing rate causes a larger depolarization of the interneuron. For suprathreshold stimuli the larger the depolarization (up to a limit) the shorter the time taken for the cell to fire. Increasing the stimulus intensity will change the amplitude of the reflex (e.g. the amount by which a limb moves) but may also alter the form of the response by recruiting additional muscles, this is called **irradiation**. The exact form of the reflex response depends on precisely which afferents are excited and so where the stimulus is applied. This is called the **local sign**.

Probably all reflexes are plastic, that is they can be modified by experience. The attenuation of a reflex by the repeated application of a constant innocuous stimulus is **habituation**. It is caused by synaptic depression. Any change to the stimulus (e.g. in its intensity) causes **dishabituation** in which the reflex returns to its baseline state. By contrast, repeated application of a noxious stimulus causes an enhancement of the reflex which may include a decrease in latency, increased amplitude and irradiation. This is known as **sensitization** and results (in invertebrates at least) from an increase in neurotransmitter release. Both habituation and sensitization are examples of nonassociative learning because only one stimulus is involved. Some reflexes are capable of the more complicated associative learning, in which a response occurs if two stimuli are paired in time. These are **conditioned reflexes**. The cellular mechanisms involved in plasticity of reflexes are reviewed in Topic Q2.

Muscle spindle reflexes

The most elementary modulation of motor unit output is made by sensory input from the **muscle spindles** which measure the length and rate of change of length (velocity) of the muscle. Any attempt to stretch the muscle rapidly, for example by suddenly loading it, is met by contraction. This is the **muscle spindle reflex** (**stretch reflex** or **myotatic reflex** are synonyms) and is a negative feedback mechanism which defends a constant muscle length in the face of external forces which act to perturb it. A stretch reflex can be elicited from any skeletal muscle by tapping it or its tendon. The resultant stretch causes the muscle to contract. The stretch reflex is most easily demonstrated by tapping the patellar ligament between its insertion into the tibia and the patella (kneecap), causing the contraction of the quadriceps femoris (the powerful group of extensor muscles on the front of the thigh). The basic circuit of this **knee-jerk reflex** is shown in *Fig. 2*.

The sensory side of the stretch reflex consists of the **muscle spindle** and its afferents. Muscle spindles lie in parallel with the standard **extrafusal fibers** so any force acting on the whole muscle acts in the same way on the spindle. Each muscle spindle is a fluid filled capsule of connective tissue, 4–10 mm long and 100 μm in diameter, containing about seven modified muscle fibers called **intrafusal fibers** (*Fig. 3*). Intrafusal fibers have contractile ends but their central

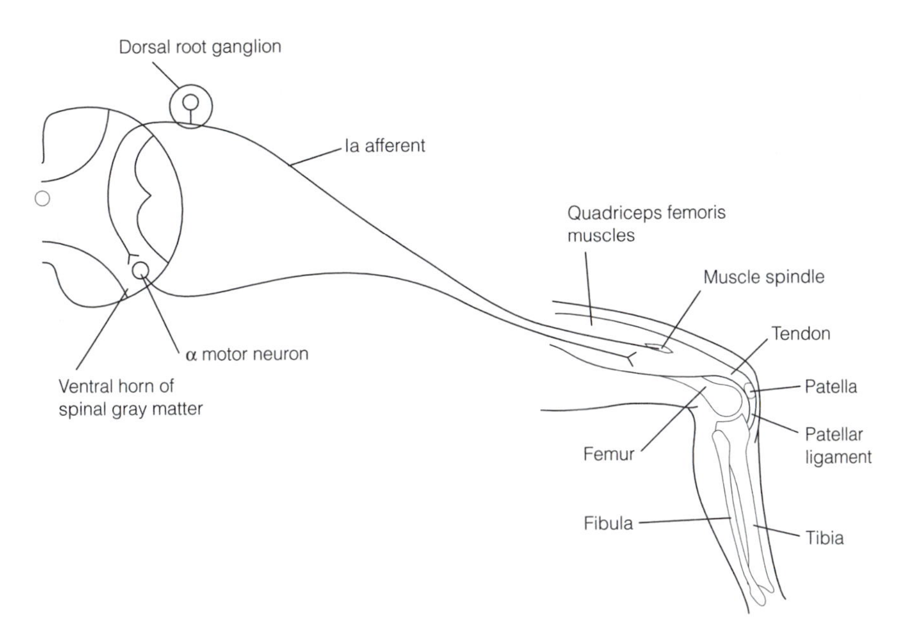

Fig. 2. Basic circuit of a stretch reflex. Striking the patellar ligament excites a few hundred Ia afferents.

regions are noncontractile. There are two types of intrafusal fiber, nuclear bag and nuclear chain.

Nuclear bag fibers (b) are swollen at their center, where the nuclei are clustered and are innervated by large diameter (~16 μm) myelinated (Ia) primary afferents, the ends of which spiral around the central region of the fiber. There are two types of nuclear bag fibers which can be recognized by whether, in addition to primary afferent innervation, they also receive secondary, group II (8 μm diameter) myelinated afferents. Those that do not are **dynamic (b_1)** those that do are **static (b_2)**. Primary afferents show dynamic responses (Topic F2) and so respond to the rate of change of length (velocity). This is because of the properties of the dynamic nuclear bag fiber. When stretched the central region elongates causing the Ia afferent to fire a volley of action potentials. Subsequently, however, the poles of the fiber elongate slowly, permitting the central region to creep back to a shorter length so that the firing rate of the Ia afferent decreases. Primary afferents also show static responses, signaling muscle length, by virtue of their innervation of static (b_2) nuclear bag fibers which are stiffer than the dynamic fibers and hence elongate in proportion to muscle stretch.

Nuclear chain fibers (c) are of uniform diameter, are about half the size of b fibers and their central region contains a line (chain) of nuclei. They are innervated by primary and secondary afferents and being stiff (like the b_2 fibers) these afferents respond to muscle length. Typically a spindle will contain one b_1, one b_2 and three to five c intrafusal fibers.

The majority of Ia spinal afferents form synapses on **homonymous** motor neurons (i.e. motor neurons going to the same muscle). However about 40% make synapses with motor neurons which go to synergistic muscles. For example, the quadriceps femoris consists of four muscles that act synergistically (they are all leg extensors). Afferents from spindles in one of them (e.g. the rectus femoris) will establish connections with the motor pool of the rectus

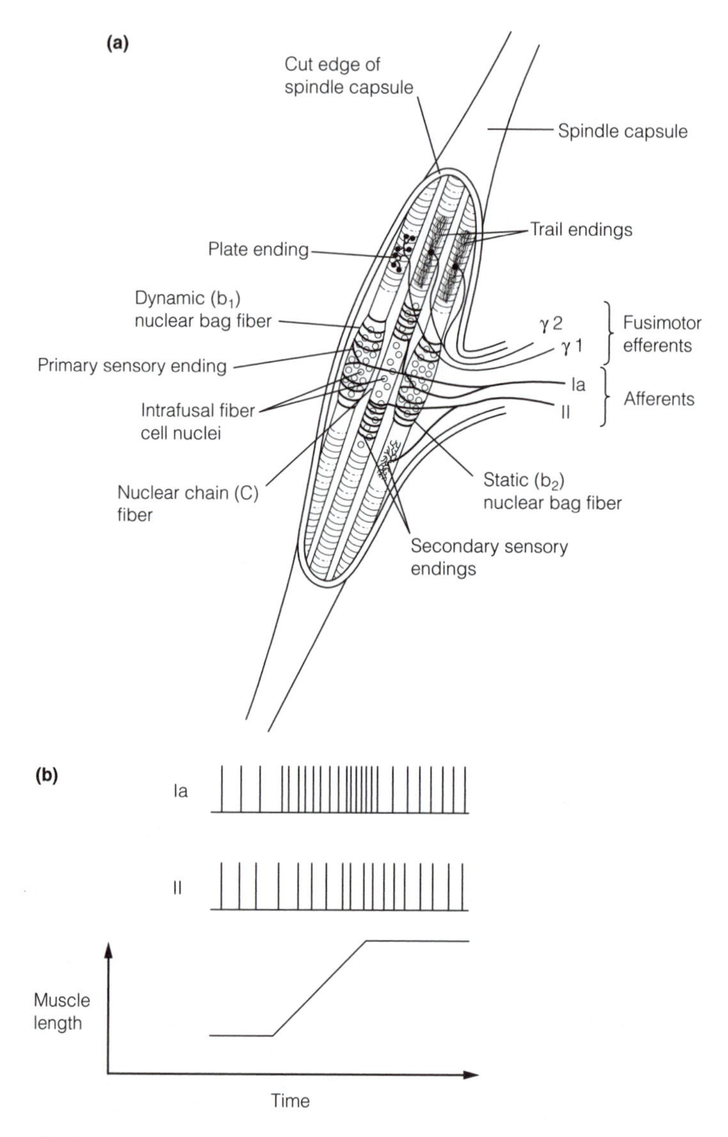

Fig. 3. Muscle spindles: (a) a spindle opened to show intrafusal fibers and their innervation. A spindle normally contains one b_1, one b_2 and several c fibers; (b) responses of Ia and II afferents to muscle stretch.

femoris, and the motor pools of the other extensors making up the quadriceps femoris.

A stretch reflex has two components. The **phasic** component is that seen by tapping the tendon of a muscle: it occurs rapidly, is brief and occurs because of the dynamic activity of the Ia afferents. The **tonic** component is the much more sustained contraction brought about by the static activity of the Ia afferents and the secondary group II afferents. This component is particularly important in maintaining posture. For standing in a moving vehicle, for example, the muscles in the legs and trunk that are stretched by the swaying will be contracted, so keeping the body upright. A sudden jolt will, of course, also trigger the phasic component.

Muscle spindles are innervated by motor neurons. The motor pool is bimodal in terms of cell size. The neurons which drive the extrafusal fibers to produce muscle contraction are Aα class (see *Table 2*, Topic E2) with a cell body diameter averaging 80 μm, usually referred to as α **motor neurons**. In addition, there is a population of smaller cells belonging to the Aβ and Aγ classes. The Aγ class, called γ **motor neurons** or **fusimotor** efferents send their axons to the muscle spindles and their influence has been studied extensively. All intrafusal fibers have their contractile ends innervated by γ motor neurons. Contraction of the ends of the intrafusal fibers keeps the central region taut so that it can respond to muscle stretch. So, one purpose of γ efferent discharge is to maintain the sensitivity of the muscle spindle to changes in length over a wide range of lengths. Without it, muscle contraction would cause the intrafusal fibers to slacken and fail to respond to stretch. This is referred to as the servo assist function of the γ efferents.

There are two categories of γ motor neuron: γ_1 (dynamic) innervate b_1 fibers forming **plate endings**; γ_2 (static) innervate b_2 and c fibers forming **trail endings** and they can be activated independently by CNS motor systems. Stimulation of γ_1 fibers increases the sensitivity of the b_1 fibers, so that the primary afferent firing rate in response to rapid stretch is higher. Stimulation of γ_2 fibers enhances firing of secondary afferents in response to constant stretch. In both cases the γ efferents are increasing the **gain** of the spindle. Firing rates of γ efferents are raised when performing movements that are particularly complex.

Stretch reflexes must be over-ridden to allow the execution of a movement since the muscle must contract isotonically and shorten. This is achieved by descending motor pathways exciting both α and γ motor neurons at the same time. This is called **coactivation**. It makes the extrafusal and intrafusal fibers shorten together in such a way that the intrafusal fibers are always sufficiently taut to respond to stretch.

Clinically, stretch reflexes are rather misleadingly termed **tendon reflexes**. A neurological examination includes eliciting stretch reflexes from several muscle groups throughout the body since abnormal or absent reflexes can reveal the level of any damage to the nervous system. An absent reflex may signify a lesion anywhere in the reflex arc: sensory or motor neuron or CNS. The stretch reflex can be studied by electrically stimulating the nerve which supplies a muscle and recording the electrical activity of the muscle with electrodes placed on the skin overlying the muscle or needle electrodes inserted into the muscle. Recording muscle activity in this way is **electromyography (EMG)**. This procedure is most easily done by stimulating the **tibial nerve** at the back of the knee and recording from the **gastrocnemius** and **soleus** (calf) muscles. Ia fibers have the lowest threshold of any nerve fibers, so low intensity stimulation of the nerve elicits the stretch reflex which can be seen as the H (Hoffman) wave on the EMG which occurs about 30 msec after the stimulation. This is the latency of the reflex. Progressively increasing the stimulus strength eventually results in exciting the α motor neurons, as well as the Ia afferent, with the appearance of the M (motor neuron) wave with a latency of only 5–10 ms. In the clinic this procedure can show whether an absent reflex is due to loss of sensory or motor function.

Inverse myotatic reflex

Located in the tendons, in series with muscle fibers, are **Golgi tendon organs (GTO)** which measure muscle tension. Increases in muscle tension activate a negative feedback reflex, the **inverse myotatic (Golgi tendon) reflex**, which opposes the increases in tension. It is brought about by GTO input activating inhibitory interneurons that synapse with α motor neurons supplying the muscle (*Fig. 4*).

The GTO is composed of the collagen fibers which join muscle fibers to tendons, interwoven through which are axon branches of a group Ib afferent neuron. The increased tension that occurs with muscle contraction stretches the collagen fibers, distorting the terminals of the Ib afferent which fires. Individual Ib afferents respond statically, reflecting the level of tension, to the activation of a single motor unit. GTOs do not measure the average tension of the muscle, only the tension developed by the muscle fibers inserted into its particular region of the tendon. Also fewer than 1% of fibers in motor units are coupled to GTOs. Because GTOs have no motor supply they cannot be adjusted during contraction. Ib afferents enter the spinal cord to synapse in the intermediate zone (laminae VI–VIII) on inhibitory neurons which then synapse with motor neurons of the homonymous and synergistic muscles. These inhibitory neurons are all specific to this disynaptic reflex pathway and so are designated **Ib inhibitory neurons (IbINs)**. Inhibition of homonymous and synergistic muscle by the inverse myotatic reflex is known as **autogenic** (self-generated) **inhibition**. Autogenic inhibition by GTOs is augmented by inputs from Ia spindle afferents, joint afferents and cutaneous mechanoreceptor afferents onto IbINs. The functional importance of these connections is not clear. Descending motor systems may either excite or inhibit IbINs.

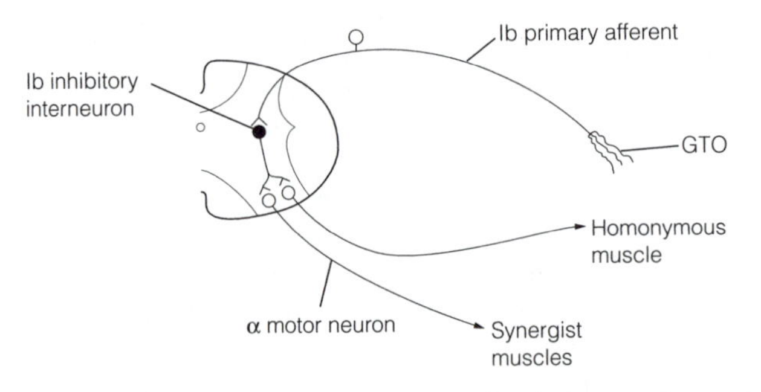

Fig. 4. Circuitry of the inverse myotatic reflex. An increase in muscle tension causes the Golgi tendon organ (GTO) Ib afferent to fire at a greater rate.

Control of muscle stiffness

The muscle spindle reflex for maintaining muscle length and the GTO reflex for keeping constant tension often work in opposition. When the loading on a muscle is altered either the muscle is stretched, or it must contract isometrically to maintain a constant length, in which case the muscle tension rises. It is impossible for both length and tension to be held constant at the same time. This suggests that overall, motor systems do not control length or tension independently, but muscle stiffness (see Topic K1). The advantage of this is that the organization of supraspinal motor systems is simpler since the nonlinearity of muscle stiffness over the normal working range of muscle can be compensated for by the operation of these reflexes.

K4 Spinal motor function

Key Notes

Elements of spinal cord motor function

Most neurons in the spinal cord are interneurons, which implies that considerable processing of sensory input takes place before it influences motor output. Much of the circuitry of the spinal cord is either organized for the execution of motor reflexes, or into central pattern generators, which generate the cycles of muscle activity involved in locomotion. Three types of inhibition are important in spinal cord function.

Reciprocal inhibition

Reciprocal inhibition occurs between mutually antagonistic muscle motor neurons. Muscle spindle Ia sensory neurons synapses with Ia inhibitory interneurons which supply the motor neurons of antagonist muscles. This circuit allows for relaxation of antagonist muscle while the agonist is contracting. Reciprocal inhibition is important in allowing movement that needs alternating activity in agonist and antagonist muscles.

Presynaptic inhibition

Primary afferents inhibit the terminals of other primary afferents via inhibitory γ-aminobutyrate (GABA)ergic interneurons that make axoaxonal synapses. The effect of GABA is to depolarize the terminal, inhibiting its release of transmitter in response to an invading action potential. Presynaptic inhibition is organized so that flexor afferents inhibit extensor afferents and *vice versa*; it is modified by descending motor systems.

Recurrent inhibition

Glycinergic Renshaw cells in the ventral horn activated by motor neurons produce inhibition of neighboring synergistic α motor neurons. This recurrent inhibition enables movements to be executed more crisply.

Flexor reflexes

Flexor reflex afferents (FRAs) elicit flexor reflexes in the ipsilateral limb, and other reflexes in the contralateral limb. The neural circuitry for these reflexes is recruited during normal limb movements, which are continuously modified by peripheral sensory input via FRAs. Flexor reflexes generated by nociceptor input enable a limb to be withdrawn from a noxious stimulus and override ongoing movement.

Central pattern generators

Locomotion – movement from place to place – involves alternate flexion and extension of limbs which are phased in specific ways depending on the manner of locomotion (e.g. walking or running). The basic rhythms of locomotion are produced by central pattern generators (CPGs), networks of spinal interneurons. Each CPG acts as an oscillator, driving a limb to flex then to extend alternately. CPG activity is regulated by a midbrain locomotor region which projects to the spinal cord via the reticulospinal tract.

Related topics

Fast neurotransmission (C2)
Elementary motor reflexes (K3)
Brainstem postural reflexes (K5)

Elements of spinal cord motor function

A dog lumbar spinal segment contains about 375 000 neuron cell bodies, the vast majority small (diameter less than 34 μm). Each dorsal root has about 12 000 sensory fibers, half of which are unmyelinated C fibers with diameters of 1 μm or less. All sensory fibers make connections in the grey matter of the spinal cord, and all send axon collaterals into pathways that ascend to supraspinal levels. Each ventral root has 6000 efferent fibers, two-thirds of which are α and β motor neurons, the remainder γ motor neurons. Hence most of the neurons in the spinal cord are interneurons which implies that massive processing of sensory input occurs before it influences motor neurons.

Two elements contribute to the organization of movement by the spinal cord, reflexes and central pattern generators (CPGs). A number of motor reflexes, including the myotatic and inverse myotatic reflexes (Topic K3) are centered on the spinal cord. Motor reflexes should not be thought of as independent but as elements which allow descending motor control of muscles to be continually modified on the basis of proprioceptor input from muscles and joints, and exteroreceptor input from skin. With the exception of a few protective reflexes such as the flexion withdrawal reflex, which serves to remove a limb from a noxious stimulus, motor reflexes are not seen in isolation under normal conditions but operate in concert to enable the smooth execution of movements.

Locomotion involves cycles of activity in which muscles groups are made to contract in a precisely timed sequence. This requires neural networks that can generate the required rhythmic output. These networks, thought to be autonomous, though modifiable by reflexes and activated by supraspinal influences, are called **central pattern generators** (**CPGs**). The presence of CPGs can be inferred by experiment in a wide variety of vertebrates including primates and humans, and putative circuits modeled using computer simulations, though the neurons that make up CPGs in mammals have not yet been identified.

Three types of inhibition contribute to spinal cord function, reciprocal, presynaptic and recurrent inhibition.

Reciprocal inhibition

Axon collaterals of Ia afferents from muscles spindles synapse with **Ia inhibitory interneurons** in lamina VII (IaINs) that use glycine as a transmitter and project to motor neurons of antagonist muscles. This disynaptic circuit allows antagonist muscles to be relaxed during agonist contraction (*Fig. 1*). The inhibition of mutually antagonistic muscle motor neurons is called **reciprocal inhibition**.

IaIN activity mediating reciprocal inhibition is modified by descending motor pathways (corticospinal, rubrospinal and vestibulospinal) and locomotor networks in the cord. Two reasons for this are:

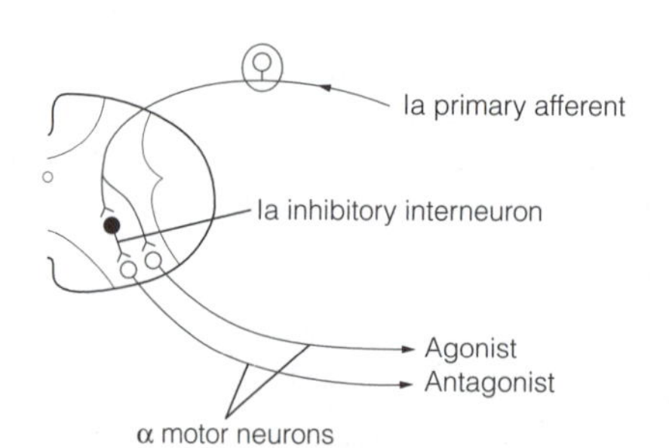

Fig. 1. A disynaptic reflex for reciprocal inhibition of antagonist muscle.

- To facilitate rapid movements. Because muscle contractions are so long lasting, muscles follow faithfully only slowly changing neural input. Rapid changes in motor neuron firing cannot be translated into corresponding alterations in muscles tension. Hence to produce fast fluctuations in tension the motor system alternates contraction in agonist and antagonist muscles. This is helped by reciprocal inhibition (*Fig. 2*).
- To adjust muscle stiffness so that it is appropriate for the load. The stiffness of a joint is increased by co-contraction of muscles with opposing actions at the joint. For example, co-contraction of the biceps and triceps increases stiffness of the elbow. Provided one muscle contracts more powerfully than the other the joint moves. Co-contraction stabilizes a joint, it provides better control when loads change unexpectedly because a given difference between expected and actual load will have a smaller effect on limb trajectory if a joint is stiffer. Co-contraction requires suppression of reciprocal inhibition.

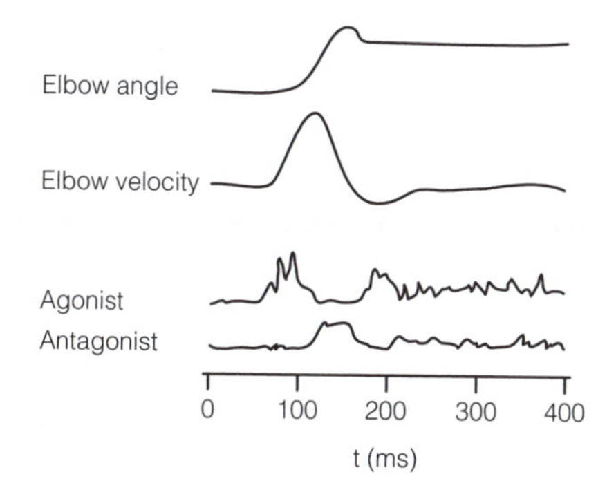

Fig. 2. Pattern of activation of agonist (biceps) and antagonist (triceps) muscles to facilitate rapid movement (elbow flexion).

Presynaptic inhibition

Reflexes dependent on sensory input via Ia, Ib and II afferents, can be modified by presynaptic inhibition from inhibitory GABAergic interneurons in the spinal cord. These interneurons make axoaxonal synapses on the afferent terminals. GABA secreted at these synapses acts on $GABA_A$ receptors to bring about depolarization because in these sensory neurons the membrane potential is more negative than the reversal potential of the chloride current through the $GABA_A$ receptor channels (see *Fig. 4*, Topic C2). The effect of this **primary afferent depolarization (PAD)** is inhibitory in that when an action potential sweeps into the terminal its amplitude gets smaller, less Ca^{2+} influx occurs, so transmitter release from the terminal is reduced.

Presynaptic inhibition of Ia terminals is organized on a reciprocal basis in which flexor afferents inhibit extensor afferents and *vice versa*. In humans this presynaptic inhibition causes a more prolonged reciprocal inhibition between antagonist muscles. It can be modified by descending motor pathways which project to one or other of the interneurons in the circuit. Normally, at the start of a movement presynaptic inhibition to Ia terminals going to agonist motor neurons is reduced, whereas inhibition of antagonist Ia terminals is increased. The effect of this is that spindle activity in the agonist reinforces its contraction whereas the antagonist muscle myotatic reflex is dampened.

Similar but separate circuitry is involved in presynaptic inhibition of Ib and II afferents, allowing control of tension and tonic length signals independently of phasic length signals.

Recurrent inhibition

A population of interneurons called **Renshaw cells** in the ventral horn are activated by axon collaterals of α motor neurons and project to neighboring, homonymous and synergistic α motor neurons. These cells fire high frequency bursts of action potentials that produce fast large inhibitory postsynaptic potentials (ipsps). The effect of the inhibition is to silence motor neurons excited weakly and firing at low rates, and to dampen the firing frequency of strongly excited motor neurons. This is a sort of lateral inhibition. It enhances contrast, and so enables economical movements. Renshaw cells are glycinergic and blockade of glycine receptors (which are ligand-gated ion channels selective for Cl^-, having some homology with $GABA_A$ receptors) with strychnine, results in convulsions due to the failure of recurrent inhibition.

Flexor reflexes

A variety of afferents, including group II and III muscle afferents, joint afferent and skin mechanoreceptor and nociceptor afferents elicit flexor reflexes in the ipsilateral limb and so are known as **flexor reflex afferents (FRAs)**. These same afferents trigger an extension of the contralateral limb, the **crossed extensor reflex** or alternative reflexes pathways. Different types of FRAs are connected to specific subsets of interneurons so the reflexes they excite differ in form and timing. Many normal limb movements consist of either flexion or alternate flexion and extension, and are brought about by the actions of supraspinal motor systems on the interneurons targeted by the flexor reflex afferents. In one model, to account for this, these interneurons are organized into sets called **half centers**, between which reciprocal inhibitory connections exist. A flexor half center gets inputs from ipsilateral FRAs and excites flexor motor neurons, while extensor half centers are activated by contralateral FRAs driving extensor motor neurons (*Fig. 3*) or alternative reflexes.

For the execution of a particular movement, descending motor axons are activated which project to the specific set of FRA interneurons that bring about the movement. The reciprocal connections ensure that the alternative set is inhibited. As the movement proceeds FRA input from muscles, joints and skin reinforces the movement and allows it to be fine tuned.

Although flexor reflexes are usually recruited as elements of normal movement, the withdrawal flexor reflex, triggered by Aδ or C fiber (group IV) nociceptors is quite distinct in character. It overrides ongoing movement, involves

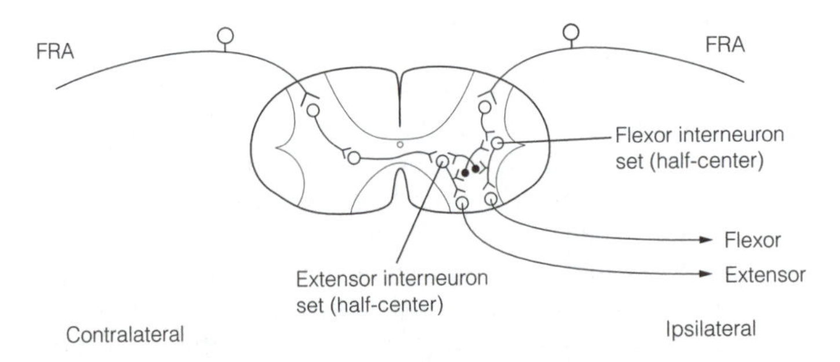

Fig. 3. Organization of flexor reflex afferent (FRA) interneurons. Excitatory neuron cell bodies are open circles, inhibitory interneuron cell bodies are filled circles.

flexor muscles throughout the limb so the response is dramatic, and it is long lasting. It serves a protective function.

Central pattern generators

Locomotion is movement from place to place. There are numerous modes of locomotion (e.g. stepping, swimming, flying) but all depend on cycles of muscle activity, alternate flexion and extension of each limb, though the **gait** or manner of locomotion (e.g. walking or running) depends on the speed. The gait adopted is the one that minimizes the energy expenditure for the desired speed.

In human stepping each leg has a **stance phase**, when extensors are the most active, and a **swing phase**, when the flexors are the most active. The same sequence in the opposite leg is out of phase, though during walking there is a brief overlap of the stance phase in both legs. As the speed increases the stance phase shortens until there is no overlap and the switch to running occurs.

The basic rhythms of locomotor activity are produced by **central pattern generators (CPGs)**, networks of interneurons in the spinal cord that generate the precisely timed sequences of α motor neuron activation without the need for sensory input.

Each limb has an array of CPGs. Each CPG is an oscillator with two half centers, one driving flexors, the other driving extensors, with reciprocal connections between them. Each half center produces rhythmic bursts of action potentials that are terminated in a time and manner determined by the intrinsic excitable properties of its constituent neurons. The cessation of firing of one half center releases its opposite number from reciprocal inhibition, allowing it to fire a burst; in this way burst firing alternates between the two half centers. Note that the interneuron sets responsible for controling alternate flexion and extension of limbs during movement (*Fig. 3* above) are half centers

The large oscillations occur because depolarization of the CPG cells activates their *N*-methyl-D-aspartate (NMDA) receptors causing a prolonged plateau depolarization which triggers rapid firing. Calcium influx through the NMDA receptor activates K_{Ca} channels, allowing K^+ efflux and hyperpolarization which terminates the burst. In addition, the NMDA depolarization excites neurons (L

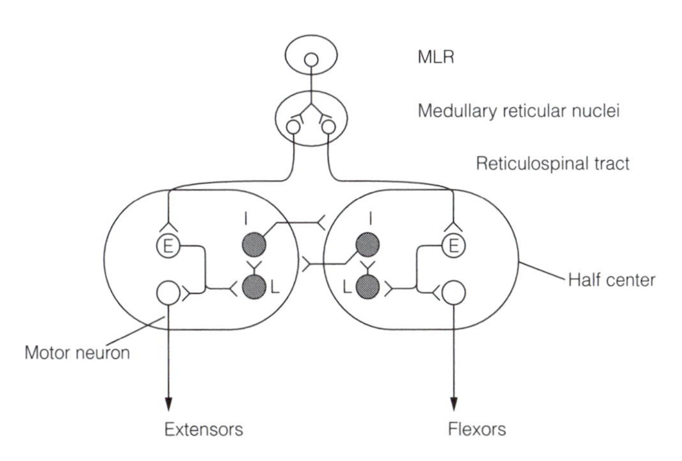

Fig. 4. A simplified hypothetical model of a central pattern generator (CPG) based on studies of the lamprey. Each of the neuron symbols represents several cells. The excitatory (E) cells use glutamate and show burst firing in response to supraspinal input. The inhibitory (filled symbol) neurons are glycinergic; contralaterally projecting interneuron (I) inhibits the opposite half center; L, lateral interneuron. MLR, mesencephalic locomotor region.

cells in *Fig. 4*) which inhibit the type 1 neurons responsible for the reciprocal inhibition. This disinhibition allows the opposite half center to depolarize, driving it to burst. CPGs have been studied in the lamprey, a primitive vertebrate, and it is likely that mammalian CPGs use similar principles. Locomotor activity is initiated by activity in the **mesencephalic locomotor region** (MLR) which projects to reticular nuclei in the medulla (*Fig. 4*). Axons from here run in the **reticulospinal** tracts to the spinal cord (see also Topic K5). These reticular nuclei are excitatory, releasing glutamate to produce a large depolarization of the CPG neurons which then produce oscillating output for as long as the MLR input continues. CPGs are interconnected so that timing of events in all limbs is coordinated. The basic locomotor rhythms of the CPGs are extensively modified by the supraspinal motor systems.

K5 Brainstem postural reflexes

Key Notes

Postural reflexes

Postural reflexes stabilize the body against forces which shift the center of mass, including limb movements. Posture is maintained largely by the action of antigravity muscles which include extensors of the back and legs (and in humans, arm flexors). Postural reflexes are organized by the brainstem in response to vestibular, proprioceptor and visual input. The precise form of postural adjustments depends on the context, such as body position and the size and direction of the unbalancing force.

Vestibular (labyrinthine) reflexes

Vestibular input is used to maintain the orientation of the head in space in the face of tilting or rotation of the head and body as a unit. These vestibular reflexes achieve this by acting on neck muscles (vestibulocollic reflexes) or limb muscles (vestibulospinal reflexes).

Neck reflexes

When the head moves with respect to the body, neck muscle stretch produces reflex contraction of neck muscles (cervicollic reflexes) and limb muscles (cervicospinal reflexes). Cervicocollic and vestibulocollic reflexes act synergistically, cervicospinal and vestibulospinal reflexes are antagonistic in some contexts and synergistic in others.

Righting reflexes

Reflexes that allow an animal to resume a normal posture are righting reflexes. They include vestibular and neck reflexes, and other brainstem reflexes that require visual or somatosensory input. These reflexes allow a cat (for example) to land on all four paws after falling from a height.

Limb placing reactions

Responses in which an animal places its feet so as to hold a stable posture – limb placing reactions – require visual or somatosensory input and involve the cerebral cortex.

Postural reflex pathways

The vestibular and reticular nuclei integrate vestibular and proprioceptor (e.g. muscle spindle) input and execute postural reflexes by way of the vestibulospinal and reticulospinal tracts of the medial motor system. Since the reticular nuclei also get input from the premotor cortex they are able to modify postural reflexes (and spinal locomotor rhythms) in the light of locomotor demands. Via their connections with the cerebellum and cerebral cortex the medial motor pathways also allow adjustments to be made to posture in advance of planned voluntary movements.

Related topics

Sense of balance (G4)
Spinal motor function (K4)
Cortical control of voluntary movement (K6)

Postural reflexes

The purpose of postural reflexes is to prevent the body from being destabilized by forces, including gravity, which tend to shift the center of mass. Postural reflexes also help to maintain the center of mass during limb movements. Muscles either oppose or assist gravity when contracting; those that oppose are described as **antigravity muscles**. Many antigravity muscles, such as the leg extensors and the short deep extensor muscles of the back (axial muscles), are involved in maintaining posture. In humans, the flexor muscles of the arms are also antigravity muscles. Since antigravity muscles are generally more powerful than muscles assisted by gravity in human limbs the strongest muscles are the leg extensors and arm flexors.

Postural reflexes are organized by the brainstem. The sensory input to the reflex circuitry is from three sources:

- Vestibular, from the otolith organs.
- Proprioceptor, from muscle spindles, Golgi tendon organs and joint receptors.
- Visual, from the superior colliculus.

These inputs are highly integrated, so as to recruit the required sequence of muscle contractions which compensate for unpredicted disturbances in body position and movement. The neural mechanisms at work here involve negative feedback (*Fig. 1*).

The exact nature of postural adjustments made by humans depends on context, that is the initial position of the body and the size and direction of the destabilizing force. Swaying produced by sudden displacement of the surface on which a person is standing will activate quite different sets of muscles depending on the direction of sway, but in general distal muscles are excited before proximal ones, with most movement occurring at the ankle joint. Rotation or tilt of the surface, however, results in bending at the hips.

Several distinct postural reflexes can be seen in animals by surgically transecting the brainstem (**decerebration**) and in humans who have suffered severe brain damage. These reflexes cannot easily be elicited in isolation in healthy behaving humans because motor functions are normally so highly integrated. They may be seen in newborn infants, in whom motor systems are immature.

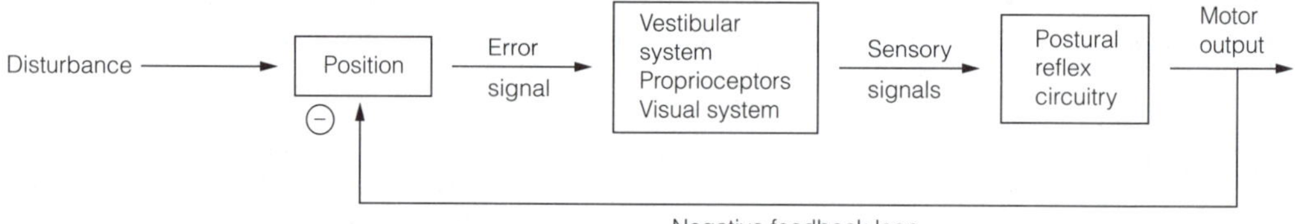

Fig. 1. Postural reflex negative feedback. The circuitry produces a motor output which reduces the mismatch between the desired position of the body and its actual position. The error is detected by the sensory systems which feed information into the reflex circuitry.

Vestibular (labyrinthine) reflexes

Vestibular (labyrinthine) reflexes stabilize the orientation of the head in space. Any tilting or rotation of the head and body as a unit activates motor neurons to muscles that maintain the head vertical with respect to gravity. These mainly tonic reflexes are driven by input from the otolith organs and semicircular ducts and have a latency of about 40–200 ms. Their output is to neck and limb muscles.

Vestibulocollic reflexes act on the neck muscles to keep the head upright. If the body sways forward the neck extensors contract bringing the head up. If the body sways backwards, neck flexors are activated. Vestibulocollic reflexes operate effectively over a wide frequency range of head oscillations (0.025–5 Hz), and act to compensate for the mechanical properties of the load represented by the head (e.g. its inertia) so that the reflex is the right amplitude and occurs with the correct timing. How the neural circuitry accomplishes this is not known.

Vestibulospinal reflexes act on limb muscles. They trigger contraction of arm extensor muscles and leg flexor muscles when falling, to reduce the impact of landing. Swaying sideways triggers extension of the ipsilateral limbs to brace against further tilt in that direction. Otolith organ signals are important in low frequency swaying but above 1 Hz signals from semicircular ducts become more important.

Neck reflexes

Turning the head relative to the body excites spindles in neck muscles and afferents from the cervical vertebral joints which evoke reflex contractions of neck muscles (**cervicocollic reflexes**) and limb muscles (**cervicospinal reflexes**). Cervicocollic reflexes contract neck muscles that are stretched and so act to reorientate the head on the body. Cervicocollic and vestibulocollic reflexes are synergistic. Cervicospinal reflexes (sometimes referred to as **tonic neck reflexes**) cause contraction of limb muscles in response to head movement which are sometimes antagonistic to vestibulospinal reflexes. In standing humans a force which throws the head backwards on the trunk activates all the limb extensors, whereas a force throwing the head forwards activates all limb flexors. Tilting the head to the side causes ipsilateral limb extension and contralateral limb flexion, like the vestibulospinal reflex. The purpose of these reflexes is to maintain the center of gravity to prevent falling or to put the limbs in a position to brace against falling. In quadrupeds, such as the cat, postural reflexes are organized a little differently. For example, the cervicospinal reflex in response to raising the head is an increase in forelimb extension tone, but a decrease in hindlimb extensor tone, and the response to sideways roll of the head is ipsilateral forelimb flexion (the opposite of the human vestibulospinal response).

Righting reflexes

If an animal is placed in an abnormal position it rapidly rights itself to assume a normal posture. The reflexes that enable this are called **righting reflexes** and include vestibular and neck righting reflexes. In addition are **optical righting reflexes** in which input from the visual cortex to the superior colliculus causes head rotation, and **body righting reflexes**, mediated by the brainstem, in which mechanoreceptor input from the flank of an animal lying on its side causes the head to be raised upright. Righting reflexes can be rapid. A cat falling from an upside down position uses a succession of optical, vestibular and neck righting reflexes to turn over in mid-air and land on all four feet in only 150 ms.

Limb placing reactions

Responses that allow an animal to place its feet appropriately to maintain a stable posture, called **limb placing reactions**, are part of the postural repertoire and although brainstem circuitry is involved in their execution they are organized primarily by the cerebral cortex.

In the **visual placing reaction** the feet are moved to a visible surface, a response requiring the visual cortex, and in **tactile placing reactions** cutaneous somatosensory input from head, whiskers, or the front of the foot touching a surface, will cause the feet to be properly placed, and limbs extended to support the animal. If the body of a standing animal is pushed horizontally, shifting its center of mass it shows a **hopping reaction** in which a leg is rapidly lifted, moved in the direction of the displacement and replaced to restore stability. The stimulus for the hopping reaction is muscle stretch, however, it is not mediated via spinal myotatic reflexes but by the motor cortex.

Postural reflex pathways

Supraspinal descending control of movement involves two sets of pathways. Those mediating postural reflexes, which go from brainstem to spinal cord, are collectively called the **medial motor pathways** to distinguish them from the **lateral motor pathways** (see Topic K6) for making voluntary movements.

The medial motor system has inputs from the cerebellum and the cerebral cortex via which postural reflexes are extensively modified. This allows anticipatory adjustments to posture to be made before execution of voluntary movements. This is feed forward control of posture in which cerebellar and cerebral cortices generate a **postural set**, the preparatory state required to maintain stability during a movement, before it is initiated.

The medial motor system involved in postural reflexes comprises the vestibulospinal (*Fig. 2*) and reticulospinal tracts (*Fig. 3*) which terminate in the intermediate and ventral gray matter of the spinal cord. Input from the vestibular labyrinth to neck muscle motor neurons is relayed by neurons in the **medial and inferior vestibular nuclei**, which give rise to the bilateral **medial vestibulospinal tract**. Both excitatory and inhibitory neurons are found in this tract, many making monosynaptic connections with neck muscle motor neurons. In general, ipsilateral motor neurons are excited whilst contralateral ones are inhibited. The pathway mediates some vestibulocollic reflexes.

Vestibular afferents to the **lateral vestibular (Deiter's) nucleus** are implicated in the control of limb muscles. The lateral vestibular nucleus projects via the

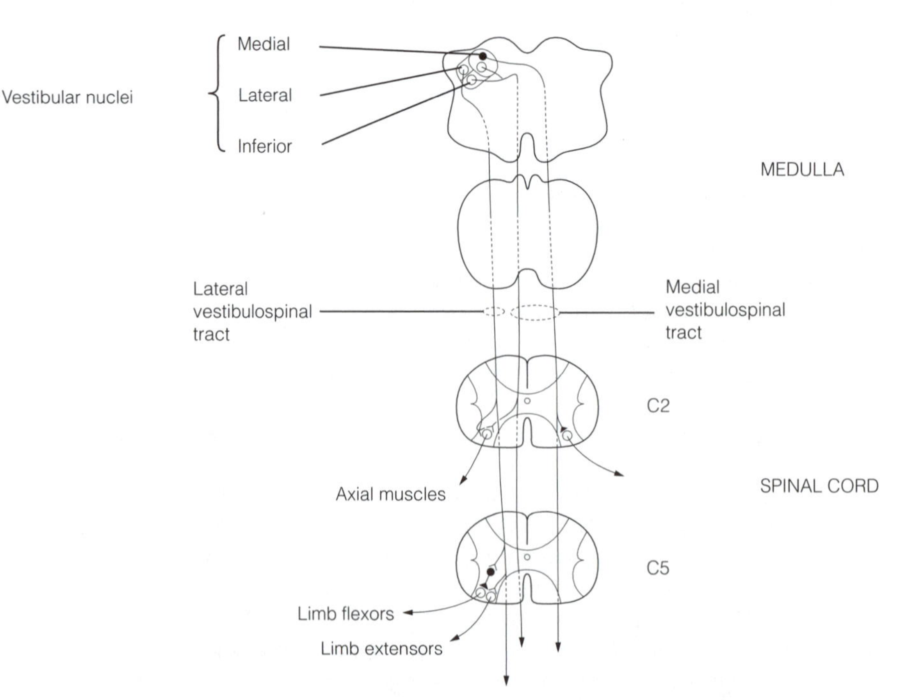

Fig. 2. The vestibulospinal tracts. Filled circles and triangles represent the cell bodies and axon terminals, respectively, of inhibitory neurons.

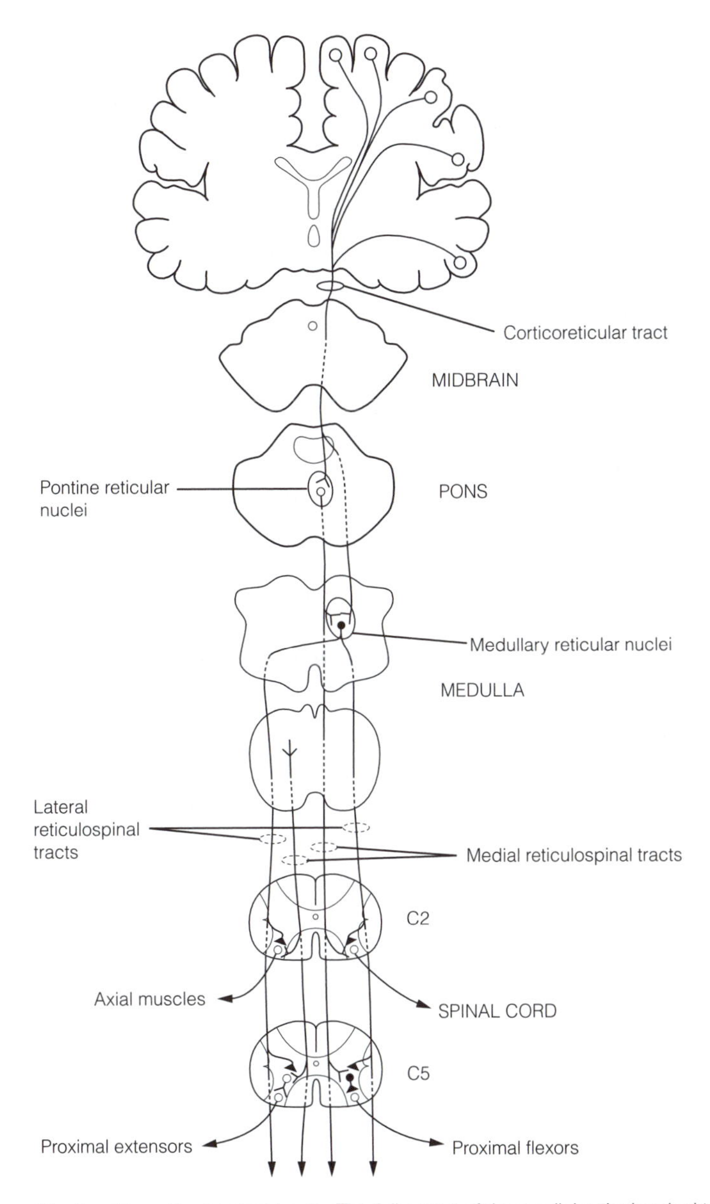

Fig. 3. The reticulospinal tracts. The full extent of the medial reticulospinal tracts is shown on one side only. The action of the tracts on extensors and flexors is shown on opposite sides for clarity. Filled circles and triangles are the cell bodies and axon terminals, respectively, of inhibitory neurons.

uncrossed **lateral vestibulospinal tract** to all segments of the spinal cord. Neurons in this pathway facilitate extensor motor neurons and inhibit flexor motor neurons. This pathway is responsible for some vestibulospinal reflexes. Input from neck muscle spindles and vertebral joint receptors is relayed via collaterals of proprioceptor afferents to vestibular nuclei which thus integrate vestibular and proprioceptor information.

There are two reticulospinal tracts. The **medial reticulospinal tract** originates from the **pontine reticular nuclei** and is ipsilateral. Its neurons terminate on and facilitate axial and limb extensor motor neurons but are inhibitory via polysynaptic pathways to limb flexors. The **medullary reticular nuclei** give rise to the **lateral reticulospinal tract** which is bilateral, and produces monosynaptic inhibition of neck and axial motor neurons and polysynaptic inhibition of proximal limb extensors, but excitation of proximal limb flexors. The medullary reticular nuclei get input from the mesencephalic locomotor region and project to the interneurons that form the central pattern generators.

The reticular nuclei receive both proprioceptor input from muscle spindles and vertebral joint receptors, and vestibular input, so reticulospinal tracts service both neck and vestibular reflexes. In fact, otolith signals for head forward and back pitching components of the vestibulocollic reflexes are not transmitted through vestibulospinal tracts but probably involve the reticulospinal tracts. In addition the reticular nuclei receive input from the premotor cortex which allows postural reflexes and central pattern generators output to be modified when the circumstances demand (e.g., when a running cat scales a wall). After lesions of the medullary reticular nuclei in cats, anticipatory adjustments are compromised and the animals lose balance temporarily when attempting to move a forelimb.

Table 1 summarizes the effects of the descending motor pathways on motor neurons.

Table 1. A summary of the major features of the descending motor pathways[a]

Motor system	Tract	Distribution	Principal effects on motor neurons	
			Excitatory to:	*Inhibitory to:*
Medial	Lateral vestibulospinal	Ipsilateral	Axial and proximal limb extensors	Axial and proximal limb flexors
	Medial vestibulospinal	Bilateral	Axial ipsilateral	Axial contralateral
	Pontine (medial) reticulospinal	Ipsilateral	Axial and proximal limb extensors	Proximal limb flexors
	Medullary (lateral) reticulospinal	Bilateral	Proximal limb flexors	Axial and proximal limb extensors
Lateral	Corticospinal	Largely contralateral	Distal limb flexors	Distal limb extensors
	Rubrospinal	Bilateral	Distal limb flexors	Distal limb extensors

[a] The lateral motor system is described in Topic K6.

K6 CORTICAL CONTROL OF VOLUNTARY MOVEMENT

Key Notes

Intentional movement

Voluntary movements are those made intentionally. They will normally be guided by perceptual (e.g., visual) input. Intentional movements are planned and then executed by output of motor commands which specify the correct sequence of muscle activation. Sensory feedback can be used to optimize performance.

Lateral motor pathways

The motor cortex gives rise to two lateral pathways, the corticospinal (pyramidal) tract and the corticorubrospinal tract, which goes to the red nucleus and then the spinal cord, which allow for the execution of intentional movement. The corticospinal tract is made up of axons of pyramidal cells in layer V of the cortex which descend through the internal capsule into the brainstem where most cross the midline, giving rise either to corticonuclear fibers that go to the motor nuclei of cranial nerves or to the lateral corticospinal tract. Corticospinal tract neurons are excitatory, make monosynaptic connections with α motor neurons to distal limb muscles and polysynaptic connections to α motor neurons to proximal and axial muscles. Fusimotor (γ) motor neurons are coactivated polysynaptically. The corticospinal tract is commonly excitatory to flexors and inhibitory (via Ia inhibitory interneurons) to extensors. The rubrospinal tract from the red nucleus is adjacent to the corticospinal tract, but is much less prominent in humans than in other mammals.

Motor cortex

The motor cortex is divided into three reciprocally interconnected areas, the primary motor cortex (MI), and the supplementary motor area (SMA) and premotor area (PM) which together constitute the secondary motor cortex (MII). The SMA forms a motor loop with the basal ganglia, while MI forms a motor loop with the cerebellum. These loops allow the cortex to recruit and coordinate specific motor programs. The motor cortex has several somatotopic maps. The map in MI has much more cortical space devoted to regions such as the hands and face where the variety and complexity of movements is greatest. MI receives a somatotopic projection from the somatosensory cortex. Part of this input is from muscle spindles and is the sensory side to a cortical loop of the stretch reflex. Neurons in MI may correlate with a wide variety of movement parameters (e.g. force, velocity, direction) and firing of an individual cell usually relates to more than one parameter. Direction of limb movement is not encoded by the behavior of a single cell, but by the average firing of a population of neurons that are distributed over the cortex.

The secondary motor cortex is involved in planning movements and its neurons may fire hundreds of milliseconds before a movement begins. The SMA has a bilateral somatotopic map and is crucial for performing complicated tasks involving both sides of the body such as two-handed tasks. The premotor area is particularly concerned with planning movements that require sensory cues.

Red nucleus

The red nucleus has a motor map, firing of rubrospinal tract neurons correlates with motor parameters in a similar manner to corticospinal tract neurons, and they are distributed to motor neuron pools in a comparable fashion. This implies that the rubrospinal and corticospinal tracts are very similar. However, the rubrospinal tract may operate when learnt, automated movements are being executed, while the corticospinal tract is active when new motor tasks are being acquired. A pathway involving the cerebellum may control which is operating depending on whether errors in performance are made.

Related topics

Elementary motor reflexes (K3)
Cerebellar function (L4)
Anatomy of the basal ganglia (L5)
Basal ganglia function (L6)

Intentional movement

Voluntary movements are those made intentionally to achieve a particular aim or reach for a target. Making an intentional movement involves several interdependent processes. Perceptual input may evoke and guide it. For example, an object to be grasped may be located visually. Many neurons in the motor system fire hundreds of milliseconds before any muscle contraction occurs showing that movements are planned. This is necessary because the same motor tasks can be performed in various ways depending on the context, a property called **motor equivalence**. For example, driving a large truck needs a different motor strategy than driving a small car. Furthermore, planning is required to organize precisely which set of postural adjustments should be recruited at the start and during the movement. The movement is executed by the output of **motor commands** which specify the correct temporal sequence of muscle activation. Finally, sensory feedback during the movement, particularly from proprioceptors such as muscle spindles and Golgi tendon organs, is used to fine tune the execution to ensure that the performance matches the desired goal. The planning of voluntary movements and the elaboration of motor commands for their execution is done by the motor cortex which has its outputs via the lateral motor pathways.

Lateral motor pathways

There are two lateral pathways for descending control of voluntary movement. Both originate in the motor cortex which lies on the frontal lobe just anterior to the central sulcus. The **corticospinal tract** consists of the axons of about 1 000 000 pyramidal cells in layer V of the cortex. Over half come from the **primary motor cortex (M1)**, Brodmann area 4, and the **supplementary motor area (SMA)** or **premotor area (PM)** in Brodmann area 6. These project to the ventral horns of the spinal cord. About 40% of corticospinal tract axons come from the somatosensory cortex (Brodmann areas 1, 2 and 3) or other regions of parietal cortex (Brodmann areas 5 and 7). These axons from the parietal lobe terminate in the dorsal horns of the spinal cord and regulate sensory input. Most of the corticospinal tract consists of fine myelinated and unmyelinated axons with conduction velocities between 1 and 25 m s^{-1}. However, there are about 30 000 extremely large (20–80 μm diameter) pyramidal cells in area 4, called **Betz cells,** with large myelinated axons that conduct with velocities of 60–120 m s^{-1}. Axons of the corticospinal tract pack tightly to pass through the internal capsule which lies between the thalamus and the lentiform nucleus (*Fig. 1*) and descend into

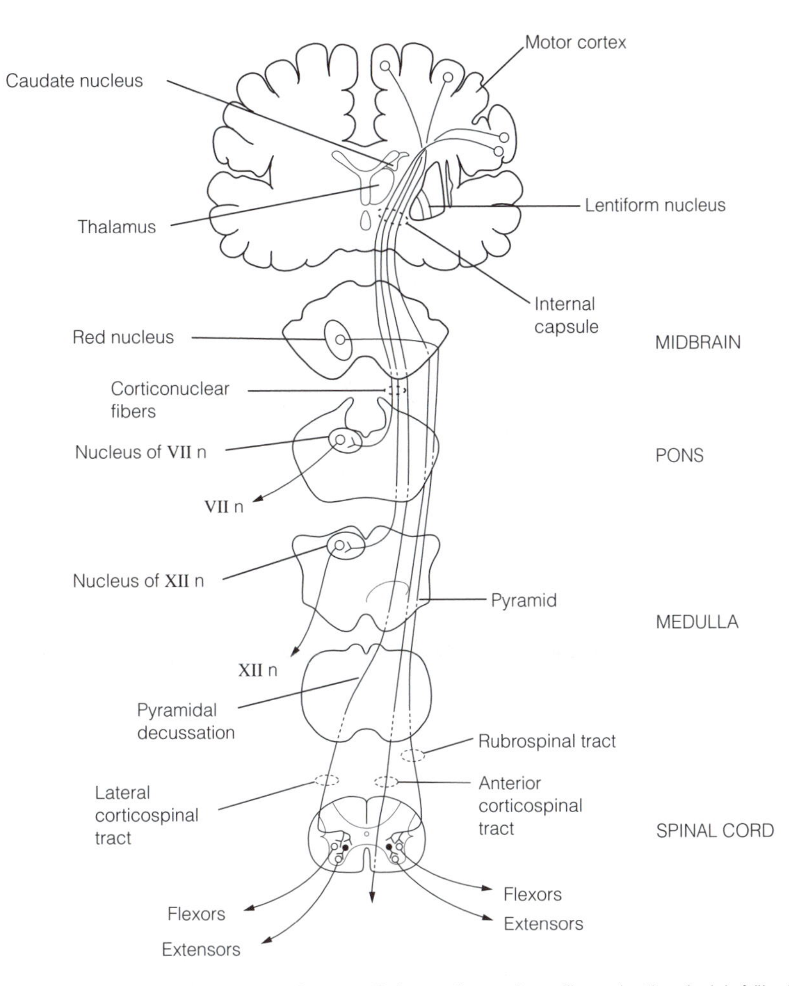

Fig. 1. The lateral motor pathways. Only corticonuclear fibers in the facial (VII n) and hypoglossal (XII n) nerves are shown. Synapses of the anterior (ipsilateral) corticospinal tract with spinal cord interneurons are not shown.

the brainstem. Here the most medial fibers peel off and cross the midline to go to nuclei [trigeminal (V), facial (VII), hypoglossal (XII) and accessory (XI)] of the cranial nerves. These are **corticonuclear (corticobulbar)** fibers and are motor to the face, tongue, pharynx, larynx and to the sternomastoid and trapezius muscles. The remaining axons descend through the medulla causing a swelling on its ventral surface, the **pyramid**, and for this reason the corticospinal tract is often referred to as the **pyramidal tract** (the term is unrelated to its derivation from pyramidal cells). At the caudal medulla 85% of fibers cross the midline as the **pyramidal decussation**, giving rise to the **lateral corticospinal tract**. The remaining ipsilateral axons form the **anterior corticospinal tract**.

Corticospinal tract neurons use glutamate as a transmitter and are excitatory. They either synapse directly with α motor neurons supplying distal limb muscles in Rexed lamina IX, or synapse with interneurons in laminae VII and VIII which make polysynaptic connections with α motor neurons of proximal limb muscles and axial muscles. Fusimotor (γ) neurons that must be coactivated with α motor neurons to override the myotatic reflex (see Topic K2) during voluntary movement are excited polysynaptically. Stimulation of the corticospinal tract is predominantly excitatory to flexors but inhibitory to extensors. The corticospinal tract inhibits motor neurons disynaptically via Ia inhibitory interneurons.

The corticospinal tract axons arising from the somatosensory cortex project to cranial nerve sensory nuclei and dorsal horns and produce presynaptic inhibition on primary afferent terminals, except those of Ia spindle afferents.

Some pyramidal cells in layer V of the motor cortex send their axons in the **corticorubral tract** to the magnocellular part of the **red nucleus** in the midbrain, which also receives collaterals from the corticospinal tract. The magnocellular part of the red nucleus gives rise to the **rubrospinal tract** which is the second of the lateral motor pathways. Some axons in this tract constitute the rubrobulbar fibers and go to cranial nerve nuclei in the pons and medulla. The tract descends as far as the lumbar cord in the macaque monkey and terminates in the dorsolateral gray matter.

Motor cortex

The motor cortex is subdivided into three areas on the basis of cytoarchitecture, connectivity and function. The primary motor cortex, M1, is **agranular cortex**, in that layer IV which receives inputs from the thalamus is very sparse, whereas layer V is well developed and contains numerous large pyramidal cells. Brodmann area 6, in which layer IV is rather better developed, contains the SMA and PM areas. The motor areas are reciprocally connected with each other (*Fig. 2*). They are also connected with subcortical structures which send inputs back to the motor cortex via the thalamus forming closed motor loops. Reciprocal back projections from the motor cortex to the thalamus also exist. These are not shown on *Fig. 2*.

The supplementary motor area is part of a motor loop with the basal ganglia. It sends output to the striatum which projects back to the SMA via its connections with the globus pallidus and ventrolateral (VL_O) thalamus. Many corticospinal tract axons from MI either terminate in the pons or give off collaterals there. These make synapses with pontine neurons which project to the cerebellum. Outputs from the cerebellum to the thalamus, which in turn projects back to MI and PM motor areas, form motor loops. These motor loops with basal ganglia and cerebellum are required for the initiation of specific motor patterns and their coordination, and their role in voluntary movement is dealt with in Section L.

Several somatotopic maps exist in the motor cortex (*Fig. 3*). Their topography is preserved in the orderly arrangement of the fibers in the corticospinal tract; axons innervating leg motor neurons are lateral, those going to the arm motor neurons being most medial. Like somatosensory maps, these motor maps are grossly distorted. Much more cortical space is devoted to the face, tongue and hands than other regions because the variety and precision of movements executed by them is so much greater. MI receives a substantial input from the somatosensory cortex and many MI neurons have sensory receptive fields. MI neuron receptive fields (RFs) are located in the position where activity of the neuron is likely to cause movement. That is, MI neurons are wired to be responsive to the sensory consequences of their actions. Those in caudal MI respond mostly to cutaneous mechanoreceptor input, those more rostral respond to proprioceptor input, particularly from muscle spindles. This contributes to a cortical loop (**long loop**) of the myotatic reflex circuitry. This long loop conveys

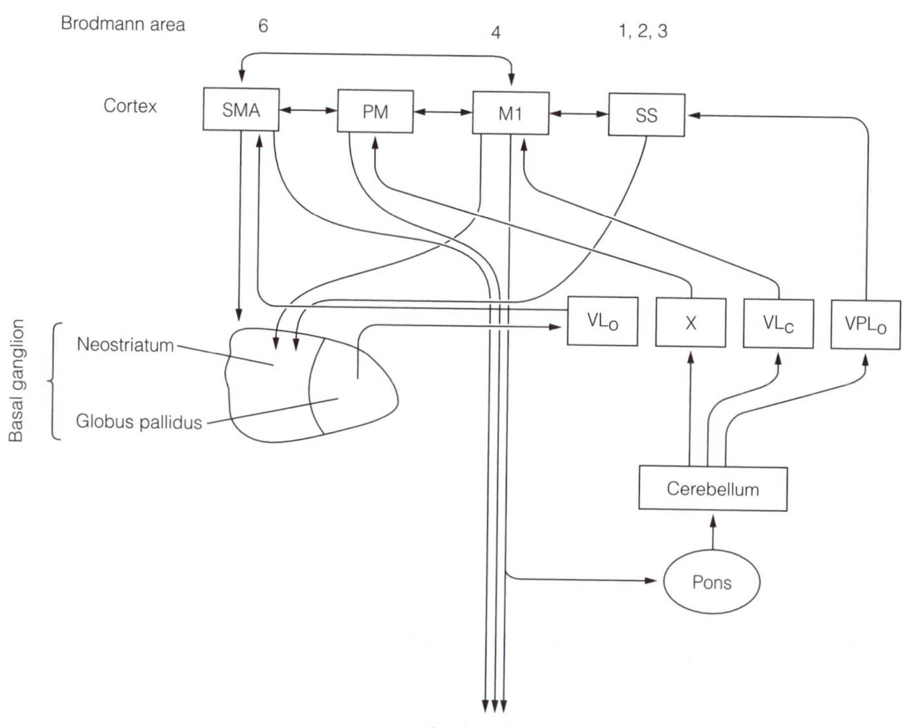

Fig. 2. The connections of the motor cortex establish motor loops with the basal ganglia and the cerebellum in which the thalamus provides the input that closes the loop. SS, somatosensory cortex. Thalamic nuclei: VL_c, ventroposterior nucleus pars caudalis; VL_o, ventroposterior nucleus pars oralis; VPL_o, ventroposterolateral nucleus pars oralis; X, nucleus X.

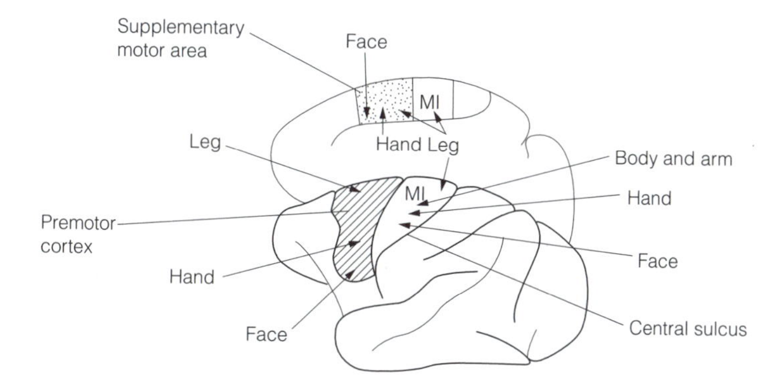

Fig. 3. Approximate somatotopic mapping in the motor cortex of the macaque. M1, primary motor cortex.

muscle spindle input to MI from where corticospinal tract axons go to influence the lower motor neurons. By informing the motor cortex about the state of muscles, the long loop circuit allows the motor cortex to rapidly modify the stretch reflex in the event of unexpected changes in load. The long loop modifies muscle contraction on a time scale slower that that of the myotatic reflex, but faster than voluntary movements.

What the somatotopic mapping in the motor cortex represents is uncertain. It is not a one-to-one mapping to individual muscles or movements. This is shown by several findings. Firstly, output from single cortical neurons diverges to several motor neuron pools. Secondly, there is convergence of outputs from quite a wide area of MI onto the motor neuron pools for muscles moving a specific body part. This is illustrated by the fact that neurons across the entire MI hand area fire during movement of a single finger. A given muscle is subserved by a region in the motor cortex which overlaps with regions controling neighboring muscles. During the execution of a movement involving the muscle, the set of neurons within the region that are activated depends on the nature of the movement; for example, its direction and force.

Recording from single cells in the motor cortex of conscious monkeys engaged in intentional limb movements shows that MI cell firing can correlate with force, rate of change of force, velocity, acceleration, direction of movement or joint position. None of these parameters is mapped in an orderly way in the cortex. Firing of an MI cell during a task is usually related to two or three of these variables so MI cells do not exclusively encode a single movement parameter. Many MI cells are rather broadly tuned for movement direction so a single MI cell cannot encode direction of movement very well. However, direction of movement is very precisely coded by the average firing of a few hundred cells. This is an example of population coding. The population of cells encoding the direction of a given movement are not localized to a discrete site, but are quite widely distributed across the cortex.

The supplementary motor area and premotor area (together called the **secondary motor cortex, MII**) contain neurons that fire in a way that correlates with the direction and force of a movement, and send axons into the corticospinal tract for the execution of movements. The SMA controls proximal limb muscles directly via its output to the corticospinal tract, but exerts control on distal limb muscles via its connections with MI.

Both the SMA and the PM have somatotopic maps. The PM actually consists of a number of discrete motor regions and so has several maps. The SMA has bilateral representations of the body and is crucial for movements involving both sides of the body, such as using both hands to perform a task or coordinating the postural responses that accompany limb movements. For this operation the motor loop involving the basal ganglia is important. Lesions of the SMA in monkeys have little effect on performance of simple motor skills but animals are impaired in their ability to perform more complex skills (such as retrieving a peanut from a small space) or in performing tasks needing both hands.

A key role of MII is in planning movements. This is shown, firstly, by the fact that neurons in MII fire a long time (possibly up to 800 ms) before a voluntary movement begins. Secondly, measurements of cerebral blood flow (cbf) in humans doing motor tasks show that a simple movement involves increased cbf in MI only, a more complex task is accompanied by increased cbf in MII as well as MI, but remarkably when subjects were required to mentally rehearse the complex task (but not execute it) an increase in cbf was seen restricted to MII.

Within MII, the SMA and PM have somewhat different roles. The SMA is involved in performing complex sequences of movements that have previously been learnt, while the PM plans movements in response to sensory (mostly visual) cues.

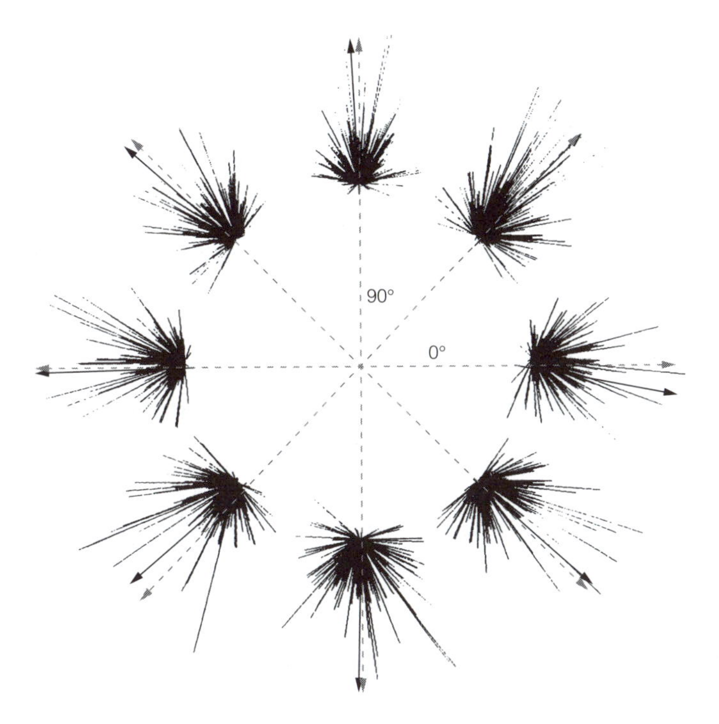

Fig. 4. Population coding of movement direction. A monkey was trained to move its arm in eight different directions. Each cluster of lines represents the activity of a population of neurons encoding that direction. The direction of each line represents the preferred direction of the cell, and the length is proportional to its firing rate. The thick arrow is the vector average for the population. In most cases it points closely in the desired direction. Reproduced from Georgopoulos, A.P. et al. *(1982)* J. Neurosci. ***2****, pp. 1527–1537, with permission. Copyright 1982, The Society for Neuroscience.*

The premotor areas receive a large input from the posterior parietal cortex (Brodmann areas 5 and 7) which in turn gets visual, somatosensory and vestibular sensory input. The posterior parietal cortex thus provides sensory input for targeted movements. Some posterior parietal neurons are context specific, firing only during goal-directed behaviour (e.g. reaching for food) but remaining silent if the limb moves in the same way in the absence of the goal. Premotor cortical neurons synapse with brainstem reticular neurons that go to axial and proximal limb muscles to provide the initial orientation of the body and limb towards a visual target. As with the SMA, the PM area influences distal limb muscles via its connections with MI.

Red nucleus

The red nucleus has a somatotopic map. Its activity precedes intentional movements and correlates with parameters such as force, velocity and direction much like corticospinal tract neurons. Furthermore, rubrospinal axons have the same distribution to proximal and distal limb motor neurons as the corticospinal tract and their activity moves individual digits. Although the two pathways appear strikingly similar, it has been suggested that they operate in different contexts. While the rubrospinal tract is active when previously learnt automated movements are executed, the corticospinal tract is required when novel movements are being learnt. Another pathway acts to switch activity between the two lateral motor systems. As a new movement is successfully learnt its execution is switched from the corticospinal tract to rubrospinal tract control. The switch operates in the opposite direction if an automatic movement needs to be adapted. It is because of the switch that each lateral motor system can compensate for the loss of the other. Corticospinal tract lesions have a more severe and protracted effect than rubrospinal tract lesions because new movements cannot be executed and only the old rubrospinal tract repertoire can be called upon. The switch pathway involves connections with the inferior olive and the cerebellum, one function of which is to detect and correct errors in motor performance (see Topic L4).

In the sequence rat–carnivores–primates–humans, the corticospinal tract becomes progressively larger, coming to dominate the rubrospinal tract. While in primates selective experimental damage to one pathway can be compensated for by the other, in humans, lesions of the corticospinal tract inevitably also damages the rubrospinal tract. This may explain why recovery from strokes is slow and incomplete in humans, in contrast with the effects of experimental lesions in nonhuman primates.

K7 Motor disorders

Key Notes

Brown–Sequard syndrome

Severing the spinal cord on one side causes paralysis and loss of touch and proprioceptor sensation below the lesion on the same side and loss of pain and temperature sensation on the contralateral side. This is the Brown–Sequard syndrome.

Decerebrate rigidity

Lesions which sever the brainstem between the red nucleus and the vestibular nuclei cause an increase in extensor tone known as decerebrate rigidity. It is caused by the loss of facilitation of flexor motor neurons by the rubrospinal tract.

Lesions of spinal motor pathways

Cutting the corticospinal tract causes an ipsilateral lesion if the transection is below the pyramidal decussation, and a contralateral lesion if above it. A pure corticospinal tract lesion in nonhuman primates causes a loss of fine movements by distal muscles. Cutting vestibulospinal and reticulospinal tracts causes deficits in posture and locomotion.

Cerebrovascular accidents

In clinical terminology, lower motor neurons are those that innervate skeletal muscles while upper motor neurons are pyramidal tract neurons. However the symptoms of an upper motor neuron lesion cannot be accounted for by lesions of the corticospinal tract alone. The major cause of an upper motor neuron lesion is a cerebrovascular accident (CVA; stroke) the commonest of which is an infarction of the internal capsule due to blockage of the artery which supplies it. The long-term symptoms are muscle weakness and spasticity on the side opposite the lesion. Spasticity is an increase in muscle tone, particularly in extensor muscles and is caused by hyperexcitability of stretch reflexes resulting from the loss of presynaptic inhibition on Ia terminals.

Related topics

Dorsal column pathways for touch sensations (G2)
Anterolateral systems and descending control of pain (G3)
Brainstem postural reflexes (K5)
Cortical control of voluntary movement (K6)
Strokes and excitotoxicity (R1)

Brown–Sequard syndrome

Brown–Sequard syndrome is a classic pattern of sensory and motor deficits seen when the spinal cord is severed on one side. On the side of the lesion there is a motor paralysis and loss of all sensation transmitted through the dorsal columns (touch and proprioception). On the contralateral side there is loss of nociceptor and thermoreceptor sensation from a few spinal segments below the lesion. This is due to the interruption of the anterolateral columns which contain spinothalamic axons that have crossed from the opposite side (*Fig. 1*).

Decerebrate rigidity

Patients in whom brain trauma or a tumor produces functional disconnection of the brainstem from the rest of the brain at the level between the red nucleus and

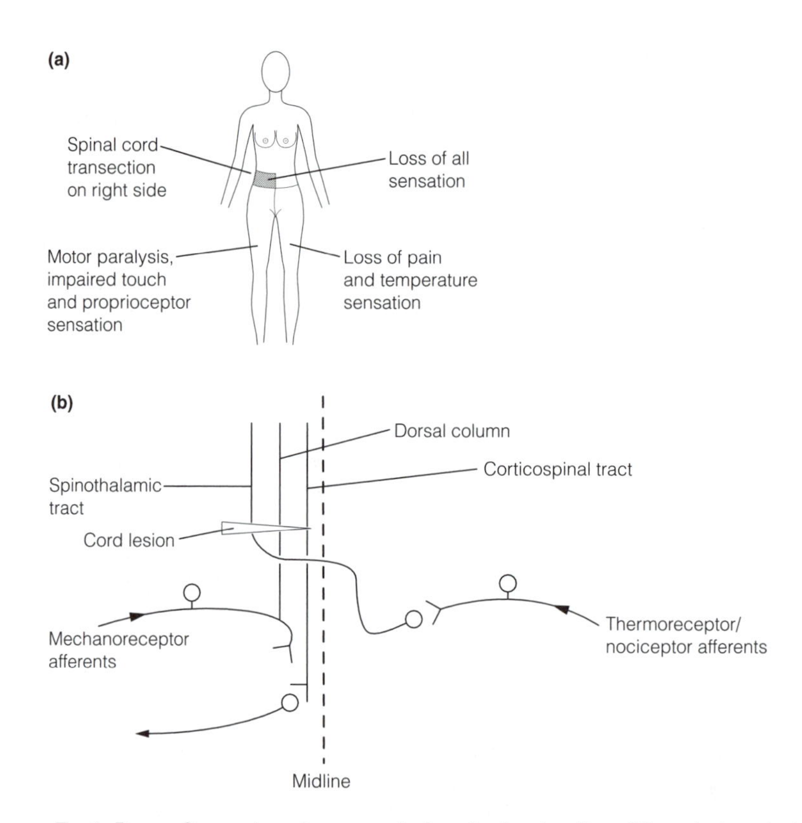

Fig. 1. Brown–Sequard syndrome results from the hemisection of the spinal cord which interrupts dorsal column input and motor output on the side of the lesion and spinothalamic input from the contralateral side. (a) Signs and symptoms; (b) lesion.

the vestibular nuclei show an increase in extensor tone called **decerebrate rigidity**. It is caused by tonic activity in the vestibulospinal and reticulospinal neurons that is no longer opposed by the powerful facilitation of flexor motor neurons by the rubrospinal tract (*Fig. 2*). The overall effect of the vestibulospinal activity is the activation of extensors. Reticulospinal inhibition and excitation of extensor motor neurons tend to cancel, but reduced reticulospinal neuron inhibition of interneurons driven by flexor reflex afferents enhances net extensor activity. The high firing rates of both α and γ motor neurons facilitated by vestibulospinal and reticulospinal inputs requires tonic muscle spindle afferent input, since cutting dorsal roots in decerebrate animals abolishes decerebrate rigidity. Ablation of the cerebellum in animal studies, by removing inhibitory input to the lateral vestibular nucleus intensifies decerebrate rigidity.

Lesions of spinal motor pathways

Experimental transection of the corticospinal tract below the pyramidal decussation in primates causes an ipsilateral motor deficit below the level of the section (see Brown–Sequard syndrome above). Corticospinal tract lesions above the pyramidal decussation cause a contralateral deficit. The deficit seen with a pure corticospinal lesion is a loss of the ability to make fine movements with distal muscles; for example, an inability to make independent finger movements needed to retrieve an object from a narrow hole. Almost complete recovery is eventually seen. A similar though less severe and transient deficit is seen if the

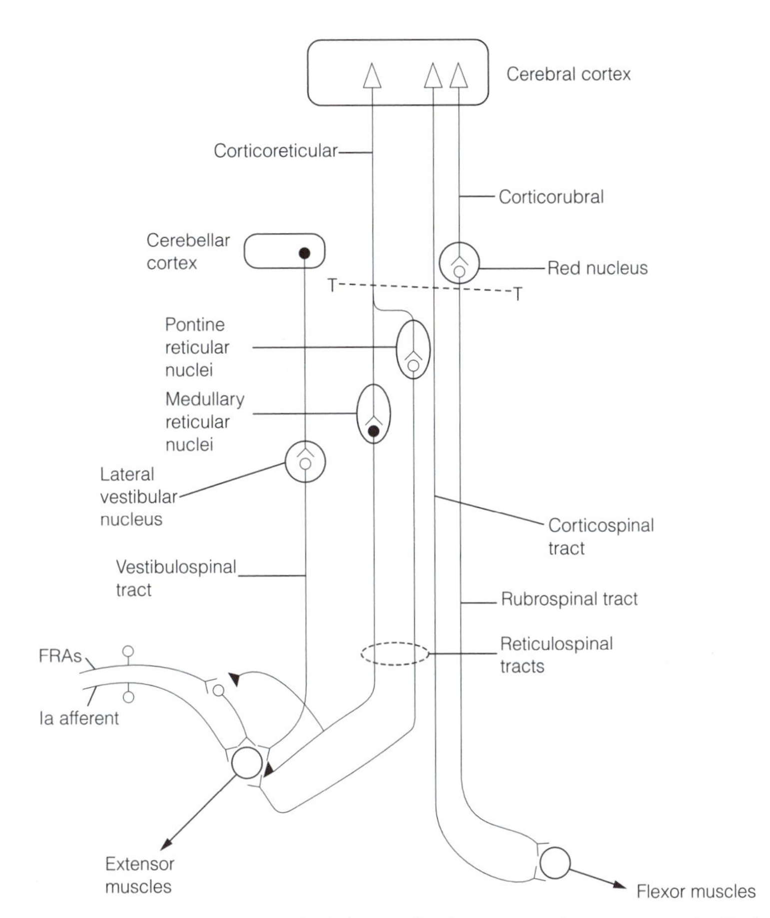

Fig. 2. Illustration of the principal descending inputs to motor neuron pools. Fusimotor fibers (not shown) are generally affected in the same manner as α motor neurons. A transection of the brainstem at the level of T–T causes decerebrate rigidity. FRAs, flexor reflex afferents.

rubrospinal tract alone is cut. However a lesion of both lateral motor pathways results in the deficit being permanent.

In contrast, experimental transection of the vestibulospinal and reticulospinal tracts which control output to proximal limb and axial muscles produces much more extensive deficits in posture, walking and climbing but leaves fine control of distal muscles intact.

Cerebrovascular accidents

Clinicians distinguish deficits due to lesions of **lower motor neurons** (brainstem and spinal cord neurons innervating skeletal muscle, see Topic K2) from those due to lesions of **upper motor neurons** (which is often taken to refer to the corticospinal and corticobulbar neurons of the pyramidal tract). Unfortunately, the symptoms attributed to upper motor neuron (pyramidal) lesions cannot be explained by damage to corticospinal or corticobulbar neurons alone. This is exemplified in strokes, the major cause of upper motor neuron lesions.

The commonest **cerebrovascular accident** (**CVA, stroke**) is caused by a thromboembolism affecting the branch of the middle cerebral artery that supplies the internal capsule. Infarction of the internal capsule produces a syndrome which does not resemble experimental lesions of the lateral motor pathways because the internal capsule also contains corticoreticular axons which drive lateral and medial reticulospinal tracts (see *Fig. 2*, Topic K6). After an initial period of flaccid paralysis and lack of reflexes on the side opposite the lesion two major deficits are seen.

(1) **Hemiparesis**. Muscle weakness on one side has a characteristic pattern because in the upper limb flexors are stronger than extensors and in the lower limbs the reverse is true. Hence the weakness is greater in arm extensors and leg flexors. If the corticobulbar fibers are affected facial muscle voluntary movements are prevented. When the weakness is so severe that paralysis results the term **hemiplegia** is used. Weakness occurs because the loss of descending excitation means that fewer motor units are recruited.

(2) **Spasticity**. An increase in muscle stiffness (**muscle tone**) is seen in the stronger (antigravity) limb muscles. It is caused by enhanced excitability of the monosynaptic stretch (myotatic) reflex, particularly the phasic component, since attempts at rapid muscle stretch are met with much greater resistance than slow stretch.

Two earlier ideas about spasticity have not been borne out. Firstly, it is not due to enhanced fusimotor activity, which is unaltered in humans with spasticity. Secondly, the sudden failure of the myotatic reflex in response to forceful attempts to stretch a muscle, the **clasp knife response**, is not due to activation of the inverse myotatic reflex by Golgi tendon organ stimulation, as was once thought, but is caused by firing of high threshold muscle afferents, distinct from either primary or secondary spindle afferents. Spasticity, in part, results from the loss of presynaptic inhibition on Ia terminals. Normally, presynaptic inhibition is brought about by the action of the reticulospinal tracts on γ-aminobutyrate (GABA)ergic Ia presynaptic inhibitory interneurons. GABA released from these interneurons acts on $GABA_B$ and $GABA_A$ receptors on the Ia terminals. $GABA_B$ receptors are metabotropic receptors and when stimulated act via G_i proteins to increase the K^+ conductance. The resulting hyperpolarization reduces Ca^{2+} influx into the primary afferent terminal thereby curtailing the release of glutamate onto the motor neurons. In spasticity, this descending reticulospinal input is lost which leads to failure of presynaptic inhibition and so hyperexcitability of the stretch reflex. Baclofen is an agonist at $GABA_B$ receptors and is used orally and intrathecally in the treatment of spasticity. Because benzodiazepines (e.g. diazepam) are agonists at the $GABA_A$ receptors involved in presynaptic inhibition, they too can be used in spasticity, but have the disadvantage of producing sedation at the doses needed to reduce muscle tone.

L1 ANATOMY OF THE CEREBELLUM

Key Notes

Functional overview of the cerebellum

The cerebellum is concerned with the execution of intentional postural and limb multijoint movements. It does this either by comparing motor commands with proprioceptor input so as to correct errors in performance, or for fast movements running programs for motor sequences that have been learnt.

Gross anatomy of the cerebellum

The cerebellum is divided into three lobes, anterior, posterior and flocculonodular, each of which are subdivided into lobules. Longitudinally it has a central vermis and two lateral hemispheres. The cerebellum is covered with a cortex and embedded within its core of white matter are deep nuclei which, with the vestibular nuclei, provide the cerebellar output. Input to the cerebellum is from the spinal cord and brainstem sensory systems, cerebral cortex and inferior olive.

Proprioceptor pathways

Sensory input from proprioceptors is used by the cerebellum to provide feedback on motor performance. Proprioceptor input to the cerebellum from the upper part of the body comes from collaterals of axons ascending in the dorsal columns which enter the accessory cuneate nucleus. This nucleus projects to the cerebellum by way of the cuneocerebellar tract. Proprioceptor afferents from the lower body terminate in Clark's column of the dorsal horn which gives rise to the dorsal spinocerebellar tract. A ventral spinocerebellar tract comes from the ventral horn and conveys information about the state of spinal circuits controling locomotion.

Related topics

Dorsal column pathways for touch sensations (G2)
Spinal motor function (K4)
Cerebellar cortical circuitry (L2)
Functional subdivisions of the cerebellum (L3)

Functional overview of the cerebellum

The cerebellum has a major role in the execution of voluntary movements, both postural and limb movements, particularly those involving several joints. By comparing command signals generated by the motor cortex and red nucleus with proprioceptor feedback from muscles and joints, it acts to correct differences that arise between motor intention and motor performance. Errors are fed back to the cortex and red nucleus to fine tune the motor commands. Movements that are too fast to be corrected by feedback are executed using programs that are based on predictions of their outcome. These predictions are based on experience. Hence the cerebellum is crucial for motor learning, by which new motor skills (e.g. driving, playing tennis) are gradually acquired. More recent evidence suggests that the cerebellum is also involved in cognitive functions such as language.

Gross anatomy of the cerebellum

The cerebellum is part of the hindbrain, in humans it constitutes about one-quarter of the mass of the brain and contains in excess of 10^{11} neurons. It is divided into three lobes, the **anterior lobe** and **posterior lobe** separated by the primary fissure and the **flocculonodular lobe**, separated from the posterior lobe by the posterolateral fissure (*Fig. 1*). The lobes are further subdivided into **lobules** which are differently named in humans and other mammals. Longitudinally the cerebellum has a central **vermis** and two **lateral hemispheres**.

Over the surface of the cerebellum lies the cerebellar cortex which is folded into coronal strips called **folia** (singular, **folium**). The cerebellum has an internal core of white matter containing **deep (intracerebellar) nuclei**. These, together

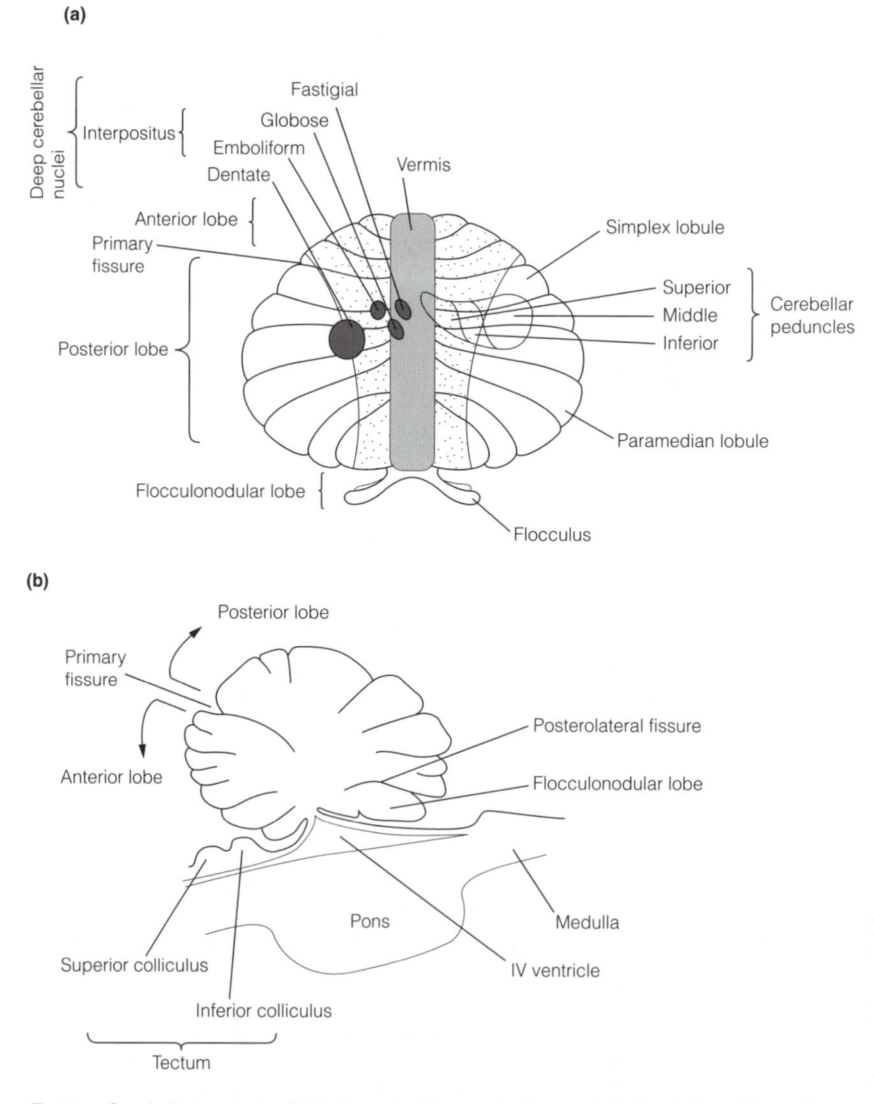

Fig. 1. Cerebellar anatomy. (a) A diagram of the cerebellum, unfurled and viewed from above. Locations of the deep cerebellar nuclei are shown on the left and of the cerebellar peduncles on the right. The medial zone is pale gray, the intermediate zone stippled and the lateral zone clear. (b) Sagittal section through the brainstem and cerebellum.

with the vestibular nuclei, provide the output of the cerebellum. Afferent and efferent connections of the cerebellum go by way of three pairs of cerebellar peduncles, the **inferior**, **middle** and **superior cerebellar peduncles**.

Input to the cerebellum is from three major sources:

- The spinal cord and brainstem, conveying sensory information from several modalities.
- The cerebral cortex, by way of pontine nuclei, in the massive corticopontine-cerebellar tract relaying motor and sensory input.
- The inferior olivary nucleus by way of the olivocerebellar tract.

Details of these inputs are given in *Table 1* and the following section.

Cerebellar output goes to three principal destinations:

- The ventrobasal thalamus which projects to the motor cortex by which it can influence the corticospinal outflow to motor neurons.
- The red nucleus, to modify the behavior of neurons projecting via the rubrospinal tract.
- The vestibular and reticular nuclei to modulate the output of the medial motor system.

Table 1. Principle inputs to the cerebellum

Tract	Origin	Peduncle	Distribution	Modality
Vestibulocerebellar	Vestibular nuclei	ICP	Crossed and uncrossed to flocculonodular lobe cortex and fastigeal nucleus	Vestibular
Trigeminocerebellar	Secondary afferents in trigeminal nerve (V nerve) nuclei	ICP	Crossed and uncrossed	Proprioceptive and cutaneous somatosensory from jaw and face
Cuneocerebellar	Accessory cuneate nucleus	ICP	Uncrossed	Proprioceptive from arm and neck
Dorsal spinocerebellar	Clark's column	ICP	Uncrossed	Proprioceptive and cutaneous somatosensory from trunk and leg
Ventral spinocerebellar	Ventral horn	SCP	Crossed and uncrossed	Proprioceptive and cutaneous somatosensory from all parts of the body
Tectocerebellar	Superior colliculi, inferior colliculi	SCP	Crossed	Visual and auditory
Pontocerebellar	Pontine nuclei	MCP	Crossed	Cognitive, motor, somatosensory and visual from cerebral cortex
Olivocerebellar[a]	Inferior olivary nucleus	ICP	Crossed to all deep cerebellar nuclei	Motor error signals

[a]The olivocerebellar input is via climbing fibers, all other afferents are mossy fibers. ICP, Inferior cerebellar peduncle; SCP, superior cerebellar peduncle; MCP, middle cerebellar peduncle.

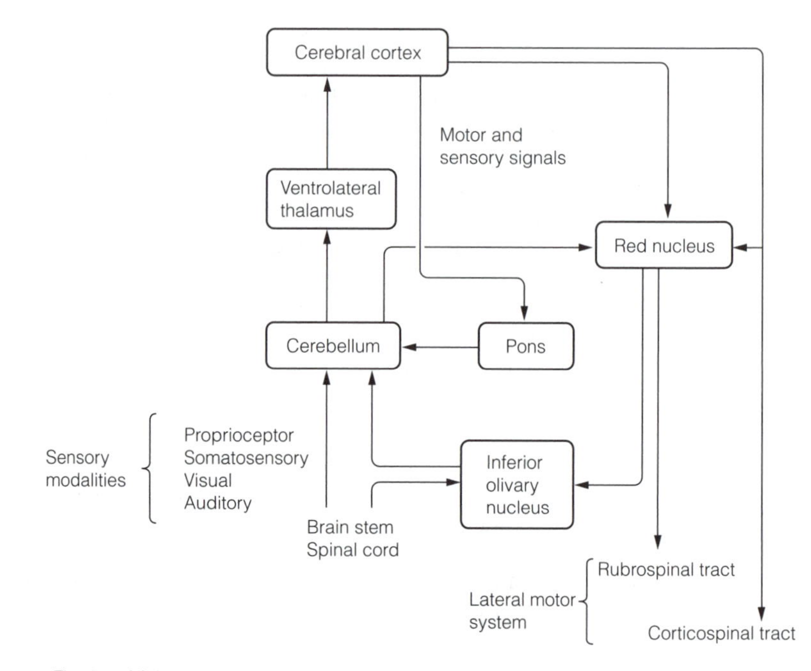

Fig. 2. Major connections of the cerebellum. Cerebellar output also goes to the medial motor system (not shown).

These general input–output relations are depicted in *Fig. 2*, although there are variations on this arrangement for different parts of the cerebellum.

Proprioceptor pathways

Proprioceptor information coming from muscle spindles, Golgi tendon organs and receptors in joints is used by the cerebellum to provide feedback on motor performance, and also provides for conscious awareness of body position and movement. Proprioceptor input from the neck, arms and upper trunk is relayed in the dorsal columns to the cuneate nucleus, from where it follows exactly the same path as touch sensation from the same areas (see Topic G2). This is the route for conscious upper body proprioception. Cerebellar input comes by way of the **cuneocerebellar tract** which arises from the **accessory cuneate (external arcuate) nucleus**. This receives the collaterals of axons ascending to the cuneate nucleus on the same side.

The proprioceptor pathway from the lower trunk and legs is distinct from that serving the upper body. Proprioceptor afferent terminals from the lower body enter the **nucleus dorsalis (Clark's column)** which is located in lamina VII of the medial dorsal horn and extends along the spinal cord between segments C8 to L3. The nucleus dorsalis houses second order neurons, axons from which ascend on the same side to form the **dorsal spinocerebellar tract (DST)**. Collaterals of the DST axons enter **nucleus Z**, which is located just above the gracile nucleus (see Topic G2) in the medulla. Nucleus Z neurons send their axons into the medial lemniscus and thereby provide input for conscious lower body proprioception.

The **ventral spinocerebellar tract (VST)** arises from the ventral horn and transmits signals reflecting the activity of spinal interneurons that are driven by motor commands regulating locomotor cycles. Hence the VST informs the cerebellum about the current state of spinal cord central pattern generators.

L2 CEREBELLAR CORTICAL CIRCUITRY

Key Notes

Inputs to the cerebellar cortex

Axons of spinal cord and brainstem neurons conveying sensory and motor information enter the cerebellum as mossy fibers to synapse with granule cells in multisynaptic complexes called glomeruli. Granule cell axons bifurcate to form parallel fibers which form synapses with thousands of Purkinje cells aligned in a row. Each parallel fiber makes just one synapse with each Purkinje cell but every Purkinje cell is contacted by 200 000 parallel fibers. Parallel fibers excite Purkinje cells to fire simple spikes. Climbing fibers from the inferior olive wrap themselves around 10 Purkinje cells making very powerful excitatory synapses with each one. Climbing fiber stimulation causes Purkinje cells to fire complex spikes.

Cerebellar cortex output

The output of the cerebellar cortex is provided exclusively by large γ-aminobutyrate (GABA)ergic inhibitory Purkinje cells and goes to the deep cerebellar nuclei.

Cerebellar cortical interneurons

There are three types of GABAergic inhibitory neuron in the cerebellar cortex. Basket and stellate cells produce lateral inhibition by inhibiting those Purkinje cells that are the immediate neighbors of those activated by parallel fibers. Golgi II cells terminate the effect of parallel fibers on Purkinje cells.

Related topic

Fast neurotransmission (C2)

Inputs to the cerebellar cortex

The cerebellar cortex has three layers and contains five cell types that are organized into a simple circuit repeated millions of times (*Fig. 1*).

The major input to the cerebellum are **mossy fibers**, axons of second order neurons from the spinal cord and brainstem conveying proprioceptor input, or the pontine–cerebellar relay from the cerebral cortex conveying sensory and motor signals. Each mossy fiber terminates in a discrete patch confined within a single lobule. Mossy fibers are glutamatergic and excitatory, and after giving off an axon collateral which goes to the appropriate deep cerebellar (or vestibular) nucleus, they synapse with granule cells in synaptic complexes called **glomeruli** (*Fig. 2*). Each glomerulus consists of the swollen terminal of a single mossy fiber which forms 15–20 synapses with the surrounding dendrites of four to five granule cells. Mossy fibers branch, so each one can excite about 30 granule cells. Every granule cell is contacted by five to eight mossy fibers. Glomeruli also contain axodendritic synapses between Golgi cells (see below) and granule cells.

Granule cells in the **granule cell layer** are small (5–8 μm diameter). Their axons ascend to the most superficial layer of the cerebellar cortex, the **molecular layer**, where they bifurcate into **parallel fibers** that in primates extend for 6 mm

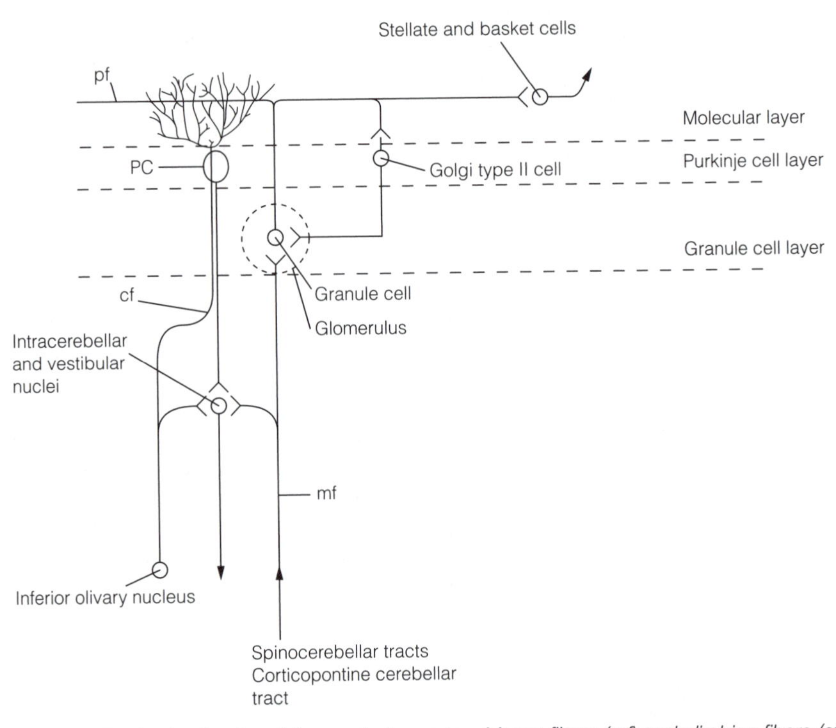

Fig. 1. The basic circuitry of the cerebellar cortex. Mossy fibers (mf) and climbing fibers (cf) are excitatory as are granule cells and the intracerebellar nuclei cells. All other cell types are inhibitory. Stellate and basket cells inhibit adjacent Purkinje cells (PCs).

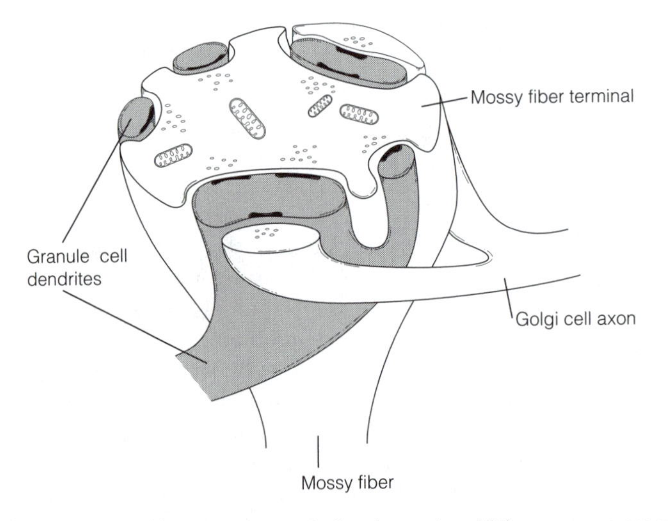

Fig. 2. The structure of a cerebellar glomerulus. All the synapses are axodendritic.

or so in each direction along the long axis of a folium. Parallel fibers intersect with the perpendicularly oriented planar dendritic trees of **Purkinje neurons**. Because of this arrangement every parallel fiber excites a longitudinal beam of 2000–3000 Purkinje cells, making just a single synapse with each one. Every Purkinje cell is contacted by about 200 000 parallel fibers. Mossy fibers, and the granule cells driven by them, have high background firing rates (50–100 Hz) that are changed by sensory input and during movements. The effect of parallel fiber activity is to cause the Purkinje cell to fire **simple spikes** repetitively (*Fig. 3a*). Purkinje cells fire at background rates between 20 and 50 Hz.

The second input to the cerebellum are the **climbing fibers** that come exclusively from the **inferior olivary nucleus** (**ION**) via the **olivocerebellar tract**. Each climbing fiber, of which there are about 15 million in humans, establishes contact with around 10 Purkinje cells, each Purkinje cell getting input from just one climbing fiber that winds its way round the soma and dendrites making about 300 powerful glutamatergic, excitatory synapses. Climbing fibers fire with a frequency of 1–10 Hz, each time causing the Purkinje cell to discharge a **complex spike** (*Fig. 3b*).

A third diffuse set of inputs into the cerebellum comes from monoaminergic cells in the brainstem. They establish sparse connections with deep cerebellar nuclei and cortex to produce modulating effects.

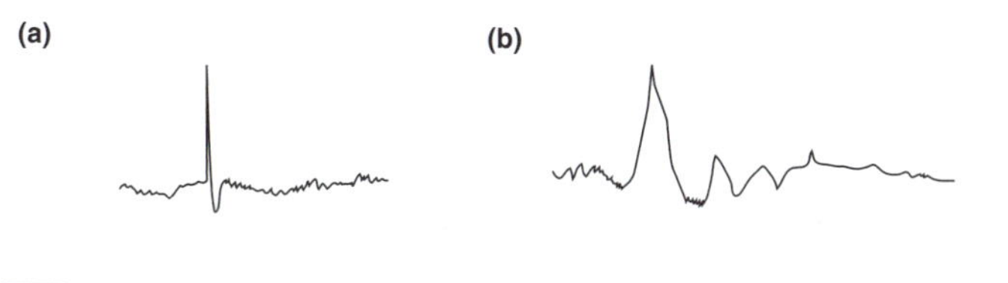

Fig. 3. Simple (a) and complex (b) spikes of Purkinje cells produced by mossy fiber or climbing fiber activation respectively.

Cerebellar cortex output

The sole output of the cerebellar cortex is via the Purkinje cells which have their large cell bodies (50 μm diameter) in the **Purkinje cell layer** of the cortex. Their extensive dendrites are aligned into a flat plane and are all oriented in the same direction, at right angles to the long axis of the folium in which they are located. The axons of the Purkinje cells go to the deep cerebellar nuclei. Purkinje cells use GABA as a neurotransmitter so the entire output of the cerebellar cortex is inhibitory.

Cerebellar cortical interneurons

The cerebellar cortex has three types of GABAergic inhibitory interneuron. **Basket cells** and **stellate cells** in the molecular layer receive input from parallel fibers and send axons to synapse with proximal or distal dendrites, respectively, of neighboring Purkinje cells. Activation of a mossy fiber excites a cluster of granule cells and hence stimulates linear arrays of **on-beam** Purkinje cells via the parallel fibers. However the basket and stellate cells inhibit surrounding **off-beam** Purkinje cells. This is a surround antagonism mechanism which produces spatial focusing of cerebellar cortex output.

Golgi II cells get input from parallel fibers and synapse with granule cells to produce feedback inhibition. In this way the Golgi cell brings about temporal focusing so that the net effect of mossy fiber input is brief firing of Purkinje cells.

In summary, the inhibitory interneurons constrain Purkinje cell output both in space and time.

L3 FUNCTIONAL SUBDIVISIONS OF THE CEREBELLUM

Key Notes

Cerebellar zones

The human cerebellum is parceled into three sagittal zones, each with distinct deep nuclei outputs and connections with the rest of the central nervous system (CNS). The medial zone (vermis) sends its output to the fastigial nucleus which relays to the vestibular nuclei. The intermediate zone output is via the interposed nucleus while the lateral zone output goes by way of the dentate nucleus. The interposed and dentate nuclei project to the red nucleus and ventrolateral thalamus.

Somatotopic mapping

Inputs to the cerebellum are organized topographically and this gives rise to somatotopic maps in the cerebellar cortex, the deep cerebellar nuclei and in their outputs to the red nucleus and thalamus. Each map represents not only sensory input but is also a motor output map.

Vestibulocerebellum

The vestibulocerebellum is the flocculonodular lobe and gets input from the vestibular system. Lesions cause defects in balance.

Spinocerebellum

The spinocerebellum is divided into a medial zone part and an intermediate zone part. The medial zone division controls postural adjustments by way of the medial motor system, using inputs from several sensory modalities. Lesioned animals cannot stand or walk. The intermediate zone part receives proprioceptor input and sensorimotor input from the cerebral cortex via the corticopontinecerebellar pathway. It exerts control over limb movements via the lateral motor system and damage disrupts the ability to make accurate limb movements.

Cerebrocerebellum

The lateral zones comprise the cerebrocerebellum which gets sensory, motor and cognitive input from the cerebral cortex. Lesions have only small effects on movements involving single joints but severe effects on more complex multijoint movements.

Related topics

Stimulus localization (F3)
Anatomy of the cerebellum (L1)
Oculomotor control (L7)

Cerebellar zones

The output of the cerebellar cortex to deep cerebellar nuclei, and climbing fiber input, is organized into parallel sagittal zones that extend the complete rostro-caudal length of the cerebellum. The exact number and arrangement of the zones depends on the species; in humans there are three. The medial zone occupies the vermis (see *Fig 1*, Topic L1) and sends its output to the **fastigial nucleus**. The cerebellum is not interrupted in the midline; parallel fibers extend across the middle of the vermis. This facilitates the coordination of movement of both halves of the body. The intermediate zone projects to the **interpositus**

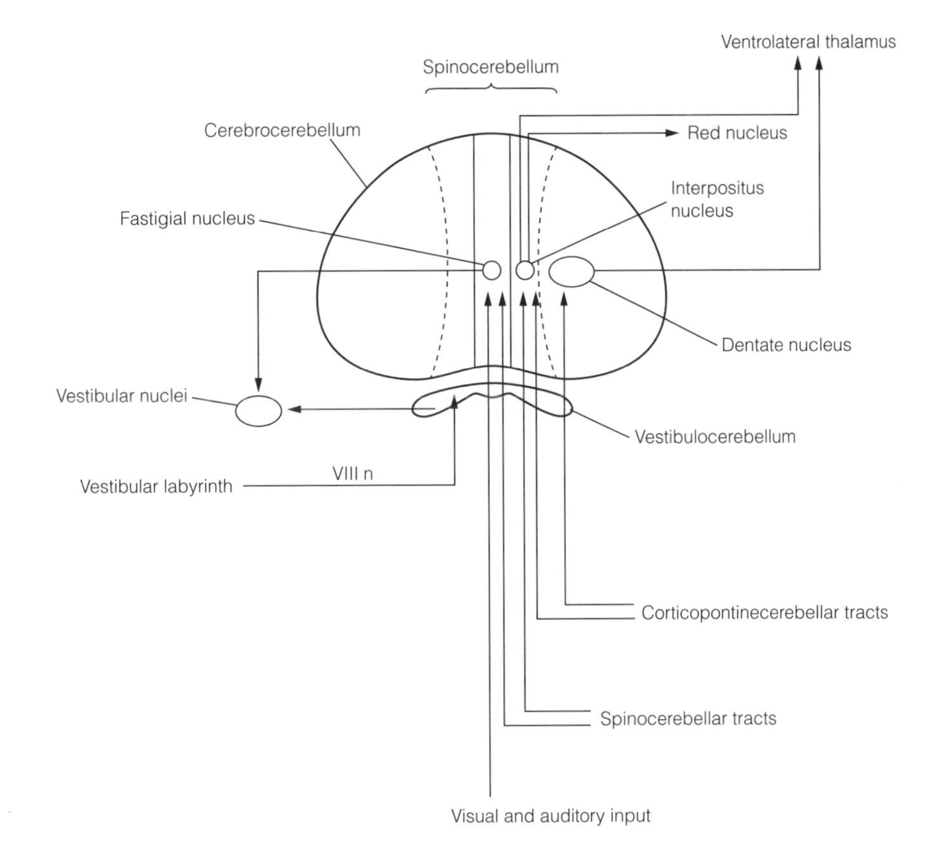

Fig. 1. The major connections and functional subdivisions of the cerebellum.

nucleus (which in humans consists of two separate **emboliform** and **globose** nuclei) while the lateral zone output is to the **dentate nucleus**. The zones, and the flocculonodular lobe, each have fairly well segregated and distinct connections with the rest of the CNS; these are summarized in *Fig. 1*.

Somatotopic mapping

Mossy fiber input and climbing fiber input are organized topographically, giving somatotopic maps in the cerebellar cortex that are retained in the deep cerebellar nuclei, and in their output to the thalamus and red nucleus. The cerebellar cortex has several maps which exhibit fractured somatotopy (see Topic F3), in which neighboring regions may get mossy fiber input from distant body parts, and there are multiple representations of body parts. A bilateral representation lies in the anterior lobe and two ipsilateral representations lie in the posterior lobe (see *Fig. 2*). The head regions of these overlap, and get visual and auditory input from the tectum as well as somatosensory input. Each of these representations is actually three maps in register. One is formed by the mossy fiber sensory input. A second map is that of the corticopontine input. The third is an output map that preserves a somatotopic projection of movements.

Vestibulocerebellum

The **vestibulocerebellum** corresponds to the flocculonodular lobe. It gets input from the ipsilateral vestibular labyrinth via the vestibulocochlear (VIII) nerve, and projects directly to the vestibular nuclei (*Fig. 3*). Lesions of the vestibulo-

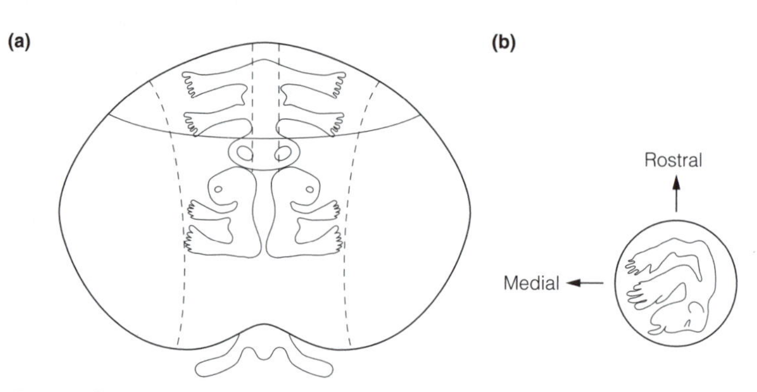

Fig. 2. Somatotopic maps in: (a) cerebellar cortex based on inputs and clinical lesions, the fractured nature of these inputs revealed by detailed studies is not shown; (b) deep cerebellar nucleus.

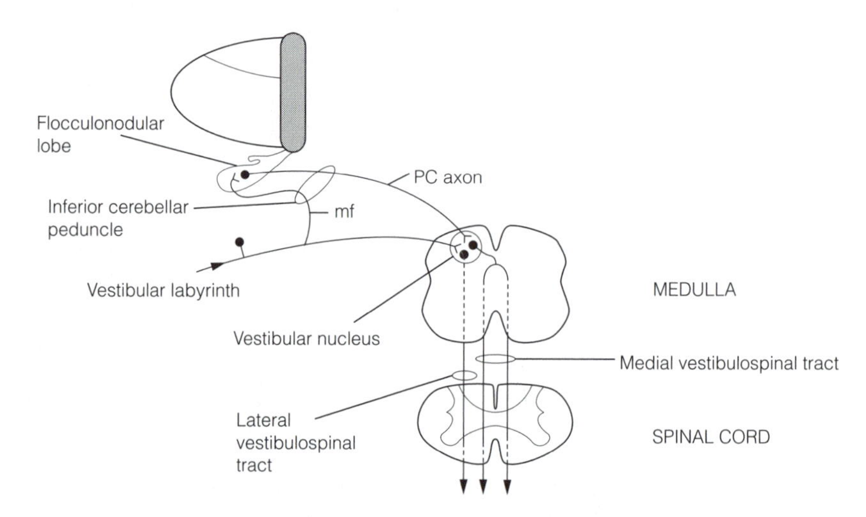

Fig. 3. Connections of the vestibulocerebellum. All cell bodies are shown as filled circles, as in other figures in this topic.

cerebellum in primates causes swaying and **ataxia** (staggering gait). If the lesion is unilateral the head is tilted to the side of the injury and **nystagmus** is seen. This is horizontal flicking of the eyes back and forth, a normal reflex response to rapid head rotation (see Topic L7), but here an abnormal response.

Spinocerebellum

The **spinocerebellum** consists of the anterior lobe and parts of the medial and intermediate zones of the posterior lobe: the vermis, simplex and paramedian lobules.

The medial zone of the spinocerebellum receives sensory input from several modalities, vestibular, proprioceptor and cutaneous somatosensory input from the trunk, together with visual and auditory input. The output from this part of the spinocerebellum goes by way of the fastigial nucleus to the vestibular nuclei (*Fig. 4*). It controls postural adjustments in response to sensory input by signaling to axial muscles via the medial motor systems. In primates, inactivation of the fastigial nucleus prevents standing and walking and the animals fall towards the side of the lesion.

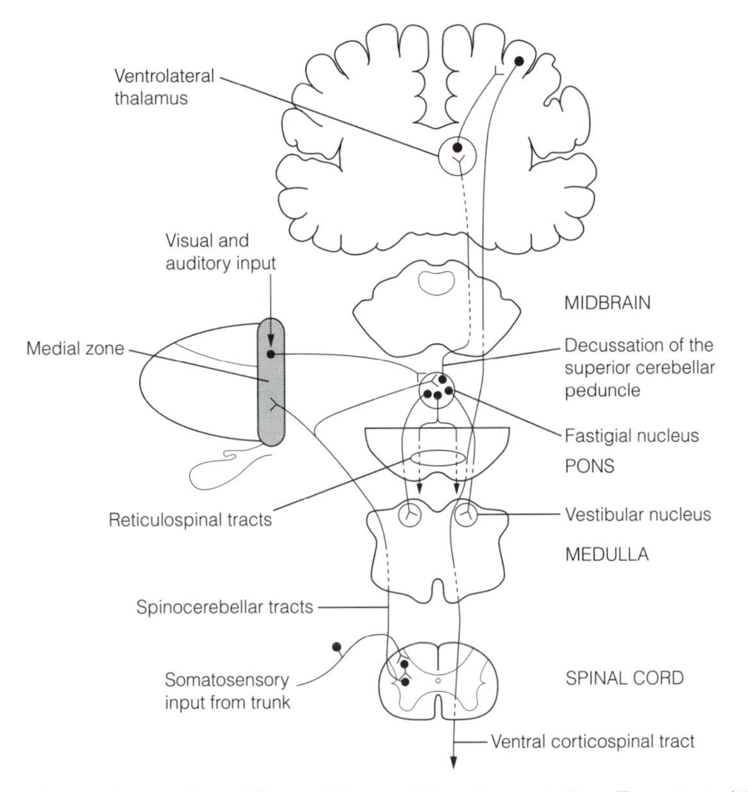

Fig. 4. Connections of the medial zone of the spinocerebellum. The output of the vestibular nuclei are as shown in Fig. 3. The inferior olivary nucleus is not shown.

In the intermediate zone regions of the spinocerebellum, input is from the cuneocerebellar and dorsal spinocerebellar tracts conveying proprioceptor and cutaneous somatosensory data and the ventral spinocerebellar tract which imparts information about the activity of spinal motor circuits (*Fig. 5*). In addition it receives inputs from the somatosensory and motor cortex via axon collaterals of the corticospinal tract that synapse with nuclei in the pons (**nuclei pontis**). These pontine nuclei give rise to fibers that cross the midline to enter the contralateral cerebellar cortex by way of the middle cerebellar peduncle. This **corticopontinecerebellar pathway**, containing 20 million axons is one of the largest tracts in the CNS, and also goes to the cerebrocerebellum.

The output of the intermediate zone region of the spinocerebellum is via the interpositus nucleus which projects to the ventrolateral thalamus and the red nucleus. By this route the spinocerebellum controls the lateral motor pathways to the limbs. Inactivation of the interpositus nucleus has little effect on standing or walking but results in a large amplitude tremor of the limbs with a frequency of 3–5 Hz when an animal attempts to reach for an object. This is known as an **action (intention) tremor** and is seen often in humans with cerebellar damage.

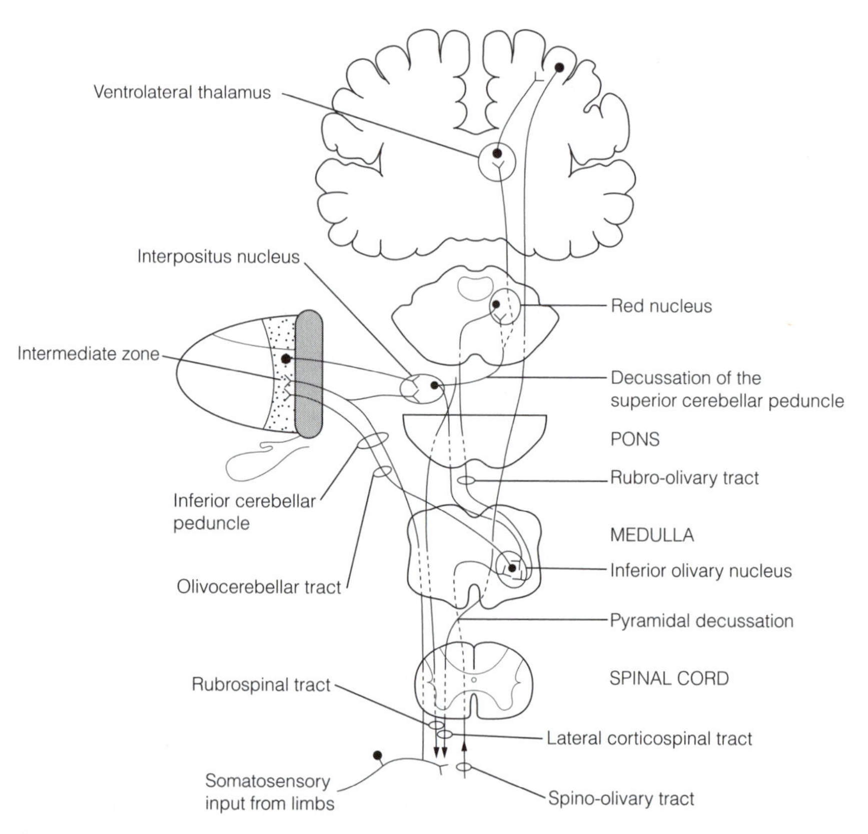

Fig. 5. Connections of the intermediate zone of the spinocerebellum. The corticopontine cerebellar tract is not shown.

Cerebro-cerebellum

The **cerebrocerebellum** corresponds roughly with the lateral zones of the posterior lobe. It receives inputs from frontal, parietal and occipital cerebral cortex by way of the corticopontinecerebellar tract that relays sensory, motor and visual information However it also gets input from prefrontal cortex concerned with cognitive not motor functions. The cerebrocerebellar outflow via the dentate nucleus goes to the ventrolateral thalamus which in turn projects to frontal cortex motor and prefrontal areas (*Fig. 6*). In addition, the dentate nucleus has reciprocal connections with the red nucleus. Lesions of the cerebrocerebellar cortex or dentate nucleus cause slight delays and modest overshooting of movements involving single joints. However, for multijointed movements the deficits are much more severe, so animals have difficulty performing complex manipulations with the fingers.

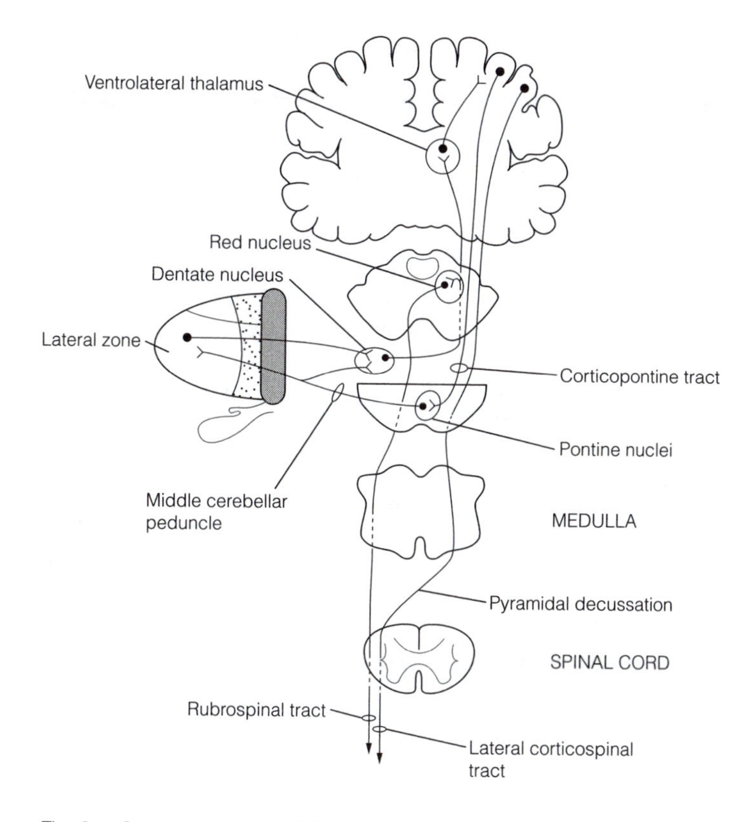

Fig. 6. Connections of the lateral zone (cerebrocerebellum). The connections of the inferior olivary nucleus are not shown.

L4 Cerebellar function

Key Notes

General principles

The cerebellum coordinates movements initiated from the motor cortex, can initiate movements itself, and learns new motor tasks. Parallel fibers excite arrays of Purkinje cells that control muscles spanning several joints. This supports the idea that the cerebellum is important for complicated multijoint movements. The cerebellum can operate in feedback or feedforward mode.

Feedback operation

Here the cerebellum compares motor intentions with motor performance. A discrepancy between these generates an error signal which is used to make the mismatch smaller. Sensory signals caused by movement errors activate mossy fibers which excite Purkinje cells. These inhibit the deep cerebellar nuclei which drive the red nucleus and thalamus and so the erroneous movement is prevented.

Feedforward operation of the cerebellum

For movements that are too fast for feedback to operate, the cerebellum runs preprogrammed sequences that have predictable effects on motor function. This feedforward works well provided that nothing unexpected happens.

Motor learning

In humans most voluntary movements must be learnt. In motor learning the cerebellum acquires a program which specifies the motor commands needed for a given movement as a consequence of errors in performance. Sensory errors are translated into motor errors by the inferior olive and sent to the cerebellum via climbing fibers. These motor error signals cause the Purkinje cells to become less responsive to the mossy fiber input occurring at the same time. Whenever the same input recurs the Purkinje cells are excited less than before the learning.

Related topics

Spinal motor function (K4)
Cortical control of voluntary movement (K6)
Cerebellar cortical circuitry (L2)
Cerebellar motor learning (Q5)

General principles

Despite functional subdivisions, the same circuit is repeated across the entire cerebellum, so it is likely that the same computations are performed by all parts of the cerebellum. Although the cerebellum is usually regarded as coordinating the execution of movements initiated by the motor cortex, the cerebrocerebellum is implicated in initiating movements, particularly in response to visual and auditory stimulation, since in movements triggered in this way the order of activation is: dentate nucleus – motor cortex – interpositus nucleus – muscle. Furthermore, central to the operation of the cerebellum is motor learning, the acquisition of new motor skills.

In primates, parallel fibers average 6 mm in length and so affect a comparable length array of Purkinje cells that lie across the cerebellum. This is sufficiently long to span an entire deep cerebellar nucleus or to bridge adjacent nuclei. Purkinje cell arrays coupling both fastigial nuclei, for example, would ensure coordination of postural muscles across the midline, which is important in gait. The Purkinje cell arrays influenced by a given set of parallel fibers span muscles over several joints. This anatomical configuration supports the results of recording and lesion studies showing that the cerebellum is much more concerned with control of movements involving many joints rather than single joints. The cerebellum is thought to operate in one of two modes, feedback or feedforward, depending on the circumstances.

Feedback operation

During the execution of well rehearsed movements that are not too fast the cerebellum acts as a feedback device to compare motor intentions with motor performance, and works to reduce any mismatch between them. For the spinocerebellum, the motor intentions are the signals relayed by the corticopontinecerebellar tract. Motor performance is monitored by proprioceptor (and other sensory) input, and by the ventral spinocerebellar signals reporting on the activity of spinal cord and brainstem motor circuits. Similarly, the cerebrocerebellum compares inputs from the supplementary motor cortex and the primary motor to produce error signals that reflect a discrepancy between motor planning and motor commands. In each case the error signals are used to correct the mismatch.

The intermediate spinocerebellum seems to be involved in correcting errors in limb movements since when a limb is perturbed by an unexpected force the order in which various neural elements fire is: muscle afferents – interpositus nucleus – motor cortex – dentate nucleus.

Feedback error correction probably works as follows: an error means that the actual position of a limb is not the intended one; this produces unpredicted muscle stretch. Precisely the same thing happens if an unexpected force is applied to a limb. In either case the muscle stretch will excite Ia and Ib afferents in the loaded muscles. These proprioceptor signals are relayed to the cerebellum by mossy fibers. The stimulated mossy fibers deliver tonic excitation to the intracerebellar (interpositus) nucleus via axon collaterals (see *Fig. 1*, Topic L2) and stimulate a group of granule cells.

The granule cell parallel fibers activate several arrays of on-beam Purkinje cells (*Fig. 1*) which strongly inhibit their target neurons in the interpositus nucleus. Normally, when mossy fibers are firing at background rates their tonic facilitation of the interpositus dominates the inhibition by Purkinje cells. Consequently interpositus neurons maintain excitation of the red nucleus and the ventrolateral thalamus. When the mossy fibers are activated during a movement however, Purkinje cell inhibition **disfacilitates** the interpositus neurons and this inhibition is transmitted downstream to the red nucleus and thalamus. In contrast, neighboring off-beam PC arrays, inhibited by the γ-aminobutyrate (GABA)ergic interneurons in the cortex allow their interpositus cells to fire at higher than background rates. Thus the pattern of activation of the deep cerebellar nucleus is a negative image of the input activation.

The overall effect, mediated by the rubrospinal and corticospinal tracts, is to correct the movement error by activating spinal reflexes that defend the correct limb position and dampen those that do not. There is some evidence that this might involve altering the firing of fusimotor efferents to muscle spindles.

The intermediate spinocerebellum seems to control the precise timing of the contraction of agonist and antagonist muscles during a movement. During

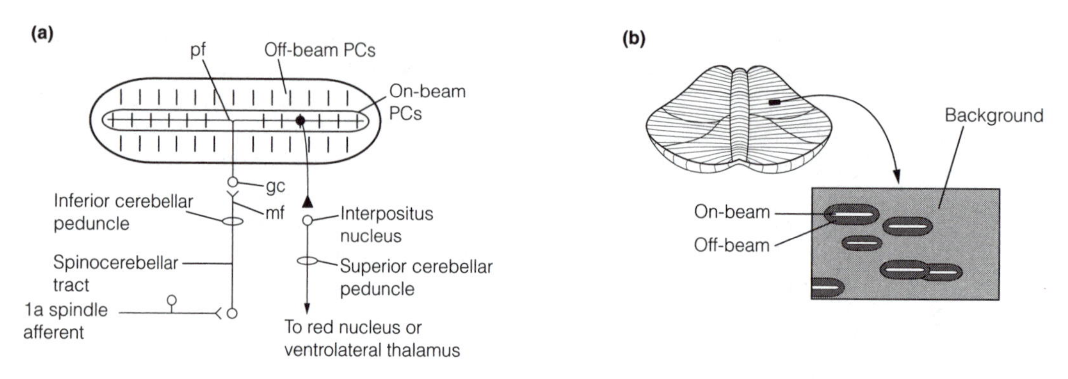

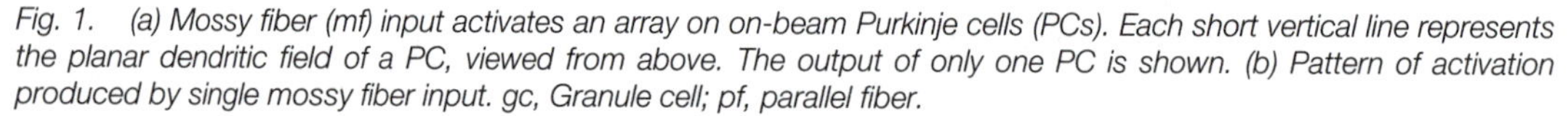

Fig. 1. (a) Mossy fiber (mf) input activates an array on on-beam Purkinje cells (PCs). Each short vertical line represents the planar dendritic field of a PC, viewed from above. The output of only one PC is shown. (b) Pattern of activation produced by single mossy fiber input. gc, Granule cell; pf, parallel fiber.

reciprocal activation of agonist and antagonist muscles, Purkinje cells responsible for controling these muscles fire alternately, driving interpositus neurons to do the same. During co-contraction, however, the Purkinje cells are silent. A role for the cerebellum in organizing these patterns of muscle activity is supported by the fact that the action tremor resulting from lesions of the interpositus nucleus or intermediate cerebellar cortex appears to be due to derangement in the timing of agonist contraction. In normal humans, a rapid wrist movement involves an initial burst of activity in the agonist, followed by a burst in the antagonist to produce braking, and finally a second agonist burst to stabilize the joint at the desired end point (see *Fig. 2*, Topic K4 for an illustration of this pattern of activity). In cerebellar tremor the start of the movement is normal but the second agonist burst is late. Consequently the antagonist burst moves the wrist beyond the end point; this sets up the tremor.

Feedforward operation of the cerebellum

For the execution of well practiced, very rapid (**ballistic**) movements (e.g. playing fast passages on a musical instrument or a tennis serve), there is insufficient time for feedback correction of errors. For these movements the cerebellum operates in a feedforward mode in which it runs a program that predicts the motor consequences of its own action. Unexpected perturbations that occur when the cerebellum is in this mode cannot be corrected for in time and so performance will be degraded.

Motor learning

The predictions inherent to feedforward operation must be learnt during numerous trials attempting to perform the task. This is motor learning and is probably important in acquiring skill in all voluntary motor tasks, including (in humans) learning to walk.

In one model of motor learning (the **cerebellar feedback–error–learning model**) the cerebellum acquires a program called an **inverse model** of the motor task, which is the transformation from the desired trajectory to the motor commands needed to bring about the movement. The inverse model is so called because when it operates on the motor system it results in an actual trajectory close to the desired one (*Fig. 2*). Errors in a movement are initially represented as sensory errors. For example, during a tennis match an incorrect arm movement will be visually obvious from the direction of the ball, and an error in

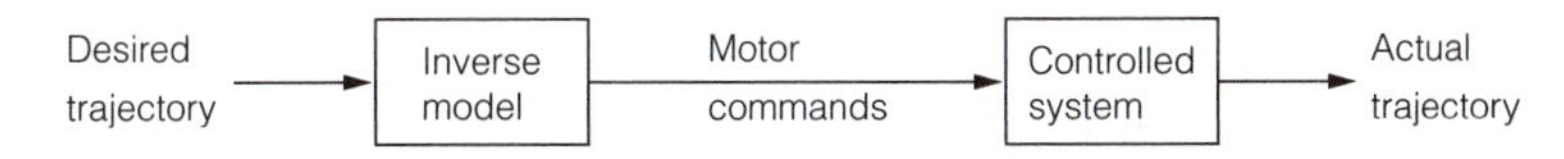

Fig. 2. The operation of the inverse model on the controlled system (motor neurons and muscles) turns a desired into an actual trajectory. The better the inverse model the closer the actual is to the desired trajectory.

playing a musical instrument will sound wrong. Sensory errors need to be converted into a pattern of nerve impulses that specify errors in motor performance, an operation of the **inferior olivary nucleus (inferior olive)**. The inferior olive sends these motor error signals to the cerebellum via the olivocerebellar tract climbing fibers (*Fig. 3*). The motor error is a feedback signal because it is produced by mistakes in carrying out the motor task. Hence, the cerebellum gets mossy fiber input that represents the desired trajectory (from motor cortex), sensory input (e.g. from visual cortex or proprioceptor pathways) and climbing fiber input from the inferior olive that represent errors in motor performance. The conjunction of climbing fiber and mossy fiber input onto a single Purkinje cell, on successive attempts to execute the task, eventually alters the output of the Purkinje cell in such a way that the motor performance is improved.

When a monkey is learning to perform a new motor task, initially it makes a large number of errors. During this time activity in climbing fibers is greatly increased so Purkinje cells show lots of complex spikes as well as simple spikes due to mossy fiber input. (When the monkey is performing routine movements there are very few complex spikes.) Gradually the number of simple spikes decreases in parallel with the improvement in performance. Eventually climbing fiber activity falls to baseline levels. It appears that the error signals carried by the climbing fibers causes concurrent mossy fiber input to become less effective in activating Purkinje cells. In other words, whenever the same pattern of mossy fiber input occurs subsequently, the Purkinje cells fire fewer simple spikes and cause less inhibition on downstream motor pathways. This corrects the motor output. The altered responsiveness of the Purkinje cells to a given pattern of mossy fiber inputs is an example of plasticity called long-term depression, the cellular events of which are dealt with in Topic Q5.

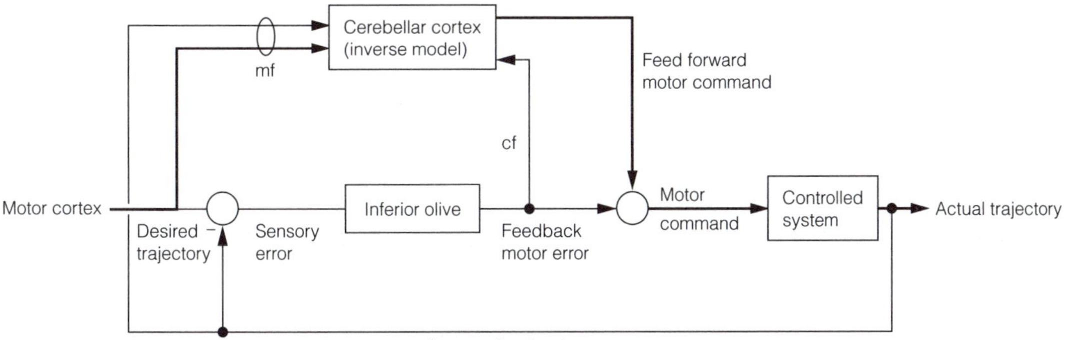

Fig. 3. A model for motor learning by the cerebellum. When the cerebellum is operating in feedforward mode the bold pathway is activated. cf, Climbing fibers; mf, mossy fibers.

L5 Anatomy of the basal ganglia

Key Notes

Overview

Several interconnected structures make up the basal ganglia: striatum, globus pallidus, substantia nigra and subthalamus. Cerebral cortical input to the basal ganglia goes to the striatum. The basal ganglia output goes from the globus pallidus and substantia nigra to the cortex via the thalamus. The basal ganglia are responsible for producing motor sequences during voluntary movement.

Striatum

The caudate nucleus and putamen together make up the striatum. Glutamatergic axons from the cerebral cortex and dopaminergic axons from the substantia nigra (pars compacta) terminate on the medium spiny neurons which are γ-aminobutyrate (GABA)ergic. There are two populations of medium spiny neurons, one is inhibited by dopamine, the other excited. The inhibitory output of the striatum goes to the globus pallidus and substantia nigra.

Output structures of the basal ganglia

Parts of the globus pallidus (pars interna) and substantia nigra (pars reticulata) send axons to specific thalamic nuclei which in turn project to particular areas of the cerebral cortex. This circuitry provides basal ganglia control of limb, facial and eye movements. Part of the globus pallidus (pars externa) projects to the subthalamic nucleus.

Subthalamic nucleus

Excitatory neurons of this nucleus are excited by the motor cortex, inhibited by the globus pallidus (pars externa) and send their axons to the globus pallidus and substantia nigra.

Parallel processing in the basal ganglia

Five circuits form loops between specific regions of the cortex and basal ganglia and each circuit seems to have a distinct functional role. Two of the circuits are concerned with motor functions, the others with aspects of memory, cognition and emotion.

Related topics

Slow neurotransmission (C3)
Cortical control of voluntary movement (K6)
Dopamine neurotransmission (N1)
Motivation (O1)

Overview

The basal ganglia consists of several extensively interconnected structures, the striatum (the caudate and putamen), the globus pallidus (pars interna and pars externa), the substantia nigra (pars compacta and pars reticulata) and the subthalamic nucleus. Most inputs to the basal ganglia are from the cerebral cortex and enter the striatum. The output of the basal ganglia emerges from the pars interna (internal segment) of the globus pallidus, and the substantia nigra pars reticulata, to go to the thalamus. The thalamus projects back to the cortex thus closing a loop. The thalamocortical axons return to the same region of cortex which gave rise to the striatal inputs (*Fig. 1*).

Basal ganglia circuitry is responsible for the execution of appropriate pre-programmed motor sequences during voluntary movements and inhibiting unwanted sequences. Firing of many neurons in the basal ganglia correlates with movement but occurs quite late, so the basal ganglia are not implicated in the initiation of movement. Classically the basal ganglia constitute the **extrapyramidal system**, on the basis that lesions of the basal ganglia produce quite different symptoms from lesions of the pyramidal (corticospinal) tract. Recently it has been proposed that the basal ganglia also have cognitive and affective functions.

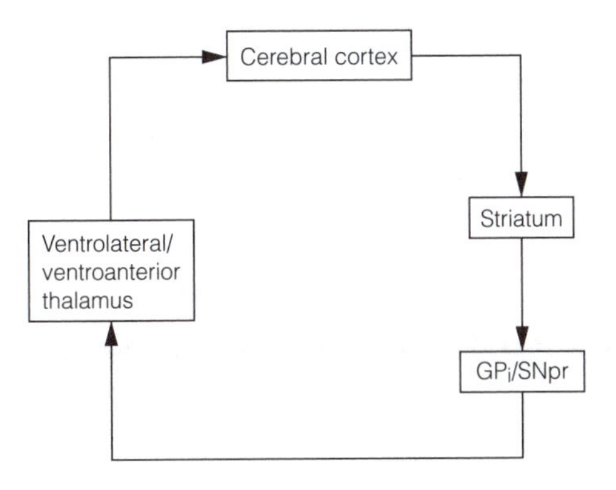

Fig. 1. Block diagram of the interface between basal ganglia and cerebral cortex. GPi, Globus pallidus pars interna; SNpr, substantia nigra pars reticulata.

Striatum

The caudate nucleus and putamen are functionally a single unit, the **dorsal striatum** (neostriatum) but split anatomically by the internal capsule. Although this is distinct from the ventral striatum which is part of the limbic system, the two have similar circuitry.

The striatum receives excitatory input from the cortex via the glutamatergic corticostriate pathway (*Fig. 2*). The input is organized topographically so that somatotopy is preserved in the projections from the somatosensory cortex and motor cortex. Corticostriate axons terminate on the major neuron type in the striatum, the **medium spiny neuron**. These make up 95% of striatal neurons, use GABA as their transmitter, and provide the inhibitory output of the striatum. Medium spiny neurons have large dendritic trees, and inhibit each other via extensive axon collaterals. There are two populations of medium spiny neuron which although morphologically indistinguishable have different connections and neurochemistry. One type has **substance P (SP)** and **dynorphin (DYN)** as cotransmitters, express dopamine D1 receptors, and projects to the globus pallidus pars interna (GPi) and substantia nigra. The second type uses **enkephalin (ENK)** as a cotransmitter, express D2 dopamine receptors, and projects to the globus pallidus pars externa (GPe).

The medium spiny neurons receive projections via the nigrostriatal pathway from the **substantia nigra pars compacta (SNpc)**. The SNpc uses dopamine as a transmitter. Because the two types of medium spiny neuron express different dopamine receptors they are differently modulated by this input. At the GABA/SP/DYN cells, dopamine acting on D1 receptors that are positively

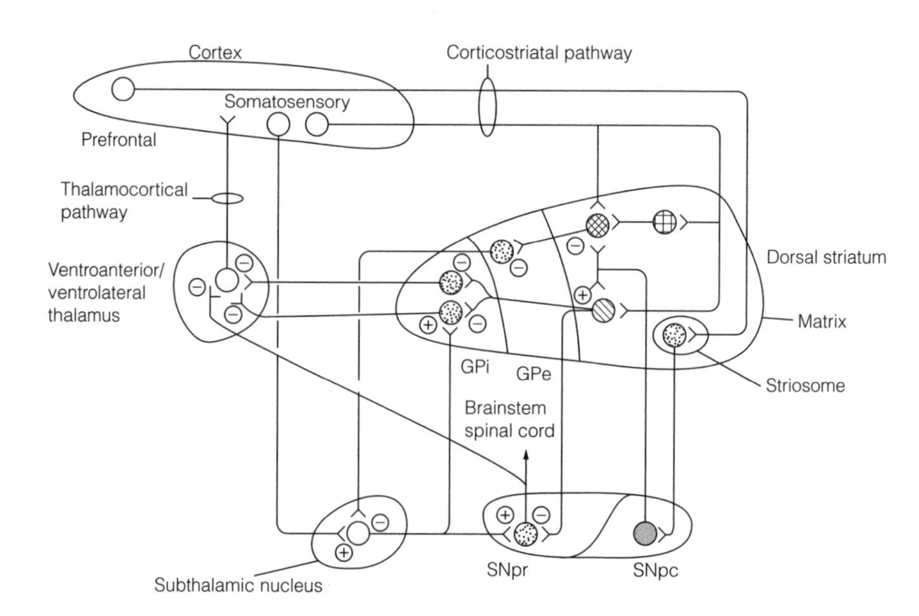

Fig. 2. Connections of the basal ganglia. GABAergic neurons in GPe also project to GPi and SNpr (not shown). GPe, Globus pallidus pars externa; GPi, globus pallidus pars interna; SNpc, substantia nigra pars compacta; SNpr, substantia nigra pars reticulata. ⊕, excitatory synapse; ⊖, inhibitory synapse. The neurons are coded by their neurotransmitter: ○, glutamate; ⊕, acetylcholine; ●, dopamine; ◎, GABA; ◍, GABA/substance P/dynorphin; ◍, GABA/enkephalin.

coupled to the adenylyl cyclase-cyclic 5′-adenosine monophosphate (cAMP) second messenger system, enhances the effect of excitatory cortical input. In contrast, the action of dopamine on the GABA/ENK cells is to reduce the effect of cortical excitation, because D2 receptors couple to G_i proteins that inhibits adenylyl cyclase (see Topic C3).

Medium spiny neurons have a third input from large **aspiny interneurons** that constitute about 2% of striatal neurons. These cells use acetylcholine as a transmitter, are excitatory, and are driven by cortical inputs.

Staining of the striatum for acetylcholinesterase shows it to be compartmented into a heavily stained **matrix** and a lightly stained three-dimensional labyrinth the **striosomes** that are about 10–20% of the striatal bulk. The connectivity of cells in these two compartments is different. The matrix gets inputs from throughout the cerebral cortex and sends outputs to the globus pallidus and **substantia nigra pars reticulata (SNpr)**, whereas the striosomes get restricted input from the prefrontal cortex and projects to the SNpc. Moreover the cortical inputs come from different layers; superficial layer 5 in the case of the matrix, deep layer 5 and layer 6 for the striosomes. These differences suggest that the matrix is concerned with sensorimotor function, while striosomes are associated with the limbic system, and may control the dopaminergic pathway from SNpc to striatum.

Output structures of the basal ganglia

Both the globus pallidus and substantia nigra are divided into two parts. The **globus pallidus pars interna (GPi)** of primates (equivalent to the **entopeduncular nucleus** in rodents) and the SNpr have very similar structures and are functionally equivalent. Both get inhibitory connections from the GABA/SP/DYN population of striatal neurons and excitatory inputs from the subthalamic nucleus, and both send GABAergic inhibitory outputs to the thalamus. The thalamus in turn projects to specific locations in the cerebral cortex. The GPi makes connections with the **oral part** of the **ventrolateral thalamus** and the **ventroanterior thalamus**, which relay to the motor cortex. The output of the SNpr goes to the **medial ventrolateral thalamus** and **dorsal medial thalamus** which sends its axons to the **frontal eye fields**, zones of the prefrontal cortex which organize gaze. The GPi and SNpr provide basal ganglia output for limb and facial movements, and part of the SNpr is concerned with eye movements.

The **globus pallidus pars externa (GPe)** receives its striatal connections from the GABA/ENK medium spiny neurons. The GPe neurons are GABAergic and go mostly to the **subthalamic nucleus**, but projections to GPi and SNpr have also been described.

Subthalamic nucleus

The subthalamic nucleus (**STN**) lies at the junction between midbrain and diencephalon, and is particularly well developed in primates. It gets excitatory input from the motor cortex and GABAergic input from the GPe. The STN neurons use glutamate as a transmitter, are excitatory, and send their axons principally to the GPi and SNpr.

A good somatotopic representation is present in all structures of the basal ganglia, except the SNpc. The output of the basal ganglia to the thalamus establishes somatotopic maps there, but they are distinct from the somatotopic maps in the thalamus produced by the cerebellum. In other words inputs of the basal ganglia and cerebellum to the thalamus are kept separate.

Parallel processing in the basal ganglia

Five parallel circuits like that illustrated in *Fig. 1* are thought to exist in the basal ganglia, each one getting corticostriatal inputs from several functionally related areas of cortex and projecting back to a restricted locus of the same area by way of specific thalamic nuclei. They are referred to as **basal ganglia–thalamocortical circuits**. A massive convergence occurs between the striatum and GPi/SNpr. This funneling of input from a large to a restricted region in the cortex by the basal ganglia serves as a way of integrating information.

Of the five, only the motor and oculomotor circuits serve the functions classically attributed to the basal ganglia; the others, dorsolateral prefrontal, orbitofrontal and anterior cingulate, are associated with memory, cognition and emotion respectively. The anterior cingulate circuit, unlike the others, goes by way of the ventral striatum (nucleus accumbens), which is part of the dopamine motivation system, rather than the caudate or putamen (see Topic O1). Since the circuits are all wired in much the same way it is likely that they all perform the same computations. The different outcomes of the operation of each of the circuits depends on the cortical regions they are connected to, and the contexts in which they are activated. Interestingly, basal ganglia disorders can produce deficits in thought processes not unlike those that affect movement. The extent to which the circuits are integrated or remain segregated is not known.

L6 Basal ganglia function

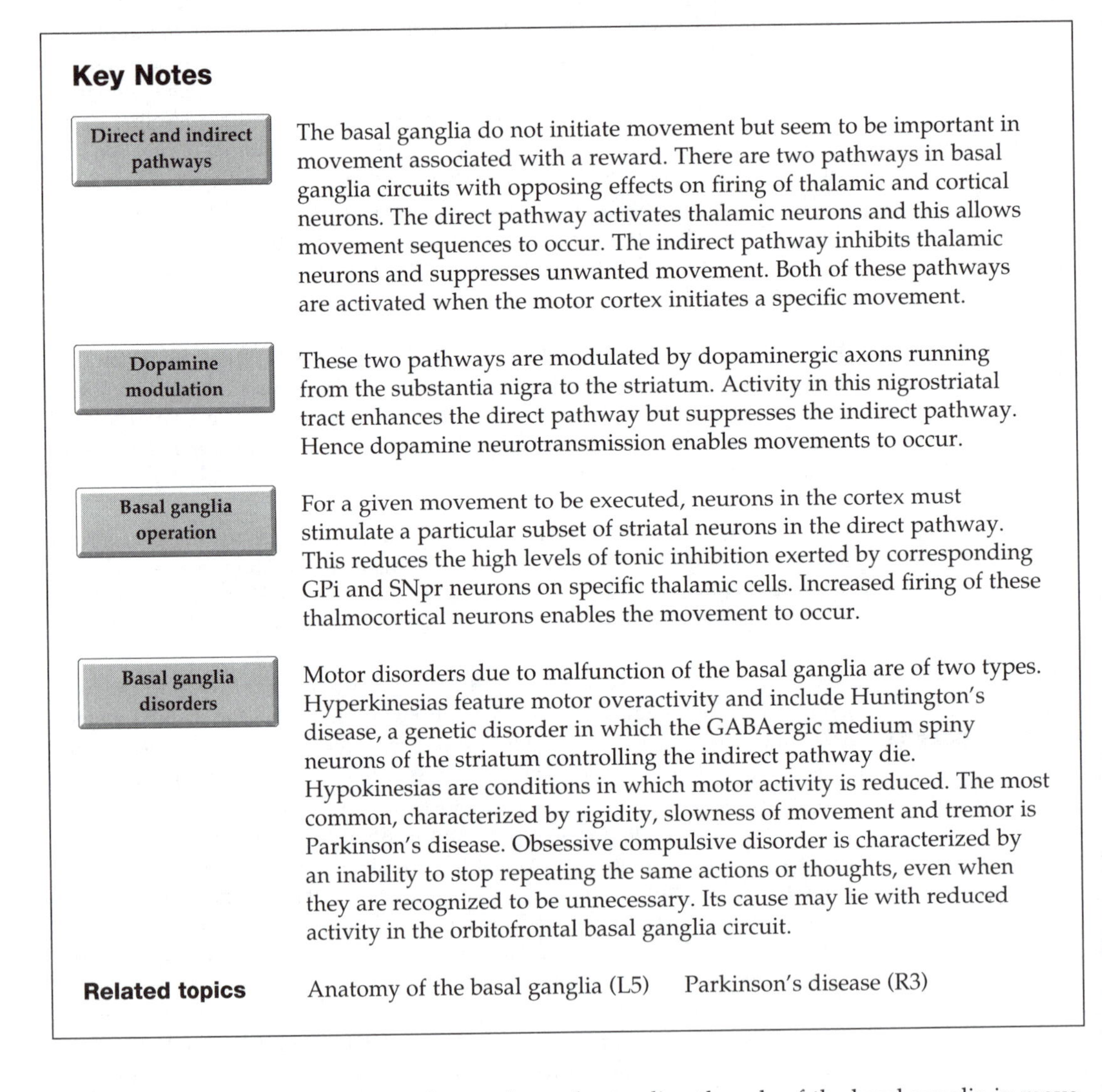

Key Notes

Direct and indirect pathways

The basal ganglia do not initiate movement but seem to be important in movement associated with a reward. There are two pathways in basal ganglia circuits with opposing effects on firing of thalamic and cortical neurons. The direct pathway activates thalamic neurons and this allows movement sequences to occur. The indirect pathway inhibits thalamic neurons and suppresses unwanted movement. Both of these pathways are activated when the motor cortex initiates a specific movement.

Dopamine modulation

These two pathways are modulated by dopaminergic axons running from the substantia nigra to the striatum. Activity in this nigrostriatal tract enhances the direct pathway but suppresses the indirect pathway. Hence dopamine neurotransmission enables movements to occur.

Basal ganglia operation

For a given movement to be executed, neurons in the cortex must stimulate a particular subset of striatal neurons in the direct pathway. This reduces the high levels of tonic inhibition exerted by corresponding GPi and SNpr neurons on specific thalamic cells. Increased firing of these thalmocortical neurons enables the movement to occur.

Basal ganglia disorders

Motor disorders due to malfunction of the basal ganglia are of two types. Hyperkinesias feature motor overactivity and include Huntington's disease, a genetic disorder in which the GABAergic medium spiny neurons of the striatum controlling the indirect pathway die. Hypokinesias are conditions in which motor activity is reduced. The most common, characterized by rigidity, slowness of movement and tremor is Parkinson's disease. Obsessive compulsive disorder is characterized by an inability to stop repeating the same actions or thoughts, even when they are recognized to be unnecessary. Its cause may lie with reduced activity in the orbitofrontal basal ganglia circuit.

Related topics

Anatomy of the basal ganglia (L5) Parkinson's disease (R3)

Direct and indirect pathways

An important feature for understanding the role of the basal ganglia in movement is the presence of two routes through the basal ganglia circuitry with opposite effects on firing of thalamic, and hence cortical neurons (*Fig. 1*). The **direct pathway** uses the GABA/substance P (SP)/dynorphin (DYN) medium spiny striatal neurons which inhibits GABAergic outflow of the globus pallidus pars interna (GPi) and SNpr to the thalamus. Cortical activation of this pathway increases the firing of thalamic neurons (since inhibiting an inhibition is excitation).

The **indirect pathway** starts with the GABA/encephalin (ENK) medium spiny neuron output to the GPe, inhibitory neurons from which go to the subthalamic nucleus (STN). The STN excites inhibitory neurons in the GPi and

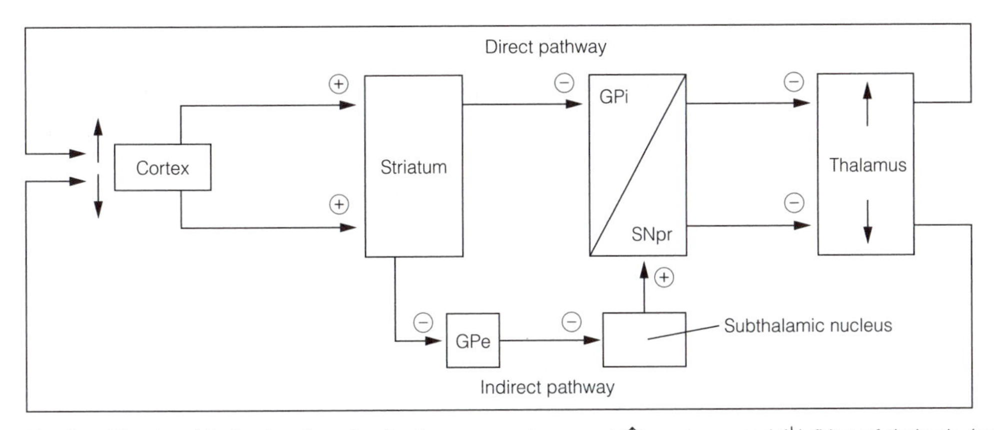

Fig. 1. Direct and indirect pathway activation causes increased (↑) or decreased (↓) firing of thalamic (and cortical) neurons respectively. GPe, Globus pallidus pars externa; GPi, globus pallidus pars interna; SNpr, substantia nigra pars reticulata.

SNpr that go to the thalamus. Corticostriate activation of the indirect pathway results in decreased firing of thalamic neurons. This dual circuitry allows the possibility that given movement sequences may be triggered or suppressed by differential activation of direct or indirect pathways respectively.

Dopamine modulation

The neurons of the substantia nigra pars compacta (SNpc) use dopamine as a transmitter. At rest they fire sporadically at low frequency and their firing is uncorrelated with movement *per se*. SNpc neurons alter their firing pattern in response to stimuli that reward a movement. They modulate the response of medium spiny neurons in the striatum to corticostriate inputs but have opposite effects on the two populations. While the GABA/SP/DYN neurons are made more excitable, the GABA/ENK cells become less excitable in the face of SNpc inputs. In summary, the direct pathway is enhanced, and the indirect pathway suppressed, by the nigrostriatal tract from the SNpc.

Basal ganglia operation

One function of the basal ganglia is to enable the execution of motor sequences. Each sequence is represented by an array of cells , a micro-loop, within the basal ganglia–thalamocortical motor or oculomotor circuit and can be either activated or (if unwanted) inhibited. While some sequences are stereotyped movements, the circuitry for which is genetically specified, many sequences are learnt, that is, particular micro-loops come to operate as a result of experience.

At rest, most medium striatal neurons fire at low frequencies (0.1–1 Hz), while GPi and SNpr neurons have high background firing rates (about 100 Hz). The current model is that movements are initiated by activity in the motor cortex which is relayed to the striatum. During a movement, striatal neurons increase firing, as a result of elevated activity of the corticostriatal neurons that drive them. Tonic inhibitory output of the GPi and SNpr at rest, which is increased about 50 ms before a movement by excitatory drive from the subthalamus, is due to the operation of the indirect pathway and results in widespread, and complete, suppression of unwanted movement sequences. Making a particular movement requires that the direct striatopallidal pathway to the GPi/SNpr cells that enable the movement (those belonging to the correct micro-loop) become activated. These GPi/SNpr cells reduce their firing, releasing their corresponding thalamocortical cells from inhibition. The nigrostriatal dopamine system acts to raise the likelihood that a movement sequence is actually made.

The similarity of basal ganglia–thalamocortical circuits implies that they all perform the same computation. If this is so, it seems likely that the cognitive circuits may operate to select particular behavioral sequences in a manner that is appropriate to the context. In the light of this it is interesting that lesions of the orbitofrontal cortex result in perseveration, in which a behavior is continued long after it is needed or appropriate.

Basal ganglia disorders

Motor disorders arising from dysfunction of the basal ganglia, whether caused by disease, or by lesions in animal studies, fall into two distinct categories, hyperkinesias and hypokinesias.

Hyperkinesias

These are disorders in which motor overactivity occurs. Characteristically these disorders consist of frequent, random, twitch-like or writhing movements, resembling fragments of normal movements, termed **choreoathetosis**. It is the principle symptom of Huntington's disease, and of **tardive dyskinesia**, an unwanted effect of the treatment of Parkinson's disease with L-DOPA (see below and Topic R3), or infarcts of the subthalamic nucleus.

Huntington's disease is a progressive neurodegenerative disorder in which symptoms (cognitive as well as motor) begin between 40 and 50 years of age. It is an autosomal dominant disease caused by an abnormality in a gene on chromosome 4 which codes for a widely distributed protein, **huntingtin**, the function of which is not known. The abnormality is an excessive number of trinucleotide (CAG) repeats which code for a string of glutamine residues near the N terminus of the protein. This causes huntingtin molecules to form aggregates in the nuclei of specific neurons. Particularly afflicted are the GABA/ENK medium spiny neurons of the striatum. Their death causes abnormal inhibition of the subthalamic nucleus, and thus increased and inappropriate firing of thalamocortical neurons. In summary, the chorea is a failure of the indirect pathway to block unwanted movement sequences.

Infarcts of the subthalmic nucleus (STN) are rare and cause a greatly exaggerated chorea characterized by large flinging movements of limbs, contralateral to the lesion, termed **hemiballismus**. Loss of the tonic excitatory drive from the STN is thought to cause GABAergic neurons in the GPi and SNpr to go into a burst firing mode. Why this happens is not known, neither is it clear what adaptation allows the recovery from hemiballismus that is seen after a few weeks both in humans and in STN-lesioned animals.

Tics are hyperkinesias in which highly stereotyped and sometimes quite complex movements of face and hands appear. Sometimes tics are associated with behavioural disorder such as in the rare **Gilles de la Tourette** syndrome where they are accompanied by involuntary uttering of sexual obscenities.

Hypokinesias

These are disorders in which motor activity is reduced. In animals, lesions of the globus pallidus results in abnormal co-contraction of agonist and antagonist muscles in the contralateral limbs. The effect of this is to raise the stiffness of joints causing **rigidity** and slowness of movement, **bradykinesia**. Bilateral lesions result in animals adopting abnormal flexed postures that they seem unable to move out of. This resembles **dystonias** seen in humans in a variety of conditions including end stage Huntington's disease, strokes, and an unwanted effect of treatment with dopamine D2 receptors such as metoclopramide. The prototypical hypokinetic disorder is **Parkinson's disease**, characterized by rigidity, bradykinesia and tremor. Its pathology and treatment are dealt with in Topic R3.

Obsessive–compulsive disorder (**OCD**) is a chronic psychiatric disorder in which a person is unable to prevent themselves endlessly repeating the same actions or thoughts. The afflicted individual may spend many hours each day acting out pointless rituals, such as hand washing because of an obsessional fear of contamination, or having to check that the front door is locked a set number of times before going out. As with other neuroses, it represents an exaggeration of normal behavior, and in OCDs this appears as excessive perseveration. Brain imaging studies show a reduction in cerebral blood flow in the orbitofrontal cortex that correlates with the severity of the disorder in patients. Lesions of the orbitofrontal cortex in primates causes perseveration. This suggests that the cause of OCD may lie with dysfunction of the orbitofrontal basal ganglia circuit.

L7 OCULOMOTOR CONTROL

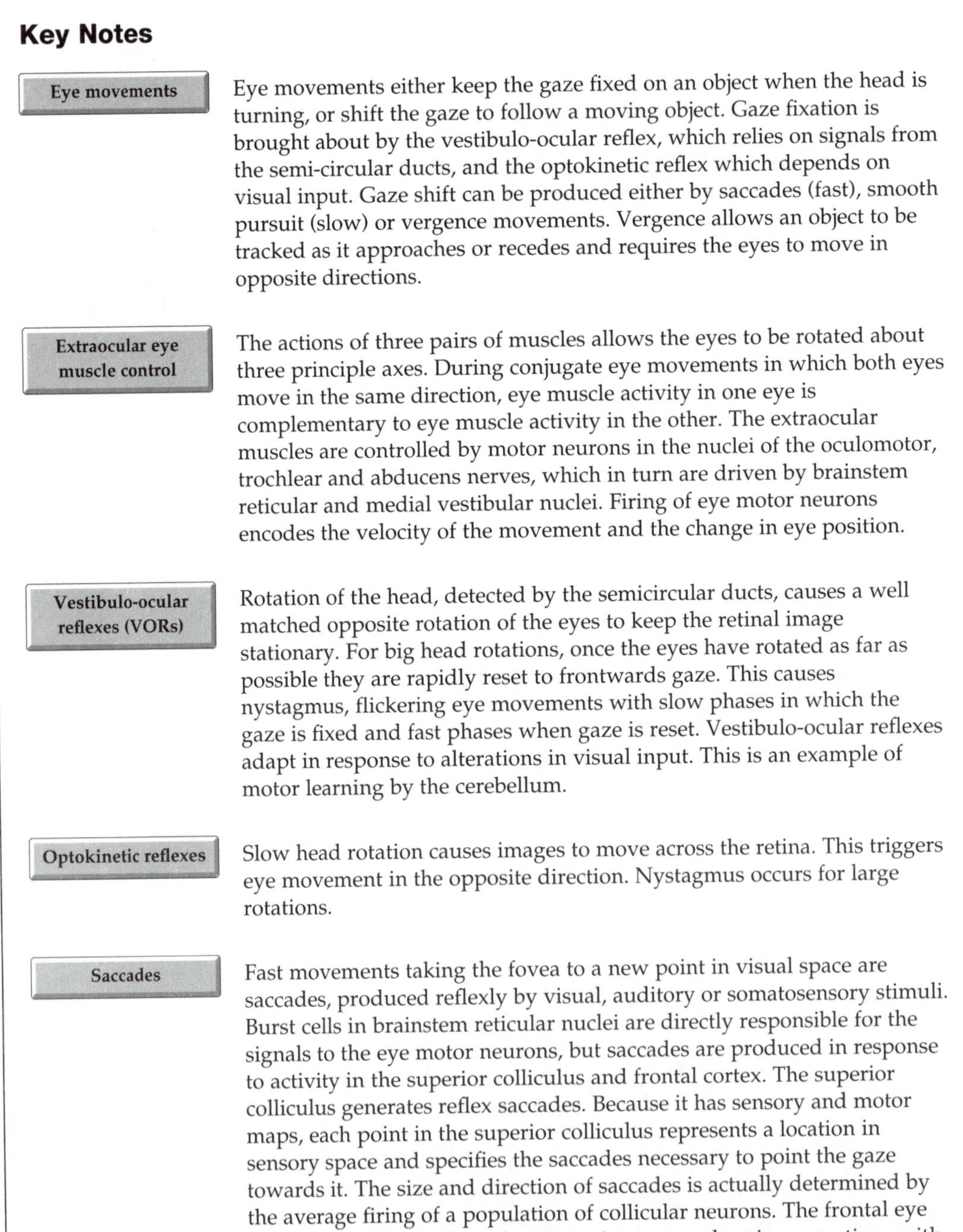

Key Notes

Eye movements

Eye movements either keep the gaze fixed on an object when the head is turning, or shift the gaze to follow a moving object. Gaze fixation is brought about by the vestibulo-ocular reflex, which relies on signals from the semi-circular ducts, and the optokinetic reflex which depends on visual input. Gaze shift can be produced either by saccades (fast), smooth pursuit (slow) or vergence movements. Vergence allows an object to be tracked as it approaches or recedes and requires the eyes to move in opposite directions.

Extraocular eye muscle control

The actions of three pairs of muscles allows the eyes to be rotated about three principle axes. During conjugate eye movements in which both eyes move in the same direction, eye muscle activity in one eye is complementary to eye muscle activity in the other. The extraocular muscles are controlled by motor neurons in the nuclei of the oculomotor, trochlear and abducens nerves, which in turn are driven by brainstem reticular and medial vestibular nuclei. Firing of eye motor neurons encodes the velocity of the movement and the change in eye position.

Vestibulo-ocular reflexes (VORs)

Rotation of the head, detected by the semicircular ducts, causes a well matched opposite rotation of the eyes to keep the retinal image stationary. For big head rotations, once the eyes have rotated as far as possible they are rapidly reset to frontwards gaze. This causes nystagmus, flickering eye movements with slow phases in which the gaze is fixed and fast phases when gaze is reset. Vestibulo-ocular reflexes adapt in response to alterations in visual input. This is an example of motor learning by the cerebellum.

Optokinetic reflexes

Slow head rotation causes images to move across the retina. This triggers eye movement in the opposite direction. Nystagmus occurs for large rotations.

Saccades

Fast movements taking the fovea to a new point in visual space are saccades, produced reflexly by visual, auditory or somatosensory stimuli. Burst cells in brainstem reticular nuclei are directly responsible for the signals to the eye motor neurons, but saccades are produced in response to activity in the superior colliculus and frontal cortex. The superior colliculus generates reflex saccades. Because it has sensory and motor maps, each point in the superior colliculus represents a location in sensory space and specifies the saccades necessary to point the gaze towards it. The size and direction of saccades is actually determined by the average firing of a population of collicular neurons. The frontal eye fields, located in the frontal cortex, trigger saccades via connections with the superior colliculus and brainstem. The frontal cortex is responsible for intentional saccades.

Smooth pursuit movements

These are used to voluntarily track an object that is moving in the visual field. The velocity of the object is signaled by the cortex of the visual 'where' system to neurons in the pons. These cells convert the velocity signals to smooth pursuit motor commands.

Vergence

Signals for vergence include blurring of the retinal image or the degree of accommodation and require the visual cortex. Fast vergence movements are made during saccades.

Related topics

Sense of balance (G4)
Eye and visual pathways (H2)
Parallel processing in the visual system (H7)
Cortical control of voluntary movement (K6)
Cerebellar function (L4)
Anatomy of the basal ganglia (L5)
Cerebellar motor learning (Q5)

Eye movements

The purpose of eye movements is either gaze stabilization, in which the eyes remain fixated on an object during rotation of the head, or gaze shifting which allows the central part of the retina, the fovea, to be brought to bear on an object, or track a moving object. Five types of eye movements, each controlled by a distinct neural system, bring about these aims.

Gaze stabilization is controlled by the vestibulo-ocular and optokinetic systems. Rapid head rotation, detected by the semi-circular ducts provides input for vestibulo-ocular reflexes (VOR), whereas optokinetic reflexes depend on visual input to monitor slow head rotations. For both systems their output causes conjugate eye movements in the opposite direction to the head rotation, so that retinal images do not shift.

Three systems organize gaze shift. The saccadic system generates extremely rapid eye movements, saccades, which move the gaze from one point in the visual field to another, bringing new targets onto the fovea. The smooth pursuit system permits gaze to follow a moving target, so that its image remains on the fovea. Finally, for animals with binocular vision, the vergence system allows the eyes to move in opposite directions (disjunctive movements): either both converge or both diverge, so that both eyes can remain directed towards an object as it gets closer or recedes.

The output of all five eye motor systems is via oculomotor neurons in the brainstem, the axons of which run in three pairs of cranial nerves to the skeletal muscles that move the eyes.

Extraocular eye muscle control

Each eye is moved by three pairs of extraocular (extrinsic) eye muscles. Two pairs of rectus muscles (superior and inferior, medial and lateral) originate from a common annular tendon attached at the back of the orbit. These muscles insert into the sclera in front of the equator of the eyeball. The third pair are the oblique muscles (superior and inferior) which insert into the sclera behind the equator of the eyeball (*Fig. 1*).

Working in concert these muscles act to rotate the eye about three principal axes (*Fig. 2*). The actions of the medial and lateral rectus muscles are simple. They cause the eye to rotate about the vertical axis so that the gaze moves horizontally. The medial rectus brings about rotation towards the midline

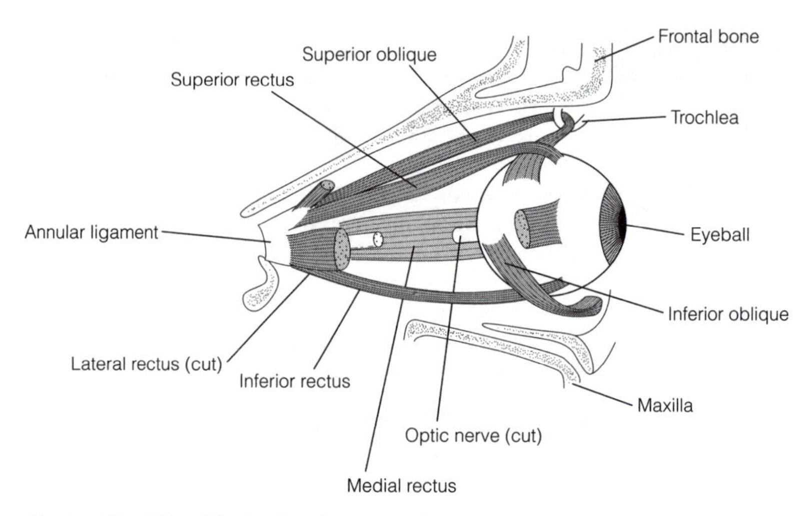

Fig. 1. The right orbit showing the extraocular muscles.

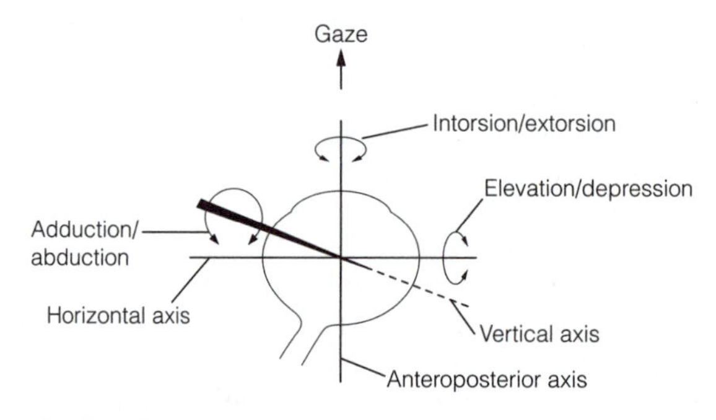

Fig. 2. Principal axis for rotation of the eye, shown for the right eye. In health, torsional movements (rotation about the anteroposterior axis) are small.

(adduction) while the lateral rectus causes lateral rotation (abduction). The other two pairs of muscles produce rotations that have components along two of the principal axes, and the components change depending on the horizontal position of the eye. These actions are summarized in *Table 1*.

Eye muscles act in complementary fashion in the two eyes during conjugate movements in which the two visual axes move in parallel. Thus, contraction of the lateral rectus in one eye is coupled with contraction of the medial rectus in the other eye for a conjugate horizontal shift in gaze (see *Table 1*).

The extraocular muscles are innervated by motor neurons in the nuclei of the oculomotor (III), trochlear (IV) or abducens (VI) cranial nerves. These neurons are the final common path for the output of all five movement systems and are driven by brainstem reticular and medial vestibular nuclei axons that run in the **medial longitudinal fasciculus**. Eye motor neurons fire both statically, in a manner relating to eye position, and dynamically, reflecting eye velocity. To hold the eye steady in a given position requires tonic discharge by a particular set of motor neurons. The set will be different for different positions. Each motor neuron fires with a frequency needed to maintain eye position, so its firing rate is linearly related to position. The difference in the discharge rate of a

Table 1. Actions of extraocular eye muscles. For the muscles moving the eye vertically the action is different depending on whether the eye is also abducted or adducted. For example, the superior rectus elevates the eye if the lateral rectus is active at the same time, but causes intorsion if the eye is adducted by the medial rectus.

Muscle	Innervation	Movement	Contralateral eye complementary muscle
Lateral rectus	Abducens (VI)	Abduction	Medial rectus
Medial rectus	Oculomotor (III)	Adduction	Lateral rectus
Superior rectus	Oculomotor (III)	Elevation and intorsion	Inferior oblique
Inferior rectus	Oculomotor (III)	Depression and extorsion	Superior oblique
Inferior oblique	Oculomotor (III)	Extorsion and elevation	Superior rectus
Superior oblique	Trochlear (IV)	Intorsion and depression	Inferior rectus

motor neuron seen for two eye positions is known as a **step**. Activity by a given neuron is not needed for all positions (e.g. sustained leftward gaze requires high firing rates of motor units in the left lateral rectus, but the left medial rectus is an antagonist of this movement so its motor neurons remain silent).

Eye movements are brought about by high frequency pulses of action potentials in oculomotor neurons. The discharge rate during a pulse is directly proportional to the velocity of the movement. Any eye movement that requires the eyes first to move, then to be held at their new positions have a **pulse–step** configuration. For the eyes to be held in their new position requires the generation of a new position signal. It is thought that integration of the velocity signal is responsible for generating this position signal, an operation carried out by the vestibulocerebellum and **prepositus nucleus** of the brainstem reticular system.

Vestibulo-ocular reflexes (VORs)

Head rotation detected by the semi-circular ducts (see Topic G5) triggers equal and opposite rotation of both eyes. For large amplitude head rotations the eyes cannot continue to rotate but must be reset to a central position by rapidly moving in the same direction as the head. This gives rise to **nystagmus**, eye movements characterized by slow phases that stabilize the retinal image, and quick phases that reset the eyes. By convention the direction of the nystagmus is the direction of the quick phase (*Fig. 3*).

The horizontal semi-circular ducts are effectively wired to the medial and lateral rectus muscles to produce the eye movements that counter the head rotation (*Fig. 4*).

The gain of the VOR (the magnitude of the eye rotation divided by the magnitude of the head rotation) is quite close to one for fast head rotations. This means there is a good match between eye and head movements which makes for a stable retinal image. The VOR can be modified by visual experience. When human subjects wear magnifying lenses which means that they should make bigger eye movements to match head rotations, the gain of their VOR increases appropriately over the next few days. The cerebellum is required for this adaptation to occur, but not for it to be maintained once established. The instability of the retinal image acts as an error signal which is relayed to the cerebellum from the inferior olivary nucleus in climbing fibers. The cerebellum learns to minimize the error and alters its drive to the extraocular muscles. This is an example of motor learning, further details of which are in Topics L4 and Q5. Lesions of the vestibulocerebellum impair the ability to maintain steady gaze, causing inappropriate nystagmus.

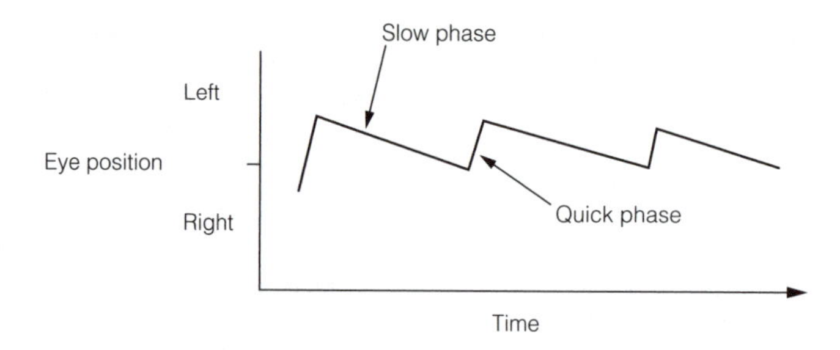

Fig. 3. Leftward nystagmus during head rotation.

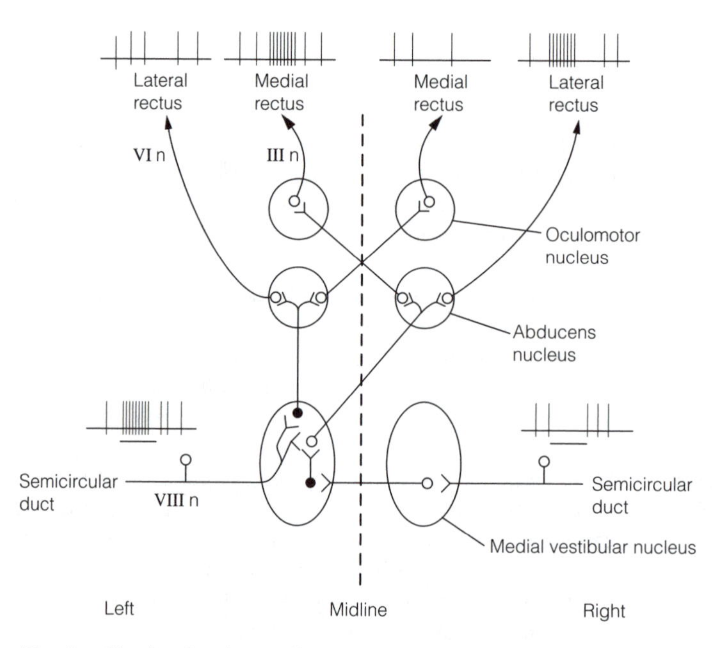

Fig. 4. Circuitry for the vestibulo-ocular reflex. Stimulation of the horizontal semi-circular ducts by the leftward rotation of the head excites the ipsilateral medial rectus and the contralateral lateral rectus and inhibits their antagonists. Excitatory neurons, open circles; inhibitory neurons, filled circles. Firing patterns are shown for cranial nerve neurons.

Optokinetic reflexes

Slow rotation of the head causes an apparent movement of the visual world in the opposite direction called **retinal slip**, This is detected by large, movement sensitive retinal ganglion cells and used to produce eye movements which are equal in speed but of opposite direction to the retinal slip. As in the VOR, nystagmus occurs for large head rotations.

Saccades

Saccades are very fast conjugate eye movements that move the fovea to target a different point in visual space. The saccade system uses visual, auditory and somatosensory input to determine the eye rotation required to realign the gaze. Horizontal saccades are controlled by the **paramedian pontine reticular formation** which lies at the midline adjacent to the nuclei of the oculomotor, trochlear and abducens cranial nerves. Vertical saccades are organized by the **rostral**

interstitial nucleus of the medial longitudinal fasciculus, situated in the midbrain rostral to the oculomotor (III) nerve nucleus. Both of these structures contain burst cells which code for the size and direction of the eye movement and produce saccades by exciting the oculomotor neurons. The signals that trigger saccades come from two sources, the superior colliculus and the frontal eye fields. Both these structures can generate saccades independently of the other. Destruction of both renders primates incapable of making saccades.

The **superior colliculus** lies in the tectum of the midbrain and is divided into superficial, intermediate and deep layers. The superficial layers receive visual information from the retina and visual cortex that gives rise to a map of the contralateral visual field. The deep layers get auditory and somatosensory input and so have two maps, an auditory map depicting the location of sounds in space, and a somatotopic map in which the body parts closest to the eyes gets the greatest representation. The intermediate layers are the site of a motor map. Neurons here are called **collicular saccade-related burst neurons** because they fire a high frequency burst of action potentials about 20 ms before a saccade. Each one has a **movement field** (the equivalent of a receptive field) that covers the sizes and directions of the saccades it participates in. These movement fields are large, in that the cells are active for many similar saccades, but they fire maximally for a preferred saccade. The broad tuning of these cells means that the direction of any given saccade is encoded by a population of neurons whose firing determines precisely the direction of the required saccade. This is exactly the way in which the primary motor cortex uses population coding to determine the direction of a movement (see Topic K6).

A key function of the superior colliculus is to turn sensory coordinates into motor coordinates. All of the four maps are in register. Each point on the superior colliculus represents a specific location in sensory space and the saccades necessary to direct gaze towards it. Visual input to the superficial layers need not lead to firing of collicular saccade-related burst neurons in the intermediate layer. This is because superficial layer cells do not synapse directly with intermediate layer cells but instead connect with them indirectly via a relay through the **pulvinar** of the thalamus and the visual cortex. The purpose of this relay may be to decide the importance (**salience**) of a particular visual stimulus and so allow saccades only to stimuli with high salience. *Fig. 5* shows the circuitry involved in saccades.

In addition to generating saccades the superior colliculus cause head rotation, by way of the **tectospinal tract**, to neck muscle motor neurons. This allows orientation towards a stimulus, so-called **orienting responses**.

The **frontal eye field (FEF)** of the frontal cortex triggers saccades via the intermediate layers of the superior colliculus, and the pontine and midbrain reticular nuclei. The FEF directly stimulates the collicular saccade-related burst neurons in the intermediate layers of the superior colliculus. In addition the FEF (and associated cortical areas) signal via the oculomotor basal ganglia–thalamocortical circuit (Topic L5) to the burst cells, releasing them from GABAergic inhibition by the substantia nigra pars reticulata.

Lesions of the superior colliculus cause temporary impairment in triggering saccades, but recovery occurs because the FEF can trigger saccades by its direct connections with the pons and midbrain. Damage to the FEF causes transient paralysis of gaze towards the opposite side, but reflex saccades soon return, produced by the superior colliculus. However, loss of the FEF prevents intentional or anticipatory saccades.

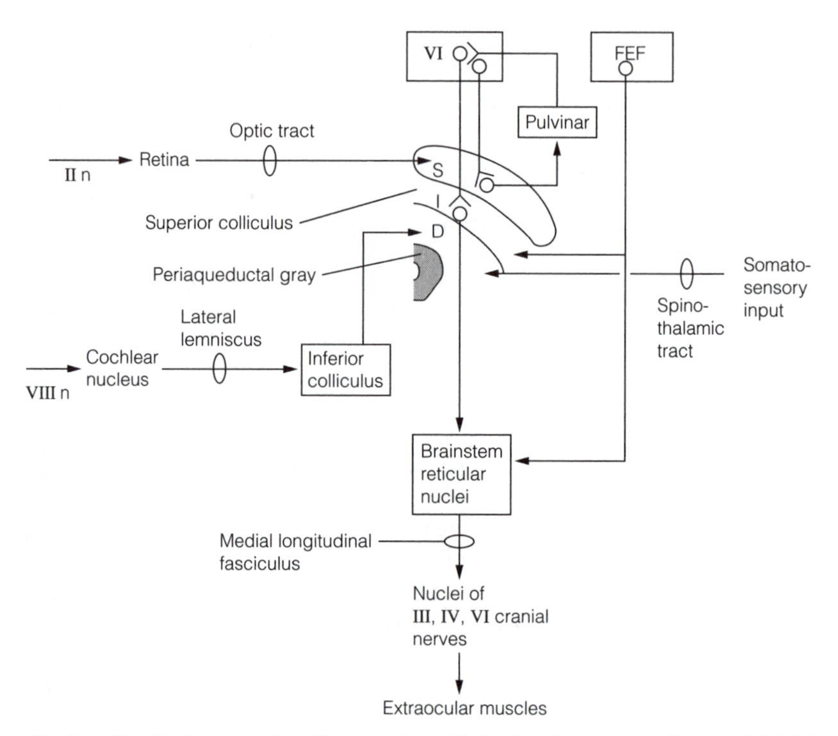

Fig. 5. Circuitry for saccades. The superior colliculus has three layers: S, superficial; I, intermediate; D, deep. FEF, frontal eye fields; VI, primary visual cortex.

Smooth pursuit movements

Intentionally tracking a moving object so that its image remains on the fovea is done by the smooth pursuit system. Smooth pursuit movements differ from optokinetic reflexes in being voluntary, and in attending to movement over a small part of visual space; optokinetic reflexes are involuntary and are responses to movement of the entire visual world.

Signals relating to the velocity (i.e. speed and direction) of the target are generated by the medial temporal cortex of the visual 'where' system (see Topic H7) which analyses motion. Lesions of this cortex impair pursuit movements. These signals are transmitted to the **dorsolateral pontine nucleus (DLPN)**, which translates the target velocity into the motor commands for the pursuit movement. The DLPN projects to the vestibulocerebellum, cells of which fire in a precisely correlated way with smooth pursuit movements. Its output to the medial vestibular nuclei drives the smooth pursuit movements.

Vergence

Vergence is the only disjunctive eye movement. For example, shifting gaze to a closer target requires adduction of both eyes, which is achieved by contracting both medial rectus muscles. Signals to produce vergence include blurring of the retinal image by large degree of retinal disparity, the extent of accommodation, or monocular cues to distance. These all require the visual cortex. Fast vergence movements occur during saccades.

M1 Anatomy and connections of the hypothalamus

Key Notes

Hypothalamic anatomy

The hypothalamus, located in the diencephalon, is composed of numerous nuclei and is implicated in sleep, appetitive behaviors, and the control of autonomic and endocrine functions, which are largely the responsibility of the paraventricular nuclei (PVN). The PVN contain large neurons that project into the posterior pituitary, via the pituitary stalk, and small neurons which have their terminals in the median eminence, which lies immediately above the pituitary stalk. The hypothalamus is divided into three zones in the mediolateral direction and four subdivisions along the rostro-caudal axis.

Connections of the hypothalamus

The hypothalamus is part of the limbic system which is concerned with emotion. It gets input from the hippocampus via the fornix, fibers of which go largely to the mammillary bodies, but also to all other areas of the hypothalamus. The output of the mammillary bodies goes to the anterior thalamic nuclei. The anterior thalamus projects to the cingulate cortex which, in turn, sends output to the hippocampus, thus closing a loop, the Papez circuit. Input from the amygdala arrives by two routes, the stria terminalis and the amygdalofugal pathway. The medial forebrain bundle, which consists of monoaminergic axons, passes through the hypothalamus making many connections *en route*.

Pituitary gland

The pituitary gland consists of a posterior lobe, which is a direct outgrowth of the hypothalamus, and an anterior lobe. It is connected to the base of the brain by a pituitary stalk. Large neurosecretory neurons in the hypothalamus send axons into the posterior lobe and release hormones there. Hormones secreted into the median eminence by small hypothalamic cells are, in contrast, carried in a vascular network, the hypothalamic–pituitary portal circulation, to the anterior lobe.

Related topics

Organization of the central nervous system (E2)
Posterior pituitary function (M2)
Neuroendocrine control of metabolism and growth (M3)
Neuroendocrine control of reproduction (M4)
Control of feeding (O2)
Sleep (O4)

Hypothalamic anatomy

The hypothalamus is part of the diencephalon and lies ventral to the thalamus. It is involved in the control of a variety of functions: sleep–wakefulness, thermoregulation, feeding and the regulation of metabolic energy expenditure, drinking and fluid homeostasis, growth and reproduction. Some of these functions are brought about by the hypothalamus regulating the autonomic nervous system or the output of hormones from the pituitary gland. This section is confined to the neuroendocrine and autonomic activities of the hypothalamus.

The hypothalamus consists of many nuclei clustered around the third ventricle (*Fig. 1*). At its rostral end lies the optic chiasm. The most caudal parts of the hypothalamus are the paired mammillary bodies. The floor of the third ventricle is a sheet of gray matter called the **tuber cinereum** extending from the optic chiasm to the mammillary bodies. At its anterior end this thickens to become the **median eminence** which projects as the **infundibular stalk** (part of the pituitary stalk) to the posterior pituitary gland. The hypothalamus is divided into three longitudinal zones, the **periventricular zone**, which surrounds the lining of the third ventricle, the **medial (intermediate) zone**, and

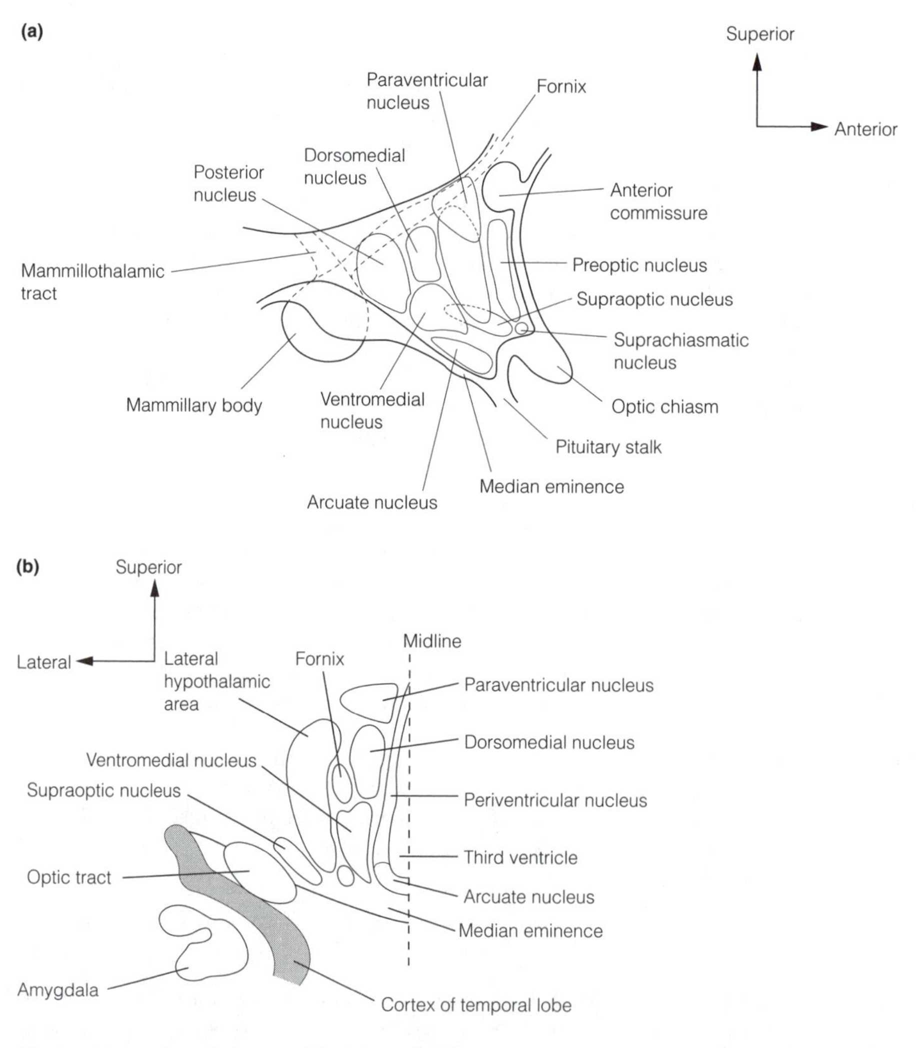

Fig. 1. Human hypothalamus of the left cerebral hemisphere shown diagrammatically: (a) midsagittal section; (b) coronal section.

the **lateral zone**. It also has four subdivisions along its rostro-caudal axis, **preoptic**, **anterior**, **tuberal** and **mammillary**. The locations of major nuclei are specified in *Table 1.* Much of the endocrine and autonomic functions of the hypothalamus involve the paraventricular nuclei. They contain many populations of neuroendocrine cells, each secreting specific peptides, but they fall into two groups. Magnocellular (large) cells which project to the posterior pituitary and parvocellular (small) cells that project to the median eminence.

Table 1. Location of some hypothalamic nuclei

	Zone		
Subdivision	Periventricular	Medial	Lateral
Preoptic		Medial preoptic nucleus	Lateral preoptic nucleus
Anterior	Suprachiasmatic nucleus Paraventricular nucleus Anterior periventricular nucleus	Anterior nucleus	Supraoptic nucleus
Tuberal	Arcuate nucleus	Ventromedial nucleus Dorsomedial nucleus	Lateral hypothalamic area
Mammillary	Posterior hypothalamic nucleus	 Medial mammillary nuclei[a]	Lateral hypothalamic area Lateral mammillary nuclei[a]

[a] Mammillary bodies.

Connections of the hypothalamus

The hypothalamus is connected with limbic structures that are concerned with emotion and its expression (*Fig. 2*). It gets input from the hippocampus by way of the **subiculum** (a region of the transitional cortex (see Topic E2)) via the **post-commissural fornix**, which projects mainly to the mammillary bodies, and by way of the **septum**, via the **precommissural fornix**, which makes connections with all three zones of the hypothalamus. Input from the amygdala arrives at

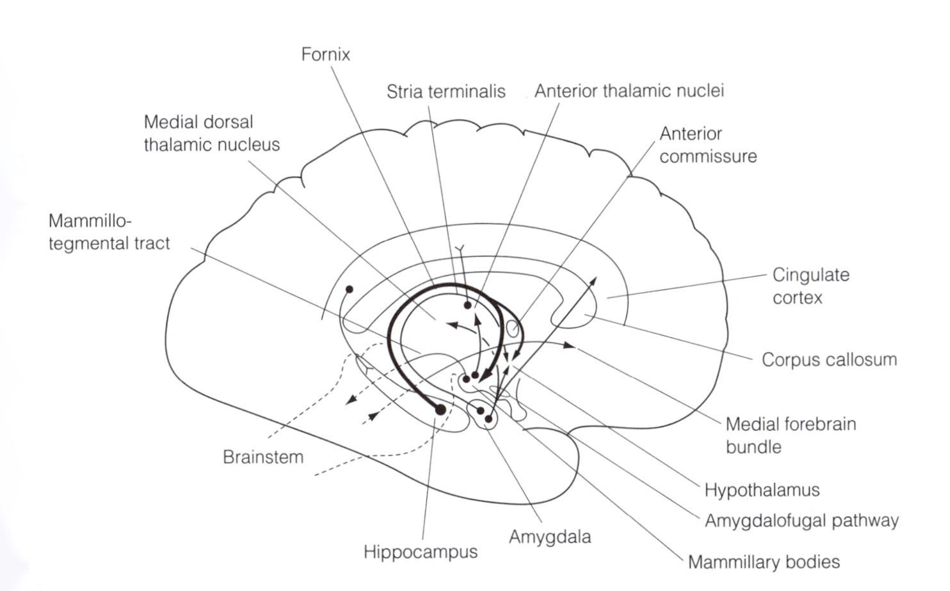

Fig. 2. Major connections of the human hypothalamus.

the hypothalamus via the **stria terminalis**, a loop that follows a similar course to the fornix, and the **amygdalofugal pathway**.

Hypothalamic output from mammillary bodies (MB) via the **mammillo-thalamic tract** goes to the **anterior thalamic nuclei** that are connected to the **cingulate cortex (CC)**. The CC projects to the hippocampus, so closing a loop (MB–ATN–CC–hippocampus–hypothalamus) termed the **Papez circuit**, originally proposed for conscious awareness of emotions and cognitive effects on emotion. The hypothalamus also projects to the prefrontal cortex. The mammillary bodies project via the mammillotegmental tract to the midbrain.

The **medial forebrain bundle** passes through the lateral hypothalamic zone. It consists mainly of monoaminergic axons ascending from brainstem nuclei. Many noradrenergic and serotonergic axons synapse with hypothalamic neurons. Dopaminergic axons from the substantia nigra and the ventral tegmentum, however, simply traverse the hypothalamus without establishing connections.

The paraventricular hypothalamus and the lateral hypothalamic area receive visceral sensory input from the nucleus of the solitary tract which is important for hypothalamic control of the ANS.

Pituitary gland

The **pituitary gland (hypophysis)** is divided into the **neurohypophysis** and the **adenohypophysis**. The neurohypophysis, which is a direct outgrowth of the hypothalamus, consists of the **posterior lobe**, the infundibular stalk and the median eminence. The adenohypophysis consists of the **anterior lobe**, an intermediate lobe which is poorly developed in humans, and the **pars tuberalis**, an extension surrounding the infundibular stalk. The pars tuberalis and the infundibular stalk are together termed the **pituitary stalk (infundibulum)** (*Fig. 3*).

There are two routes by which the hypothalamus controls hormone output from the pituitary. The functional connection between the hypothalamus and the posterior lobe is *neural*. Large **neurosecretory neurons** with their cell bodies

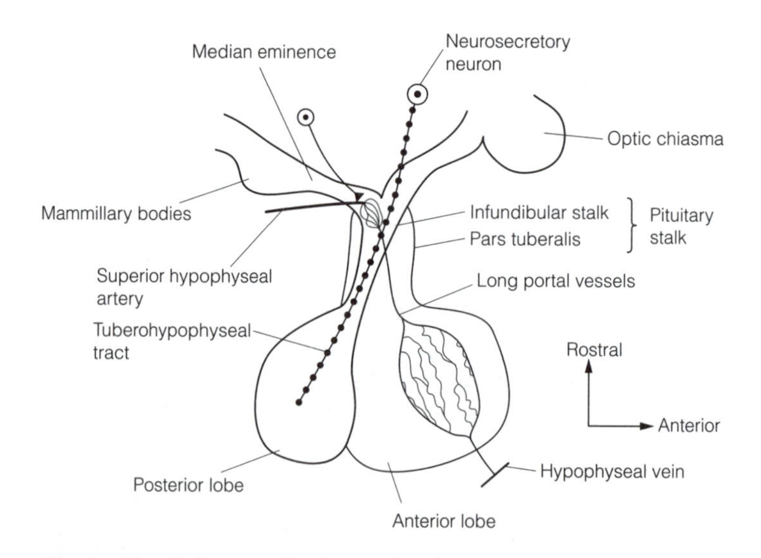

Fig. 3. The pituitary showing its neural and vascular connections with the hypothalamus. In humans the tuberohypophyseal tract contains about 100 000 axons.

in the hypothalamus send their axons through the median eminence down the infundibular stalk into the posterior lobe as the **tuberohypophyseal tract**. Hormones released from the posterior lobe are actually made in the cell bodies of the neurosecretory cells and secreted from their axons. By contrast, the functional connection between the hypothalamus and the anterior lobe is *vascular*. The superior hypophyseal artery forms a plexus of capillaries in the medial eminence which drains into **long portal vessels** that descend to the anterior lobe. The portal vessels open into venous sinusoids which supply the cells of the anterior pituitary with blood. The function of this **hypothalamic–pituitary portal circulation** is to deliver hormones, secreted by small hypothalamic neurons into the median eminence, to the anterior lobe.

M2 Posterior pituitary function

Key Notes

Posterior lobe hormones

Neurons in the supraoptic nucleus (SON) and the paraventricular nucleus (PVN) send their axons into the posterior lobe and release two peptide hormones, arginine vasopressin (AVP) and oxytocin.

Arginine vasopressin

AVP is secreted from the posterior pituitary in response to a rise in the osmolality of the extracellular fluid or a fall in blood volume. It acts to restore these by increasing the reabsorption of water by the nephrons of the kidney. Changes in osmolality are detected by neurons in a circumventricular organ which synapse on the PVN and SON. Alterations in blood volume are detected in two ways. Firstly, as changes in mean arterial blood pressure that are signaled by baroreceptors, afferents of which go to the nucleus of the solitary tract (NST). The NST relays the baroreceptor signals to the PVN and SON. Reduced blood pressure causes an increase in AVP secretion. Secondly, a fall in blood volume is detected by the nephron, which responds by secreting renin. This enzyme triggers a cascade that generates angiotensin II (AII). AII stimulates both AVP secretion and drinking.

Oxytocin

Oxytocin, released from cells in the PVN and SON different from those that secrete AVP, stimulates contraction of smooth muscle. Suckling stimulates reflex milk ejection by release of oxytocin, and uterine contractions during childbirth occur in response to oxytocin release triggered reflexively by the pressure of the fetus on the neck of the uterus.

Related topics

Blood–brain barrier (A5)
Autonomic nervous system function (M6)

Posterior lobe hormones

Magnocellular (large) neurosecretory neurons in the **supraoptic nucleus (SON)** and **paraventricular nucleus (PVN)** send their axons into the posterior lobe. These neurons secrete the nonapeptides **arginine vasopressin (AVP)**, also known as **antidiuretic hormone**, and **oxytocin**. Both are synthesized as prohormones in the cell bodies of the neurons and packed into large (120 nm) neurosecretory vesicles which are delivered to the terminals by axoplasmic transport. The prohormones are cleaved in the vesicles to give both the hormone and a cleavage product, a **neurophysin**.

Arginine vasopressin

AVP is secreted from the posterior lobe into the systemic circulation in response to an increase in extracellular fluid osmolality or reduced blood volume. It increases the water permeability of the collecting ducts of the nephron, thereby promoting water reabsorption. This has the effect of reducing extracellular fluid

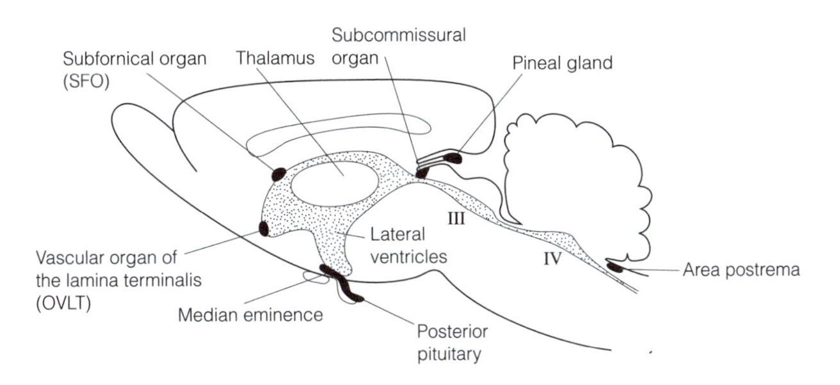

Fig. 1. Location of the circumventricular organs (CVOs, shaded) in the rat brain (midsaggital section); the ventricles are stippled.

osmolality and urine output (an antidiuretic effect) and restoring blood volume. Thus, AVP acts as a negative feedback regulator, defending set points in body fluid osmolality and blood volume.

The stores of AVP in the posterior lobe are large, sufficient to maintain maximum antidiuresis during several days of dehydration. The osmoreceptors which respond to changes in osmolality are in the **vascular organ of the lamina terminalis (OVLT)**. The OVLT is one of the circumventricular organs (CVO) of the brain which lie on the blood side of the blood–brain barrier (see *Fig. 1* and Topic A5) situated at the anterior end of the hypothalamus. Osmolality-sensitive neurons in the OVLT synapse with the PVN and SON cells (*Fig. 2*), increasing their background discharge when osmolality rises. There is a linear relationship between plasma osmolality and AVP secretion.

A reduction in blood volume greater than about 10% stimulates AVP secretion. This is seen in dehydration from any cause (e.g. water deprivation, vomiting or diarrhoea) or with hemorrhage. Two mechanisms operate to trigger AVP release in **hypovolemia** (reduced blood volume).

(1) Hypovolemia lowers mean arterial blood pressure. This is detected by stretch receptors (**baroreceptors**) in the walls of the carotid sinus and aorta. The afferents of these pressure sensors run in the glossopharyngeal (IX) and vagus (X) cranial nerves to the **nucleus of the solitary tract (NST)** in the medulla. The NST activates noradrenergic neurons in the ventrolateral medulla which project to the PVN and SON to bring about AVP release. A reduced blood pressure causes decreased firing of the baroreceptor afferents and hence disinhibition of the circuitry triggering AVP secretion. This is shown in *Fig. 2*.

(2) Activation of the renin–angiotensin cascade (*Fig. 3*). Renin is produced by granular cells in the **juxtaglomerular apparatus (JGA)** of the kidney. It is secreted in response to several factors contingent on a fall in blood volume:

- a drop in perfusion pressure at the renal afferent arterioles
- increased sympathetic stimulation of β receptors on the granular cells (see Topic M5).
- decreased Na^+ delivery to the JGA.

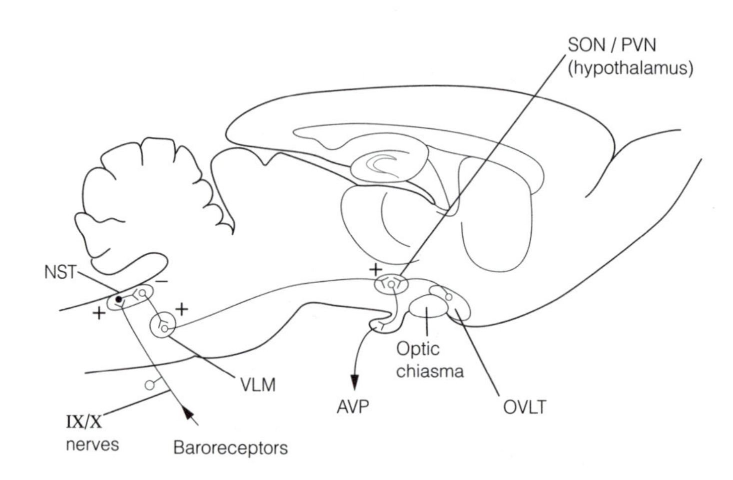

Fig. 2. A model for neural control of arginine vasopressin (AVP) secretion. Increased osmolality detected by the vascular organ of the lamina terminalis (OVLT) stimulates supraoptic and paraventricular nuclei (SON and PVN) cells to secrete AVP. Reduced arterial blood pressure is signaled via the nucleus of the solitary tract (NST) and the ventrolateral medulla (VLM) to the SON/PVN.

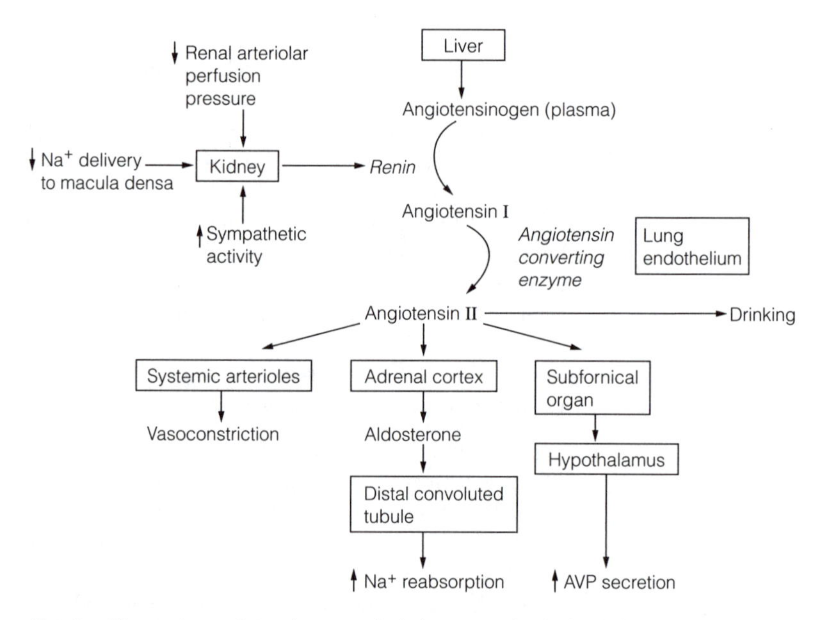

Fig. 3. The renin–angiotensin cascade helps to maintain body fluid osmolality and blood volume.

Renin is a proteolytic enzyme which cleaves a plasma protein substrate, **angiotensinogen**, to yield a decapeptide, **angiotensin I**. This is further cleaved by angiotensin converting enzyme, expressed on pulmonary endothelial cells, to the octapeptide, **angiotensin II (AII)**. AII stimulates another CVO, the subfornical organ, neurons of which stimulate AVP secretion. In addition, AII produces a profound vasoconstriction and stimulates aldosterone secretion by the adrenal cortex. Vasoconstriction has the immediate effect of raising blood pressure, while aldosterone promotes Na^+ reabsorption by the nephron, so producing a delayed expansion of blood volume. Furthermore, AII stimulates drinking, so all the actions of AII are homeostatic in that they restore volume and blood pressure.

In dehydration due to water deprivation, about 70% of the fluid restoration is brought about by osmoreceptor-driven AVP secretion and the remainder is due to responses to reduced blood volume.

Failure or impairment of AVP secretion or inability to respond to the hormone causes **diabetes insipidus (DI)**, characterized by very high urine output, **polyuria** (10–20 l/day), and excessive drinking, **polydipsia**. DI is most commonly caused by destruction of the magnocellular cells of the PVN and SON by tumors or autoimmune disease. Mutations in the vasopressin gene account for genetic DI in humans and in the *Battleboro* rat. Defects in the vasopressin (V2) receptor accounts for **nephrogenic** DI in which patients have reduced responsiveness to AVP. Long half-life analogs of AVP can be used to treat DI.

Oxytocin

Oxytocin stimulates the contraction of smooth muscle. This underlies its actions as a hormone mediating the **milk ejection** reflex in lactating females, and maintaining uterine contractions during **parturition**. Suckling is the most potent stimulus for milk ejection. Primary afferents from the areolar and nipple skin relay with spinothalamic tract neurons in the dorsal horn of the spinal cord. Spinothalamic input causes oxytocin secretion via an undefined neural pathway from midbrain to the PVN and SON. Neurons that secrete oxytocin are distinct from those that secrete AVP. Suckling causes burst firing of oxytocinergic cells, each burst causing the release of a pulse of oxytocin.

Oxytocin does not itself trigger the onset of parturition. However, at term a fall in the concentration of progesterone in the face of a maintained high concentration of estradiol is associated with an upregulation of the number of oxytocin receptors in uterine smooth muscle which becomes very sensitive to oxytocin. Once parturition is established, pressure of the fetal head on the cervix evokes the secretion of oxytocin from the posterior pituitary via a reflex pathway similar to that for milk ejection. The oxytocin stimulates contractions of the uterine smooth muscle. This is the **Ferguson reflex**, and is a positive feedback mechanism since the oxytocin-induced contractions further increase the pressure of the fetus on the cervix. However, since parturition can be normal in spinally transected women, or those with DI who lack oxytocin, additional mechanisms must also be important.

M3 Neuroendocrine control of metabolism and growth

Key Notes

Hypothalamic–anterior pituitary axes

The hypothalamus and anterior pituitary, acting in concert, control five endocrine axes that regulate aspects of metabolism, reproduction, development and growth. Hypothalamic neurons secrete hormones that either stimulate or inhibit the anterior pituitary secretion of trophic hormones. Trophic hormones released into the circulation in turn stimulate target tissues (e.g. adrenals, thyroid and gonads) to secrete their hormones. The secretion of hypothalamic hormones and hence of the trophic hormones is pulsatile. The size and period of the pulses varies cyclically over 24 h, and sometimes over longer times also. Secretion from the endocrine axes is under negative feedback regulation which maintains set point concentrations of hormones. By varying the set point the endocrine axes can change their hormone output.

Hypothalamic–pituitary–adrenal (HPA) axis

The HPA axis controls the secretion of glucocorticoids by the adrenal cortex. Paraventricular neurons secrete corticotrophin releasing hormone (CRH) which causes a population of anterior pituitary cells to release adrenocorticotrophic hormone (ACTH) into the circulation. ACTH is the trophic hormone that stimulates the adrenals to release glucocorticoids. The output of CRH, ACTH and so glucocorticoids varies on a daily basis, typically being highest in the early morning. Glucocorticoids act at two types of receptor that belong to a superfamily of intracellular steroid receptors. Steroids easily diffuse across the plasma membrane, bind to the steroid receptors which consequently translocate to the nucleus, binding hormone responsive elements on DNA to alter gene transcription. Type I receptors are high affinity and are found in limbic structures. Type II receptors are of low affinity, so only bind glucocorticoids when they are in high concentration.

Stress

Stress can be defined as a state in which there is a protracted elevation in concentrations of ACTH and glucocorticoids. Glucocorticoids promote the synthesis of glucose from noncarbohydrate precursors (derived from fats and proteins) and the storage of glucose as glycogen. This transfer of long-term to short-term energy stores is adaptive in stress. The HPA is activated in stress by catecholaminergic neurons concerned with arousal, or hunger and thirst sensations, by brainstem cholinergic neurons conveying visual, auditory and somatosensory input associated with the stressor and from other hypothalamic nuclei relaying limbic system information about the stressful situation.

Hypothalamic–pituitary–thyroid (HPT) axis

Neurons in the paraventricular nucleus (PVN) release thyrotrophin-releasing hormone (TRH) which causes anterior pituitary cells to secrete thyroid stimulating hormone (TSH), a trophic hormone which stimulates growth of the thyroid gland and release of the thyroid hormones (T3 and

T4). There is a daily rhythm in thyroid hormone output, being greatest when it is dark. Thyroid hormone receptors are members of the intracellular steroid receptor superfamily, but unlike steroid receptors, they are bound to nuclear DNA in the absence of hormone. On binding T3 the receptor activates gene transcription. The output of thyroid hormone is regulated by negative feedback modification of both TRH and TSH secretion. Cold exposure excites neurons in the preoptic hypothalamus that activate the HPT axis. The increased secretion of thyroid hormones raises the metabolic rate helping to maintain core temperature. Thyroid hormones are required for fetal brain development, and maternal thyroid hormone deficiency, due to lack of dietary iodide, can cause neurological cretinism in infants.

Growth hormone (GH)

GH released from the anterior pituitary stimulates cell division and growth of many tissues, and mobilizes fatty acids as energy substrates. It is secreted in increased amounts during exercise, stress and fasting. GH secretion is stimulated by growth hormone releasing hormone (GHRH), produced by neurons in the arcuate nucleus, and inhibited by somatostatin from the anterior periventricular nucleus. At both hypothalamus and pituitary, GH stimulates the production of a mediator, insulin-like growth factor, either in the brain or by peripheral tissues. Insulin-like growth factor exerts negative feedback control on the release of GH. GH itself stimulates somatostatin secretion and this also contributes to negative feedback inhibition of GH release. GH release is pulsatile and much higher at night. Several neurotransmitter systems and hormones modify GH output; sex steroids particularly stimulate the high GH output responsible for the growth spurt of puberty.

Related topics

Anatomy and connections of the hypothalamus (M1)
Neuroendocrine control of reproduction (M4)
Brain biological clocks (O3)
Sleep (O4)

Hypothalamic–anterior pituitary axes

Acting through the anterior lobe of the pituitary gland the hypothalamus controls five endocrine **axes**. Between them these neuroendocrine axes regulate key aspects of metabolism, reproduction, development and growth. The five axes share many common features. Neurons, located in several hypothalamic nuclei, send their axons to the external zone of the median eminence and the **tuberoinfundibular tract**. These axons secrete **hypophysiotropic hormones** into the hypothalamic–pituitary portal circulation which carries them into the anterior lobe. Each hypophysiotropic hormone acts on a particular population of cells in the anterior lobe, either exciting or inhibiting their secretion of a specific **stimulating (trophic)** hormone. Hypophysiotropic hormones that excite secretion are termed **releasing** hormones, those that inhibit are called **release inhibiting** hormones. Trophic hormones of the anterior pituitary are secreted by specific cell types into the systemic circulation and have endocrine effects on target tissues, particularly endocrine glands (*Table 1*).

Secretion of hypothalamic hormones is pulsatile with a period of 60–180 min. This drives pulsatile release of anterior pituitary hormones. The amplitude and period of the pulses is varied on a circadian basis and in some cases on longer

Table 1. Five hypothalamic–anterior pituitary neuroendocrine axes

Hypophysiotropic hormone				
Releasing hormone	Release-inhibiting hormone	Anterior pituitary stimulating (trophic) hormone [anterior pituitary cell type]	Target tissue for trophic hormone	Secreted hormone
Corticotrophin releasing hormone	–	Adrenocorticotrophic hormone [corticotroph]	Adrenal cortex	Glucocorticoids
Thyrotrophin releasing hormone	–	Thyroid stimulating hormone [thyrotroph]	Thyroid	Triiodothyronine (T3) and thyroxine (T4)
Gonadotrophin releasing hormone	–	Follicle stimulating hormone Luteinizing hormone [gonadotroph]	Gonads	Sex steroids: estrogens, progestogens and androgens
Growth hormone releasing hormone	Somatostatin	Growth hormone (GH, somatotrophin) [somatotroph]	Liver, fibroblasts, myoblasts, chondrocytes, osteoblasts and others	Somatomedins (insulin-like growth factors)
Prolactin releasing factor[a]	Dopamine (acting at D2 receptors)	Prolactin [lactotroph]	Mammary glands	

[a] The molecule responsible for stimulating prolactin release has not been unambiguously identified.

timescales. Secretion from the neuroendocrine axes is modulated by feedback acting at several levels which tends to defend a set point in the concentration of the end product (*Fig. 1*).

Negative feedback is an extremely common homeostatic principle in biology. Its purpose is to hold some variable at a constant level, the **set point**. In *Fig. 1*, if the concentration of the end product hormone exceeds the set point, more receptors are activated in the hypothalamus and anterior pituitary which consequently reduce their output of hormones. The effect is that, after some delay, the concentration of end product hormone falls. If it falls below the set point the hypothalamus and pituitary secrete more hormones, provoking an increase in the synthesis of the end product. **Autofeedback inhibition** is a special case of negative feedback in which a substance directly inhibits its own synthesis. Several mechanisms exist to alter the set points of physiological systems so that hormone concentrations can be varied as circumstances change. For example, in many endocrine systems hormone concentrations fluctuate in a cyclical manner during the day because set points are adjusted by biological clocks in the brain (see below and Topic O4).

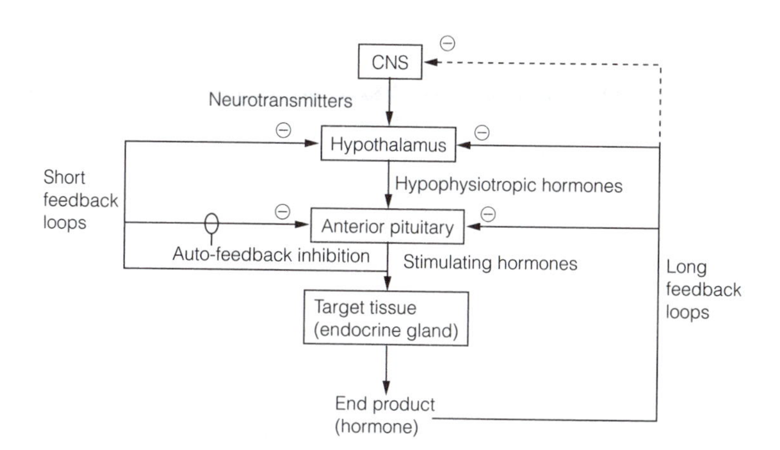

Fig. 1. Negative feedback loops that control neuroendocrine secretion.

Hypothalamic–pituitary–adrenal (HPA) axis

The HPA axis regulates the synthesis and secretion of **glucocorticoids**, a group of steroid hormones which help to control the metabolism of energy substrates. The most important glucocorticoid in humans is **cortisol.** Cells in the paraventricular nucleus (PVN) of the hypothalamus secrete **corticotrophin releasing hormone (CRH)**, a 41 amino acid residue peptide, which together with arginine vasopressin (AVP) acts synergistically to stimulate corticotrophs to release **adrenocorticotrophic hormone (ACTH)**. This is synthesized from a large precursor, **pro-opiomelanocortin (POMC)** which, in corticotrophs, is cleaved to produce **β-endorphin** in addition to ACTH. In response to ACTH, cells in the adrenal cortex synthesize and secrete glucocorticoids. Negative feedback by glucocorticoids at the hippocampus, hypothalamus and pituitary regulates secretion of the steroids (*Fig. 2*).

A circadian rhythm in glucocorticoid output is driven by the suprachiasmatic nucleus (Topic O4) acting on CRH secreting cells. In humans ACTH pulses are greatest early in the morning and decline through the day to reach a low point

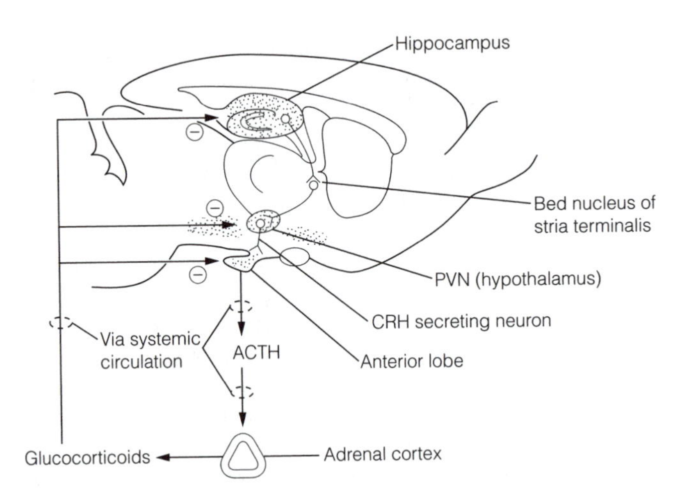

Fig. 2. Feedback in the hypothalamic–pituitary–adrenal axis. Stippled regions harbor glucocorticoid receptors. ACTH, Adrenocorticotrophic hormone; CRH, corticotrophin releasing hormone; PVN, paraventricular nucleus.

around midnight. Glucocorticoid secretion follows a similar pattern with a delay of about 30 minutes. This daily rhythm is influenced by the timing of light–dark cycles, sleep and meals.

The effects of glucocorticoids are mediated via two distinct receptors, coded for by separate genes: the high affinity mineralocorticoid receptor (MR, Type I) and the low affinity glucocorticoid receptor (GR, Type II), which has a 10-fold lower affinity for cortisol. Mineralocorticoid receptors are in greatest numbers in limbic structures. Glucocorticoid receptors are more widespread, and expressed in glia as well as neurons; the receptors colocalize in the same cells in the hippocampus. At basal levels of cortisol secretion MRs are largely occupied, while occupation of significant numbers of GRs occurs only when the cortisol concentration is high such as during the early morning circadian peak.

MRs and GRs are members of a nuclear receptor superfamily that includes receptors for other steroids (estrogen receptors, progesterone receptors, androgen receptors), thyroid hormone receptors, and receptors for vitamin D3 and retinoic acid. Glucocorticoids are lipophilic and so diffuse readily across plasma membranes of cells. MRs and GRs are present in the cytoplasm, complexed with **heat shock proteins** that act as molecular chaperones, stabilizing the receptors into their functional configuration. Binding of the ligand to the receptor causes it to translocate into the nucleus where it binds to specific sequences of DNA, **hormone responsive elements**, which results in increased or decreased transcription of specific genes (*Fig. 3*).

The corticotrophs of the anterior pituitary, CRH-secreting neurons of the PVN and hippocampal neurons express GRs. When the concentration of glucocorticoids is high these GRs are activated and they inhibit transcription of the genes for CRH and AVP. This is one mechanism for negative feedback control of glucocorticoid concentrations.

Stress

The HPA axis is activated by **stress**. No satisfactory comprehensive definition of stress exists. Physiological stressors such as hunger, thirst, exercise or trauma are universal, threaten homeostasis and the coordinated physiological responses

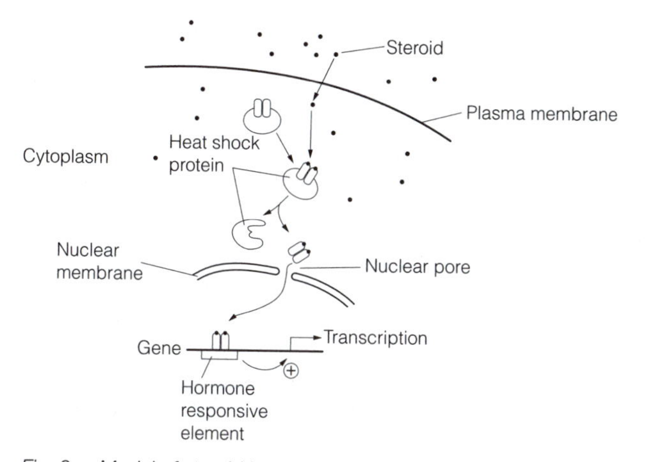

Fig. 3. Model of steroid hormone receptor action to modulate gene transcription.

that occur (including HPA activation) are adaptive in that they tend to maintain and restore homeostasis. Psychological stressors do not directly derange homeostasis, do not affect individuals equally, and stress responses to them may be learnt. Psychological stress often arises from the uncertainty inherent in social interactions or in situations from which it is difficult to escape or over which there is little control. Psychological stressors produce emotional (**affective**) states: anxiety, fear, anger, frustration, depression and so on, the nature and intensity of which depends on the individual's evaluation of the situation in the light of prior experience. A frequently adopted operational definition of stress is a state in which there is a prolonged rise in ACTH and glucocorticoid concentrations.

The increased secretion of glucocorticoids is useful in stress. Glucocorticoids mobilize fats and amino acids from fat and muscle cells respectively. These are used as substrates for **gluconeogenesis** in the liver. Much of the new glucose is then converted to glycogen by **glycogenesis** (see *Instant Notes in Biochemistry*, 2nd edn, for more details on glucose metabolism) and stored. The overall effect is to harness long-term energy substrates, triglycerides and proteins, and convert them to readily available substrates, glycogen and glucose. The early morning peak in glucocorticoids, in humans, is timed to correspond with what is generally the longest interval each day without food. In addition, glucocorticoids potentiate the effects of catecholamines.

Activation of the HPA axis in stress occurs through inputs from a variety of sources converging on the CRH-secreting cells of the PVN.

- Stress-evoked arousal activates noradrenergic neurons in the locus ceruleus which project to the PVN.
- Visceral sensations associated with thirst and hunger are transmitted via the glossopharyngeal (IX) and vagus (X) nerves to the nucleus of the solitary tract and adjacent regions of the medulla. These structures project catecholaminergic axons to activate the PVN.
- Inputs from the vascular organ of the lamina terminalis and subfornical organ (SFO), which respond to rises in osmolality and angiotensin II, go to the CRH-secreting cells of the PVN providing a route to activate the HPA during dehydration.

- Neurons in the midbrain and pons, many cholinergic, project to the PVN and are thought to transmit visual, auditory and somatosensory (including nociceptive) input associated with stressful situations (e.g. a loud noise that produces a startle response).
- Most hypothalamic nuclei project to the PVN and these connections presumably funnel information about stressful situations from prefrontal cortex and limbic structures such as the amygdala or hippocampus.

The high concentration of glucocorticoids seen in stress produce 50% occupancy of glucocorticoid receptors in the PVN and hippocampus and hence terminates the stress response via negative feedback. The HPA is much more sensitive to stress activation and negative feedback inhibition at times when blood glucocorticoid concentrations are at their lowest.

Protracted activation of the HPA by chronic stress has deleterious effects. High concentrations of corticosteroids acting via glucocorticoid receptors enhances excitatory amino acid transmission and increases the calcium influx through voltage-dependent Ca^{2+} channels into hippocampal cells, thereby killing them (see Topic R1). This may account for the finding of reduced numbers of hippocampal pyramidal cells in aged rats, with a corresponding reduction in glucocorticoid receptors which blunts the effectiveness of glucocorticoid negative feedback; in both aged rats and humans corticosteroid concentrations take longer to return to basal levels after stress compared with younger individuals.

Persistently high concentrations of glucocorticoids suppress the function of immune system cells. High levels of stress increase the risk of infection, and incidence of cancers.

Hypothalamic–pituitary–thyroid (HPT) axis

Thyroid hormones, amongst other functions, regulate basal metabolic rate, increasing metabolic heat production by increasing the synthesis of a protein that uncouples oxidative phosphorylation by mitochondria. Thyroid hormone secretion is regulated by the hypothalamus and pituitary, and is influenced by several factors, such as ambient temperature.

Thyrotrophin releasing hormone (TRH) is a tripeptide synthesized, as part of a precursor, in small neurons of the PVN of the hypothalamus, The axons of these cells form part of the tuberoinfundibular tract that goes to the median eminence. TRH secreted here reaches the anterior pituitary by way of the hypothalamic–pituitary portal system and stimulates thyrotrophs to secrete **thyroid stimulating hormone (TSH)**. TSH is a glycoprotein consisting of two chains, α and β. It is liberated into the systemic circulation and stimulates division and growth of cells in the thyroid gland, and the synthesis and secretion of the thyroid hormones. There are two thyroid hormones, **thyroxine (T4)** and **triidothyronine (T3)**, so named because of the number of iodine atoms they contain. All but about 1% of thyroid hormone in the blood is bound to **thyroxine binding globulin** and other plasma proteins. Only the free hormone is directly available for receptor binding.

Thyroid hormone receptors are members of the steroid receptor superfamily. They form heterodimers together with retinoid X receptors, and differ from glucocorticoid receptors in that the heterodimer binds to hormone responsive elements in the DNA in the absence of ligand. These receptors have a higher affinity for T3 than T4. T4, which forms the bulk of the secreted hormones, is a prohormone which is converted to T3 by the neuronal cytosolic enzyme, 5′-deiodinase II. On binding of T3 the thyroid hormone receptor activates gene transcription.

Thyroid hormone output is controlled by negative feedback acting at several levels of the HPT axis. A decrease in thyroid hormone concentration causes the increased secretion of TSH by thyrotrophs of the anterior pituitary. This occurs because lack of T3 causes enhanced transcription of the genes for the TRH receptor and for TSH, so not only do thyrotrophs become more sensitive to hypothalamic TRH but their capacity for producing TSH is elevated. TRH secretion from the hypothalamus is also subject to feedback inhibition by both T4 and T3.

Pulses of TRH secretion drives pulsatile TSH output. The frequency and amplitude of the pulses is entrained into a circadian rhythm by the suprachiasmatic nucleus, rising throughout the night (the **nocturnal TSH surge**), falling during the morning and remaining low throughout the afternoon. This circadian rhythm is sensitive to light–dark cycles, but unaffected by sleep patterns.

Thyroid hormone secretion is increased in cold exposure. Temperature sensitive neurons in the preoptic hypothalamus, which get input from skin thermoreceptors, project to brainstem noradrenergic neurons. These, in turn, synapse with the TRH secreting cells of the PVN. Cold exposure activates the noradrenergic neurons, provoking a rise in TRH secretion. The resulting increase in the concentrations of thyroid hormones enhances metabolic rate, helping to maintain core temperature. The effect is rapid, elevation of thyroid hormone output is seen within 30 min of cold exposure. Cytokines such as interleukin-1 inhibit TRH gene transcription, so during infection or serious illness in which cytokines are elevated, the HPT axis is downregulated. The low concentrations of thyroid hormones help to conserve metabolic substrates.

Thyroid hormones are crucially required for human brain development from very early pregnancy, long before the fetal thyroid begins to function at about 17 weeks gestation. Before 17 weeks, brain development is driven by maternal T4 which crosses the placenta and is converted to T3 in the fetal brain. After 17 weeks, T4 entering the placenta is deiodinated by a placental 5-deiodinase to L-3,3′,5′-triidothyronine (reverse T3, rT3) which has no endocrine activity. However, the liberated iodide is used by the fetal thyroid to synthesize its own hormones. Maternal **hypothyroxinemia**, usually caused by a lack of dietary iodide, can result in **neurological cretinism** in infants, in whom reduced thyroid hormone compromises synaptogenesis, myelination and axoplasmic transport, particularly in the cerebral and cerebellar cortices. About half a billion women live in iodine deficient regions; dietary iodide supplements are an effective remedy for hypothyroxinemia and so have a prophylactic value against cretinism.

Growth hormone (GH)

Growth hormone (**somatotrophin**) stimulates cell division and growth of many tissues, particularly during the perinatal period and the adolescent growth spurt, enhancing protein synthesis by increasing transcription and translation. GH stimulates lipolysis, mobilizing fatty acids as energy substrates. This is adaptive during exercise, stress and fasting, three major physiological variables which increase GH secretion.

GH secretion from somatotrophs of the anterior pituitary is regulated by two hormones, **growth hormone-releasing hormone** (**GHRH**) and **somatostatin** (also known as **somatotrophin release-inhibiting hormone**). GHRH is a 44 amino acid residue peptide synthesized from a precursor by neurons in the **arcuate nucleus** of the hypothalamus, the axons of which terminate in the median eminence. Secreted GHRH is transported to the anterior pituitary via the hypothalamic–pituitary portal system. Somatostatin is an important transmitter throughout the central nervous system (CNS), but the somatostatin containing cells responsible for inhibiting the secretion of GH are restricted to the **anterior periventricular nucleus**, and their axons project to the median eminence. Somatostatin is a 14 amino acid residue peptide synthesized from a precursor. GHRH and somatostatin exert their opposing effects on GH secretion via metabotropic receptors coupled to the cyclic adenosine monophosphate (cAMP) second messenger system. GHRH receptors enhance cAMP concentrations via G_s proteins while somatostatin receptors reduce cAMP by coupling to G_i proteins. The cAMP second messenger system modulates GH secretion by altering Ca^{2+} influx in the somatotrophs.

The secretion of GH is circadian and pulsatile, driven mostly by pulses of GHRH from the hypothalamus. The pulses are much bigger at night and triggered by deep (stages 3 and 4) slow wave sleep (see Topic O5). This nocturnal GH secretion is greatest in children and declines with age. It is brought about by a serotonergic pathway from the brainstem to the hypothalamus. GHRH secretion is also stimulated by dopaminergic, noradrenergic and enkephalinergic pathways in the brain. GH secretion is greatly influenced by other hormones. The GH gene contains transcriptional response elements for thyroid hormones and for glucocorticoids. Thyroid hormones are required for normal levels of synthesis and secretion of GH, and basal concentrations of glucocorticoids enhance, while high concentrations inhibit, GH synthesis.

Negative feedback control of GH secretion occurs at the pituitary by suppression of the synthesis and secretion of GH, and at the hypothalamus by reduction of GHRH secretion. GH also stimulates the secretion of somatostatin. These negative feedback effects are exerted by **insulin-like growth factor** (**IGF-1**), one of a group of peptides called **somatomedins** which mediate the effects of GH. IGF-1 is produced either in the brain, or peripherally, in response to GH.

A rapid rise in the rate of growth, the **growth spurt**, occurs during puberty. In girls it occurs, on average, between 11 and 14 years of age and in boys it occurs 2 years later. During puberty, gonadal secretion of sex steroids rises and both androgens and estrogens stimulate the high secretion of GH that is responsible for the growth spurt. The concentration of GH is higher during puberty than at any other time. In women, the secretion of GH is briefly enhanced by the high estrogen concentrations that are seen just before ovulation.

M4 Neuroendocrine control of reproduction

Key Notes

The hypothalamic–pituitary–gonadal (HPG) axis

Gonadotrophin releasing hormone is synthesized by neurons in several hypothalamic nuclei. It stimulates the anterior pituitary to release two gonadotrophins, follicle stimulating hormone (FSH) and luteinizing hormone (LH). These trophic hormones stimulate the gonads to produce sex steroids. Gonadotrophin secretion is pulsatile. In males the period of the pulses is constant, but in females the period depends on the phase of the reproductive cycle.

Feedback in males

Gonadotrophins stimulate the testis to produce testosterone and inhibin, which both produce negative feedback suppression of the HPG axis. Testosterone acts at both the hypothalamus and the anterior pituitary but the effects of inhibin are confined to the suppression of anterior lobe secretion of FSH. Testosterone output is pulsatile and greatest between midnight and noon.

Feedback in female reproductive cycles

In women, gonadotrophins stimulate the ovarian follicles to grow, producing estradiol and inhibin during the first half of the menstrual cycle (the follicular phase). After ovulation, the follicle becomes a corpus luteum which secretes progesterone through the second half of the cycle (the luteal phase) in response to FSH and LH. During most of the menstrual cycle sex steroids exert a negative feedback suppression of gonadotrophin output. In the follicular phase it is mediated by estradiol (and inhibin) whereas in the luteal phase it results from both estradiol and progesterone. However, just before ovulation the high concentrations of estradiol produced by the mature follicle cause the HPG to switch briefly into a positive feedback mode. Now, estradiol generates a midcycle surge of gonadotrophin which triggers ovulation.

Puberty and menopause

Puberty is due to the activation of the previously quiescent HPG axis. What initiates puberty is not known but it is thought to be a metabolic signal of growth or body mass. The menopause results from ovarian failure, the HPG remains functional.

Prolactin (PRL)

Secreted by the anterior pituitary, PRL stimulates breast tissue development during pregnancy and is responsible for reflex synthesis and secretion of milk by suckling in lactating women. PRL secretion, like that of growth hormone (GH), is subject to dual regulation by the hypothalamus. Dopamine acts on the anterior pituitary to inhibit PRL release. Several peptides can stimulate PRL release, but which one does so *in vivo* is uncertain. High levels of PRL secretion, arising as a result of lactation or pathology causes infertility by suppressing LH secretion.

Related topics	Anatomy and connections of the hypothalamus (M1) Neuroendocrine control of metabolism and growth (M3)	Dopamine neurotransmission (N1)

The hypothalamic–pituitary–gonadal (HPG) axis

The hypothalamic–pituitary–gonadal axis is central to the control of reproduction. In primates, neurons scattered widely throughout the hypothalamus (preoptic area, arcuate nucleus, periventricular nucleus and lateral hypothalamus) synthesize a decapeptide, **gonadotrophin releasing hormone (GnRH**, also referred to as **luteinizing hormone releasing hormone**) from a large precursor. GnRH is secreted from axon terminals in the median eminence into the hypothalamic–pituitary portal circulation. GnRH stimulates gonadotrophs of the anterior pituitary to secrete two gonadotrophins, **follicle stimulating hormone (FSH)** and **luteinizing hormone (LH)**, into the systemic circulation which carries them to the gonads.

The gonadotrophins are large glycoproteins each consisting of two peptides, an α chain and a β chain. The α chains of FSH and LH are identical (and very similar to the α chain of thyroid stimulating hormone) but the β chains are distinct and confer the specificity of the hormones. Gonadotrophins stimulate the gonads to produce **sex steroids** and have effects on gamete development. Gonadotrophin secretion is cyclical in females but not in males. The secretion of gonadotrophins is pulsatile, as with other anterior lobe hormones, and is driven by bursts of GnRH from the hypothalamus. In males, the pulses are regular, spaced about 3 h apart, but in females the period varies between 1 and 12 h depending on the phase of the reproductive cycle. Experimentally replacing pulsatile with continuous GnRH delivery in female rhesus monkeys abolishes gonadotrophin secretion, showing that pulsatile GnRH delivery is mandatory for proper HPG axis function.

Feedback in males

At the testis, LH stimulates **Leydig cells** to synthesize and secrete androgens, principally **testosterone**. FSH, together with testosterone, acts on **Sertoli cells** to organize the development of spermatozoa and secrete a glycoprotein, **inhibin**.

Gonadotrophin secretion in the male is subject to negative feedback control by both of the secretions from the testis. Testosterone acts at the hypothalamus and the anterior pituitary, both express androgen receptors. At the hypothalamus, testosterone decreases the frequency of the episodic GnRH bursts, while the anterior pituitary becomes less responsive to GnRH. Inhibin specifically suppresses only FSH secretion and inhibin must act at the anterior lobe, since the hypothalamus has no inhibin receptors. Humans have a circadian rhythm in testosterone secretion but show no longer-term cycles in testosterone output. Testosterone secretion is pulsatile and the amplitude and duration of the pulses are largest between midnight and noon. Positive feedback effect of testosterone on gonadotrophins is not seen in males.

Feedback in female reproductive cycles

In females, the situation is more complicated since here the role of the HPG axis is firstly to stimulate the growth of a group of ovarian follicles (one of which develops to maturity), secondly to produce cyclical changes in sex steroid output, which prepares the reproductive tract for fertilization and implantation, and thirdly to trigger ovulation at the appropriate time, which in humans occurs on day 14 of the 28 day cycle.

The first half of the cycle (days 1–14) is called the **follicular phase**, since it is dominated by the growth of the ovarian follicle which secretes **estradiol** and **inhibin**. The second half of the cycle is the **luteal phase** (days 15–28), since after ovulation the follicle becomes a **corpus luteum** which secretes **progesterone**.

The extent to which feedback by steroids acts at the anterior pituitary or hypothalamus depends on species. In primates, the anterior lobe is the predominant mediator. What follows relates to humans.

Feedback in females depends on the phase of the menstrual cycle. For most of the follicular phase the low or moderate levels of estradiol and inhibin exert negative feedback effect on gonadotrophin secretion (*Fig. 1*).

Estradiol causes the anterior pituitary to become less sensitive to the effects of GnRH with the effect that the amplitude of the LH pulses is reduced. Hence during the follicular phase LH pulses are high frequency and low amplitude. The reduced GnRH sensitivity may be due to down regulation of GnRH receptors on the gonadotrophs.

However, by about day 14, levels of estradiol become high enough to flip the HPG axis into a positive feedback mode. This stimulates a rise in LH and FSH secretion, which triggers ovulation, and switches the steroid metabolism of the ruptured follicle so that it synthesizes and secretes progesterone (*Fig. 2*). The rise

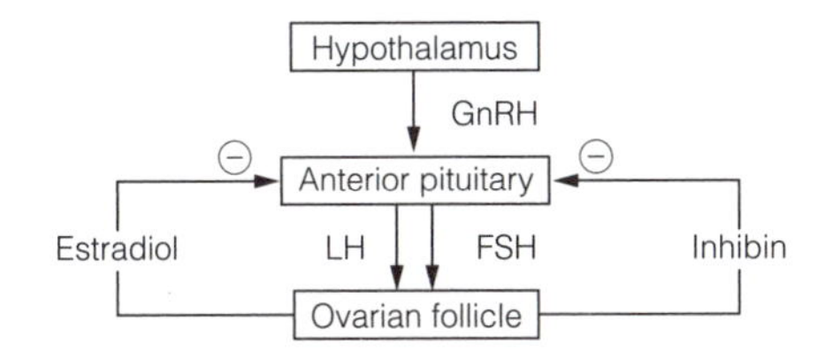

Fig. 1. Negative feedback inhibition during the follicular phase of the menstrual cycle. Negative feedback in males is very similar except that luteinizing hormone (LH) is regulated by testosterone from the testis. FSH, follicle stimulating hormone; GnRH, gonadotrophin releasing hormone.

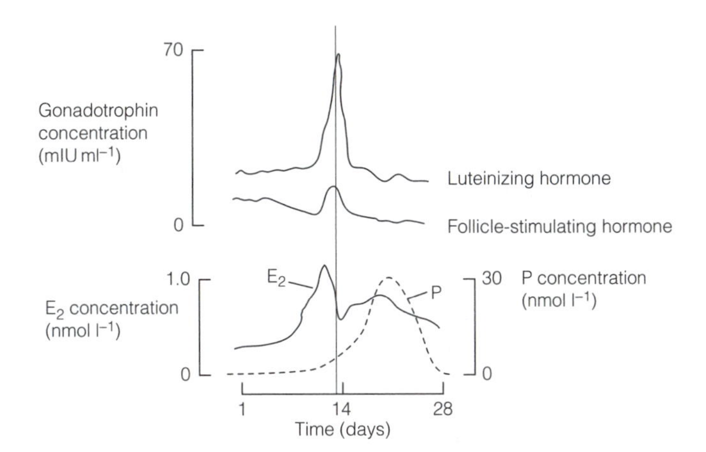

Fig. 2. Pattern of hormone secretion during the human menstrual cycle. Ovulation, triggered by the estradiol (E_2) evoked surge in luteinizing hormone occurs around day 14. P, progesterone; mIU, milli-international units.

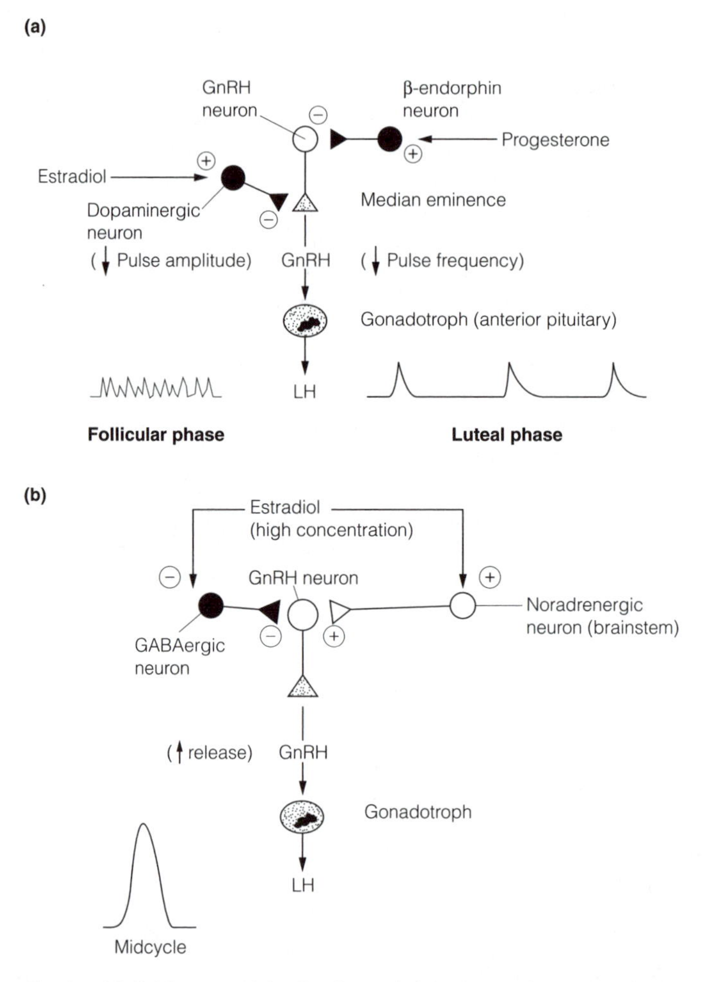

Fig. 3. Model for steroid feedback regulation of gonadotrophin releasing hormone (GnRH) producing neurons: (a) negative feedback by estradiol (left) and progesterone (right); (b) positive feedback by estradiol works by inhibiting γ-aminobutyrate (GABA) inhibition and stimulating noradrenergic excitation of GnRH release. Excitatory neurons (○) inhibitory neurons (●). LH, luteinizing hormone.

in progesterone secretion at the start of the luteal phase terminates the positive feedback LH surge by inhibiting GnRH secretion, and the gonadal steroids once again curtail the secretion by negative feedback. *Fig. 3* shows the feedback mechanisms operating during the follicular and luteal phases (a) and mid-cycle (b).

Puberty and menopause

Except for a brief postnatal period in primates, the HPG axis is quiescent until puberty so circulating levels of gonadal steroids are low. Inactivity of the HPG axis results from a tonic γ-aminobutyrate (GABA)ergic inhibition on the GnRH neurons, and the onset of puberty is accompanied by decreased GABA inhibition. During puberty there is a dramatic circadian rhythmicity in gonadotrophin output: LH pulses are much larger during sleep. The precise trigger for puberty is not known, but a metabolic signal reflecting growth or body mass is probably involved. In girls, a critical mass of 30 kg appears necessary for puberty to commence, and needs to reach about 47 kg before menstrual cycles begin. Female dancers, athletes and anorexics fail to menstruate if their body mass falls too low.

The end of reproductive life, the menopause, is characterized by ovarian failure; the hypothalamus and pituitary continue to function.

Prolactin (PRL)

Prolactin is secreted by lactotrophs of the anterior lobe. Together with estrogen it stimulates the growth of the alveoli and ducts of the breasts during pregnancy. Suckling produces a reflex secretion of prolactin which stimulates the synthesis and secretion of milk. PRL is a glycoprotein with a similar amino acid sequence to growth hormone. It is secreted in pulsatile fashion and the pattern of secretion varies with time of day (highest between midnight and 9 a.m.), sex (higher in females as secretion is stimulated by estrogens) and stage of the menstrual cycle. Like GH, both exercise and stress stimulate prolactin release. PRL concentrations are high during pregnancy and lactation.

PRL secretion is regulated by the hypothalamus, by dopaminergic neurons which are inhibitory, and by a number of peptides which activate secretion, a pattern similar to control of GH secretion. The dopaminergic neurons (A12 group, see Topic N1) have their cell bodies in the arcuate nucleus of the hypothalamus and their axons run in the tuberoinfundibular pathway to terminate in the median eminence. Here the dopamine is released directly into the hypothalamic–pituitary portal circulation, carrying it to the anterior lobe. Lactotrophs express D2 dopamine receptors that are negatively coupled to adenylyl cyclase. The fall in cAMP concentration reduces transcription of the prolactin gene.

PRL release is stimulated by:

- thyrotrophin releasing hormone (see Topic M3);
- vasoactive intestinal peptide secreted into the hypothalamic–pituitary portal circulation;
- oxytocin from the posterior pituitary, which gains access to the anterior lobe via tiny blood vessels called the **short portal vessels.**

However there must be other prolactin releasing factors to account for secretion under all physiological conditions. With frequent suckling (every 2–3 h) the amount of prolactin released is sufficient to block ovulation by suppressing LH secretion, so lactating women are relatively infertile.

Excessive secretion of prolactin (**hyperprolactinemia**) may occur with some pituitary tumors or as a result of drugs which interfere with dopamine inhibition of prolactin secretion, e.g. dopamine receptor antagonists. Hyperprolactinemia causes infertility in males and females because of its suppression of LH secretion, and inappropriate secretion of milk in both sexes.

M5 SMOOTH AND CARDIAC MUSCLE

Key Notes

Smooth muscle

Smooth muscle is unstriated, composed of single cells and located mostly in hollow viscera, blood vessels, and airways. It is a major target for the autonomic nervous system (ANS). Although it contracts slowly, smooth muscle can develop large forces for long periods on a low oxygen consumption. There are two types. Single unit smooth muscle cells are electrically coupled by gap junctions and are pacemakers, firing Ca^{2+} action potentials automatically. Single unit smooth muscle is regulated largely by hormones, local factors and intrinsic neurons; gut smooth muscle is of this type. Multi-unit smooth muscle has its tonic activity maintained by the release of the ANS transmitters that are either excitatory or inhibitory. These produce small junction potentials (similar to synaptic potentials) on the muscle cells, which usually do not fire; vascular smooth muscle is of this type.

Smooth muscle contraction is triggered by a rise in cytoplasmic Ca^{2+} concentration brought about by action potentials, or stimulation of receptors that are linked by inositol trisphosphate (IP_3) production to liberation of Ca^{2+} from internal stores, or that are coupled to Ca^{2+} channels. Ca^{2+} causes contraction by activating myosin light chain kinase. Inhibition of the same enzyme by cyclic adenosine monophosphate (cAMP) promotes relaxation.

Cardiac muscle

Cardiac muscle cells are striated, linked by gap junctions, and have pacemaker properties caused by slow depolarization at rest. Excitation–contraction coupling resembles that of skeletal muscle in that Ca^{2+} is released from the sarcoplasmic reticulum. However Ca^{2+} entry via L-type Ca^{2+} channels is also important. Increased cAMP by β1 adrenoceptors raises the rate and force of the heart by opening Ca^{2+} channels. Acetylcholine (ACh) acting on muscarinic receptors reduces the heart rate by opening potassium channels.

Related topics

Overview of synaptic function (C1)
Slow neurotransmission (C3)
Skeletal muscles and excitation–contraction coupling (K1)
Autonomic nervous system function (M6)

Smooth muscle

A major target for the postganglionic fibers of the ANS is smooth muscle, so called because it lacks the striations seen in skeletal and cardiac muscle under the light microscope. Smooth muscle is located in the hollow viscera (e.g. gut, urinary bladder, uterus), in major airways, blood vessels (except capillaries), ducts of exocrine glands and eyes. Smooth muscle cells are spindle shaped, have a central nucleus, and vary in length (15–300 μm) and diameter (2–10 μm) depending on location. Within them, thick and thin filaments, containing actin and myosin respectively, provide the contractile force, but smooth muscle is far more economical than striated muscle in terms of power generated per mole of adenosine 5′ triphosphate (ATP) hydrolyzed. So, although smooth muscle metabolizes aerobically, its oxygen demands are low. Smooth muscle contracts slowly, but force can be maintained for long periods without fatigue, and the forces generated (per cross sectional area of muscle) can be as large as those of skeletal muscle. Thin filaments are attached to **dense bodies** within the cytoplasm that serve the same function as Z discs in striated muscle. Each smooth muscle cell has several dense bodies linked by intermediate filaments that distribute the forces throughout the cell to **desmosomes** of the plasma membrane which mechanically weld adjacent cells together. In this way mechanical force is transmitted through a bundle of smooth muscle cells.

Two distinct types of smooth muscle can be identified, though many smooth muscles share characteristics of both. **Single unit** smooth muscle has gap junctions (Topic C1) which electrically couple adjacent cells together. This allows all the linked cells to respond as a whole to an input. Smooth muscle producing phasic (rhythmic) output, such as that in the gut responsible for peristalsis, is of this type. Single unit smooth muscle cells have pacemaker properties. Their resting membrane potential is not stable but instead slowly depolarizes automatically until it eventually triggers a burst of Ca^{2+} action potentials and the cell contracts. Hyperpolarization (due to the opening of Ca^{2+}-sensitive K^+ channels) silences the cell briefly until slow depolarization causes the cycle to repeat. Since the pacemaker activity is a result of intrinsic membrane properties of the muscle it is called **myogenic activity** and continues after cutting the autonomic nerves.

Often single unit smooth muscle receives only a sparse extrinsic ANS supply, its activities being coordinated by neurons intrinsic to the tissue, as in the case of the gut. Some single unit smooth muscle has no neural input and its activity is altered solely by circulating hormones or local factors. Stretch triggers contraction of single unit smooth muscle by opening ion channels sensitive to mechanical stress increasing the rate of pacemaker depolarization. This underlies autoregulation of cerebral blood vessels (see Topic E6).

Multi unit smooth muscle cells are not linked by gap junctions, so each acts autonomously. Multi unit smooth muscle is in a continual state of contraction modified as a result of neural input and circulating hormones. The neurotransmitters and hormones have either excitatory or inhibitory effects on the muscle, generating junction potentials that are analogous to synaptic potentials. Tonic activity like this is typical of smooth muscle found in arteries and sphincters and is maintained by low frequency firing of autonomic axons which results in depolarization of the smooth muscle, but not action potentials. Axons of postganglionic autonomic neurons have, spaced at intervals along their lengths, numerous swellings called varicosities, each about 1.5 μm in diameter and containing synaptic vesicles. These form synapses with the smooth muscle cells called neuroeffector junctions, with synaptic clefts that vary between 20 and 200 nm across. The larger gaps are seen in tissues that have the fewest terminals and may allow room for secreted neurotransmitter to diffuse widely, so as to affect several cells. There is little specialization of the smooth muscle cell membrane postsynaptically though it contains neurotransmitter receptors.

The resting membrane potential of smooth muscle is low relative to that of skeletal muscle (V_r about –60 mV). Contraction of smooth muscle is brought about by an increase in the concentration of free Ca^{2+} in the cytoplasm. This can be achieved in several ways.

(1) By action potentials generated myogenically or by neural depolarization. Smooth muscle cells have no T tubules but instead their plasma membranes have L-type Ca^{2+} channels activated by depolarization, and influx of Ca^{2+} through these channels is responsible for action potentials which lasts for between 10 and 500 msec. Action potentials are important in single unit (phasic) smooth muscle.
(2) Via neurotransmitters or hormones acting on receptors coupled to the phosphoinositide second messenger system (see Topic C3). By stimulating the synthesis of IP_3 they trigger the release of Ca^{2+} from the sarcoplasmic reticulum (SR). This mechanism operates in multi-unit (tonic) smooth muscle cells. For example, norepinephrine released from sympathetic nerve terminals acting on α1 adrenoceptors of vascular smooth muscle. It is also how ACh secreted by parasympathetic terminals, acting at muscarinic M3 receptors causes contraction of gut smooth muscle.
(3) G protein-linked receptors can also control Ca^{2+} entry by a direct action of their G proteins on **receptor operated Ca^{2+} channels**.

Calcium causes smooth muscle contraction by binding to calmodulin (CaM), a cytosolic protein with sequence similarities to troponin (see Topic K1). The Ca^{2+}–CaM complex activates a kinase which phosphorylates myosin, **myosin light chain kinase (MLCK)**, allowing it to interact with actin and so bring about contraction. Relaxation in smooth muscle can be achieved in two ways. One is by active transport mechanisms which, by sequestering Ca^{2+} back into the SR or extruding Ca^{2+} from the cell, reduce the cytosolic Ca^{2+} concentration. The other is by neurotransmitters or hormones acting at receptors which inhibit the contraction biochemistry. This is the case with β adrenoceptors that are coupled to the cAMP second messenger system (see Topic C3). Cyclic AMP activates protein kinase A which then inactivates MLCK. β2 receptors mediate relaxation of airway and gut smooth muscle in this fashion. Regulation of smooth muscle contraction is shown in *Fig. 1*.

Cardiac muscle

Cardiac muscle cells are striated, but like single unit smooth muscle are coupled together via gap junctions (permitting rapid spread of action potentials through the heart) and are myogenic. The pacemaker activity of heart muscle cells comes about by slow depolarization at rest caused by activation of:

- specific Na^+ channels that differ from the normal fast voltage-dependent sodium channels;
- T-type Ca^{2+} channels.

This depolarization is partly offset by the efflux of K^+ through muscarinic cholinergic activated K^+ channels (K_{ACh}). The depolarization eventually reaches threshold and a Na^+ spike occurs (*Fig. 2*).

Excitation–contraction coupling in cardiac muscle is similar to that in skeletal muscle in that the Na^+ spike invades T tubules in the muscle membrane to trigger the release of Ca^{2+} from SR via ryanodine receptors (see Topic K1). However cardiac muscle cells also have L-type voltage-dependent Ca^{2+} channels which are activated by the Na^+ spike and this generates a prolonged Ca^{2+} spike so that the cardiac action potential lasts for about 250 msec. Hence the calcium responsible for triggering the contraction of cardiac muscle comes from two

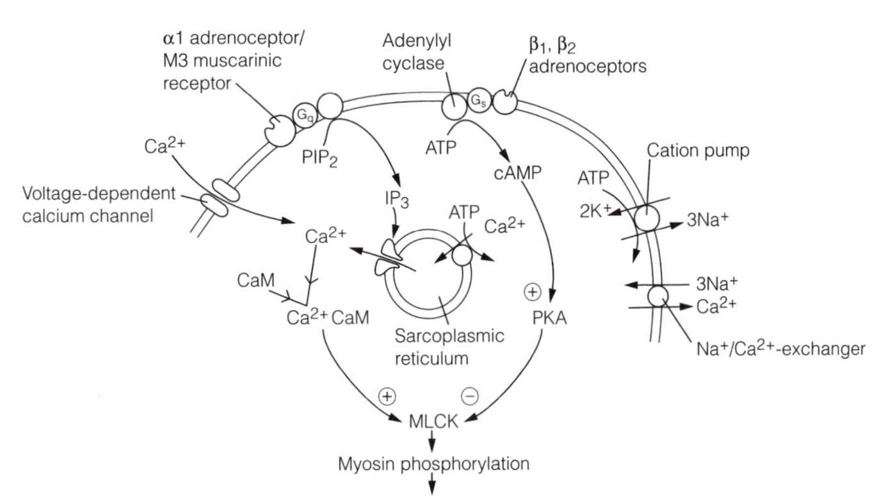

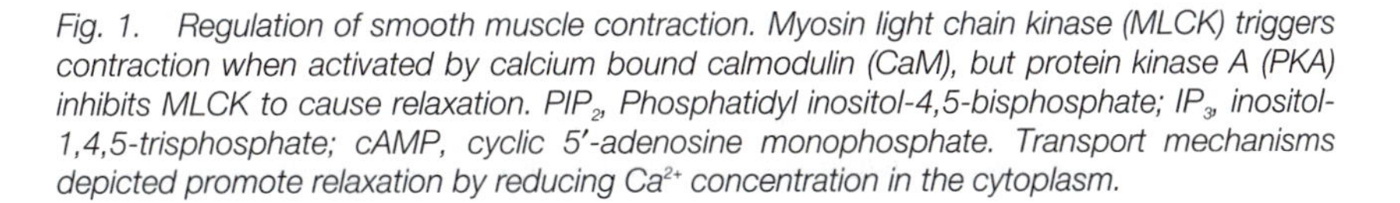
Fig. 1. Regulation of smooth muscle contraction. Myosin light chain kinase (MLCK) triggers contraction when activated by calcium bound calmodulin (CaM), but protein kinase A (PKA) inhibits MLCK to cause relaxation. PIP_2, Phosphatidyl inositol-4,5-bisphosphate; IP_3, inositol-1,4,5-trisphosphate; cAMP, cyclic 5′-adenosine monophosphate. Transport mechanisms depicted promote relaxation by reducing Ca^{2+} concentration in the cytoplasm.

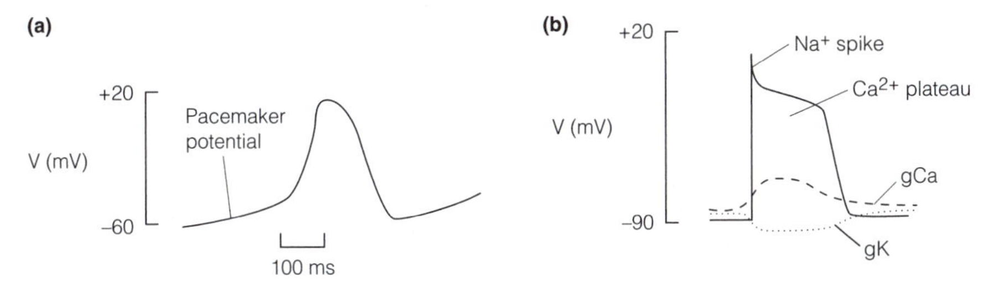

Fig. 2. Cardiac action potentials: (a) in the cardiac pacemaker (sinoatrial node), note the slow depolarizing pacemaker potential; (b) in the ventricular muscle. gCa, Calcium conductance; gK, potassium conductance.

sources, an internal store (the SR), and from outside (via voltage-dependent calcium channels). As the Ca^{2+} channels inactivate the muscle cells begin to repolarize, allowing K^+ efflux which reinforces the repolarization. Observe in *Fig. 2b.* that during a cardiac action potential the potassium conductance (gK) decreases when the cells are being depolarized by Na^+ and Ca^{2+} influx. Potassium channels that show decreased current flow with depolarization are **inward rectifiers**. Note that this is the opposite of the behavior of the outward rectifier K^+ channels during a neuronal action potential (see Topic B2).

Activation of β1 receptors on cardiac muscle cells increases intracellular cAMP. By phosphorylation of the Na^+ and T-type Ca^{2+} channels this makes the pacemaker depolarization more rapid, which raises heart rate. By phosphorylating the L-type Ca^{2+} channels it increases calcium influx, strengthening the force of contraction. ACh acting on M2 muscarinic receptors slows the pacemaker depolarization by activating the K_{ACh} channels. A G_I protein directly couples the receptor to the channels.

M6 Autonomic nervous system function

Key Notes

Overview of the autonomic nervous system (ANS)

The ANS acts on smooth muscle, cardiac muscle and glands to keep key physiological variables at a level appropriate to an animals' activity and its environment. Much ANS regulation is by negative feedback in that it operates to defend a set point in some variable (e.g., mean arterial blood pressure). In other circumstances, autonomic mechanisms are homeostatic in the sense that variables change to meet altered demands. The sympathetic division is activated in stress to produce 'fright, fight or flight' responses, whereas the parasympathetic division is more active in 'rest and digest' situations. The two divisions may exert opposite effects on an organ (e.g. the heart). In a few instances ANS mechanisms are positive feedback in that they drive a physiological system away from its normal stable state (e.g. sexual responses).

ANS physiology

At autonomic ganglia, divergence of preganglionic input to postganglionic output occurs, which is much greater for the sympathetic division. Convergence is also seen and this allows numerous weak inputs to sum so as to fire postganglionic cells. Acetylcholine (ACh) is the transmitter of autonomic ganglia and acts on both nicotinic (nAChR) and muscarinic (mAChR) receptors. The nAChR mediate fast ACh transmission. Activation of the mAChR greatly prolongs the time for which the postganglionic cells fire in response to the nAChR stimulation. Postganglionic sympathetic axons are usually noradrenergic and postganglionic parasympathetic terminals are invariably cholinergic. The adrenal medulla chromaffin cells are effectively postganglionic sympathetic cells and secrete mostly epinephrine. Postganglionic terminals corelease a variety of peptide transmitters which modulate, or have additional actions to, the primary transmitters.

An example of coordinated autonomic control: the urinary bladder

The urinary bladder has a dual autonomic innervation. Tonic sympathetic activity permits the bladder to fill. Urination is achieved by a spinal reflex in which bladder wall stretch receptors activate parasympathetic relaxation of the internal sphincter of the bladder. This reflex is normally under conscious control exerted via neurons in the pons. After severing the spinal cord, urination becomes entirely reflexive.

Axon reflexes

Some visceral and nociceptor afferents, when excited by tissue damage propagate action potentials in the 'wrong' direction along axon collaterals to release peptides from the terminals which have inflammatory actions. This is the axon reflex.

Enteric nervous system (ENS)

The myenteric and submucosal plexus coordinate gut motility and secretion respectively. Myenteric motor neurons exert tonic excitatory or inhibitory actions on gut smooth muscle. They are driven, via cholinergic interneurons, by sensory neurons that respond to stretch or chemical signals in the gut lumen. Usually sensory neuron firing stimulates motility in the oral direction, but inhibits it in the anal direction. Submucosal reflexes increase glandular secretion both directly and by increasing local blood flow.

Related topics

Slow neurotransmission (C3)
Organization of the peripheral nervous system (E1)
Smooth and cardiac muscle (M5)
Central control of autonomic function (M7)

Overview of the autonomic nervous system (ANS)

The autonomic nervous system adjusts the contraction of smooth muscle and heart muscle and controls glandular secretion so that key physiological variables (e.g. core temperature, cardiac output, blood pressure, blood glucose) are maintained at levels appropriate to an animals' activity or the environment in which it finds itself.

The term autonomic (self-governing) is apt since the ANS usually operates without conscious awareness and has no cognitive component. The term involuntary – sometimes applied to the ANS – is misleading since individuals can be trained using **biofeedback** techniques to control (within limits) variables regulated by the ANS (e.g. blood pressure).

To a first approximation much of the activity of the ANS is concerned with homeostatic regulation of physiological variables. For example, mean arterial blood pressure is kept fairly constant despite changes in posture which cause large swings in the hydrostatic pressure of blood. When suddenly standing from a lying position the tendency for blood to pool in the legs due to the force of gravity is offset by autonomic reflexes which monitor the drop in pressure and elicit constriction of arterioles and venules in the legs. This is a classic negative feedback mechanism.

Many autonomic adjustments are not negative feedback, since they do not defend a set point in some variable, but are homeostatic in that they change physiological variables so as to cope with altered demands. In response to a wide variety of stressors, activation of the sympathetic nervous system (SNS) to targets such as the heart, blood vessels, airways and liver results in increased cardiac output, regional alterations in blood flow, raised airflow through the lungs and elevations in blood glucose concentrations, all adaptations which improve the chances of surviving the stress unscathed. In general the SNS mediates the response of 'fright, fight and flight'. In contrast, parasympathetic nervous system (PSNS) activation is seen when the body is in 'rest and digest' mode; the PSNS generally stimulates exocrine gland secretion and promotes anabolic processes.

The sympathetic and parasympathetic divisions of the ANS can have opposing effects, for example on pupil diameter (see Topic H2) or heart rate, and it is the balance of activities in the two divisions that achieves the appropriate outcome; that is the SNS and PSNS work in concert.

In a few situations the ANS works by positive feedback. Sexual responses in humans require autonomic reflexes (both sympathetic and parasympathetic) in which the motor response (enlargement of the penis or clitoris by vasocongestion) increases the firing of the same visceral afferents which drive the reflex response. This is positive feedback because it carries the system away from its usual stable state.

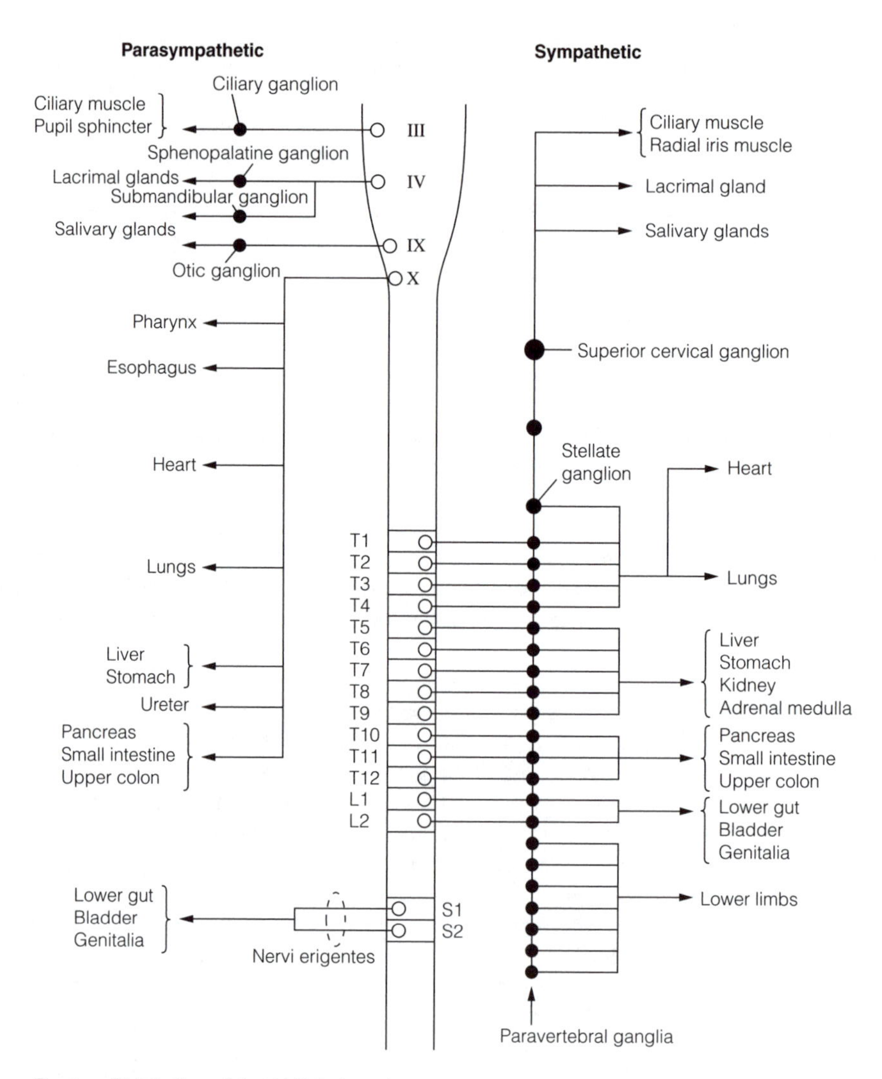

Fig. 1. Distribution of the ANS to target organs.

The classical view of the ANS (given in Topic E1) is that it is exclusively a visceral motor system, the activities of which can be altered by sensory input. An alternative view is that the ANS should include visceral afferents since these run in the same nerve trunks as visceral efferents, they can be distinguished by their neuropeptide transmitters from somatic afferents and it brings both sensory and motor components into the same part of the nervous system. The distribution of the ANS to target organs is shown in *Fig. 1.*

ANS physiology

The anatomy of the ANS was reviewed in Topic E1. The distribution of the ANS to target organs is shown in *Fig. 1.* Autonomic ganglia are a site for divergence, in which preganglionic axons branch to establish connections with several postganglionic cells, thus spreading neural activity over a wider target area. Convergence, in which several preganglionic axons form synapses on a given postganglionic neuron, also occurs. The preganglionic to postganglionic ratio is 1:3 in the PSNS and 1:200 in the SNS. Convergent connections vary in strength, most inputs are weak and summation of many preganglionic inputs are required to fire the postganglionic cell.

Acetylcholine (ACh) is the major neurotransmitter at all autonomic ganglia. ACh released by stimulation of preganglionic neurons acts on nicotinic cholinergic receptors (nAChR) (see Topic C4) rapidly causing a fast excitatory postsynaptic potential (epsp), which if sufficiently large, makes the postganglionic cell fire. In addition, the ACh acts on M1 muscarinic receptors which greatly prolongs the firing of the postganglionic cell by an action on a population of K^+ channels, K_m channels. K_m channels are both voltage-dependent and ligand-gated. They are activated by depolarization, and the efflux of K^+ that results tends to hyperpolarize the cell. Hence normally K_m channels tend to stabilize the membrane potential against depolarizing influences. M1 receptor activation causes the K_m channels to close, producing a slow epsp so that the postganglionic cell continues to fire for many seconds.

Almost all terminals of the sympathetic postganglionic axons secrete norepinephrine (the only important exception being the cholinergic sympathetic supply to sweat glands). The adrenal medulla chromaffin cells are regarded as a postganglionic component of the SNS and secrete epinephrine (and norepinephrine) directly into the blood as a result of activity in the preganglionic sympathetic fibers that supply it. There are four major types of adrenoreceptors that mediate the various effects of sympathetic stimulation. They are all G protein-linked receptors and their properties are summarized in *Table 1*.

All parasympathetic postganglionic axons release ACh, the effects of which are brought about by muscarinic receptors (mAChR). There are several subtypes of mAChR, all G protein-linked, which account for the diverse effects of parasympathetic activity (*Table 1*).

Table 1. Properties of adrenoceptors and muscarinic receptors and their principal actions at ANS targets.

Receptor	G protein	Second messenger	Major tissues	Effect
α1	Gq	IP_3/DAG	Vascular smooth muscle Sphincters[a]	Contraction
α2	Gi	↓ cAMP	Adrenergic terminals (presynaptic)	↓ NA release
β1	Gs	↑ cAMP	Cardiac muscle	↑ Force of contraction ↑ Rate
β2	Gs	↑ cAMP	Airway smooth muscle Gut smooth muscle[b] Liver	Relaxation Gluconeogenesis Glycogenolysis
β3	Gs	↑ cAMP	Fat cells	Lipolysis
M1	Gq	IP_3/DAG	Autonomic ganglia	Close K_m channels
M2	Gi and Go	↓ cAMP opens K^+ channels	Cardiac muscle Sphincters[a] Gut smooth muscle[b] Airway smooth muscle	↓ Rate Relaxation Contraction
M3	Gq	IP_3/DAG	Exocrine glands Endothelium	↑ Secretion NO release

[a] Includes gut and genito-urinary sphincters. [b] Except sphincters. IP_3, Inositol triphosphate; DAG, diacylglycerol; NA, norepinephrine; cAMP, cyclic adenosine monophosphate; NO, nitric oxide.

Autonomic nerve terminals, in addition to secreting norepinephrine or ACh, also release adenosine 5′ triphosphate (ATP) and peptides, which act as cotransmitters. ATP released from sympathetic nerve terminals, for example, acts on the smooth muscle of blood vessels to produce fast excitatory postsynaptic potentials and rapid contraction. This is followed by a slower response due to norepinephrine. Peptide cotransmitters include neuropeptide Y (NPY) and vasoactive intestinal peptide (VIP). They prolong and modulate the effects of the primary transmitter. For example, NPY in sympathetic terminals enhances the vasoconstrictor response to norepinephrine. VIP from parasympathetic terminals on salivary glands causes vasodilation which enables ACh to produce a greater salivary secretion.

An example of coordinated autonomic control: the urinary bladder

The urinary bladder receives a dual autonomic innervation (*Fig. 2*). The sympathetic supply to the bladder wall causes relaxation of the detrusor smooth muscle via β2 adrenoceptors. By contrast the internal sphincter smooth muscle has α1 adrenoceptors and is contracted by sympathetic activity. Hence tonic sympathetic activity allows the bladder to fill and prevents voiding. Contraction of the striated muscle of the external sphincter by somatic motor neurons in the sacral spinal cord also helps to maintain continence. **Micturation** or urination requires the activation of a parasympathetic reflex. Stretch receptors in the bladder wall signal via Aδ and C visceral afferents which enter the sacral spinal cord to synapse with preganglionic parasympathetic neurons. The parasympathetic supply causes contraction of the detrusor muscle but relaxation of the internal sphincter. In addition, collaterals of the afferents establish a long reflex arc via the pons which inhibits both the sympathetic input to the bladder and the motor neurons to the external sphincter. Conscious control over micturation is exerted by descending pathways acting on the pontine neurons. This is lost in patients in whom the spinal cord is severed; in these people micturition is controlled entirely by the spinal reflex and this can be trained to produce urination 'on demand'.

Axon reflexes

Some visceral afferents, particularly those associated with the sympathetic nervous system, respond to local tissue damage by generating action potentials which are propagated **antidromically** (away from the cell body) along axon collaterals. The collateral terminals release a class of peptide transmitters called **tachykinins** (substance P, neurokinin A and neurokinin B) that relax arteriolar smooth muscle, producing vasodilation that increases local blood flow. This sequence is called the **axon reflex**, the resulting response is termed **neurogenic inflammation**. Axon reflexes are also mediated by C fiber nociceptor afferents in the skin where they are responsible for the **flare** of the **triple response**.

Enteric nervous system (ENS)

The ENS, often regarded as the third division of the autonomic nervous system, is responsible for coordinating gut motility (myenteric plexus), secretion and absorption (submucosal plexus). Most myenteric neurons are motor neurons, unipolar cells supplying the smooth muscle of the longitudinal and circular muscle layers. Excitatory motor neurons are cholinergic and corelease tachykinins and NPY. Inhibitory motor neurons release VIP and NO. The motor neurons are tonic and fire volleys of action potentials for as long as they are stimulated.

The sensory neurons are multipolar cells that respond either to stretch, or have neurites in the mucosa that can respond to chemical signals in the intestinal lumen or distortion of the mucosa, associated with the presence of partially digested food (**chyme**). The sensory neurons are normally phasic. In

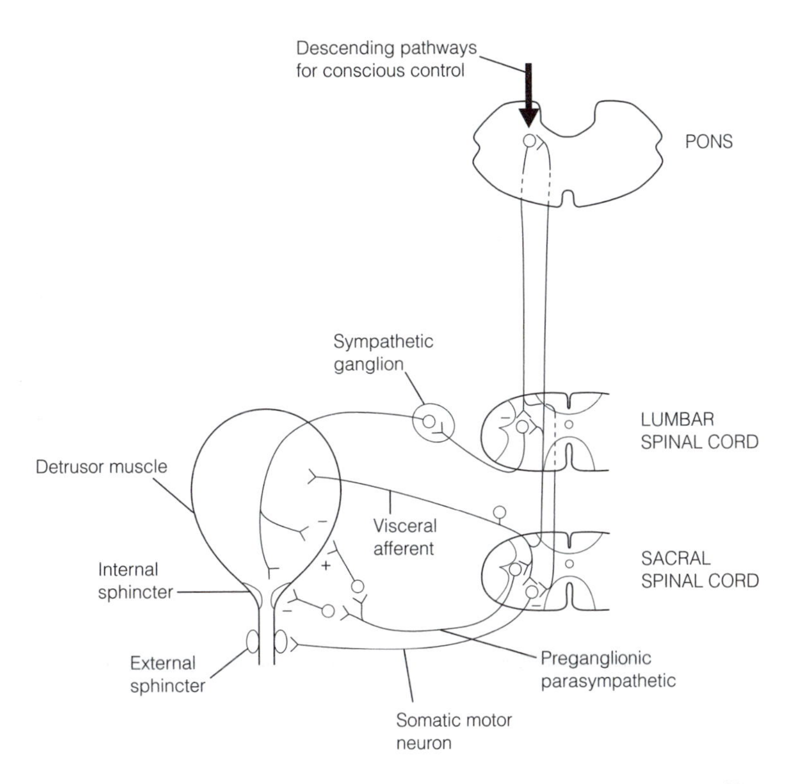

Fig. 2. Control of the urinary bladder. +, Excitatory; –, inhibitory; synapses without a symbol are excitatory.

response to sustained stimulus they fire just a few action potentials, because each action potential has a large after-hyperpolarization caused by the activation of Ca^{2+}-dependent K^+ channels. However their activity can be modulated by input from other neurons, the transmitter of which (VIP, substance P or serotonin) causes a slow epsp by closing K^+ channels. This renders the sensory neurons more sensitive to incoming stimulation.

Most myenteric interneurons which couple sensory and motor neurons are cholinergic, and the receptors on motor neurons (and other interneurons) are nicotinic. Hence transmission between interneuron and motor neuron is fast. Typically, activation of a sensory neuron by stretch, caused by a bolus of chyme, stimulates contraction of circular layer smooth muscle on the oral side, but inhibits it on the anal side. Furthermore, in the longitudinal muscle layers smooth muscle is relaxed on the oral side but contracted on the anal side. The effect of this coordinated smooth muscle activity, **peristalsis**, is to move chyme in the anal direction. *Fig. 3* shows the circuitry involved.

The submucosal plexus contains sensory neurons sensitive to chemical stimuli or mechanical distortion of the mucosa. These connect via cholinergic interneurons with stimulatory **secretomotor neurons** which release ACh and VIP. They increase secretion by a direct effect on the glands, and by relaxing smooth muscle of gut arterioles, produce vasodilation that increases local blood flow. Inhibition of the secretomotor neurons by enkephalinergic neurons within the ENS, or by noradrenergic sympathetic nervous system input, acts to promote absorption.

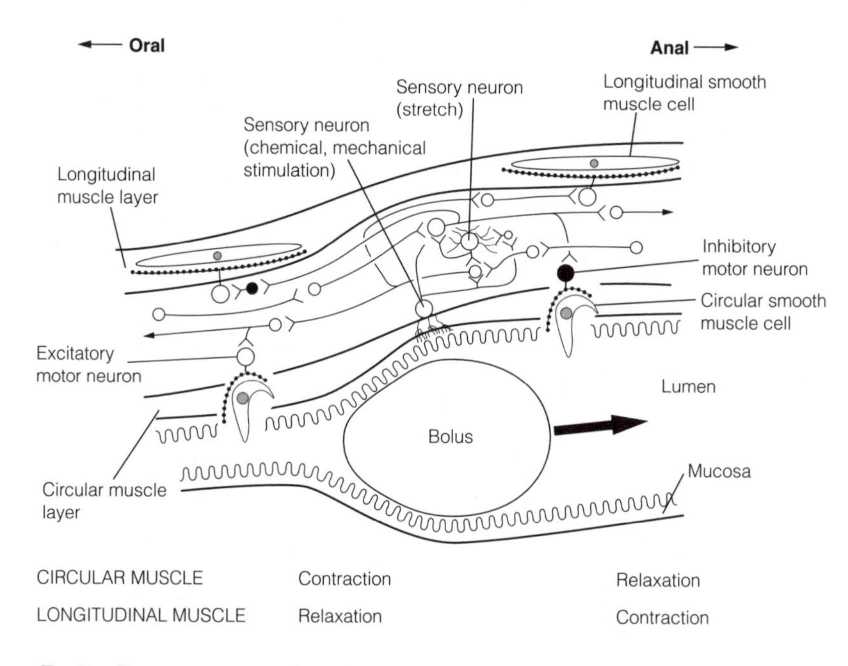

Fig. 3. Elementary gut motility reflex circuitry of the myenteric plexus.

M7 CENTRAL CONTROL OF AUTONOMIC FUNCTION

Key Notes

Thermoregulation

Both behavior and physiological mechanisms allow core temperature to be held at a roughly constant 37°C. Outside a narrow thermoneutral window in which an individual is comfortable, heat gain or heat loss is initiated by negative feedback processes. Small shifts in ambient temperature cause altered sympathetic tone to skin arterioles, producing either cutaneous vasodilation (in the warm), or vasoconstriction (in the cold). Bigger changes in the temperature of the surroundings also trigger sweating, mediated by sympathetic cholinergic stimulation of sweat glands, or shivering, a rapid contraction of muscles driven by the somatic motor system. The posterior hypothalamus integrates signals from internal warm receptors in the hypothalamus and the spinal cord, and signals from skin thermoreceptors. The result of this integration is an appropriate thermoregulatory response. The set point signal ('thermostat') is provided by temperature insensitive hypothalamic interneurons. The set point falls at night, and is raised by progesterone and immune responses to infection (fever).

Cardiovascular regulation

The autonomic nervous system (ANS) is vital for short-term negative feedback control of mean arterial blood pressure (MAP). This is achieved by a combination of modifying the cardiac output and peripheral resistance: increasing either raises MAP. Adjustments to the tonic output of sympathetic and parasympathetic supply to the heart alters its rate and force, increasing them when the sympathetic dominates. Tonic sympathetic discharge to vascular smooth muscle controls blood vessel diameter and so peripheral resistance. Increased activity causes vasoconstriction which elevates the resistance. Mean arterial pressure is monitored by baroreceptors in the aorta and carotid arteries, afferents from which go to the nucleus of the solitary tract (NST) which, by way of other medullary nuclei, controls preganglionic autonomic neurons. An increase in MAP excites baroreceptors, reflexly activating parasympathetic, but inhibiting sympathetic neurons. The consequent fall in heart rate and force, and peripheral resistance, restores the blood pressure. Numerous inputs to this circuitry from other brain regions are responsible for alterations to the cardiovascular system in exercise and emotions.

Control of breathing

Although breathing uses the somatic motor system and skeletal muscles, the circuitry involved gets inputs from visceral afferents and is interconnected with central autonomic neurons. Respiratory muscles (e.g. the diaphragm) are driven by rhythmic activity in motor neurons in the cervical spinal cord. Input to these respiratory motor neurons comes from the ventral respiratory group (VRG) in the medulla. A network of VRG cells, some of which have intrinsic pacemaker activity, acts as a central pattern generator to produce the respiratory rhythm. Inputs to the VRG modify breathing. Several types of receptors in the airways bring about reflex inhibition of inspiration, as does baroreceptor stimulation. Peripheral chemoreceptors in arteries that respond to reduced blood oxygen concentration, and central chemoreceptors in the brain that are responsive to an increase in CO_2 or H^+ concentration, stimulate breathing. The basic respiratory rhythm is modified in many activities by other brain regions

Related topics

Anterolateral systems and descending control of pain (G3)
Spinal motor function (K4)
Smooth and cardiac muscle (M5)
Autonomic nervous system function (M6)

Thermoregulation

A core temperature of around 37°C is defended homeostatically by behavioral and physiological mechanisms. Behavior, which includes seeking sun or shade, curling up into the fetal position when cold (to minimize the surface area for radiation), wearing clothes and building shelter, is particularly useful for reducing the effects of extreme environmental temperatures.

Physiological heat loss or heat gain mechanisms are activated whenever the ambient temperature moves outside the **thermoneutral zone**, a window about 1°C wide in which an individual feels comfortable. The position of the thermoneutral zone depends on humidity, wind velocity and clothing. For naked humans in still air at 50% relative humidity it is 28°C. The first response to ambient temperature moving outside the thermoneutral zone is adjustment in sympathetic tone to smooth muscles of skin arterioles. In heat stress, reduced tone results in a fall in norepinephrine evoked vascular smooth muscle contraction and so **cutaneous vasodilation** occurs. This warms the skin, increasing heat loss by radiation. In the cold, increased sympathetic activity causes **cutaneous vasoconstriction**. Larger excursions from the thermoneutral zone evoke either sweating or shivering. Sweat glands are innervated by sympathetic neurons that are atypical in secreting acetylcholine (ACh) rather than norepinephrine. ACh acts on muscarinic receptors to trigger sweat production which causes skin cooling by evaporation. Shivering is the almost simultaneous contraction of agonist–antagonist muscle pairs. It starts in masseter (jaw) muscles in humans and spreads to the trunk and proximal limb muscles. Shivering is brought about by activation of brainstem reticular neurons that synapse with γ-fusimotor neurons. The contraction of intrafusal fibers excites stretch reflexes (see Topic K3). So, shivering is mediated peripherally by the somatic not the autonomic nervous system. Muscle contraction, both in shivering or exercise, generates heat.

A further heat gain mechanism, **nonshivering thermogenesis**, is particularly important in human babies. It is caused by increased sympathetic activity to brown adipose tissue (BAT), mostly located in the neck and between the shoulder blades. Released norepinephrine acts on β3 adrenoceptors to stimulate a rise in cAMP. This activates lipolysis, liberating free fatty acids which are metabolized by β-oxidation in BAT mitochondria, and at the same time uncoupling oxidative phosphorylation in the mitochondria, generating heat (see *Instant Notes in Biochemistry*, 2nd edn).

Thermoregulation depends on the integration of signals from two classes of thermoreceptors. Cutaneous warm and cold thermoreceptor afferents, conveying information about skin temperature, run in the spinothalamic tract. **Internal warm thermoreceptors** monitor core temperature and are located in the preoptic area of the hypothalamus and the cervical spinal cord. Both cutaneous and internal thermoreceptor afferents go to the posterior hypothalamus, the region of the CNS responsible for driving thermoregulatory responses. Thresholds for activating either sweating or shivering depend on both core temperature and skin temperature. For example during exercise, sweating, triggered by internal thermoreceptors as core temperature rises, is reduced linearly the colder the skin temperature.

The core temperature maintained by thermoregulation is the **set point**. It is defined as the temperature at which neither heat loss nor heat gain mechanisms are activated. The neural signal that acts as the set point 'thermostat' is provided by the integrated activity of interneurons in the hypothalamus that are not temperature sensitive. These interneurons are regulated by catecholaminergic neurons in the pontine reticular formation, and the set point is not constant. It shows a circadian rhythm, falling about 0.5°C during sleep and is increased by progesterone during the luteal phase of the menstrual cycle by about the same amount. Chronic exposure to hot or cold environments causes gradual long term shifts (adaptation) of the set point.

During infections, bacterial endotoxins stimulate macrophages to secrete **interleukin I (IL-1)** while virus-infected cells produce **interferons**. These **cytokines** (signaling molecules of the immune system) act at the hypothalamus to raise the set point, causing fever.

Cardiovascular regulation

Long-term regulation of blood pressure does not require the ANS and relies on control of blood volume and osmolality via vasopressin and the renin–angiotensin–aldosterone cascade (see Topic M2). However, the ANS is crucial for the short-term regulation of blood pressure.

Mean arterial blood pressure (MAP) is controlled by the ANS. At rest the ANS operates to maintain a roughly constant MAP by negative feedback. Mean arterial pressure is the product of cardiac output, the volume output of the left ventricle per minute, and the peripheral resistance, which is related to the radius of the arterioles. Now, the cardiac output is in turn a product of **stroke volume**, the volume ejected from the left ventricle per beat, which is determined by the contractile force of the heart, and **heart rate**. Hence cardiac output can be raised (lowered) either by increasing (decreasing) stroke volume or heart rate or both.

The ANS regulates cardiac output via both sympathetic and parasympathetic supply to the heart (see Topic M5, *Fig. 1* and *Table 1*). Both are tonically active at rest and increases in cardiac output are achieved by increasing sympathetic activity and reducing parasympathetic activity. The effect is to elevate both force of contraction and heart rate. Peripheral resistance is controlled solely by altering the tonic firing rates of sympathetic neurons going to vascular smooth muscle. Increased firing frequency causes vasoconstriction which raises peripheral resistance.

Circuitry for the negative feedback regulation of MAP resides in the medulla (*Fig. 1*). Baroreceptors are stretch receptors located in the carotid sinus and the aortic arch that are sensitive to rapid alterations in MAP. Their afferents run in the glossopharyngeal (IX) and vagus (X) cranial nerves respectively and terminate in the NST, a structure involved in a wide variety of visceral reflexes (e.g. swallowing, chemoreceptor responses). The NST projects to the dorsal vagal

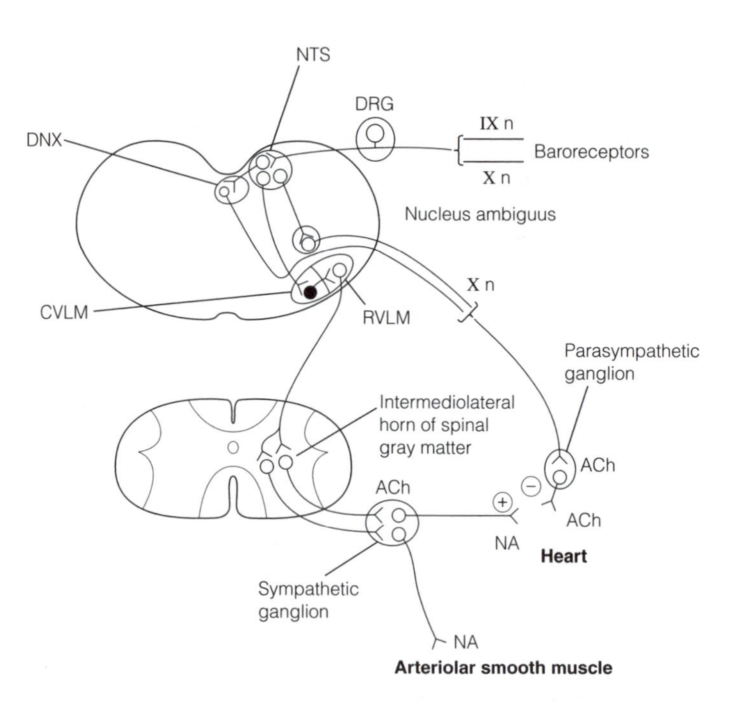

Fig. 1. Brainstem circuits controlling mean arterial blood pressure. DRG, Dorsal root ganglion; DNX, dorsal vagal nucleus; CVLM, caudal ventrolateral medulla; RVLM, rostral ventrolateral medulla; NST, nucleus of the solitary tract.

nucleus and to the **nucleus ambiguus**, both of which give rise to preganglionic parasympathetic axons that project via the vagus (X) nerve to the heart. The NST controls sympathetic outflow to heart and blood vessels by input to the **caudal ventrolateral medulla (CVLM)**. This contains GABAergic inhibitory neurons which synapse in the **rostral ventrolateral medulla**, axons of which run down the spinal cord, terminating on preganglionic sympathetic neurons. A rise in MAP increases the firing rate of baroreceptor afferents, and this directly activates the parasympathetic innervation to the heart, slowing its rate. However, the presence of inhibitory neurons in the CVLM means that baroreceptor discharge suppresses sympathetic outflow to the heart, reducing its rate and force of contraction, and arterioles, which reduces peripheral resistance. The net effect is a fall in blood pressure back to the set point. Responses occur in the opposite direction to an initial drop in MAP.

If blood pressure is altered persistently for any reason then the set point is adjusted and the baroreceptor reflex will now defend the new set point. In **hypertension**, defined as a MAP chronically raised above 140/90 mmHg, baroreceptor reflexes are reset so as to maintain the abnormally high pressure.

Arterial pressure is altered to match circumstances. The **defense reaction**, a stereotyped autonomic response seen in animals faced with sudden danger that includes **tachycardia** (a rise in heart rate), widespread vasoconstriction and a sharp rise in MAP, is organized by a **defense area** in the anterior hypothalamus. Stimulation of the defense area results in inhibition of those neurons in the NST that are driven by baroreceptor afferents. Cardiovascular changes that occur during exercise involve the cerebellar and cerebral cortex which act to modify hypothalamic autonomic regulation. Similarly cardiovascular responses seen in emotional states require elements in the limbic system such as the amygdala and the cingulate cortex.

Control of breathing

The diaphragm and muscles of the chest wall used in breathing are skeletal muscles supplied by motor neurons of the somatic nervous system. However, the central circuitry which regulates breathing gets sensory input from visceral afferents, and is interconnected with central autonomic circuits controlling the cardiovascular system. This is illustrated by sinus arrhythmia, a change in heart rate with the phase of respiration: during inspiration heart rate rises, during expiration it falls. It is caused by neurons responsible for inspiration inhibiting the preganglionic parasympathetic neurons, in the nucleus ambiguus, that supply the heart.

Breathing results from the rhythmic discharge of spinal motor neurons which supply ventilatory muscles. Axons of motor neurons in spinal segments C3–C5 run in the phrenic nerves to the **diaphragm**, contraction of which increases chest volume during **inspiration**. Motor neurons in C4–L3 supply neck muscles and external intercostal muscles that aid inspiration and internal intercostal muscles and abdominal muscles responsible for **expiration**. Most of the muscles used for breathing are also important in other functions; for example abdominal muscles are needed to increase intra-abdominal pressure for defecation and vomiting, and during locomotion.

The spinal motor neurons are driven by premotor neurons located in the **ventral respiratory group (VRG)** of the ventrolateral medulla. It is here that the respiratory rhythm is generated. Both inspiratory and expiratory VRG neurons are found and their axons make excitatory (glutamatergic) connections with the motor neurons. Inspiratory neurons are also found in the NST. These receive sensory input largely via the vagus (X) nerve from pulmonary receptors which are sensitive to the state of the lungs, from baroreceptors, and from peripheral chemoreceptors in the carotid body and aortic arch that monitor blood concentrations of O_2. In addition are central chemoreceptors, sensitive to brain extracellular fluid concentrations of CO_2 and H^+, located in the NST. Inputs from all these sources modify the basic rhythm of breathing (*Fig. 2*).

The respiratory rhythm comes from a network of cells associated with the VRG called the **pre-Botzinger** complex. This contains several populations of neurons, each of which fires a burst of action potentials at a specific phase during the respiratory cycles. Together neurons of the pre-Botzinger complex act as a central pattern generator producing the oscillating output that drives respiration. Exactly how the circuit produces oscillations is uncertain. Two mechanisms shown to be operative in the brainstem may be responsible and they are not mutually exclusive. Reciprocal inhibition, via GABAergic interneurons, between populations of neurons active at different phases ensures that only one population fires at a time. Pacemaker cells which fire in an oscillatory manner probably act to generate the rhythm because blocking reciprocal antagonism with GABA receptor antagonists in an *in vitro* preparation fails to abolish it.

Lung inflation reflexly inhibits inspiration and prolongs subsequent expiration. This is called the **Hering–Breuer reflex** and is mediated by slowly adapting pulmonary stretch receptor afferents which activate neurons in the NST. The NST neurons synapse with neurons in the VRG, terminating inspiration and triggering a protracted expiration. Rapidly adapting stretch receptors respond to irritation of the airways and trigger the cough reflex via the NST and VRG

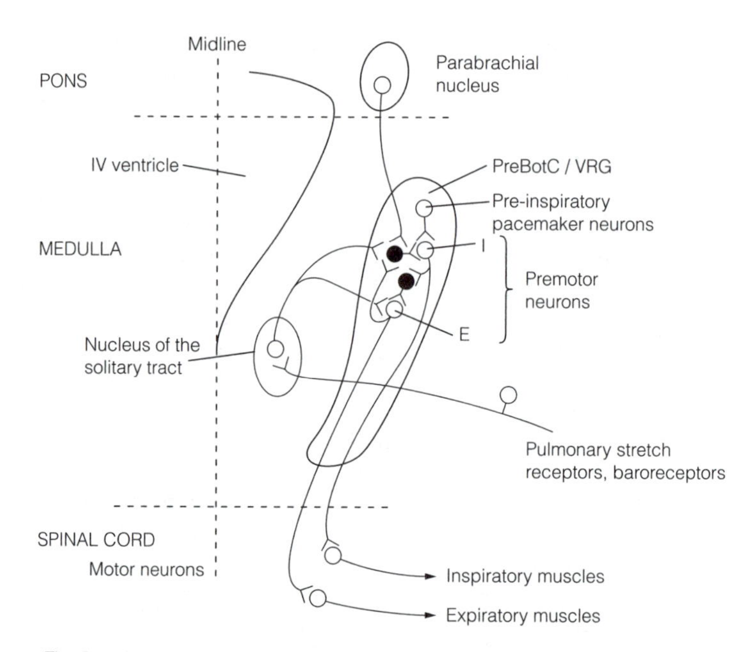

Fig. 2. A highly simplified model of the central control of respiration. E, Expiratory neurons; I, inspiratory neuron; PreBotC, pre-Botzinger Complex; VRG, ventral respiratory group.

circuitry. These same receptors detect the increased lung stiffness due to collapse of alveoli and stimulate a large inspiratory sigh which reopens alveoli. Small diameter C fiber input from lung irritant receptors into the NST is responsible for the breath holding (**apnea**) or fast shallow respiration that occurs when forced to breathe noxious gases.

Baroreceptor discharge inhibits inspiration. So if MAP falls (as a result of hemorrhage, for example) depth of inspiration increases. **Peripheral chemoreceptors** activated principally by a reduced partial pressure of O_2 (**hypoxia**) and **central chemoreceptors** stimulated by raised partial pressure of CO_2 (**hypercapnia**), and fall in pH, drive increased depth of breathing via neurons in the NTS.

Glutamatergic neurons in the parabrachial nucleus (NPBM) of the pons help to maintain tonic inhibition on inspiratory neurons. Lesions of the NPBM in animals which also have cut vagus nerves, causes **apneusis**, abnormal breathing characterized by sustained inspiration punctuated by brief expirations. The brainstem central pattern generator producing the respiratory rhythm is altered in many situations (e.g. sleep, exercise, emotional states and speech) so it is clear that many other brain regions are involved in the control of breathing.

N1 DOPAMINE NEUROTRANSMISSION

Key Notes

Dopaminergic pathways

The major dopaminergic pathways arise from the midbrain and go to the forebrain. The nigrostriatal tract from the substantia nigra to the striatum contains most of the brains' dopamine neurons and is involved in movement. Dopaminergic neurons in the ventral tegmentum project to limbic structures via the mesolimbic pathway and to the cortex by way of the mesocortical pathway. These form a motivation system. Dopamine cells in the hypothalamus control pituitary hormone secretion.

Dopamine synthesis

The catecholamines (dopamine, noradrenaline, adrenaline) are synthesized from tyrosine. The first, rate-limiting step which generates L-DOPA is catalyzed by tyrosine hydroxylase. This enzyme is inhibited by catecholamines. This end point inhibition is one method by which the synthesis of catecholamines is controlled. L-DOPA is decarboxylated to give dopamine.

Inactivation of dopamine

Synaptic dopamine is taken back into nerve terminals by a high affinity dopamine transporter. This process is inhibited by amphetamines and cocaine. Dopamine which escapes reuptake is catabolized to homovanillic acid by catechol-O-methyltransferase then monoamine oxidase (MAO). Dopamine free in the cytoplasm is converted to dihydroxyphenyl acetic acid by mitochondrial MAO.

Dopamine receptors

The five metabotropic receptors for dopamine fall into two families. The D1 receptor family (D1 and D5) increase cAMP concentrations, whereas the D2 receptor family (D2, D3 and D4) decrease cAMP concentrations. In general, D1 receptors are postsynaptic, D2 receptors are localized both pre- and postsynaptically.

Related topics

Slow neurotransmission (C3)
Anatomy of the basal ganglia (L5)
Basal ganglia function (L6)
Motivation (O1)
Parkinson's disease (R3)

Dopaminergic pathways

Dopamine neurons are widely distributed in the nervous system, being found in the retina (amacrine cells), olfactory bulb, adjacent to the ventricles of the brain and in autonomic ganglia. Most dopaminergic neurons, however, are confined to a few nuclei in the brainstem but send their axons to many regions of the forebrain including the cerebral cortex. These major pathways are illustrated in *Fig. 1*.

About 80% of dopamine neurons are in the zona compacta of the substantia

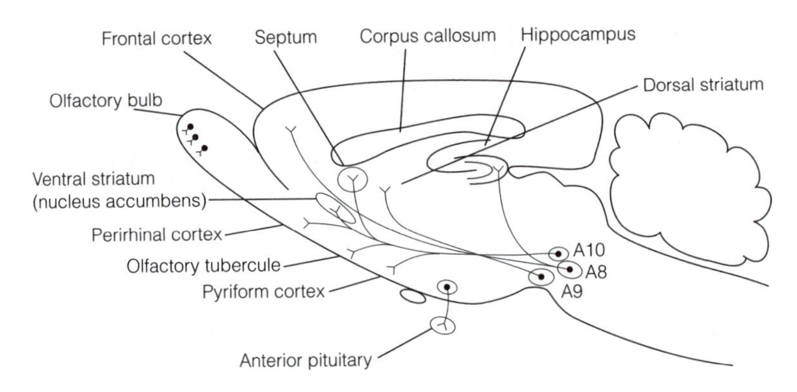

Fig. 1. Major dopamine pathways in a sagittal section of the rat brain. The A8 and A10 group of dopamine neurons give rise to the mesolimbic and mesocortical tracts. The nigrostriatal tract originates in the substantia nigra (A9). A12 neuron axons run in the tuberoinfundibular pathway.

nigra (SNpc), which constitutes the A9 group of catecholaminergic cells. (These groups range from A1 to A16, the higher the number the more rostrally they are located). SNpc neurons project to the striatum as the nigrostriatal pathway. These cells are involved in basal ganglia regulation of movement (see Topics L5, L6) and their loss results in Parkinson's disease (see Topic R3). Dopamine cell clusters (groups A8 and A10) in the ventral tegmentum of the midbrain project to limbic structures (amygdala, septum and nucleus accumbens) or to associated cortical areas (medial prefrontal, cingulate and entorhinal cortex), giving rise to the **mesolimbic** and **mesocortical** systems respectively. These are implicated in motivation, drug addiction (see Topic O1) and in schizophrenia (see Topic R5). Several small groups of dopaminergic cells in the hypothalamus (groups A11, A12 and A13) project axons to the pituitary to inhibit the secretion of prolactin (see Topic M4) or growth hormone. This is the tuberoinfundibular pathway.

Dopaminergic neurons are small (12–30 μm diameter) and have three to six large, long dendrites. The axon arises from one of the dendrites. It is unmyelinated and about 0.5 μm across and bears numerous varicosities along its length. Action potentials in dopamine neurons are long lasting (2–5 ms) and propagated very slowly (0.5 m s^{-1}).

Dopamine synthesis

The precursor for all catecholamine transmitters (dopamine, norepinephrine, epinephrine) is the amino acid, L-tyrosine. This is hydroxylated by **tyrosine hydroxylase (TH)** to give 3,4-dihydroxy phenylalanine (L-DOPA) which is rapidly decarboxylated by the nonspecific enzyme **L-aromatic amino acid decarboxylase** to give dopamine (see *Fig. 2*).

Tyrosine is actively transported into the brain, and the brain concentration of tyrosine is normally enough to saturate TH, so administration of tyrosine cannot alter the rate of dopamine synthesis. The hydroxylation of tyrosine is the rate limiting step for catecholamine synthesis under basal conditions and TH is subject to regulation by:

- increased expression of TH genes, leading to *de novo* synthesis of the enzyme;
- phosphorylation by protein kinases which increases its activity;
- inhibition by catecholamines. This is an example of **end point inhibition**.

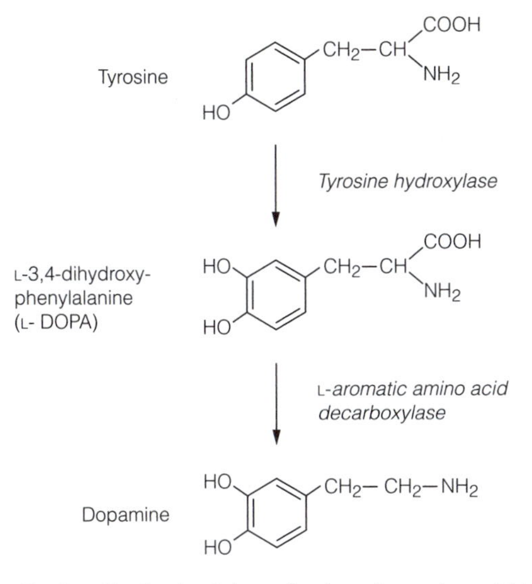

Fig. 2. Synthesis of dopamine from the amino acid tyrosine.

Dopamine is taken into vesicles by a vesicular monoamine transporter (VMAT) which actively transport catecholamines and serotonin using the efflux of protons from the vesicle to provide the energy (see Topic C5). VMATs are blocked by the drug **reserpine** which, by preventing vesicular storage, drastically impairs monoamine neurotransmission. Reserpine has proved to be a useful research tools for investigating the contribution of monoamines to behavior and psychiatric disease.

Inactivation of dopamine

Three mechanisms act to lower the concentration of dopamine after its release into the synaptic cleft. Initially, diffusion carries the dopamine away from the synaptic region. After this, dopamine is taken back into the axon by a high affinity Na^+-, Cl^--dependent **dopamine transporter**. Note that neurons of the tuberoinfundibular pathway that release their dopamine into the hypothalamic–pituitary portal system do not have the dopamine transporter. The dopamine transporter is competitively inhibited by amphetamines and by cocaine, which thus potentiate the effects of dopamine at the synapse. This mechanism may underlie the powerful reinforcing properties of these drugs which make them addictive (see Topic O1).

Two major enzymes are involved in catecholamine catabolism, although catabolism is not important in inactivating dopamine action in the synapse. The primary dopamine metabolites in the central nervous system are **homovanillic acid** and **dihydroxyphenyl acetic acid** (**DOPAC**). In primates, the major catabolic route is via homovanillic acid and this is the fate of dopamine released into the cleft which escapes reuptake. It requires the sequential action of **catechol-O-methyl transferase** (**COMT**) and **monoamine oxidase** (**MAO**), both of which are present in neuronal membranes (*Fig. 3*). Cytoplasmic dopamine that is not transported into vesicles and hence remains free in the axon is catabolized by MAO located on the outer membrane of mitochondria, then by **aldehyde dehydrogenase**, a soluble cytosolic enzyme, to DOPAC.

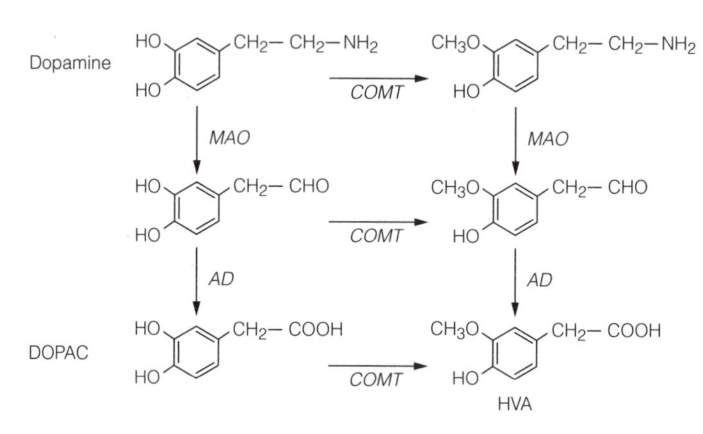

Fig. 3. Metabolism of dopamine. DOPAC, Dihydroxyphenyl acetic acid; HVA, homovanillic acid; COMT, catechol-O-methyl transferase; MAO, monoamine oxidase; AD, alcohol dehydrogenase.

Dopamine receptors

Five dopamine receptors have been identified and sequenced by genetic engineering techniques. All are metabotropic G protein-linked receptors and fall into two groups. Members of the D1 family are coupled to G_s and activate adenylyl cyclase to increase cAMP synthesis; this family consists of two members, D1 and D5. The D2 family consists of D2, D3 and D4 receptors; these are coupled to G_i and inhibit adenylyl cyclase to reduce cAMP synthesis. Both D1 and D2 receptors are located postsynaptically (e.g. in the striatum). In addition, D2 receptors are autoreceptors on dopamine neurons in the substantia nigra and ventral tegmentum where they help to regulate dopamine synthesis. When occupied by dopamine they reduce cAMP concentrations and so phosphorylation of tyrosine hydroxylase by protein kinase A falls. This reduces the synthesis of dopamine. Presynaptic D2 receptors on the terminals of the corticostriatal neurons modulate their glutamate release. D3 receptors are presynaptic autoreceptors. By closing presynaptic Ca^{2+} channels they reduce dopamine release. The mesocortical pathway differs from the nigrostriatal pathway in terms of its complement of dopamine receptors. Firstly the mesocortical neurons have no autoreceptors which means they lack the normal regulation of synthesis and release of dopamine. Secondly the cortex, but not the striatum expresses D4 receptors. These differences are important for the treatment of schizophrenia (see Topic R5).

N2 Norepinephrine neurotransmission

Key Notes

Noradrenergic pathways

Noradrenergic neurons are located in the pons and medulla. The largest group is the locus ceruleus. Noradrenergic axons project via the medial forebrain bundle to most forebrain structures including the cortex, forming wide synapses which allow considerable diffusion of the transmitter.

Norepinephrine and epinephrine synthesis

Dopamine-β-hydroxylase catalyses the synthesis of norepinephrine (noradrenaline, NA) from dopamine. In adrenergic neurons in the brain and chromaffin cells of the adrenal medulla, NA is metabolized further to epinephrine.

Norepinephrine inactivation

A high affinity transporter is responsible for reuptake of NA from the synaptic cleft. The transporter is inhibited by tricyclic antidepressant drugs. Compounds structurally related to NA (e.g. tyramine) are taken up by the transporter and they enhance NA release (indirect sympathomimetics) or are subsequently metabolized to weak adrenergic agonists and then released as false transmitters. The enzymes monoamine oxidase (MAO) and catechol-O-methyl transferase (COMT) are responsible for NA catabolism producing 3-methyl-4-hydroxyphenyl glycol which is then excreted.

Adrenergic receptors

Adrenoceptors are metabotropic receptors activated by NA and epinephrine. α1 receptors are typically postsynaptic and coupled to the IP_3/DAG second messenger system. α2 receptors are presynaptic and reduce cAMP. All β adrenoceptors are coupled to G_s proteins and raise cAMP levels.

Arousal

Noradrenergic neurons in the locus ceruleus (LC) fire in a manner that correlates with an animal's level of arousal and in response to the appearance of stimuli to which they have been trained to attend. The effect of norepinephrine secreted throughout the brain is to enhance the response of neurons to specific excitatory and inhibitory inputs.

Related topics

Slow neurotransmission (C3)
Anterolateral systems and descending control of pain (G3)
Autonomic nervous system function (M6)
Sleep (O4)

Noradrenergic pathways

Cell bodies of noradrenergic neurons are located in the pons and medulla (cell groups A1–A6, except A3). The most caudal groups, A1 and A2, send their axons into the spinal cord where they form synapses with the terminals of primary afferents. The others project in two bundles, a dorsal or a ventral bundle, which

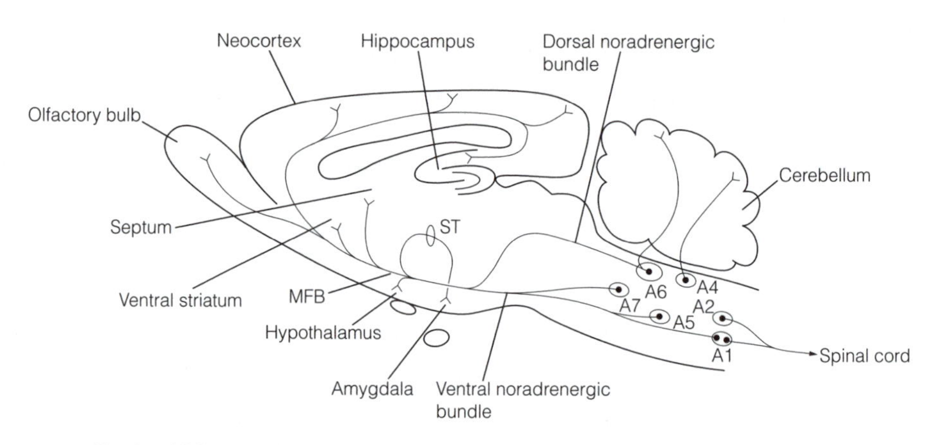

Fig. 1. Major noradrenergic pathways in a sagittal section of the rat brain. A6 is the locus ceruleus. MFB, medial forebrain bundle; ST, stria terminalis.

unite to form the **medial forebrain bundle** that ascends to supply the hypothalamus, amygdala (via the stria terminalis), thalamus, limbic structures, hippocampus and neocortex. The major noradrenergic cell group is the **locus ceruleus** (**LC**, group A6) which contributes most of the axons of the dorsal noradrenergic bundle and projects to the cerebellum. In the rat, the LC contains some 200 000 neurons (see *Fig. 1*). Noradrenergic neurons are small with fine, highly branched axons that ramify widely. The axons bear varicosities along their length, but they do not form close synaptic contacts, so release NA some distance from their targets. This and the wide distribution of its terminals has led to the description of noradrenergic transmission as being a 'neural aerosol'.

Norepinephrine and epinephrine synthesis

The first steps in NA synthesis require the synthesis of dopamine from tyrosine. **Dopamine-β-hydroxylase (DβH)**, an enzyme present in the synaptic vesicle membrane then catalyzes the synthesis of norepinephrine (NA; see *Fig. 2*). NA is actively taken into synaptic vesicles by the vesicular monoamine transporter (see Topic C5) where it is stored bound to a protein, chromogranin (reducing its osmotic activity) and adenosine 5'-triphosphate. These are coreleased with NA. Since a small amount of soluble DβH is also coreleased with NA and is not subject to metabolism or reuptake it can be used as a marker of the activity of noradrenergic neurons.

For noradrenergic neurons the reaction stops at this point. However, for the relatively few neurons in the hindbrain that are adrenergic (and for the chromaffin cells of the adrenal medulla), the enzyme **phenyletholamine N-methyltransferase** catalyzes the N-methylation of norepinephrine to epinephrine.

High activity by LC neurons results in increased expression of the tyrosine hydroxylase (TH) genes and *de novo* synthesis of the enzyme so that the demand for NA synthesis can be met. The effect of this is that DβH becomes the rate limiting enzyme rather than TH; thus dopamine and its metabolites may be coreleased with NA.

Norepinephrine inactivation

Diffusion and reuptake are the key mechanisms removing NA from the synapse. The NA transporter is a saturable Na^+-, Cl^--dependent transporter expressed in noradrenergic neurons, it shares homology with the dopamine

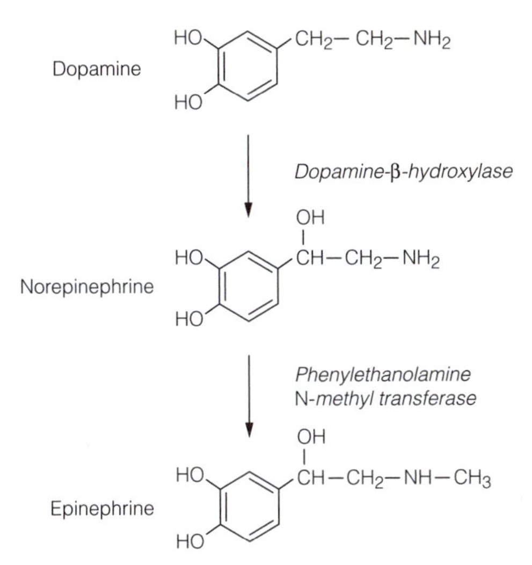

Fig. 2. Synthesis of norepinephrine and epinephrine. These catecholamines, like dopamine, are derived from tyrosine. Early synthetic steps are shown in Topic N1, Fig. 2.

transporter. The NA transporter is inhibited by the tricyclic antidepressant group of drugs.

The NA transporter does not show a high degree of substrate specificity. Amphetamines, tyramine and other compounds structurally related to NA are taken up by the NA transporter. This inhibits reuptake of NA itself, so its effect at the synapse is prolonged. Furthermore, the compounds displace stored NA from vesicles into the cytoplasm where some of it is degraded by mitochondrial MAO and the rest is release into the cleft via the NA transporter operating in reverse. Because they enhance NA transmission they are referred to as **indirectly acting sympathomimetics**. As repeated doses of amphetamine produce ever greater depletion of stored NA, the quantity required to produce a given effect gets progressively larger. This is an example of tolerance (see Topic O1). Some substrates for the NA transporter are metabolized in the terminal and the metabolites stored in the vesicles, when released, exert weak effects on adrenoceptors. These are **false transmitters**.

The metabolic degradation of NA is not important for its inactivation and occurs via different routes in the periphery and central nervous system (CNS). In the CNS, MAO catalyzes the formation of 3,4-dihydroxy phenylglycoalde-hyde which is then reduced to the corresponding alcohol, **3,4-dihydroxy phenylglycol** (**DOPEG**). Finally this is methylated by COMT to give **3-methoxy, 4-hydroxy phenylglycol** (**MOPEG**) which is excreted in the urine. MOPEG excretion has been used as a measure of NA turnover in the CNS.

Adrenergic receptors

Adrenoceptors are metabotropic receptors that are activated by both NA and epinephrine. *Table 1* summarizes the G proteins and second messenger systems that are linked to the different receptors. In the CNS, α2 receptors are pre-synaptic autoreceptors, where they reduce NA release by reducing cAMP-mediated phosphorylation of N-type Ca^{2+} channels, thus suppressing Ca^{2+} entry. Presynaptic β receptors are also found on noradrenergic terminals in the brain. These facilitate NA release by increasing cAMP-mediated phosphorylation and

Table 1. Adrenergic receptors

Receptor	G protein	Second messenger/effector
α1	Gq	IP_3/DAG
	Go	↓ gK
α2	Gi	↓ cAMP
		↑ gK ↓ gCa
β1	Gs	↑ cAMP
		↑ gCa
β2	Gs	↑ cAMP
β3	Gs	↑ cAMP

IP_3, Inositol trisphosphate; DAG, diacylglycerol; cAMP, cyclic adenosine monophosphate.

opening of Ca^{2+} channels. Both excitatory and inhibitory effects of NA release are seen postsynaptically in CNS neurons.

Arousal

Arousal is regulated by the widely distributed noradrenergic neurons. Activity of these cells causes a globally synchronized release of norepinephrine (NA) throughout much of the brain which acts to modulate neuron responses to input from other transmitters. Recording from the LC in behaving animals shows that the firing rate is low during sleep and increases with the level of arousal. In wakefulness, the frequency of firing rises when animals switch from low vigilance behavior (i.e. grooming) to orienting towards a stimulus. LC neurons will fire in response to the presentation of stimuli that animals have been trained to pay attention to. The firing is not related to the sensory attributes or motor requirements of the task.

The neuromodulatory effect of NA on cells in the cerebral or cerebellar cortex is to augment the effects of excitatory (glutamatergic) or inhibitory (GABAergic) inputs relative to the basal firing rate. This is interpreted as an improvement in the **signal-to-noise ratio** of neurons by noradrenergic inputs.

N3 SEROTONIN NEUROTRANSMISSION

Key Notes

Serotonergic pathways

Serotonin neurons are located in the raphe nuclei which lie close to the midline throughout the brainstem. Some axons descend into the spinal cord and inhibit nociceptor input into the spinothalamic tract. Axons run in the medial forebrain bundle to most forebrain structures including the choroid plexus and cerebral blood vessels.

Synthesis of serotonin

Serotonin (5-HT) is synthesized from tryptophan, the plasma concentration of which can affect serotonin levels in the brain. The rate limiting step in 5-HT synthesis is the hydroxylation of tryptophan catalyzed by tryptophan hydroxylase. The activity of this enzyme increases with neuron firing rate so that transmitter synthesis keeps pace with neural activity.

Inactivation of serotonin

Reuptake of serotonin by a transporter terminates its transmitter action. The transporter is inhibited by tricyclic antidepressants and selective serotonin reuptake inhibitors (e.g. Prozac). 5-HT is catabolized by monoamine oxidase (MAO) to 5-hydroxyindoleacetic acid.

Serotonin receptors

Of the many subtypes of 5-HT receptor all but 5-HT_3 receptors are metabotropic. 5-HT_3 receptors are ligand-gated nonspecific cation channels. Most 5-HT receptors are postsynaptic but the 5-HT_{1A} subtype is a presynaptic autoreceptor that inhibits serotonin release.

Related topics

Slow neurotransmission (C3)
Anterolateral systems and descending control of pain (G3)
Brain biological clocks (O3)
Sleep (O4)

Serotonergic pathways

Clusters of serotonin neurons (designated B1–B9) are scattered throughout the brainstem, mostly towards the midline in the **raphe nuclei**. Projections into the spinal cord that terminate in the dorsal horn are important in pain sensation by reducing nociceptor input into the spinothalamic tract (Topic G3). Other serotonergic spinal cord axons synapse with preganglionic autonomic neurons. Forward projections run into the medial forebrain bundle to go to the hypothalamus, amygdala, striatum, thalamus, hippocampus and neocortex (*Fig. 1*). Most brain structures have a serotonergic innervation, including the choroid plexus and cerebral blood vessels, where it regulates cerebrospinal fluid (CSF) secretion and cerebral blood flow respectively.

Synthesis of serotonin

The precursor for serotonin is the amino acid tryptophan. The plasma concentration of tryptophan, which varies according to dietary intake, can alter brain

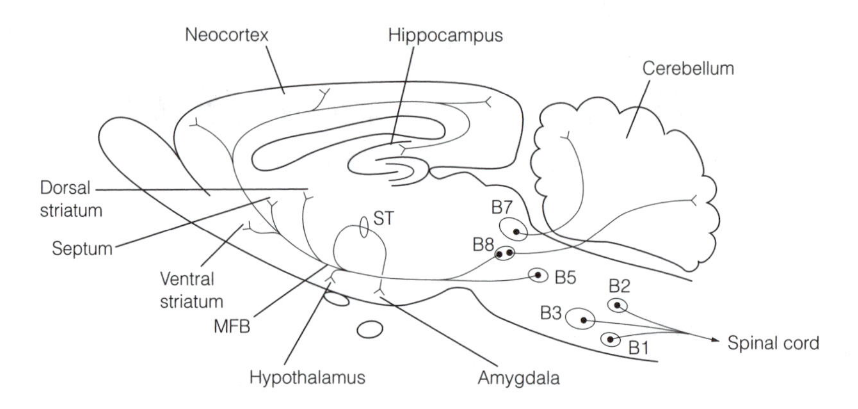

Fig. 1. Major serotonin (5-HT) pathways in a sagittal section of the rat brain. The cell groups B1–B8 correspond to the 5-HT containing raphe nuclei (except B4 and B6). MFB, Medial forebrain bundle; ST, stria terminalis.

serotonin levels. Serotonin is hydroxylated by **tryptophan hydroxylase** to give 5-hydroxytryptophan (5-HTP) and this reaction is the rate limiting step in serotonin synthesis. Decarboxylation of 5-HTP by L-aromatic amino acid decarboxylase (the same enzyme found in catecholaminergic neurons) gives **serotonin**, also referred to as **5-hydroxytryptamine (5-HT)**, which is an indolamine (*Fig. 2*).

Serotonin synthesis is matched to the firing frequency of the neuron. Higher firing rates allow increased Ca^{2+}-dependent phosphorylation of tryptophan hydroxylase, the activity of which goes up.

Inactivation of serotonin

Diffusion and reuptake via a saturable Na^{+}-, Cl^{-}-dependent transporter are the prime means by which the action of serotonin in the synapse is terminated. The transporter is inhibited by tricyclic antidepressants and the relatively new selective serotonin reuptake inhibitors that include fluoxetine (Prozac). Oxidative deamination of serotonin by MAO yields its principle metabolite, **5-hydroxyindoleacetic acid (5-HIAA)**.

Serotonin receptors

There are numerous subtypes of serotonin receptor, all except one are metabotropic. The 5-HT_3 receptor, however, belongs to the ligand-gated ion channel superfamily. *Table 1* summarizes the G protein and second messenger

Table 1. Serotonin receptors

Receptor	G protein	Second messenger/effector	Actions
$5\text{-HT}_{1A,B,D–F}$	Gi	↓ cAMP	Slow inhibitory transmission Presynaptic inhibition
$5\text{-HT}_{2A–C}$	Gq	IP_3/DAG	Slow excitatory transmission
5-HT_3	–	Ligand-gated channel (nonselective cation conductance)	Fast excitatory transmission
5-HT_4	Gs	↑ cAMP	
$5\text{-HT}_{5A,B}$	?		
5-HT_6	Gs	↑ cAMP	
5-HT_7	Gs	↑ cAMP	

5-HT, serotonin; IP_3, inositol trisphosphate; DAG, diacylglycerol; cAMP, cyclic adenosine monophosphate.

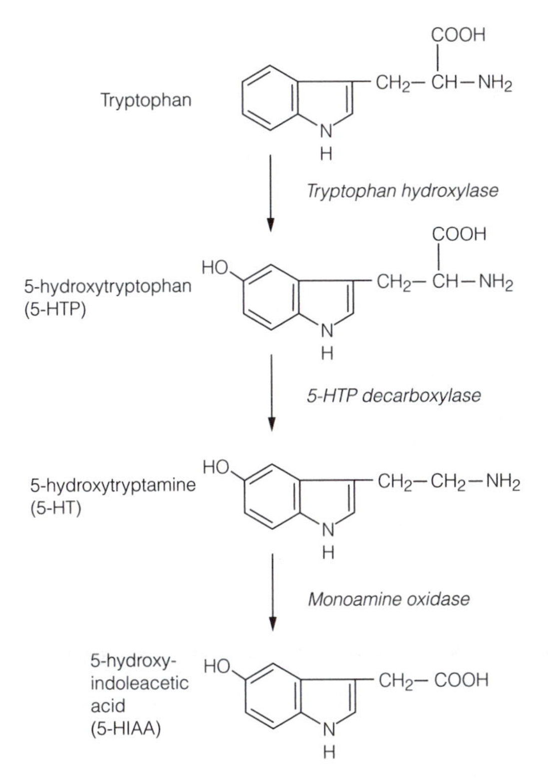

Fig. 2. Synthesis of serotonin from the amino acid tryptophan.

coupling of 5-HT receptors. Presynaptic autoreceptors of the 5-HT_{1A} subtype, inhibit the release of serotonin. They do so by a direct action of their associated G protein, opening K^+ channels which hyperpolarizes the cell membrane. Most 5-HT_1, 5-HT_2 and 5-HT_3 receptor subtypes are located postsynaptically.

N4 ACETYLCHOLINE NEUROTRANSMISSION

Key Notes

Cholinergic pathways

Somatic and autonomic preganglionic motor neurons that project from the brainstem and spinal cord are cholinergic. Central cholinergic projections come from three principal sources. The pontine reticular formation sends axons to spinal cord or forward to forebrain structures. Basal forebrain nuclei make massive connections with the cortex and the septum projects to the hippocampus.

Acetylcholine synthesis

Acetylcholine (ACh) is produced from acetyl coenzyme A (CoA) and choline by choline acetyl transferase, a marker enzyme for cholinergic neurons.

Acetylcholine inactivation

ACh is hydrolyzed to choline and acetate in the synaptic cleft by acetylcholinesterase (AChE) which terminates its transmitter action. Choline is taken back into the nerve terminal by a Na^+-dependent choline transporter.

Acetylcholine receptors

Nicotinic receptors (nAChR) are ligand-gated ion channels and muscarinic receptors (mAChR) are metabotropic receptors. In the central nervous system (CNS) nAChRs seem largely confined to Renshaw cells in the spinal cord. Central mAChRs are widely distributed with M1 receptors being postsynaptic and M2 receptors being presynaptic. In the periphery, nAChRs mediate fast transmission in autonomic ganglia and at the neuromuscular junction of skeletal muscle. mAChRs are present in smooth muscle, cardiac muscle and glands and respond to ACh released from the autonomic nervous system (ANS).

Central cholinergic function

Basal forebrain neurons in primates fire in response to presentation of reinforcing stimuli, and their effect is to cause a long-term facilitation of cortical neurons. Hence they produce cortical arousal in response to stimuli that the animal has learnt are rewarding.

Related topics

Autonomic nervous system function (M6)
Sleep (O4)
Memory circuitry in mammals (Q3)
Alzheimer's disease (R4)

Cholinergic pathways

Motor neurons in motor nuclei of the cranial nerves and ventral horn of the spinal cord are cholinergic, as are preganglionic autonomic neurons; the axons of all these cells project into the peripheral nervous system. Three regions within the brain contain cholinergic neurons that project centrally. Most caudal are the neurons of the **laterodorsal tegmental** and **interpeduncular nuclei** (part of the **pontine reticular formation**) that send axons into the spinal cord or

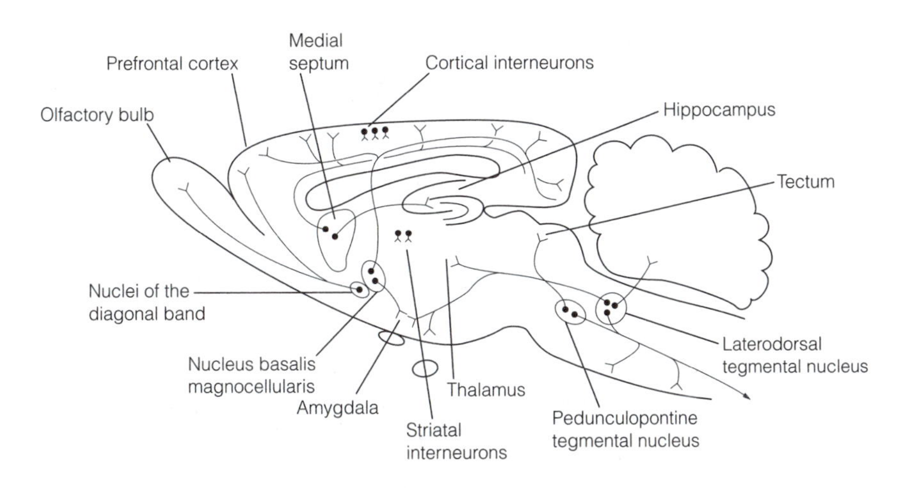

Fig. 1. Major cholinergic pathways in a sagittal section of the rat brain. The nucleus basalis magnocellularis of the rat is known as the nucleus basalis of Meynert in primates.

forward to the amygdala, thalamus and basal forebrain. A second region, the basal forebrain, contains the **magnocellular forebrain nuclei**, including the **nucleus basalis of Meynert (NBM)** and the nuclei of the **diagonal band (NDB)** which project extensively to the cerebral cortex. The third region, the **medial septum**, gives rise to the **septohippocampal pathway** (*Fig. 1*). Loss of the central cholinergic pathways is an inevitable finding in Alzheimer's dementia (Topic R4). Cholinergic interneurons are present in the striatum and the nucleus accumbens.

Acetylcholine synthesis

Synthesis of ACh from choline and acetyl CoA is catalyzed by **choline acetyl transferase (ChAT)**, a cytoplasmic enzyme. Acetyl CoA is derived from glycolysis and must be transported out of the mitochondria of cholinergic neurons. This supply of acetyl CoA, rather than ChAT activity, is thought to be rate limiting for ACh synthesis. Cholinergic neurons express a Na^+-dependent choline transporter which is saturated at plasma choline concentrations and is responsible for the uptake of choline into neurons. ChAT is a reliable marker of cholinergic neurons. ACh is loaded into vesicles by a transporter that is related to the vesicular monoamine transporters (see Topic C5).

Acetylcholine inactivation

ACh is the only neurotransmitter to have its synaptic action terminated by enzyme degradation. ACh is hydrolyzed in the cleft to choline and acetate by **acetylcholinesterase (AChE)**. The liberated choline is recovered by the Na^+-dependent choline transporter. AChE can be secreted in a Ca^{2+}-dependent manner and may act as a neuromodulator in the substantia nigra and cerebellum, where in addition to its catalytic activity it enhances the responses of cerebellar neurons to glutamate.

Acetylcholine receptors

Receptors for ACh are either ligand-gated ion channels, **nicotinic receptors (nAChR)** or metabotropic, G protein-linked **muscarinic receptors (mAChR)** (*Table 1*).

Table 1. Muscarinic receptors

Receptor	G protein	Second messenger/effector
M1	Gq	IP_3/DAG
M2	Gi	↓ cAMP
	Go	↑ gK
M3	Gq	IP_3/DAG
M4	Gi	↓ cAMP

IP_3, Inositol trisphosphate; DAG, diacylglycerol; cAMP, cyclic adenosine monophosphate.

Nicotinic receptors are extremely sparse in the CNS and the Renshaw cells of the spinal cord are the only sites where their participation in fast excitatory transmission is certain. In the CNS postsynaptic mAChR are commonly M1. Presynaptic autoreceptors inhibit the release of ACh but have no effect on the synthesis of ACh. Autoreceptors are M2 (and possibly M4) subtypes.

In the peripheral nervous system, both nicotinic and muscarinic receptors are involved in cholinergic transmission in autonomic ganglia, whereas muscarinic receptors only are found at the neuroeffector junctions of the ANS, on smooth muscle, cardiac muscle and glands. Nicotinic receptors mediate transmission at the neuromuscular junction between somatic motor neurons and skeletal muscle.

Central cholinergic function

In primates, cholinergic neurons of the basal forebrain (the nucleus basalis of Meynert) show brief changes in firing rate during behavioral tasks, particularly when positively (or negatively) reinforcing stimuli are presented, or when stimuli which consistently precede reinforcers appear. ACh produces long-term facilitation of neurons in the neocortex and hippocampus. By protracted decrease of a Ca^{2+}-activated K^+ current, ACh reduces the after-hyperpolarization that follows action potentials, rendering cortical neurons more likely to fire in response to excitatory inputs. Hence the forebrain cholinergic system may be a selective arousal system, activated by rewarding or salient events, and may also act to facilitate associative learning (see Topic Q3). Central cholinergic neurons in the brainstem are important in regulating sleep and wakefulness. This is dealt with in Topic O4.

O1 Motivation

Key Notes

Motivated behavior

Motivated behavior is directed to achieving a specific goal, and is driven by internal states (e.g. hunger) and external cues. Physiological deficits motivate appetitive behavior (e.g. search for food) and consummatory behavior (e.g. eating). A stimulus that increases the chance of a motivated response is a positive reinforcer; if it decreases the probability it is a negative reinforcer. The reinforcing quality of a given stimulus is species specific and depends on the context.

Dopamine reward system

The ascending dopaminergic pathways are thought to activate motivated behavior. Evidence for this comes from studying the contexts in which mesolimbic neurons fire, from the effect on behavior of pharmacological manipulation of the mesolimbic system, and from the fact that electrical stimulation of the mesolimbic system appears inherently highly rewarding in rats. The nucleus accumbens (ventral striatum), a limbic structure that integrates dopaminergic motivational input and information from the amygdala about reinforcers, probably determines whether motivation gives rise to action.

Drug addiction

Addictive drugs are positive reinforcers that take the place of natural reinforcers (e.g. food, sex) in driving the brain dopaminergic reward system. There are three aspects to addiction: tolerance (repeated doses of a drug become progressively less effective); dependence (normal functioning is only possible in the drugged state); and withdrawal (unpleasant effects result if the drug is not taken). Tolerance can occur without dependence ensuing, but is a necessary prerequisite for dependence. Operant studies in animals show that tolerance is greater when a drug is given in the context in which it is normally administered, than when supplied in a novel situation; it indicates the importance of learning in addictive behavior. Addictive drugs activate the mesolimbic system and various measures of addiction are prevented by destroying dopamine neurons or blocking dopamine receptors. Transmitter systems other than dopamine are also implicated in addiction to particular agents. During withdrawal the reward system is hypoactive and corticotrophin-releasing hormone is secreted by the hypothalamus and amygdala, activating stress responses.

Related topics

Anatomy of the basal ganglia (L5)
Neuroendocrine control of reproduction (M4)
Dopamine neurotransmission (N1)
Control of feeding (O2)
Types of learning (Q1)

Motivated behavior

Behavior that is driven by internal states or external cues and events and which is directed to achieve a particular goal is **motivated** or **goal-directed** behavior. Some motivated behavior occurs in order to satisfy physiological needs. Deficits in water or energy substrates (e.g. glucose or lipids) generate neural signals which give rise to the conscious perceptions of thirst and hunger. These drive **appetitive** or goal-seeking behaviors such as the search for a source of water, or foraging for food, and subsequently **consummatory** behavior; drinking and eating. Any stimulus that may act to trigger motivated behavior is an **incentive**. An incentive acts as a **positive reinforcer** if it increases the probability of a response occurring. An animal will work to get access to a positive reinforcer. By contrast a reinforcer is said to be negative if the animal works to avoid the stimulus, in which case it is displaying **aversive behavior**. Incentives are species specific. Food-deprived cats or cows presented with both grass and meat display very clear preferences and usually reject the alternative even if it is the only food available.

The reinforcing quality of a stimulus depends on context. For example, food is a powerful reward to a hungry person but its positive reinforcing quality diminishes as they become sated. However, a particular food may still be a positive reinforcer if it is novel and sufficiently delicious even if the person is not hungry. Hence the motivation to eat is a complex interplay of internal state, external cues and memory. Much motivated behavior occurs in the absence of any clear physiological deficit. This is the case with reproductive behaviors such as finding a sexual partner, copulating, nest building or rearing of young. These will be facilitated by internal brain states – brain biological clocks (see Topic O4), hormones – and by external events that may be readily identifiable. The triggers for motivating other activities such as listening to music, exploring a forest path, engaging in a sport or academic study, are far from obvious.

Dopamine reward system

Motivated behaviors may be regarded as having two components, **activation** and **direction**. By activation is meant the intensity of the behavior whereas direction is the type of behavior (drinking, feeding, copulating etc.). Considerable evidence suggests that ascending dopaminergic neurons are responsible for activating motivated behavior. The nigrostriatal pathway is concerned with motivated locomotion (see Topic L5) including that involved in consummatory activity. The mesolimbic system controls behavior motivated by external positive reinforcers. The evidence for this comes from several approaches.

(1) Activity of neurons in the mesolimbic system is increased by the presence of natural reinforcers such as food.
(2) Injections of amphetamine into the nucleus accumbens, an important target of the ascending mesolimbic system, increases motivated behavior in a manner which suggests that enhancing dopamine transmission heightens the positive reinforcing properties of stimuli.
(3) Surgical or chemical lesion of the dopamine mesolimbic pathway, with the toxic dopamine analog 6-hydroxy dopamine (6-OHDA), reduces appetitive behavior, including curtailing the activity of food-deprived rats presented with food and reducing proceptive ('predatory') sexual behavior in female rats.
(4) From **intracranial self stimulation (ICSS)** in which electrodes are chronically implanted into the brain and the animals (usually rats) learn to press a lever to deliver a small stimulating current through the electrode. With the electrode in the medial forebrain bundle (MFB), through which run the mesolimbic axons, rats lever press up to 100 times per minute in order to

stimulate their own MFB. Given the choice between food and ICSS, food-deprived rats choose ICSS, which implies that it is a very powerful reinforcer. Many areas of the brain will support high rates of ICSS, most are sites of catecholaminergic neurons, and pharmacological studies suggest that dopamine is the important transmitter.

These findings suggest that the mesolimbic neurons constitute a **brain reward** or **motivation system**. It signals the hedonic (pleasurable or positively reinforcing) qualities of a stimulus such as food or drink and so activates the appropriate goal-seeking behavior.

A major destination of the mesolimbic system is the **nucleus accumbens** (**ventral striatum**, so-called because structurally it greatly resembles the neostriatum). This forms part of the anterior cingulate basal ganglia circuit (see Topic L5) which is thought to translate motivation into appropriate motor activity. Furthermore, the nucleus accumbens gets input from the amygdala, a cluster of nuclei situated in the temporal lobe, involved in appetitive and aversive learned behaviors. Lesions of the amygdala impair the ability of animals to learn associations between stimulus and reward.

The nucleus accumbens integrates a dopamine motivation system, with glutamatergic inputs from the amygdala which carry information that has been acquired about the context in which reinforcers appear. The result of this integration may determine the extent to which motivation translates into action.

Drug addiction

All addictive drugs have positively reinforcing properties and these are primarily responsible for drug-seeking behavior in addicts. The current theory is that addictive drugs 'hijack' the brain reward system that is normally responsible for motivating behavior to seek natural reinforcers.

Addiction is characterized by three features, tolerance, dependence and withdrawal. **Tolerance** occurs to a drug when it is given repeatedly and it becomes progressively less effective with each administration, so that the dose has to be increased to achieve the original effect. Many distinct physiological mechanisms have been proposed for tolerance that have different time courses and depend on the specific drug. They include the induction of enzymes, changes to receptor numbers and alterations to transduction mechanisms. With some drugs the physiological changes brought about by tolerance are such that normal functioning can only occur when the drug is present. This is termed **dependence** and is the hallmark of drug addiction. If the drug is withheld after dependence is established a **withdrawal** (**abstinence**) syndrome results, the symptoms of which are always unpleasant and include **anhedonia** (loss of feelings of pleasure), depression, insomnia, anxiety and agitation. This lasts until the physiological changes of tolerance revert back to their normal state.

It is important to recognize that dependence cannot occur without tolerance, but tolerance can be present without dependence. For example, individuals who drink alcohol regularly but in moderation show a measure of tolerance to ethanol, but are not dependent and suffer no ill effects if deprived.

Animal operant learning studies (see Topic Q1) have proved useful in investigating both behavioral and physiological aspects of addiction. The positive reinforcing properties of a drug can be assessed by measuring the extent to which animals (usually rats or monkeys) will press a lever to obtain an oral or intravenous dose of the drug. Crudely speaking, the more positively reinforcing the drug, the harder animals will work to get it. **Drug discrimination studies** work by training animals to learn that of two levers, one is associated with the delivery of a food reward plus an intravenous drug dose while the other is associated with a food reward alone. Animals can discriminate these two conditions with greater than 90% accuracy. It is then possible to test whether other substances are like the drug on which the animals have been trained by substituting them and examining their effects on the rates of lever pressing.

The context in which drugs are taken is important. This is illustrated by **conditioned place preference** in animals. Here, firstly animals are exposed to one environment when drugged and to a different environment when in the undrugged state. Next, the animals are given a choice between the two environments (they can move freely between them) and the time spent in each is recorded. With positively reinforcing drugs, animals spend more time in the environment they experienced in the drugged state. Moreover they also reveal that tolerance depends on context; that is, tolerance is greater when a drug is given in an environment or situation where it is normally administered than in a novel situation. This is an example of **context-dependent learning** and illustrates the importance that learning has in addictive behavior. The link between addictive drugs and the mesolimbic brain reward system comes largely from animal experiments.

Addictive drugs decrease the threshold for ICSS. Cocaine increases the locomotor activity of rats, an action blocked by the previous destruction of dopamine terminals in the nucleus accumbens by injecting the toxic dopamine analog 6-OHDA. *In vivo* microdialysis, which allows continuous sampling of brain extracellular fluid from specific locations in the brain and on-line measurement of neurotransmitter output, shows that dopamine is released by the nucleus accumbens during intravenous self-administration of cocaine and by low oral doses of ethanol. Furthermore, 6-OHDA lesions of the nucleus accumbens produce long-lasting reductions in the self-administration of both cocaine and amphetamine, though not of ethanol.

The effects of dopamine (D1) receptor antagonists certainly highlights a role for dopamine neurotransmission in the action of addictive drugs. D1 receptor blockers decrease:

- self-administration of ethanol in rats;
- cocaine induced locomotor activity in rats;
- cocaine self-administration by primates;
- craving for cocaine in human addicts.

The reinforcing properties of addictive drugs are not necessarily mediated only by the mesolimbic dopamine system. Although the mesolimbic system probably has a role in the rewarding properties of ethanol and opiate drugs, its destruction by 6-OHDA does not prevent the self administration of these agents. Other mechanisms must also be at work. Ethanol, benzodiazepines and barbiturates are all positively reinforcing at least partly because they all reduce anxiety. They share in common the action of increasing GABAergic inhibition by binding to $GABA_A$ receptors and enhancing the Cl^- ion influx. $GABA_A$ receptor antagonists injected into the amygdala and related structures decreases ethanol self-administration in rats. Not surprisingly, the reinforcing properties of opiate drugs such as heroin and morphine are mediated via opioid (μ) receptors. Mu opioid receptor antagonists block heroin self-administration.

The tolerance, dependence and withdrawal seen in drug addiction arise as the result of adaptive changes in the nervous system. In the case of cocaine one possibility is that tolerance arises from the down regulation of dopamine receptors at synapses in the nucleus accumbens. Cocaine inhibits the dopamine transporter, blocking reuptake into the presynaptic terminal so raising the dopamine concentration in the cleft. With chronic administration there is a down regulation of postsynaptic (D1) receptors, which means that greater dopamine concentrations in the synapses are needed for the same degree of postsynaptic response.

In opiate addiction, there is no change in the number of μ opioid receptors but they couple less effectively to G_i proteins so, with repeated use, opiates fail to inhibit adenylyl cyclase as much as normal. Withdrawal from opiates leads to a rebound activation of adenylyl cyclase. This occurs because of the inability of the decoupled μ opioid receptors, now stimulated only by endorphins, to offset the activation of adenylyl cyclase by receptors coupled to G_s proteins in the same cells.

In general, the physiological responses seen on withdrawal of a drug are the opposite of those seen with acute administration of the drug. The threshold for ICSS is increased and there is a decrease in the dopamine release from the nucleus accumbens (as shown by microdialysis) immediately after withdrawal from cocaine, alcohol and opiates. During the withdrawal syndrome the activity of the reward system is suppressed in relation to its normal state. This is presumably because of the receptor down regulation.

Withdrawal from drugs stimulates the release of corticotrophin releasing hormone from the hypothalamus and amygdala. This causes the endocrine and behavioral stress responses (Topic M3) of the withdrawal syndrome. Rebound from neuroadaptive changes in GABAergic transmission could account for the anxiety.

There is no satisfactory explanation for the intense craving experienced by addicts during periods in which they are not physiologically dependent, nor suffering withdrawal. Learning is undoubtedly important, and it has been proposed that it may be explained by a long-lasting drug-induced alteration of the mesolimbic system that makes it hypersensitive to subsequent drug administration. This is an important issue to resolve since it is the compulsion for continuing use that is a major barrier for addicts who want to refrain permanently from drug use.

O2 Control of feeding

Key Notes

Two center hypothesis

Lesions of the ventrolateral hypothalamus (VLH) inhibit feeding, while lesions of the ventromedial hypothalamus (VMH) cause overeating. From this the two center hypothesis proposed that the VLH responds to hunger signals to initiate feeding, whereas the VMH responds to satiety signals and terminates feeding. This hypothesis is no longer accepted because VLH lesions cause many other deficits in addition to that in feeding, and they occur because of the destruction of the nigrostriatal tract that passes through the VLH. Furthermore, VMH lesions produce their effects by altering the autonomic nervous system regulation of insulin secretion. Oversecretion of insulin does elevate food intake, but only by increasing the frequency of meals, and the obesity arises by insulin stimulated fat synthesis.

Brainstem and peripheral factors

Meal size is controlled by neural and blood borne satiety signals detected by the nucleus of the solitary tract (NST) and the area postrema (a circumventricular organ). Eating is inhibited by stomach distention, cholecystokinin secreted by the duodenum in response to the products of digestion, and by the rise in plasma osmolality that occurs on feeding. The inhibition of feeding is mediated by oxytocinergic neurons in the paraventricular hypothalamus that project to the NST.

Adiposity signals: insulin and leptin

Adiposity signals released in proportion to the size of the body fat store exert homeostatic control over body mass in the long term by balancing food intake and energy expenditure. Both insulin and leptin are thought to be adiposity signals; their plasma concentrations reflect the amount of body fat and they cross the blood–brain barrier. Insulin is released from β cells of the endocrine pancreas and reduces food intake. Leptin is released from fat cells and reduces food intake and increases energy expenditure. Both of these peptides act on the ventral hypothalamus to inhibit the action of neuropeptide Y (NPY), a powerful stimulator of feeding.

Obesity

Obesity is produced by a mismatch between food intake and energy expenditure. Obese individuals probably respond more to external cues than internal (physiological) cues for feeding and have poorer capacity for energy expenditure. Mutations of the leptin gene, or its receptor, cause obesity in mice. Obese humans have high plasma leptin, reflecting their high body fat. Their obesity may be because the leptin molecule is abnormal, or because the leptin receptor is defective.

Anorexia nervosa

Anorexics self-starve and indulge in excessive physical activity. The anxiety hypothesis argues that when eating, anorexics have elevated secretion of corticotrophin releasing hormone (CRH) which both promotes anxiety and inhibits feeding. The reward hypothesis postulates that the increased glucocorticoid concentration that comes from CRH activation of the pituitary and adrenal glands by self-starvation and exercise is positively reinforcing (pleasurable).

Related topics

Anatomy and connections of the hypothalamus (M1)
Posterior pituitary function (M2)
Neuroendocrine control of metabolism and growth (M3)
Motivation (O1)

Two center hypothesis

Electrolytic lesions of the **ventrolateral hypothalamus (VLH)** result in **aphagia** (refusal to feed) whereas similar lesions of the **ventromedial hypothalamus (VMH)** cause **hyperphagia** (overeating) and obesity. These findings led to the notion that the VLH was a 'hunger center', the destruction of which prevented the animal from responding to hunger signals, whereas the VMH was a 'satiety center' which, once lesioned, resulted in an animal that could not respond to satiety signals. In this **two center hypothesis**, the VLH normally initiated feeding whilst the VMH terminated it. This hypothesis is no longer tenable for the following reasons:

(1) Lesions of the VLH produce, in addition to aphagia, **adipsia** (failure to drink), **akinesia** (greatly reduced motor activity) and **sensory neglect**, which is failure to respond to sensory (somatosensory, olfactory and auditory) stimulation. Both akinesia and sensory neglect are seen in Parkinson's disease (see Topics L6 and R3) that is caused by a loss of nigrostriatal dopaminergic neurons. The nigrostriatal pathway runs through the VLH. Destruction of this pathway by the injection of 6-hydroxydopamine (6-OHDA) into the substantia nigra (see Topic L5) causes effects that are very similar to those found with electrolytic lesions of the VLH. Hence VLH lesions destroy dopaminergic neurons and so have general effects on motivation not specific effects on feeding mechanisms.
(2) Animals with VMH lesions become hyperphagic and obese by eating more often but the size of individual meals stays the same. This is because VMH lesions damage the central control of the autonomic nervous system which in turn alters the endocrine regulation of metabolism.

Normally the secretion of insulin from the β cells of the endocrine pancreas is under autonomic control and is excited by the parasympathetic and inhibited by the sympathetic system. Secretion of insulin is also triggered by the increase in blood concentrations of glucose and amino acids that takes place shortly after feeding. Insulin promotes the uptake of glucose and amino acids into muscle, liver and fat cells and the synthesis of triglycerides from fatty acids by fat cells (**lipogenesis**, see *Instant Notes in Biochemistry*, 2nd edn).

In VMH-lesioned animals the activity of the parasympathetic nervous system is increased but that of the sympathetic system is decreased. This results in oversecretion of insulin, promoting lipogenesis and the storage of triglycerides in fat cells leading to obesity. Moreover, the elevated insulin concentration causes rapid clearance of glucose and amino acids from the circulation. According to the **glucostatic hypothesis** a transient fall in blood glucose (always seen a few minutes before eating in rats) is responsible for initiating feeding. This fall occurs earlier in VMH-lesioned animals because of their higher insulin concentration, and so they feed sooner; in other words the intervals between meals are shortened. In summary, normally the VMH is responsible for the timing of meals, by producing output appropriate to the glucose concentration that results from current feeding.

Brainstem and peripheral factors

The brainstem regulates meal size in response to **satiety signals**. Both humoral and neural satiety signals exist to inhibit feeding. The nucleus of the solitary tract (NST) and neighboring circumventricular organ, the **area postrema** (**AP** see *Fig. 2* of Topic M2) act as a functional unit to integrate input from the periphery. Afferents from taste buds and pharynx (Topic J2), stomach, intestine and liver run to the NST, and the AP contains chemosensory neurons which are able to respond to blood-borne factors. Stomach distention excites stretch receptors, afferents from which run in the vagus (X) nerve to the NST. Ablation of the AP results in rats eating larger than normal meals, but their total food intake remains normal because they eat less frequently, which suggests that longer-term controls on feeding are unimpaired.

Cholecystokinin (CCK) is a peptide secreted by the duodenal mucosa in response to the presence of fatty acids, monoglycerides and certain amino acids. Aside from its actions on the digestive system, CCK inhibits feeding in part by stimulating the vagal afferents which convey gastric distention signals. Hence gastric stretch and CCK act synergistically to limit meal size. Ingestion of food increases plasma osmolality (e.g. NaCl) which normally stimulates drinking (see Topic M2). If water is unavailable the rise in osmolality is detected by osmoreceptors in the walls of the hepatic portal vein with the end result that feeding is curtailed. This can be viewed as a mechanism for controlling plasma osmolality by restricting ingestion of osmotically active solutes. Note that it uses a different set of osmoreceptors than those involved in drinking; hypo-osmolality inhibits drinking, it does not stimulate feeding.

The inhibition of food intake by gastric distention, CCK and hyperosmolality is mediated by the paraventricular nucleus (PVN) of the hypothalamus which receives input from the NST. In turn, parvocellular (small) neurons in the PVN which use oxytocin as a transmitter, project back to the NST which controls reflex aspects of food intake such as swallowing. This control is very specific. In decerebrate rats (lacking a functional forebrain) the brainstem triggers swallowing of food, but not water, in food-deprived animals. This response is inhibited by a full stomach, intravenous CCK or hypertonic saline.

Stress reduces feeding by the activation of neurons in the PVN that release CRH; this acts locally to excite the oxytocinergic neurons of the PVN that are responsible for the inhibition of feeding (*Fig. 1*).

Adiposity signals: insulin and leptin

Long-term control of body mass is very precise and thought to be achieved by defending a stable level of body fat. This requires an **adiposity signal** which reflects the size of the fat store and acts to maintain it constant homeostatically by balancing energy input and output. Two peptide hormones have been proposed for this role, **insulin** and **leptin**. The plasma concentrations of both correlate well with body fat content, and both cross the blood–brain barrier via specific, saturable, transport mechanisms.

Insulin concentration in the cerebrospinal fluid (CSF) correlates with plasma insulin, but only in the long term. CSF insulin is unaffected by transient alterations in blood insulin concentration that depend on feeding or fasting. Insulin acts at the ventral hypothalamus to reduce food intake.

Leptin is secreted exclusively from adipose cells and acts at the ventral hypothalamus, reducing food intake and increasing energy expenditure. Leptin

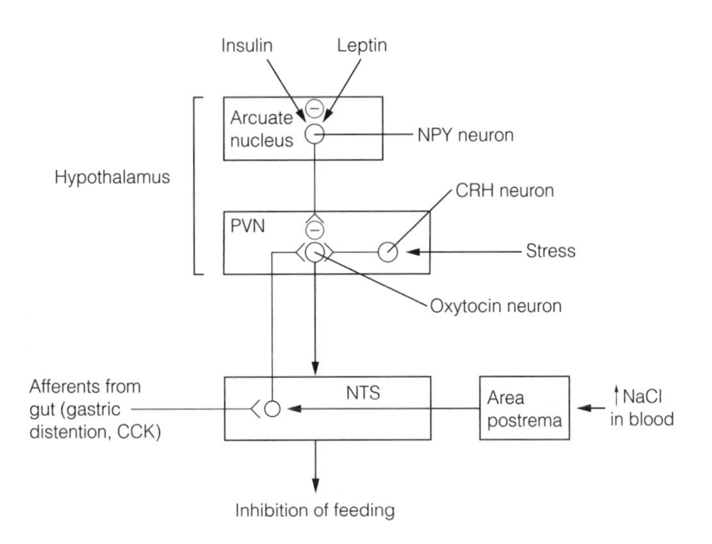

Fig. 1. Circuitry implicated in the inhibition of feeding by a variety of factors. Inhibitory actions, – . CRH, Corticotrophin releasing hormone; NPY, neuropeptide Y; NST, nucleus of the solitary tract; PVN, paraventricular nucleus; CCK, cholecystokinin; ⊖, inhibition.

production is elevated by the rise in insulin that occurs on feeding and is suppressed by the glucocorticoid secretion that accompanies fasting, so leptin regulates fat mass by means of a negative feedback loop (*Fig. 2*).

The effect of leptin and centrally acting insulin on food intake is brought about by a decrease in the secretion of **neuropeptide Y (NPY)**, which is both the most potent stimulator of feeding known, and the most abundant neuropeptide in the brain. The NPY pathway implicated in regulation of feeding runs from the arcuate nucleus of the hypothalamus to the PVN (see *Fig. 1*); a fall in NPY secretion in the PVN allows an increase in activity of the oxytocinergic neurons that inhibit feeding. The effects of leptin on energy expenditure are brought

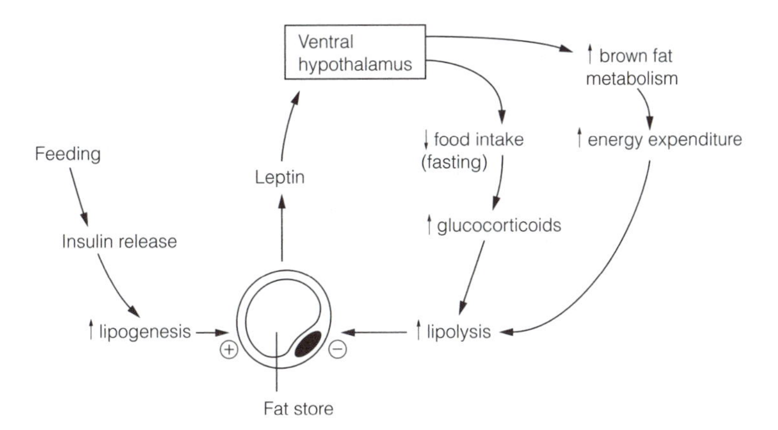

Fig. 2. Negative feedback control of fat stores by leptin. An increase in the size of the fat store causes a rise in leptin release. This activates processes which result in increased fat breakdown.

about by the hypothalamus driving increased sympathetic activity to brown adipose tissue (BAT). When activated by norepinephrine acting at β3 adrenoceptors BAT generates metabolic heat (see *Instant Notes in Biochemistry*, 2nd edn).

Obesity

Disorders of feeding and body mass regulation are prevalent in developed societies. One-third of North Americans are more than 20% overweight (the clinical definition of **obesity**) and obesity is a key risk factor for cardiovascular disease, diabetes and other disorders. Obesity can arise in a number of different ways that are poorly understood. Experiments in which obese individuals must work to obtain food suggest that they respond more to external cues (how appetizing a food seems) but less to internal cues (hunger and satiety) than lean individuals. Although it has proved difficult to show differences in activity between obese and lean individuals, obesity is associated with lower sympathetic activity and reduced amounts of BAT, so overall energy expenditure may be less.

In mice, mutations of either the leptin gene (*ob* gene) or the gene for the leptin receptor (*db* gene) causes obesity, in the first case because the leptin itself is defective and in the second case because the hypothalamus is unable to respond to available leptin (**leptin resistance**). In general, mice with obesity, whether genetic or acquired, have high plasma concentrations of leptin which reflects their adiposity. Obese humans also have high plasma leptin concentrations and in some cases the obesity can be attributed to leptin resistance, the leptin molecule itself being normal. Weight loss due to dieting is associated with a fall in leptin concentration. If leptin is acting as a satiety signal, this fall is likely to result in increased hunger and decreased energy expenditure. This highlights the pointlessness of dieting alone (i.e. in the absence of exercise) in attempts to reduce weight.

Anorexia nervosa

Anorexia nervosa is a disorder most commonly, but not exclusively, of young women, characterized by reduced food intake and enhanced physical activity. It has a poor prognosis. Several hypotheses have been proposed to explain it. Any hypothesis is blighted by the difficulty in deciding whether the neurochemical findings are the cause or an effect of the starvation. The **anxiety hypothesis** starts from the premise that when anorexic women eat they become anxious, and the well documented finding that anorexics show enhanced secretion of CRH and glucocorticoids (reflecting activation of the hypothalamic–pituitary–adrenal axis, see Topic M3). Feeding stimulates the secretion of CCK which apart from being a satiety factor also excites CRH secretion. CRH is both **anxiogenic** (promotes feeling of anxiety) and inhibits eating (see above). The **reward hypothesis** proposes that glucocorticoids are positively reinforcing; that is, they produce euphoria. Animals will self administer corticosteroids, and glucocorticoids increase the release of dopamine from terminals of the mesolimbic pathway. The argument is that the reduced food intake and physical exercise which are associated with raised glucocorticoid concentrations become secondary reinforcers, that is they become rewarding in their own right. Neither of these hypotheses accounts for how body mass homeostasis by leptin is subverted in anorexia, nor do they explain the psychological states that give rise to food aversion in the first place.

O3 Brain biological clocks

Key Notes

Intrinsic rhythms

Many physiological parameters (e.g. sleep/waking, core temperature, secretion of anterior pituitary hormones) show circadian (about a day) variation. In the complete absence of external time cues, the most important of which is light, circadian rhythms free run initially with a period of 25 rather than 24 h. After a couple of weeks, sleep/waking cycles decouple from other functions to free run with a period of 30 h. Hence, circadian clocks must exist that are entrained by external events.

Suprachiasmatic nucleus

Stimulation, lesion and transplant experiments show that the circadian clock regulating sleep/waking cycles is located in the suprachiasmatic nucleus (SCN) of the anterior hypothalamus. An intrinsically active neural oscillator in the SCN fires with a frequency that varies sinusoidally, peaking in the daytime. The SCN outputs to other parts of the hypothalamus to regulate sleep/waking cycles, autonomic and endocrine functions. Light signals arrive at the SCN by way of the retinohypothalamic tract (RHT), to synchronize it to a 24 h cycle.

Pineal gland

The pineal gland, a circumventricular organ, secretes a hormone, melatonin, during the hours of darkness. Melatonin in the blood crosses the blood–brain barrier to gain access to the SCN. The duration of the melatonin pulse is a direct measure of the length of the night, and so signals the time of year for animals living at latitudes away from the equator. For seasonal breeders, melatonin secretion acts at the SCN to control reproductive cycles. Light inhibits melatonin synthesis by means of a pathway from the SCN (which receives retinal input) via the sympathetic nervous system to the pineal. There are melatonin receptors in the SCN and melatonin will reduce symptoms of jet-lag.

Related topics

Neuroendocrine control of metabolism and growth (M3)
Neuroendocrine control of reproduction (M4)
Sleep (O4)

Intrinsic rhythms

Many body functions vary cyclically with a period of about a day; that is, in a **circadian** manner. Functions regulated in this way include sleep/wakefulness, core temperature, secretion of anterior pituitary hormones, autonomic mechanisms (e.g., regulation of skin blood flow) and urinary K^+ excretion. Humans isolated from all external time cues – for example living in deep caves, sleeping, eating, switching lights on and off whenever they wish – show intrinsic circadian rhythms with a period of about 25 h initially. This decoupling of circadian rhythms from the normal 24 h period is called **free running** and shows that there exist intrinsic **circadian clocks** that are usually entrained by environmental cues, **zeitgebers** (the German word for time-giver). Zeitgebers include light, exercise, social interactions and work schedules. Light is the most powerful zeitgeber. A powerful light pulse given during subjective night, produces shifts in the circadian cyclicity. In humans with normal sleep patterns the nadir in core temperature occurs at about 5 a.m. A light pulse given during the night before this time causes circadian rhythms to be delayed (**phase delay**) whereas a light pulse after this time causes **phase advance**.

After 1–2 weeks free running, physiological variables often desynchronize from each other. For example, typically, fluctuations in core temperature, secretion of adrenocorticotrophic hormone (ACTH) and glucocorticoids, and rapid eye movement (REM) sleep continue with a period of about 25 h, but the cycles of sleep/wakefulness and secretion of growth hormone (GH) lengthen to more than 30 h. This suggests that there are two circadian clocks, both of which are normally entrained by light–dark cycles of day and night.

Suprachiasmatic nucleus

The circadian clock that regulates sleep/waking cycles resides in the **suprachiasmatic nucleus** (**SCN**) of the anterior hypothalamus. Electrical stimulation of the SCN shifts circadian rhythms in predictable ways, destruction of the SCN produces permanent abolition of circadian cyclicity (but does not prevent sleep), and transplanting a fetal SCN into a host animal with its own SCN destroyed restores circadian rhythmicity.

The pacemaker (clock) function of the SCN is due to the presence of individual neurons in the ventrolateral **core** division of the SCN which fire with a frequency that varies in a circadian fashion. These neurons are circadian oscillators and maintain their rhythmic activity even when isolated from the rest of the nervous system in cell culture. The firing frequency of SCN neurons varies sinusoidally with a period of 24 h, peaking during the day, and dropping to its lowest rate during the night. The SCN projects largely to other hypothalamic structures to regulate sleep–wake cycles, autonomic and endocrine functions, but also sends output to the thalamus and basal forebrain (e.g. septal nucleus) which probably accounts for circadian variation in memory and cognitive functions. Most SCN neurons are GABAergic and corelease peptides, and are assumed to be inhibitory to their targets.

Light signals encoding total luminance but not color, form or movement, are relayed to the SCN by the **retinohypothalamic tract** (**RHT**). This pathway consists of the axons of a population of small retinal ganglion cells driven by cone photoreceptors over a wide area, which synapse directly with neurons in the core of the SCN. The RHT uses glutamate as a transmitter.

Pineal gland

The pineal gland is a circumventricular organ which secretes **melatonin** into the blood during the hours of darkness, and the duration of the pineal melatonin pulse is a direct measure of the length of night, and hence also of day length, the **photoperiod**. Melatonin secreted into the blood is transported across the blood–brain barrier to act on the SCN. For animals living at latitudes other than the equator, day length varies during the year, so melatonin secretion acts as a signal which codes for the time of the year. For seasonal breeders, the length of the melatonin pulse regulates the hypothalamic–pituitary–gonadal axis of both males and females via its action on the SCN. For example, in sheep the longer melatonin signals produced in the shorter photoperiods of November (in the Northern hemisphere) activates estrous cycles in ewes, and testicular growth – with a consequent rise in testosterone secretion and spermatogenesis – in the rams. Although photoperiodic control of melatonin secretion is not important

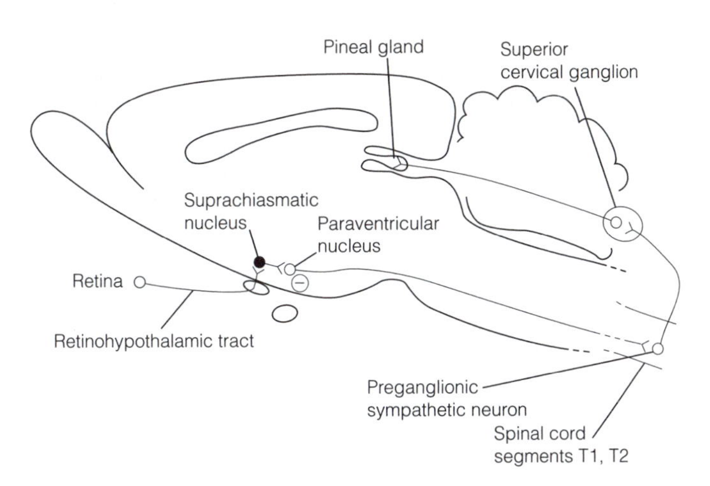

Fig. 1. The brain circuitry by which light inhibits the secretion of melatonin from the pineal gland.

for reproduction in humans, it does exert feedback effects on the function of the SCN circadian clock and so affects sleep–wake cycles.

The pathway by which light inhibits melatonin synthesis is circuitous (*Fig. 1*). Neurons in the core of the SCN (which get retinal input from the RHT) inhibit the autonomic division of the paraventricular nucleus (PVN) of the hypothalamus. The PVN sends axons through the brainstem to synapse with preganglionic sympathetic neurons in the intermediolateral horn of spinal cord segments, T1 and T2. These autonomic neurons project to the superior cervical ganglion (SCG), the postganglionic cells of which innervate the pineal gland. At night the activity of SCG neurons is increased, and the secretion of norepinephrine from sympathetic terminals acts on β adrenoceptors of **pinealocytes** to stimulate melatonin synthesis. The biosynthetic pathway is illustrated in *Fig. 2*.

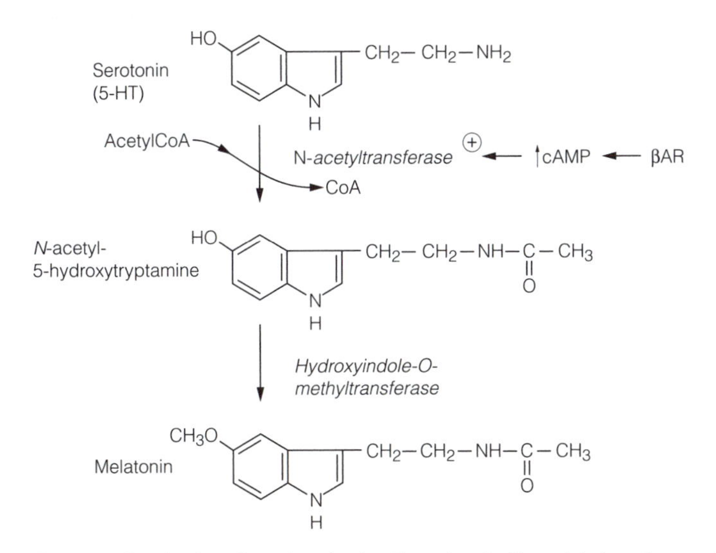

Fig. 2. Synthesis of melatonin in the pineal. N-acetyl transferase is activated by norepinephrine, secreted from sympathetic terminals, acting at β adrenoceptors (βAR). cAMP, cyclic adenosine monophosphate.

Melatonin interacts both with metabotropic receptors, linked to G_i proteins which bring about inhibition of adenylyl cyclase, and intracellular receptors that are members of the steroid receptor superfamily called **RZRβ** receptors. Both receptors are found in the SCN. Melatonin can entrain the circadian clock in the SCN, reset sleep–wake cycles in animals and humans, and reduce the symptoms of **jet-lag**, the sleep disturbance that arises when light–dark cycles and circadian rhythms are suddenly desynchronized by air travel over several time zones.

O4 Sleep

Key Notes

States of sleep

In electroencephalography (EEG), in which the electrical activity of the brain is recorded via scalp electrodes, high frequency, low voltage waveforms are seen in individuals who are awake. EEG and other physiological measures distinguish two states of sleep. Slow wave (or nonrapid eye movement, NREM) sleep has low frequency, high voltage EEG, is characterized by a drop in cerebral blood flow and brain glucose utilization and increased growth hormone output. Muscle tone is retained. As a person falls asleep they drift through four stages (1–4) of NREM sleep, each with a lower frequency than the last. After about 90 min they enter rapid eye movement (REM) sleep which has a similar EEG waveform to the awake state. Muscle tone is absent apart from rapid eye movements and brief twitches of limb muscles. Autonomic functions become irregular. Dreaming occurs during REM sleep.

Reticular system

The reticular system that runs through the brainstem consists of large cells that get input from many sensory modalities and project to the thalamus, and small cells that provide diffuse aminergic projections to the forebrain. Wakefulness requires the midbrain reticular projections to the thalamus. Monoaminergic neurons active during wakefulness are silent during sleep. Cholinergic neurons normally quiet in the awake state become active in REM sleep.

NREM sleep mechanisms

The thalamus and cortex are reciprocally interconnected, and when awake thalamic neurons fire steadily. In stage 2 NREM sleep the thalamic neurons fire bursts called sleep spindles but in deep (stages 3 and 4) sleep the thalamic neurons fall silent and cortical neurons fire in isolation with their own intrinsic rhythm. This disconnection of thalamus from cortex occurs because of the loss of excitatory drive from the preoptic area of the hypothalamus to the thalamus.

REM sleep mechanisms

In REM sleep, the thalamocortical neurons are excited by the preoptic hypothalamus. Moreover in REM sleep, unlike NREM sleep, cholinergic pathways from the pons to the thalamus become highly active, which in turn excites the cortex. This activity is recorded as pontine–geniculate–occipital (PGO) spikes in the EEG. The cells in the pons that trigger PGO spikes are also responsible for the rapid eye movements and autonomic irregularities. Activity of the medullary and pontine reticular system blocks sensory input and motor output, effectively disconnecting the brain from the external world during REM sleep.

Functions of sleep

Three broad types of hypothesis attempt to explain the purpose of sleep. Ecological hypotheses argue that sleep keeps animals quiet and hidden at times when they are at greatest risk of predation. Metabolic hypotheses suggest that sleep corrects some chemical disturbance that accrues during waking hours; this is based on the fact that sleep deprived rats die because of failure of their thermoregulatory and immune systems, and on the discovery of sleep promoting factors. Learning hypotheses postulate either that sleep is needed to unlearn spurious associations ('false' memories) or to consolidate. Learning theories, unlike the others, attempt to account for dreaming.

Related topics

Norepinephrine neurotransmission (N2)
Serotonin neurotransmission (N3)
Acetylcholine neurotransmission (N4)
Hippocampal learning (Q4)
Epilepsy (R2)

States of sleep

Recording the net electrical activity of the brain by means of surface electrodes attached to the scalp is termed **electroencephalography (EEG)**. When awake, the EEG waveforms are of low amplitude and high frequency (*Fig. 1*) and described as **desynchronized**. On the basis of the EEG and many other physiological measures, two sleep states can be distinguished, **slow wave sleep** and **rapid eye movement (REM) sleep**.

Slow wave sleep, otherwise known as **nonREM (NREM) sleep**, has high amplitude, low frequency (**synchronized**) EEG waveforms. During NREM sleep, muscle tone is retained and postural adjustments (turning over in bed) are occasionally made. Respiration rate, heart rate, and mean arterial blood pressure all fall, though gastrointestinal motility increases. Most growth hormone secretion occurs during NREM sleep. At the beginning of a period of

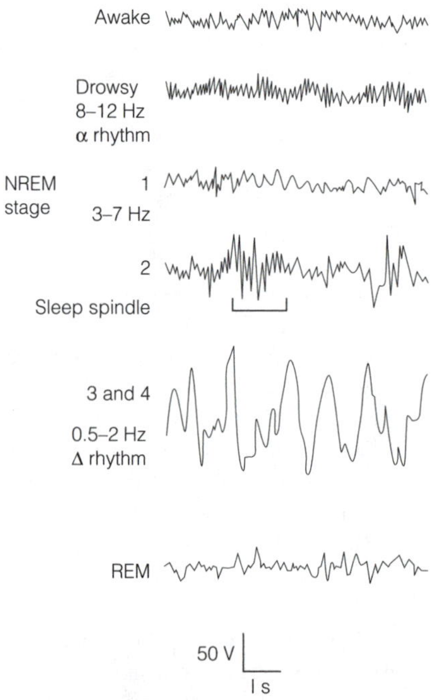

Fig. 1. EEG waveforms recorded from the human brain when awake and asleep.

sleep the EEG changes from the fully awake to the NREM sleep state by a progressive decrease in EEG frequency through four stages of NREM sleep (stages 1–4). Stage 2 is characterized by higher frequency bursts called **sleep spindles**. Stages 3 and 4 are often collectively referred to as **delta (Δ) sleep** because of its 1–4 Hz delta waves. In passing from stage 1 through to 4 the frequency of the EEG falls (and amplitude rises) and it becomes increasingly difficult to arouse the sleeper. People awakened from NREM sleep are confused, cannot do cognitive tasks and rapidly go back to sleep if left. Positron emission tomography scanning shows that cerebral blood flow and glucose utilization fall by as much as 40% in NREM sleep.

REM sleep is sometimes described as **paradoxical** sleep since its EEG resembles that seen in the awake state. Muscle tone is absent except for transient contractions of extraocular eye muscles, which produces the rapid movements of the eyes for which this stage of sleep is named, together with brief contractions of middle ear muscles and distal limb muscles. Respiration rate, heart rate, mean arterial blood pressure and core temperature become irregular. Penile erections occur during REM sleep, and their absence distinguishes physiological from psychological causes of impotence. People aroused from REM sleep usually report that they were dreaming. At this time the dream content is in short term memory (Topic Q1) since unless the details are immediately rehearsed they are forgotten within a minute or two.

During a typical night's sleep (*Fig. 2*) adults drop rapidly into deep (stage 4) NREM sleep and then REM and NREM sleep alternate about every 90 min with increasingly longer periods of REM sleep as the night progresses. The proportion of REM sleep changes dramatically during development. In humans, 10 week premature neonates spend 80% of their sleep time in REM sleep. This drops to 50% for babies born at term, 35% for 2-year-old infants and stabilizes to adult levels of about 25% by the age of 10 years. A similar pattern is seen in other mammals. In humans, total sleep time falls rapidly, from 24 h at 3 months gestation to 12 h per day at 1 year and then gradually declines for the rest of life.

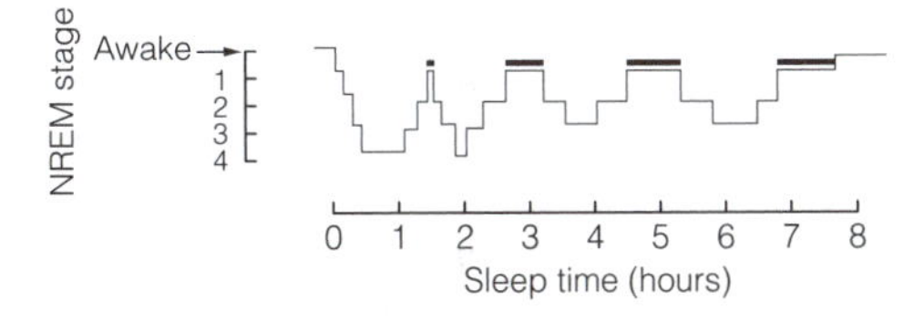

Fig. 2. Distribution of sleep stages during a typical nights sleep. Dark bars are rapid eye movement sleep periods. NREM, Nonrapid eye movement sleep.

Reticular system

Early studies established that the brainstem regulates the sleep–wake cycle. The brain of a cat in which the spinal cord is cut at C1 continues to show EEGs characteristic of normal sleep–wake cycles. However transecting the midbrain results in an isolated forebrain in which the EEG is permanently synchronized in slow wave sleep activity. More selective lesions showed that the loss of wakefulness was caused by severing the thalamic projections of the midbrain reticular formation, not by cutting the classical sensory pathways to the thalamus. High frequency electrical stimulation of the midbrain in intact cats causes arousal and desynchronized EEG. Hence the midbrain has neurons that are required for the awake state.

The reticular formation, implicated in sleep–wake cycles by the above studies, runs through the entire brainstem, projects to the thalamus, and consists of two broad functional populations of neurons. Large cells (diameter 50–100 μm) form circuits integrating sensory input from several modalities (visual, vestibular, proprioceptor somatosensory) for reflex control of eye movements and postural reflexes. They fire at high frequencies during wakefulness, have high conduction velocities and use glutamate and GABA as transmitters.

The second population are small (10–20 μm diameter) aminergic neurons that cluster in brainstem nuclei and project widely and diffusely to the forebrain (Section N). These cells regularly fire at low rates (1–10 Hz) and have very low conduction velocities. Many aminergic neurons are pacemakers which fire spontaneously in a way that correlates with behavioral state. Noradrenergic neurons, required for arousal, fire during wakefulness but are quiescent during sleep. Firing of serotonergic neurons is highest during locomotion but they are silent in REM sleep. The cholinergic basal forebrain arousal neurons (see Topic N4), episodically active in the awake state, fall silent during sleep, but the cholinergic neurons of the brainstem that are quiet in the awake state become active in REM sleep.

NREM sleep mechanisms

The neocortex and the thalamus are massively and reciprocally interconnected. Thalamic cortical neurons have two modes of activity depending on their resting potential (V_r). During the awake state (normal V_r) they show tonic (nonburst) firing. During NREM sleep the V_r becomes more hyperpolarized and the thalamic neurons fire short bursts of action potentials. Burst firing occurs because hyperpolarization removes inactivation from T type Ca^{2+} channels, allowing Ca^{2+} depolarization to trigger action potentials. Bursting of thalamocortical neurons drives bursting of cortical neurons. These bursts are sleep spindles (stage 2 NREM sleep). With the deeper stages 3 and 4, the thalamocortical neurons become so hyperpolarized they become silent and the cortical neurons now fire with their own intrinsic (delta) rhythm.

The triggers for NREM sleep are not known, but the hyperpolarization of thalamocortical neurons in NREM sleep is due to the loss of excitatory drive from the hypothalamus. At the start of NREM sleep the **ventrolateral preoptic area (VLPO)** of the hypothalamus is activated. GABAergic VLPO neurons inhibit nearby histaminergic neurons in the tubero-mammillary nucleus that are excitatory to the thalamus and cortex. Furthermore, as NREM sleep deepens there is a progressive loss of input from large reticular formation neurons to the thalamus, and a reduction in norepinephrine and serotonin modulation of the thalamus and cortex.

REM sleep mechanisms

During REM sleep as in the waking state, the thalamocortical neurons are once again excited by the histaminergic pathways from the hypothalamus. However, whereas during waking, modulation of thalamocortical neurons is predominantly by NA and 5-HT, with cholinergic neurons being quiet or only episodically active, during REM sleep NA and 5-HT neurons are silent, but brainstem cholinergic neurons are highly active. This shift between waking through NREM to REM sleep might be brought about by a reciprocal arrangement between brainstem aminergic cells (*Fig. 3*). The cholinergic cells active during REM sleep are **REM-on** cells, aminergic cells silent during REM are **REM-off** cells.

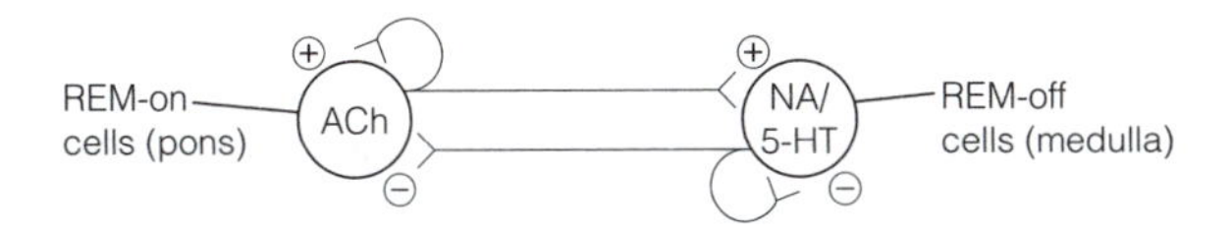

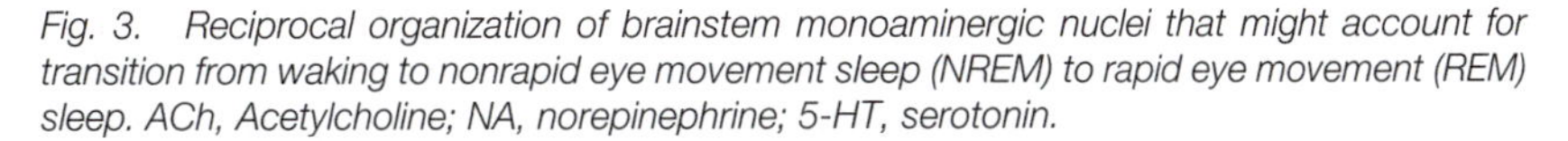

Fig. 3. Reciprocal organization of brainstem monoaminergic nuclei that might account for transition from waking to nonrapid eye movement sleep (NREM) to rapid eye movement (REM) sleep. ACh, Acetylcholine; NA, norepinephrine; 5-HT, serotonin.

The REM-on cells are probably located in the **pedunculopontine** and **laterodorsal tegmental nuclei** in the pons because:

- Lesions in these areas markedly reduce REM sleep.
- Injection of cholinergic agonists (or cholinesterase inhibitors) into their vicinity produce enhancement of REM sleep in behaving animals.
- REM sleep is immediately preceded by increased firing of the neurons in these nuclei.

These nuclei project to the medial pontine reticular nuclei which act to desynchronize cortical activity when activated by REM-on cells. REM-off cells are located in the locus ceruleus and the raphe nuclei and when active during wakefulness inhibit the REM-on neurons.

A major feature of REM sleep are periodic **pontine–geniculate–occipital (PGO) spikes** in the EEG. These originate in the peribrachial pons and other brainstem regions. Phasic activity of cells here drives vestibular and reticular neurons that excite oculomotor neurons (causing the rapid eye movements) and other cells to produce the phasic alterations in respiration, heart rate, blood flow and muscle twitches seen in REM sleep. PGO bursting neurons in the pons also initiates the spread of activity to the lateral geniculate nucleus and visual cortex that is responsible for the PGO spike. In fact, PGO spiking reflects generalized activation of thalamus and cortex that takes place in REM sleep.

A second REM sleep feature is that sensory input and motor output are both powerfully inhibited. This is due to activity of the pontine and medullary reticular system. Presynaptic GABAergic inhibition by reticular neurons on afferent terminals is responsible for the blockade of sensory input, while glycinergic postsynaptic inhibition of motor neurons is the route by which muscle **atonia** (loss of muscle tone) is brought about. Thus, during REM sleep the brain is effectively uncoupled from the external world; it is 'off-line'. Lesions of the pons which prevent the muscle atonia produces animals which express stereotyped behaviors during REM sleep. This suggests that during normal REM sleep motor patterns are generated but cannot be executed; we cannot act out our dreams!

Functions of sleep

Numerous hypotheses have been proposed for the function of sleep and modern ideas fall into three broad categories.

Ecological

Sleep has evolved to ensure that animals remain silent and hidden during times when they are at risk from predation. It might be expected then that the amount of time an animal spends asleep would correlate with its risk of predation.

Metabolic

Metabolic hypotheses postulate that during waking metabolic disturbances occur which are corrected during sleep. Sleep deprived rats suffer anorexia (even though their food intake increases), motor weakness and evidence of stress (adrenal hypertrophy and gastric ulceration). Eventually they lose the ability to thermoregulate, becoming hypothermic, and their immune systems fail, so that they die of infection after about 4 weeks. Hence, sleep may have anabolic functions which conserve energy stores, and core temperature. In this context it is interesting that the same region of the hypothalamus, the preoptic area (POA), is involved in both thermoregulation and triggering NREM sleep, that local warming of the POA in behaving animals can trigger or prolong NREM sleep, and that heat loading during waking hours causes a rise in the amount of delta sleep taken during the following night. Support for the homeostatic hypothesis comes from the identification of **sleep promoting factors**, compounds present in sleep deprived animals which induce sleep when given to awake animals. One of these is **interleukin-1**, a cytokine synthesized by glia and macrophages that stimulates immune function. It promotes NREM sleep. Maybe sleep is needed to maintain the competence of the immune system.

Learning and memory

Two rather different hypotheses suggest a role for sleep in learning and memory. Learning occurs by adjustments to the strength of selected synapses in a network of neurons in response to a given input (see Topic Q4). Whenever the same input is reapplied the trained synapses are activated, giving appropriate recall.

The first idea is that although cortical networks are 'off line' during REM sleep they are excited periodically by internally generated PGO spikes which acts as random noise. Any recall activated by this will not be specific learned associations but spurious associations, and a mechanism exists to weaken the synaptic links which allow them. Here, REM sleep is needed to unlearn 'false' memories. This could be important in preventing neural networks from becoming saturated.

The second hypothesis is that during sleep, recent memories are consolidated. Based mainly on studies of maze learning in rats, it proposes that during REM sleep, synapses activated in the hippocampus during the maze training are strengthened. This involves bursts of activity called theta (θ) rhythm (frequency 4–10 Hz), which is generated in the septum and transmitted to the hippocampus in the cholinergic septohippocampal pathway. Theta rhythm is seen both during the original training (Topic Q4) and in subsequent episodes of REM sleep.

In these learning hypotheses of sleep, dreams are viewed as fragments of 'false' memories being eliminated or as a consequence of memory consolidation. The other hypotheses do not account for dreaming.

P1 EARLY PATTERNING OF THE NERVOUS SYSTEM

Key Notes

Neural tube formation

The early human embryo consists of two layers, ectoderm and endoderm. Ectodermal cells in the midline of the posterior end form the mesoderm and notochord which lie between the other layers. Ectoderm above the notochord becomes the neural plate, which rolls up to become a closed neural tube by 28 days. Cells of the dorsal neural tube migrate away to become the neural crest which forms the peripheral nervous system. The neural tube becomes the central nervous system.

Neural induction

Signals for inducing the neural plate in the frog come from an organizer in the early mesoderm. When transplanted, the organizer can induce neural tube formation. The organizer in mammalian embryos is Hensen's node.

Formation of the anteroposterior axis

The neural tube is organized along three axes, anteroposterior, dorsoventral and radial. In frogs, the organizer secretes signals that are able to begin patterning in the anteroposterior direction. The neural tube comes to be divided into a series of compartments. This is brought about by the sequential activation of segmentation and *Hox* genes. These highly conserved genes code for transcription factors. The expression of specific *Hox* genes in particular narrow bands along the anteroposterior axis is controlled by a gradient of retinoic acid secreted from Hensen's node.

Formation of the dorsoventral axis

Sonic hedgehog (SHH) from the notochord induces the floor plate in the ventral neural tube. The floor plate then secretes its own SHH which induces the differentiation of motor neurons. Epidermal ectoderm induces a roof plate in the dorsal neural tube by secreting bone morphogenetic proteins (BMPs). The roof plate then generates its own BMPs, promoting the differentiation of dorsal horn cells. SHH and BMPs cause dorsoventral patterning by activating *Pax* genes that encode transcription factors that play a similar role to *Hox* gene products.

Related topics

Organization of the peripheral nervous system (E1)
Organization of the central nervous system (E2)
Cell determination (P2)

Neural tube formation

At 11 days post fertilization, the human embryo is a two-layered circular disc. The **endoderm** is a single layer of flat cells forming the roof of the yolk sac. The **ectoderm** is 2–4 cells thick and continuous at its edges with the amniotic membrane. In the caudal midline, ectodermal cells proliferate to form a **primitive streak**, at the front end of which a knot of cells forms called **Hensen's node**

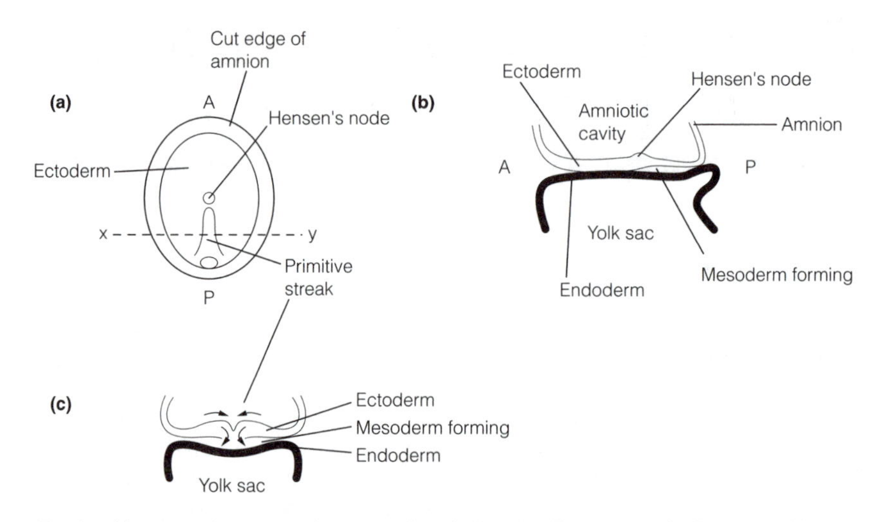

Fig. 1. Human embryo at 13 days gestation. A, Anterior; P, posterior. (a) Plan view; (b) longitudinal section; (c) transverse plane section at level x–y in (a), arrows depict migration of cells to form mesoderm.

(*Fig. 1*). The primitive streak gives rise to mesoderm which spreads laterally and forward between the endoderm and ectoderm, transforming the embryo into a three-layered structure. The morphological change that shapes the trilaminar embryo is called **gastrulation**. From Hensen's node, on day 5, a rod of cells grows forward in the midline between the ectoderm and endoderm, the **notocord**, the forerunner of the vertebral column.

Hensen's node and the notocord induces the formation, within the ectoderm above, of a **neural plate**. A groove develops along the midline of the **neural plate**, which deepens and the neural folds at the margins of the groove close over to form the neural tube. Closure of the neural tube starts in the middle and progresses in both rostral and caudal directions to be complete by the end of the fourth week (*Fig. 2*). Formation of the neural tube is called **neurulation**.

Cells on the dorsal side of the tube migrate laterally to form the neural crest which eventually gives rise to the peripheral nervous system. The neural tube becomes the central nervous system. Failure of the neural tube to close causes **anencephaly** (the fetus lacks much of the forebrain and cranium, dying shortly after birth) and **spina bifida**, in which the lumbosacral tube fails to close. In the most severe cases, **meningomyelocele**, elements of the spinal cord, cauda equina and meninges herniate through a defect in the vertebral column. More commonly, less serious varieties occur in which the defect is restricted to meninges or bone.

Neural induction

In all vertebrates an **organizer** region is responsible for triggering the differentiation of ectoderm into neural tissue, **neural induction**. In bird and mammalian embryos the organizer is Hensen's node. In frog embryos (the usual subjects for studying neural induction) the organizer is the mesoderm of the **dorsal blastopore lip** (**DBL**, *Fig. 3*).

DBL transplanted from a donor embryo to the opposite pole of a recipient embryo (see *Fig. 3a* for transplant site) causes the formation of a secondary embryonic axis, oriented in the same direction as the primary embryonic axis,

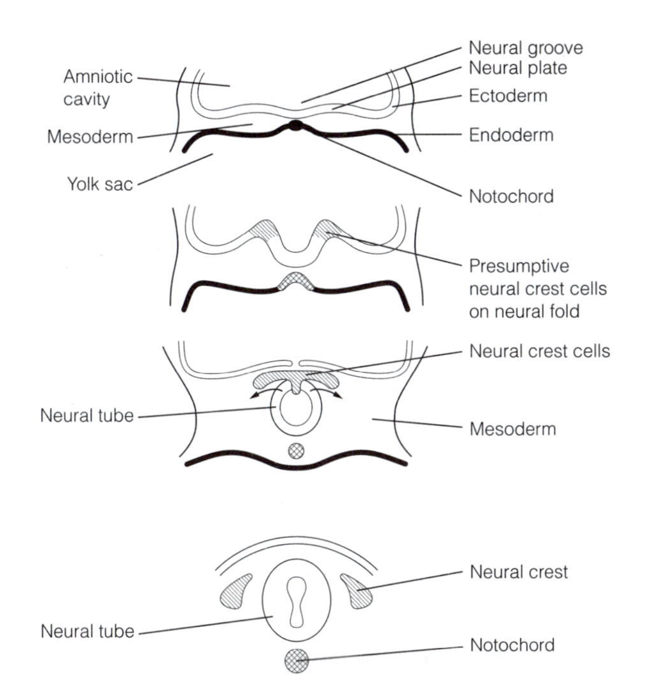

Fig. 2. Neurulation shown in midtransverse sections of the human embryo from day 20 to day 24. Closure of the rostral and caudal ends of the neural tube is not complete until day 28.

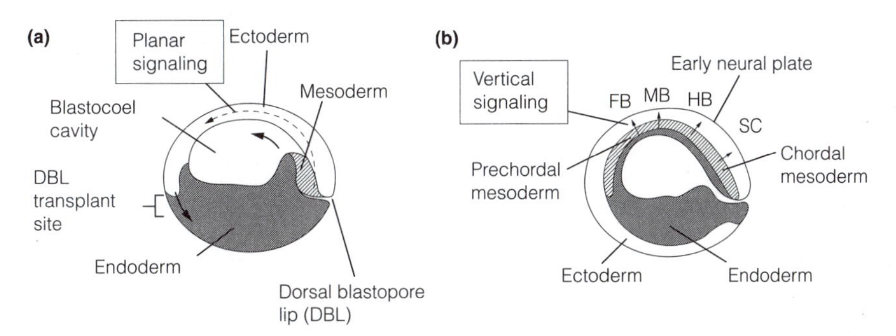

Fig. 3. Neural plate induction during gastrulation in the frog. (a) 9 hours, thick arrows show the directions of cell migrations; (b) 12 hours, note the neural plate. FB, Forebrain region; HB, hindbrain region; MB, midbrain region; SC, spinal cord region. Thin arrows depict the directions of inducing signals (see text for details).

but in which the neural tube is derived from the host ectoderm. This show that the DBL produces the chemical signals responsible for neural induction. These have not been unambiguously identified. A likely candidate is the protein, **chordin**, secreted by Hensen's node and primitive streak cells in chick embryos, which causes neural induction of frog embryos. It works by binding to polypeptide growth factors called **bone morphogenetic proteins (BMPs)** produced endogenously in the ectoderm, which normally act to transform it into epidermis. By binding BMPs, chordin inhibits their normal epidermalizing effect, with the result that ectoderm becomes neural tissue.

BMPs are one of a large group of growth factors called the **transforming growth factor β** superfamily, many of which are implicated in neural development. They bind to the extracellular face of membrane receptors, which dimerize, switching on serine/threonine kinase activity in the cytoplasmic domain of the receptor. The kinase controls downstream proteins which regulate the cell cycle.

Formation of the anteroposterior axis

The neural tube is organized in three directions:

- the anteroposterior (rostrocaudal) axis or **neuraxis** which is aligned along the long axis of the body;
- the dorsoventral (top to bottom) axis;
- the radial axis, which extends from the ventricular system out to the pial surface.

Early nervous system patterning along the anteroposterior axis comes about during gastrulation. Experiments in the frog embryo show that two directions of signaling are involved. Very early signals spread from the organizer across the ectoderm moving anteriorly. Regions of the ectoderm closest to the organizer become spinal cord, those furthest away become brain. This is called **planar signaling** (*Fig. 3a*). As gastrulation proceeds, mesoderm with organizing properties comes to lie beneath and in contact with the entire AP length of the neural plate. Now **vertical signaling** (*Fig. 3b*) by the **chordal mesoderm** at the caudal end causes overlying neural plate to become spinal cord, hindbrain and midbrain, while the **prechordal mesoderm** at the anterior end promotes the formation of forebrain.

The neuraxis of the early embryo comes to be divided into a series of compartments, **neuromeres**. Those in the hindbrain, **rhombomeres**, show repeat patterns of cellular organization (**metamerism**). Clones of cells derived from a single precursor usually remain within the confines of a given rhombomere. Early neurons appear in stripes in alternate even-numbered rhombomeres, followed by similar neurogenesis in odd-numbered rhombomeres. Segmentation of the forebrain also occurs though it is not metameric; each compartment has its own individual developmental sequence. Segmentation is intrinsic, since if induced ectoderm is removed from a gastrula and cultured a normally segmented neural tube develops.

Segmentation of the nervous system is due to the sequential activation of several sets of **segmentation genes**, each of which is expressed in restricted domains along the neuraxis. As a result of the pattern of gene expression, cells in particular compartments come to express particular cell surface proteins (cell adhesion molecules) and secrete signaling molecules which essentially code the position of the cell and specify how it interacts with its neighbors. It is the interactions between cells which determines how they migrate with respect to each other, what connections are made between them and even which cells survive, all of which shapes development into the adult nervous system.

Numerous genes have been identified in *Drosophila* by studying mutations affecting early fly development. Many of these genes have vertebrate homologs that are involved in early patterning of the neuraxis. One such group, the ***Hox* family** of **homeotic genes** is concerned with hindbrain and spinal cord segmentation. Mutations of homeotic genes produce individuals in which one body part abnormally develops in a similar way to another (homeosis); e.g. the *Antennapedia* mutant fly develops legs where it should have antennae. *Hox* genes code for transcription factors which control the expression of other genes. All *Hox* genes contain a highly conserved region, the **homeobox**, which encodes a sequence of 60 amino acids, the **homeodomain**, a part of which is responsible for binding to specific regions in the DNA of the genes they regulate. *Drosophila* has a single cluster of homeotic genes, but vertebrates have four clusters (*Hox* A–D), each located on a different chromosome. Each cluster has 13 regions, most of which correspond to identified genes. In vertebrates, genes in corresponding positions in each cluster are homologous and probably arose by gene duplication from a single ancestor cluster. Most vertebrate *Hox* genes have *Drosophila* equivalents. *Hox* genes are expressed in narrow bands across the AP axis and their effect is to help specify the compartmentation into rhombomeres.

Extraordinarily, the sequence of *Hox* genes along the chromosome exactly matches the order in which they are expressed along the neuraxis. The closer the *Hox* gene is to the 3′ end of the coding strand of DNA the closer to the anterior end it is expressed. This correspondence is called **colinearity** and probably arises because *Hox* genes must be activated sequentially in a particular order.

Activation of the correct *Hox* genes in cells along the AP axis requires positional signals. One candidate for such a signal is retinoic acid (RA) because:

- either deficiency or excess of RA cause severe malformations of the hindbrain;
- manipulating RA concentrations alters the expression of *Hox* genes; adding RA to embryos *in utero* shifts the anterior limits of the *Hox* gene expression more rostrally so that hindbrain development expands at the expense of forebrain;
- RA is produced in Hensen's node.

Cells move from the primitive streak into the node where they proliferate and migrate forward as the **head process**. The first cells to move through the node in this way get exposed to a brief, low concentration pulse of RA which activates *Hox* genes at the 3′ end of the DNA. These genes determines the most anterior structures. Later, as RA production by the node increases, cells moving through the node are exposed to high concentrations of RA for longer. This activates progressively more posterior determining (i.e. towards the 5′ end) *Hox* genes (*Fig. 4*). RA acts on members of the intracellular steroid receptor superfamily (**retinoic acid receptors**) which are regulators of gene expression. Promotors of *Hox* genes harbor retinoic acid response elements which bind the RA receptors.

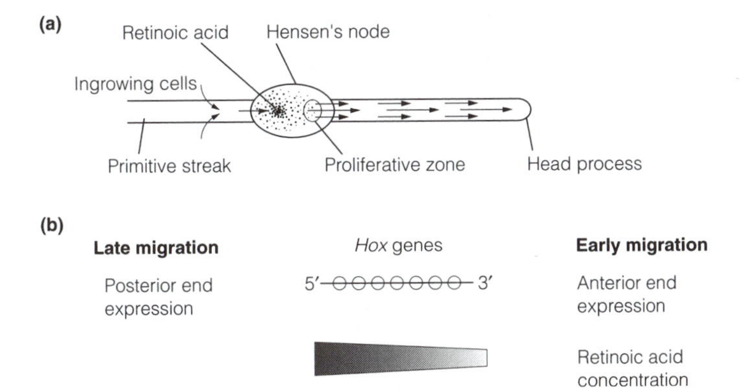

Fig. 4. Retinoic acid (RA) as a positional signal. (a) Cells are exposed to RA as they migrate through Hensen's node. (b) Early migration exposes cells to low RA concentrations which switches on anterior Hox *genes.*

Whereas the *Hox* gene expression that causes compartmentation of the hindbrain occurs in the cells of the hindbrain itself, the AP patterning of the spinal cord is brought about by *Hox* gene activation in the **paraxial mesoderm** that runs alongside the cord, which consequently secretes signals to affect cord development.

The forebrain appears to be divided into six **prosomeres**, three corresponding to the diencephalon and three to the telencephalon. It is thought that homeodomain containing proteins encoded by genes highly conserved in both vertebrates and invertebrates are crucial to brain development since they are expressed in specific domains at precise times.

Formation of the dorsoventral axis

A top–bottom differentiation of the neural tube is particularly obvious in the hindbrain and spinal cord. The notochord which lies immediately beneath the infolding neural tube expresses a gene called *sonic hedgehog* (*shh*) which encodes a protein of the same name (SHH) that is a member of the **transforming growth factor-β** superfamily of growth factors. SHH induces the formation of the **floor plate**, a narrow strip of glial cells along the ventral midline of the neural tube. The floor plate then becomes self-inducing by expressing *shh* itself. The SHH protein, first from the notochord and later from the floor plate, induces the differentiation of motor neurons in the ventral spinal cord (*Fig. 5*).

Although the floor plate secretes SHH along its whole length, in the hindbrain and midbrain regions of the neural tube it induces differentiation of serotinergic and dopaminergic neurons respectively. Hence the type of cell produced in response to the SHH signal depends on its AP position along the neuraxis.

As the neural tube closes, **bone morphogenetic proteins (BMPs)** in the epidermal ectoderm induce the formation of a dorsal roof plate, which then produces its own BMPs. These locally generated BMPs in turn induce the differentiation of the interneurons that will become dorsal horn cells.

The SHH and BMPs act as positional signals that are believed to produce their effects on dorsoventral patterning by activating families of ***Pax* genes** encoding transcription factors that serve a role similar to *Hox* genes.

In the interface region between epidermal and neural ectoderm the neural crest forms. Once the neural tube has closed, cells from here migrate laterally. Those at the caudal end will become sensory neurons of the dorsal root ganglia, sympathetic neurons (including the chromaffin cells of the adrenal medulla),

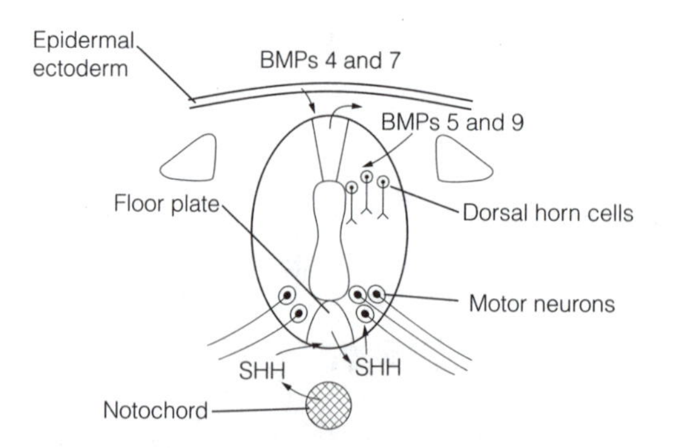

Fig. 5. Dorsoventral patterning of the neural tube. SHH, Sonic hedgehog; BMP, bone morphogenetic protein.

enteric neurons and Schwann cells. At the rostral end, neural crest cells form the sensory nuclei of the cranial nerves, parasympathetic neurons and contribute to the mesoderm of the brachial arches. The identity of neural crest cells appears to be predetermined by the expression of *Hox* genes, so when they migrate away from the neural tube they go to their appropriate destinations.

P2 Cell determination

Key Notes

Overview of cell determination

Cell differentiation is driven by an interaction between extracellular signals to which the cell is exposed and the pattern of gene expression within the cell. Extracellular signals can alter gene expression and some genes code for extracellular signals.

Neural fate determination

Epithelial cells lining the neural tube give rise to neuroblasts which proliferate and eventually give rise to postmitotic cells which differentiate into neurons. In *Drosophila*, neuroblast fate of epithelial cells requires expression of proneural genes, then neurogenic genes. Those cells which express neurogenic ones a little earlier become neuroblasts and their neurogenic gene products inhibit proneural gene expression in surrounding cells, which thus become epidermal.

Individual neural lineages

The position of a neuroblast in the neural tube determines its pattern of gene expression and the extracellular signals to which it is exposed. This specifies the type of neuron it will become.

Motor neuron specification

Motor neuron differentiation is initiated by sonic hedgehog (SHH) protein. Motor neurons then induce neighboring cells to become interneurons. Shortly after motor neurons are born the destination of their axons is unspecified by the position of the cell body in the neural tube, but within a few hours the motor neurons become committed so that their axons grow to particular muscles.

Glial cell lineages

Single progenitors can give rise to neurons or glial cells. In the optic nerve (part of the CNS) O-2A cells proliferate to give birth to oligodendrocytes in response to mitogenic signals from type I astrocytes. After about 7 days the type I astrocyte produces other signals which cause the O-2A cells to differentiate into type 2 astrocytes. In the peripheral nervous system neuroblasts produce a growth factor that induces glial cell differentiation from less committed cells.

Related topics

Glial cells and myelination (A4)
Blood–brain barrier (A5)
Early patterning of the nervous system (P1)
Neurotrophic factors (P6)

Overview of cell determination

Neuroectoderm gives rise to a large number of cell types, both neurons and glia, each with their own identity or **phenotype**. In development, the future phenotype of an undifferentiated cell is its **fate**. The process by which cell precursors are transformed into mature cells is **differentiation** and the sequence of cell types that lead from precursor to mature cell is the **cell lineage**. Differentiation occurs through an interplay between:

- Extrinsic signaling molecules (diffusible, or bound to the cell surface or extracellular matrix) in the cells' surroundings that have been generated by other cells.
- A timed sequence of gene expression mediated by intracellular signals, usually transcription factors. Some of these intrinsic signals are inherited from the cell lineage, others are generated in response to second messenger cascades activated by the extrinsic signals.

Neural fate determination

In vertebrates, some of the epithelial cells lining the neural tube become **neuroblasts** which divide mitotically. After a time, which varies depending on the fate of the cell, neuroblasts produce daughter cells that are no longer able to undergo mitosis. These migrate away from the epithelium towards the surface of the neural tube, where they continue to differentiate into neurons. In *Drosophila*, clusters of epithelial cells called **proneural clusters** become neuroblasts as a result of the activation of a group of **proneural genes**, for example, the *achaete–scute* genes. Activation of specific proneural genes in particular domains along the neuraxis depends on the pattern of expression of segmentation and homeotic genes (Topic P1) that are responsible for specifying the anteroposterior axis. Specific proneural gene expression is also regulated in the dorsoventral axis. Cells deprived of proneural gene function either develop as epidermoblasts or embark on a neural fate which is aborted by apoptosis (programmed cell death).

Neurogenic gene expression inhibits neighboring cells (including those in the same proneural cluster) from becoming neuroblasts. Destruction of a particular neuroblast in an early *Drosophila* embryo results in its place being taken by another cell that would normally have become epidermal.

Two neurogenic genes that have been extensively studied are *notch* (*N*) and *delta* (*Dl*) which encode membrane proteins, Notch and Delta. Loss of function mutations of either *notch* or *delta* causes expansion of neuroblasts at the expense of epidermoblasts. A gain of function mutation in which the Notch protein has its signaling mechanism permanently switched on even in the absence of Delta, causes all proneural cluster cells to become epidermoblasts (*Fig. 1*).

Notch protein is a membrane bound protein with a large extracellular domain containing numerous epidermal growth factor (EGF)-like repeats that act as a receptor for the smaller **Delta protein**, also a transmembrane protein with extracellular EGF-like repeats. Upon binding Delta (*Fig. 2*), the Notch receptor undergoes proteolytic cleavage releasing its intracellular domain that translocates to the nucleus. In the nucleus the intracellular domain of Notch forms a complex with a DNA binding protein **Suppressor of Hairless** which

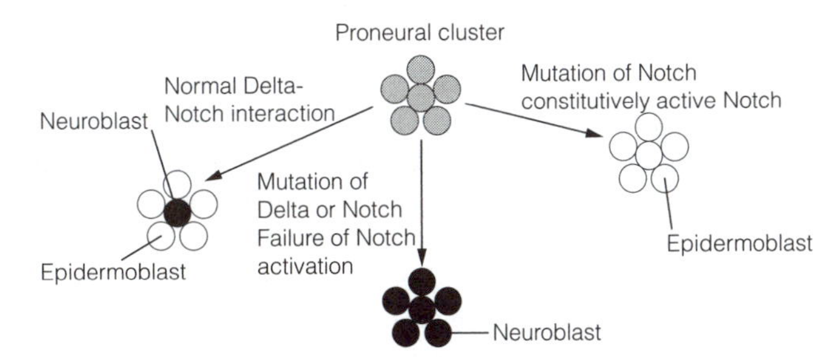

Fig. 1. The role of the neurogenic genes delta *and* notch *in neurogenesis in* Drosophila.

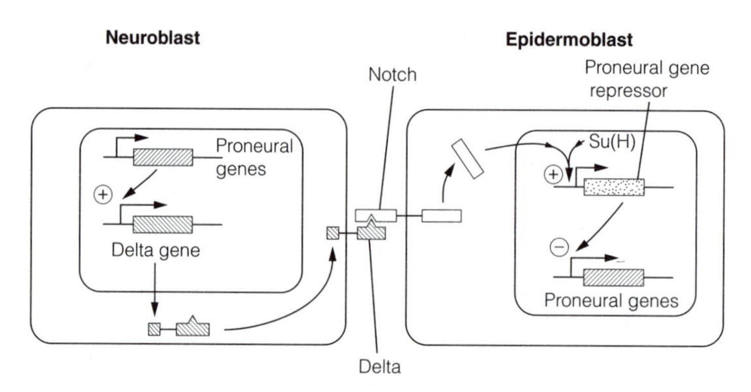

Fig. 2. Delta–Notch signaling in neural determination. Cells in a proneural cluster express proneural genes. These activate transcription and translation of Delta (and Notch) proteins. Those cells producing the highest amount of Delta cause suppression of proneural genes in neighboring cells by activating a Notch signaling pathway involving two sets of transcription factors; Su(H), Suppressor of Hairless.

causes transcription of a repressor of proneural genes. Those cell producing the highest amount of Delta will cause the greatest suppression of their neighbors' proneural genes; they will win the 'race' to become neuroblasts.

Many of the genes extensively studied in *Drosophila,* including the *achaete–scute* complex, and *notch* and *delta,* have homologs in vertebrates, where they seem to have similar functions.

Individual neural lineages

A single cell in each proneural cluster eventually becomes a fully committed neuroblast. Its subsequent fate is now determined by the activation of additional subsets of genes: which subset depends on the history of the neuroblast and its surroundings (i.e. its position). Many of these genes code for transcription factors that contain homeodomains. The particular combination expressed in a neuroblast is assumed to specify what sort of neuron results. Although many of these genes were first identified in invertebrates, homologous genes occur in vertebrates which appear to have similar roles. At present, the mechanisms by which a neuroblast is transformed into a particular type of neuron are understood (rather poorly) in only a few examples.

Motor neuron specification

When SHH from the floor plate induces motor neurons in the ventral neural tube it does so by directly activating the expression of two transcription factors and repressing two others. When this is investigated in the zebrafish it is found that the *Drosophila* hedgehog (*hh*) gene can substitute perfectly for Zebrafish *shh* gene in motor neuron induction. This illustrates how close the homologies can be between invertebrate and vertebrate development. Prospective motor neurons undergo their last mitotic division and then express a sequence of homeobox genes, the first of which (*Isl-1*) is used as an early marker of the motor neuron fate of the cell. The motor neurons, in turn, secrete proteins which induce neighboring, more dorsal cells, to become interneurons.

In zebrafish, each spinal segment initially has just three primary motor neurons so their fates are easily followed (*Fig. 3*). The rostral primary motor neuron axon grows to the lateral muscle, the middle primary motor neuron innervates the dorsal muscle, while the caudal primary projects to the ventral muscle. When these motor neurons are transplanted to different locations in the

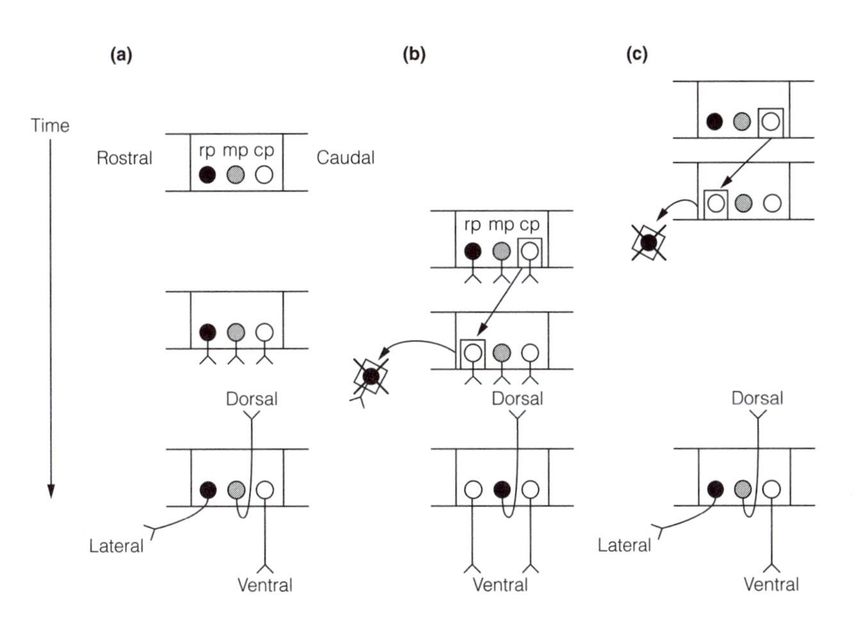

Fig. 3. Motor neuron differentiation: (a) normal development; (b) a caudal primary neuron (cp) is transplanted into the position normally occupied by the rostral primary neuron (rp) at the time of axonogenesis; (c) a cp is transplanted to the rp position before axonogenesis. mp, Middle primary neuron.

spinal segment at the time when their axons are beginning to grow, they project to their original muscles. Hence, at this time the neurons are committed to their particular fates and continue to develop independently. If the neurons are transplanted earlier, shortly after they are born, they take on the characteristics expected of the cell at their new position. Thus a caudal primary neuron transplanted to the rostral end of a spinal segment comes to innervate lateral (not ventral) muscle. This suggests that early development is dictated by the position of the cell along the spinal segment.

Glial cell lineages

Single progenitors in the neural tube or neural crest can give rise to either neurons or glial cells. Glial cell differentiation has been studied in the rat optic nerve, which is functionally CNS tissue.

The optic nerve has three kinds of glial cell. Type 1 astrocytes form contacts with blood vessels and help ensure the integrity of the blood–brain barrier, type 2 astrocytes contact nodes of Ranvier, and oligodendrocytes form the myelin sheaths around the axons of the retinal ganglion cells. These glial cell types can be identified by their expression of specific antigens.

Glial cell differentiation comes about by an interplay of several cell types and chemical signals. Type 2 astrocytes and oligodendrocytes differentiate postnatally from a common precursor called the O-2A progenitor, born on embryonic day 17 (E17). Type 1 astrocytes are derived from another lineage at the same time. The O-2A cell divides mitotically during the first postnatal week, giving rise to oligodendrocytes (*Fig. 4*). After a set number of divisions, the O-2A cells then differentiate into type 2 astrocytes. Both the division of the O-2A cells and their differentiation are controlled by signals from type 1 astrocytes; when O-2A cells are cultured in the absence of type 1 astrocytes they

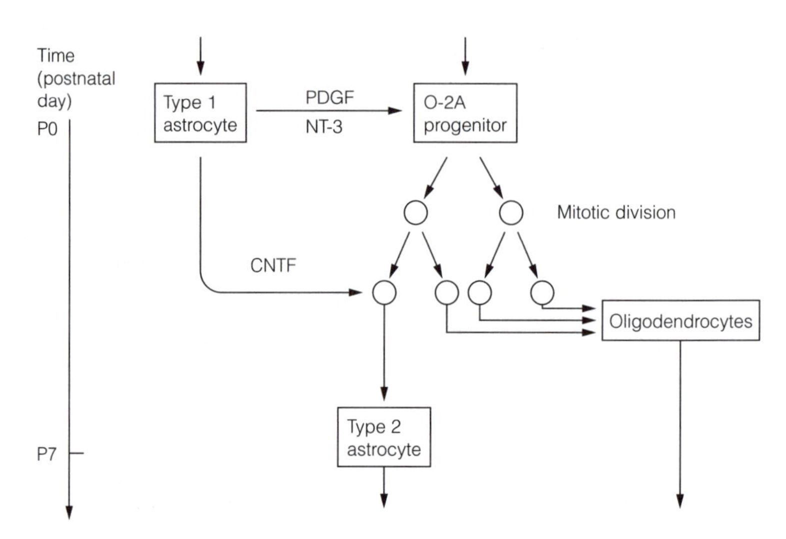

Fig. 4. CNS glial cell differentiation. CNTF, ciliary neurotrophic factor; PDGF, platelet derived growth factor; NT-3, neurotrophin 3.

immediately stop dividing and differentiate into oligodendrocytes. The mitogenic signals produced by type 1 astrocytes are **platelet-derived growth factor (PDGF)** and **neurotrophin 3 (NT-3)**. One signal for differentiation of O-2A cells into type 2 astrocytes is **ciliary neurotrophic factor (CNTF),** generated by type I astrocytes, but other unidentified factors are also required.

In the peripheral nervous system, once neural crest cells have migrated to form early ganglia, neuroblasts express **glial growth factor 2 (GGF2)** which promotes the emergence of glial cell phenotype by suppressing neuronal differentiation. Microglial cells are derived from non-neuronal mesodermal progenitors which invade the developing nervous system.

P3 Cortical development

Key Notes

Cerebral cortex development

The ventricular zone (VZ) of the neural tube contains stem cells which continually divide, giving rise to neurons and glial cells. Neurons migrate away from the VZ in such a way that the neural tube develops six radially organized layers. The two most superficial layers eventually become the neocortex, the VZ becomes the ependyma and the overlying subventricular zone retains the capacity to generate glial cells into adult life. Neuron migration occurs along radial glial cells, the processes of which extend the full thickness of the neural tube. Large projection neurons are born first and migrate the shortest distance. Smaller cells, born later, migrate through zones containing older cells to more superficial positions.

Cerebellar cortex development

The VZ gives rise to an internal granular layer which generates the neurons of the deep cerebellar nuclei and Purkinje cells which migrate superficially. VZ cells of the anterior rhombic lip migrate to form an external granular layer which generates huge numbers of granule cells that migrate down radial glial cells to lie deep to the Purkinje cells.

Related topics

Organization of the central nervous system (E2)

Cerebellar cortical circuitry (L2)

Cerebral cortex development

Early on, the walls of the three vesicles (*Fig. 1*) at the anterior end of the neural tube are a single layer of pseudostratified columnar neuroepithelium. These cells give rise to neurons and glia and initially the neuroblasts divide more frequently than the glia. In humans, up to about 6 weeks gestation cell division in the neuroepithelium causes it to increase in area but not in thickness. After 6 weeks, the walls of the neural tube go from being a single layer to several layers. This represents **radial patterning** of the neural tube. At first the neural tube becomes differentiated into an inner **ventricular zone (VZ)** in which cells proliferate and a **marginal zone (MZ)** containing radially projected cell processes. Cells that are not actively dividing (i.e. those in the interphase of the cell cycle)

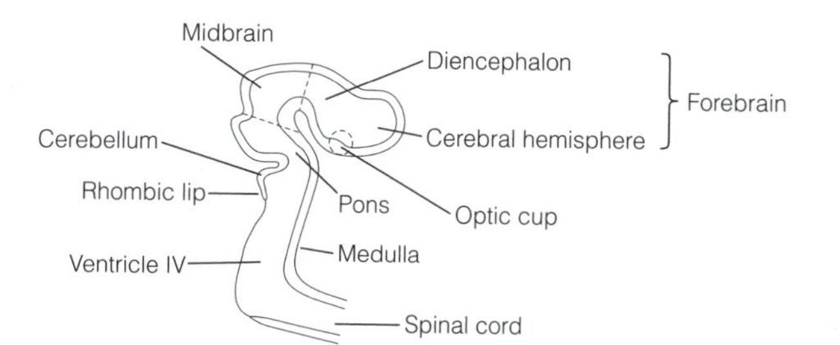

Fig. 1. Neural tube of a human embryo at about 5 weeks gestation; midsagittal section.

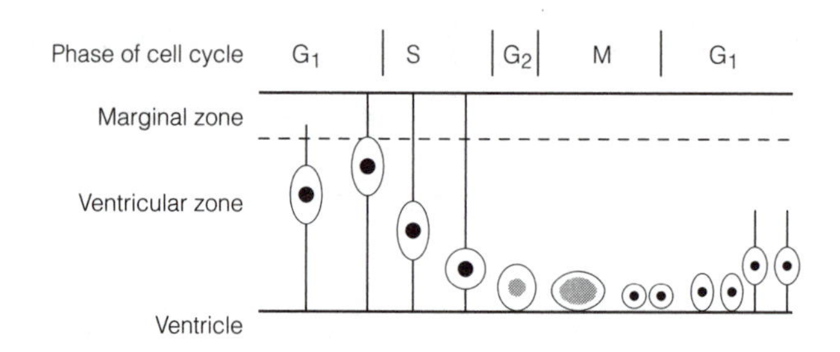

Fig. 2. Interkinetic movements of proliferating cells in the neural tube. G_1, Resting; S, DNA synthesis; G_2, preparation for mitosis; M, mitosis. (For further details of the cell cycle see Instant Notes in Biochemistry, *2nd edn.)*

are bipolar and extend their processes through the full thickness of the neuroepithelium. Progenitor cells undergoing mitosis retract their processes and sit on the wall of the ventricle. These cells divide symmetrically so that the parent produces one **stem cell** (which subsequently continues to divide) and a postmitotic cell which migrates away from the ventricle (*Fig. 2*). The periodic changes in shape and positions of progenitor cells as they go through successive cell cycles are called **interkinetic movements**.

Migrating neurons leave the VZ to form a single layer called the **preplate** (*Fig. 3*). Between the preplate and the VZ is an **intermediate zone** in which is found axons of the preplate neurons and axons growing into the cortex from developing subcortical structures, such as the thalamus. As neurons continue to be generated in the VZ they migrate through the intermediate zone to form the **cortical plate** which splits the preplate into the marginal zone above and a **subplate** region that lies below. The subplate is the site of intense **synaptogenesis** (synapse formation). As neurons migrate through the subplate they await the arrival of afferents from other areas of cortex and subcortical structures to make contact, then continue into the cortical plate. Neurons that remain in the subplate suffer apoptosis so this layer is transient.

During the later phases of neurogenesis a second region of dividing cells develops, the **subventricular zone**. This generates mostly small interneurons, a process which in humans continues into the second postnatal year. The mature cerebral cortex is formed mostly from the cortical plate, with the marginal zone being layer 1 (see Topic E2). The intermediate zone becomes the subcortical

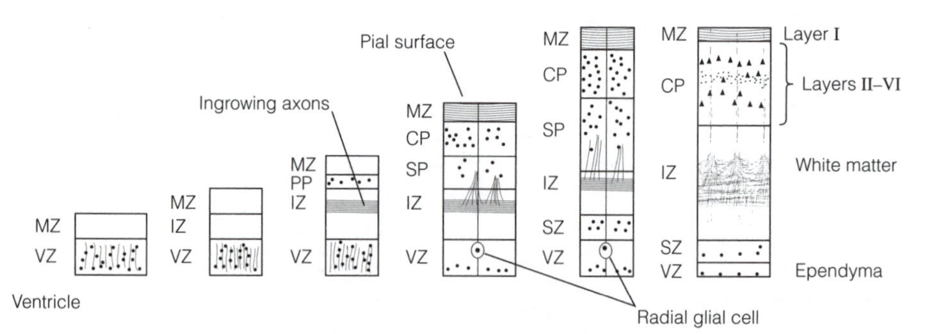

Fig. 3. Development of the neocortex. CP, Cortical plate; IZ, intermediate zone; MZ, marginal zone; SP, subplate; SZ, subventricular zone; VZ, ventricular zone.

white matter. The original ventricular zone becomes the ependymal layer, above which a thin subventricular zone contains cells which continue to proliferate to provide glia. Until recently it was thought that new neurons are not produced in the mature CNS. Recent evidence, however, shows that the ependyma of the adult brain retains stem cells capable of dividing to generate neurons.

Neurons migrate out of the VZ along **radial glial cells** by ameboid movements. Radial glial cells differentiate early and have their cell bodies in the VZ. They are bipolar and one end of their long process is attached to the ventricular cell while the other end is fixed to the pial cell basement membranes. This arrangement provides a scaffold for neuron migration. Neurons moving along a single glial cell generally remain close together after migration and this forms the columnar structure of the cortex. When migration is complete, most radial glia differentiate into astrocytes. In two brain regions they remain virtually unaltered, as Müller cells in the retina and Bergmann glia in the cerebellum.

The birthday of a neuron is the day on which it loses the ability to undergo further mitotic division. Postmitotic neurons subsequently migrate. The birthday of a neuron and its migration can be tracked by autoradiography of embryos at different times after giving a brief pulse of [^{3}H]thymidine. The isotope labels all cells synthesizing DNA (i.e. in the S phase of the cell cycle) at the time of the pulse because these cells will incorporate the label into their growing DNA molecules. This type of experiment reveals general features of neuron development. Firstly, larger projection neurons are born earlier than small neurons or interneurons. Secondly, younger neurons migrate through zones occupied by older neurons. Hence neurons with late birthdays migrate furthest and come to occupy a more superficial position. In consequence, the mature cerebral cortex has small cells in superficial layers and larger ones in deeper layers. In addition, the birthday of a neuron determines its subsequent fate; that is, what type of cell is formed and in which layer it ends up. Young progenitor cells transplanted into host embryos of a different age switch their fates to be appropriate to the developmental age of the host. However, cells about to divide for the last time (i.e. just about to give rise to a specific neuron type) retain their donor fate. This shows that extrinsic cues, operating at about the time when a neuron is born, determine its fate. The extrinsic cues are thought to come from the cell type born previously. For example, layer VI cortical pyramidal cells are born first and as they differentiate they produce signals which switch the VZ neuroblasts into generating layer V precursors.

Cerebellar cortex development

The primordial cerebellum consists of pseudostratified neuroepithelium showing interkinetic movements which soon develops into ventricular, intermediate and marginal zones, as elsewhere in the neural tube. As the cerebellum thickens, the VZ gives rise to uncommitted progenitors which form the **internal granular layer (IGL)** and **external granular layer (EGL)**. Cells which form the EGL migrate from the VZ in the anterior rhombic lip (the rostrodorsal edge of ventricle IV) (*Fig. 4*).

The IGL gives rise to nuclear neuroblasts and Purkinje neuroblasts. The nuclear neuroblasts form the neurons of the deep cerebellar nuclei and remain embedded in the intermediate zone, the future white matter of the cerebellum. The Purkinje neuroblasts migrate superficially towards their final position, trailing one neurite (the future axon) which lengthens as the cell body migrates. Finally, the IGL produces Golgi neuroblasts which migrate a short distance. The EGL initially gives rise to basket neuroblasts and subsequently its cells undergo

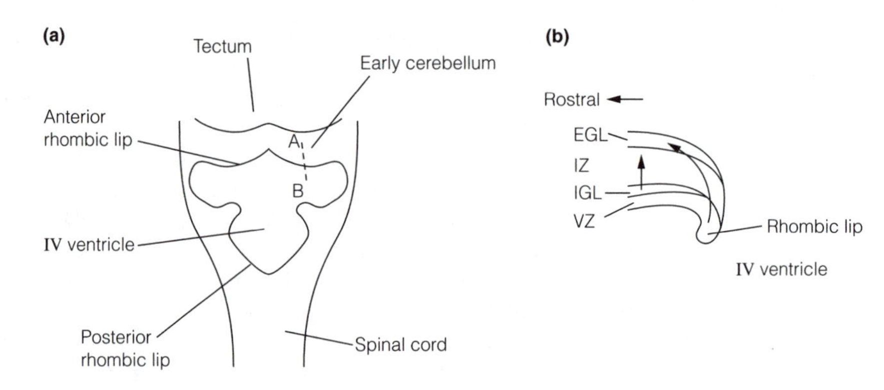

Fig. 4. Early cerebellar development: (a) Dorsal view of early hindbrain. The rhombic lip surrounds an opening in the neural tube which will eventually be covered by the cerebellum to form the IV ventricle. The posterior rhombic lip gives rise to the choroid plexus. (b) Section A–B through the rhombic lip and early cerebellum. The anterior rhombic lip generates cells which migrate to the external granule layer (EGL) to become granule neuroblasts. The internal granule layer (IGL) produces Purkinje neuroblasts (straight arrow). IZ, Intermediate zone.

an intense and prolonged proliferation generating vast numbers of granule cell neuroblasts and, at about the same time, stellate neuroblasts.

Radial glial cells (**Bergmann glia**) provide a scaffold for the migration of granule neuroblasts from the EGL through the Purkinje cell layer to the IGL. As the granule cells migrate they trail their axons behind. These axons bifurcate, giving rise to the parallel fibers (*Fig. 5*).

The importance of glial cells is illustrated by *weaver* mutant mice. Animals homozygous for the *weaver* mutation suffer disastrous motor incoordination. Bergmann cells are misaligned, and granule cell migration stops at the molecular layer, disrupting further cerebellar cortex development. Impaired signaling between glial and granule cells causes the defect.

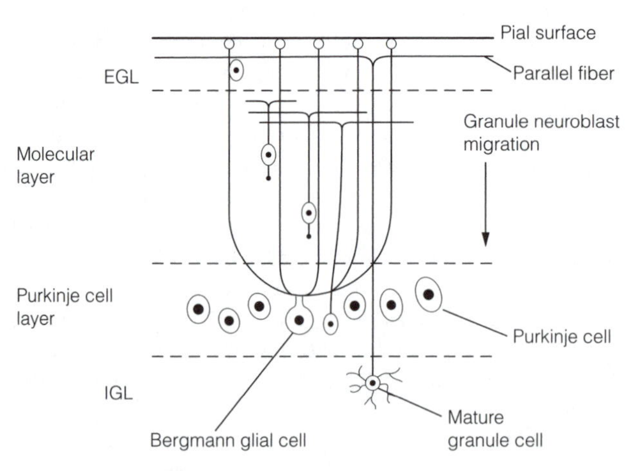

Fig. 5. Migration of granule neuroblasts through cerebellar cortex.

P4 Axon pathfinding

Key Notes

Growth cones

The tips of growing axons form growth cones. These express cell surface receptors which recognize short- or long-range signals that guide their axon, either by attraction or repulsion. Growth cones crawl with ameboid motion over the surface. Binding of growth cone receptors to molecules in the surface on which it rests causes polymerization of actin at the leading edge of the cone, which consequently advances. Polymerization of microtubules in the body of the growth cone is needed for sustained progress, partly by facilitating fast axoplasmic transport.

Axon guidance molecules

Cell adhesion molecules (CAMs) are integral membrane proteins, or extracellular matrix proteins (e.g. laminin or fibronectin) that mediate interaction between cells, or between a cell and the extracellular matrix. The interactions can be either attractive or repulsive and mediate short range interactions. There are three groups of cell surface CAMs: integrins, the immunoglobulin (Ig) superfamily and cadherins. The Ig superfamily members in the nervous system, N-CAMs, mediate cell adhesiveness either by binding to identical or different N-CAMs in adjacent cells. Cell adhesiveness causes axons to form bundles, and guides axons in specific directions through the nervous system. Netrins and semaphorins are responsible for long-range axon guidance. Netrins attract axons of dorsal neural tube cells to grow ventrally into the floor plate. One semaphorin is repulsive to axons, but attractive to dendrites of pyramidal cells, and so determines the direction in which these neurites grow in the cortex.

Formation of topographic maps

Studies in frogs and chicks of the manner in which retinal ganglion cells form precisely ordered connections with the tectum suggest that relative position on the tectum is specified by gradients of signaling molecules across anteroposterior and dorsoventral axes of the tectum that are detected by incoming growth cones. Position on the anteroposterior axis is specified by both the gradient of a ligand across the tectum and a gradient for its receptor, expressed on retinal axons.

Related topics

Stimulus localization (F3)
Eye and visual pathways (H2)
Neurotrophic factors (P6)

Growth cones

Axons emerge from neuroblasts, elongate and grow towards their proper targets. The tip of the growing axon is termed the **growth cone**. It expresses a sequence of cell surface receptors which recognize either short-range (local) cues, such as specific marker molecules on the extracellular matrix or on the surface of particular cell types, or long-range (secreted) cues, each of which can be either attractive or repulsive on axon growth. Interactions between the growth cone and the markers guide the axon to its correct destination.

The growth cone (*Fig. 1*) consists of a flattened fan-shaped cytoplasmic

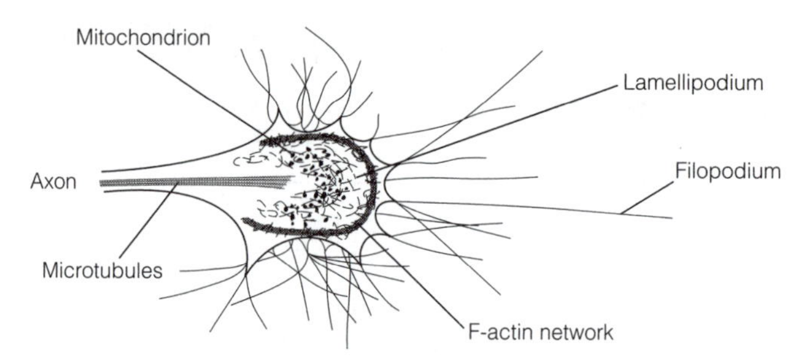

Fig. 1. The structure of a growth cone.

extension called a **lamellipodium**, rich in mitochondria and stacks of membrane bound vesicles, the membranes of which are incorporated into the surface of the advancing growth cone. The plus ends of microtubules extend from the axon into the lamellipodium. Tubulin polymerization occurs at the plus end to lengthen the microtubule. At the leading edge of the lamellipodium is a dense meshwork of **filamentous actin** (**F-actin**). Projecting from the lamellipodium are numerous spiky **filipodia**. These contain bundles of F-actin.

Growth cones advance by crawling with an ameboid motion over the surface (**substratum**) until receptors in the growth cone membrane find and bind to molecules on the substratum. This interaction causes actin, which is coupled to the receptor, to polymerize so the leading edge of the cone extends. F-actin further back depolymerizes, shrinking proximal regions of the cone. Actin monomer released by the depolymerization is recycled to the leading edge. The advance of the tip of the growth cone occurs in the following way: polymerized stretches of actin at the leading edge are driven in a retrograde direction by interaction with myosin in a process reminiscent of muscle contraction. This generates tension, and results in filipodia poorly attached to the substratum to retracting. Strong attachment of a filopodium to the substratum opposes this tension. Hence the growth cone progresses in the direction in which it can fix itself most avidly to its surroundings. Often, growth cones are guided into the appropriate place by molecules secreted either locally or at a distance, which are either attractive or repulsive to the cone. Polymerization of tubulin into microtubules is required for the sustained advance of growth cones. In part, this is because of the role microtubules play in fast axoplasmic transport which brings proteins synthesized in the cell body to the cone.

Axon guidance molecules

Several large families of molecules mediate the guidance of axons. Those that mediate short-range (local) interactions are collectively referred to as **cell adhesion molecules** (**CAMs**) which may be either integral membrane proteins or in the extracellular matrix. When the interaction is attractive the term **contact guidance** is used, when repulsive, **contact inhibition**. Extracellular matrix CAMs include **laminin**, **fibronectin** and **tenascin-C**. Laminin and fibronectin stimulate axon outgrowth and line corridors which channel axons to grow in the appropriate direction. Tenascin-C is stimulatory to axon growth in some neurons, but inhibitory in others.

Cell surface CAMs important in the nervous system fall into three broad classes, the integrins, the immunoglobulin (Ig) superfamily, and the cadherins which have a minor role.

The growth cone receptors for extracellular matrix molecules are the **integrin** family of CAMs. Integrins are integral membrane proteins made up of heterodimers consisting of an α and a β chain. There are several isoforms of each chain (15 α and eight β isoforms have been identified to date) and different combinations of isoforms generates a large number of integrins, each with preferential binding affinities for particular extracellular matrix CAM. This gives the potential to encode a large number of axon guidance pathways.

The **immunoglobulin (Ig)** superfamily is huge. Those that are expressed in the nervous system are called **N-CAMs**. They have a single transmembrane segment and a small intracellular terminal. The extracellular region is large and consists of a variable number of Ig domains and fibronectin type III domains. Many members of this group can bind themselves (**homophilic binding**) or bind other family members (**heterophilic binding**) and in doing this they are able to mediate cell adhesion. The adhesiveness of N-CAMs is regulated by long chains of sialic acid, a negatively charged sugar. The more sialic acid residues an N-CAM possesses the weaker the adhesiveness. Embryonic N-CAMs are highly sialylated so that cell–cell contacts are readily broken. This allows extensive sorting to occur before irrevocable contacts are forged by differentiated cells, which have N-CAMs that are much less sialylated, and are hence more adhesive.

Ig-CAMs called **fasciculins** are responsible for the cell adhesion often seen between axons from several neurons (either of the same or different type) which causes them to aggregate into bundles or **fasciculi**. In this situation axons are acting as guides for each other. Axons with the incorrect CAMs are excluded.

Heterophilic binding between two Ig-CAMs is needed for guidance of the axons of commissural neurons in the dorsal neural tube. These extend axons ventrally, then cross the floor plate to the opposite side. The commissural cells express axonin-1 while the floor plate cells have Nr-CAM on their surface. These CAMs bind one another. Antibodies to either axonin-1 or Nr-CAM injected into the neural tube results in about half of commissural axons failing to cross the floor plate.

Long-range interactions are brought about by two families of molecules, netrins and semaphorins. **Netrins** are large, secreted proteins but they bind to the extracellular matrix where they interact with the N-CAMs on the surface of the growth cone. The ventral growth of the commissural axons in the dorsal neural tube (see above) is mediated by a netrin expressed in the floor plate. This netrin is chemoattractant for commissural neurons. Remarkably, as the axons grow through the floor plate they become desensitized to netrin, allowing the axons to continue to extend. Netrins can be chemorepellent for some neurons. Motor neurons of the trochlear (cranial IV) nerve are located in the ventral neural tube. Their axons grow dorsally, away from the floor plate. It has been demonstrated that a netrin can be both attractant and repellent by acting on different N-CAMs.

Semaphorins are a large family of secreted and membrane-associated factors. Semaphorin 3A (sem 3A) repels primary sensory axons in the peripheral nervous system. It is also important in the cerebral cortex. Sem A is repulsive to axons of cortical pyramidal cells, but attractant to their dendrites. Since there is a gradient for sem A in the cortex – high at the pial surface and low at the subcortical white matter – the axons of pyramidal cells grow toward the white matter, while their dendrites grow in the opposite direction.

Formation of topographic maps

Axons need to make connections with the appropriate target cells. For those axons that establish topographic connections, neighboring axons must synapse with neighboring target cells in a very precise way in order to produce a smooth map. The process by which neural connections form topographic maps has been extensively studied in the **retinotectal pathway** of the frog and chick. This pathway maps visual input from the retina to the optic tectum of the midbrain. If the optic nerve of a frog is cut and the eye rotated through 180º, the optic nerve regenerates. However the frog now behaves as if its visual world has been rotated (*Fig. 2a*), and since it never learns to compensate for this deficit, the retinotectal pathway must be specified by developmental mechanisms to be immutable (*Fig. 2b*). If, in a normal frog, half of the retina is removed, axons from the remaining half eventually expand to cover the whole tectum (*Fig. 2c*). If half of the tectum is removed (*Fig. 2d*) retinal axons resort so that eventually a complete retinal map exists on the remaining hemitectum. These experiments show that it is relative, not absolute, position that is encoded. Gradients of signaling molecules across the anteroposterior and dorsoventral axes of the tectum, detected by receptors in the arriving growth cones are thought to account for the positional information.

The molecular basis for retinotectal specificity has been studied in the chick. In this animal, the nasal retina projects to the posterior tectum and the temporal retina projects to the anterior tectum (*Fig. 3*). Nasal growth cones are unaffected by incubation with anterior or posterior tectal membranes. Temporal growth cones are not influenced by anterior tectal membranes, but collapse and fail to advance when exposed to posterior tectal membranes. The tectum produces a

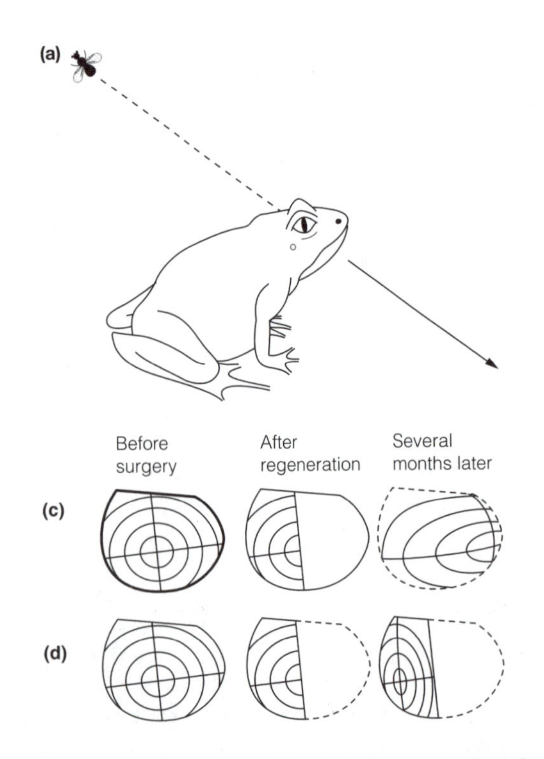

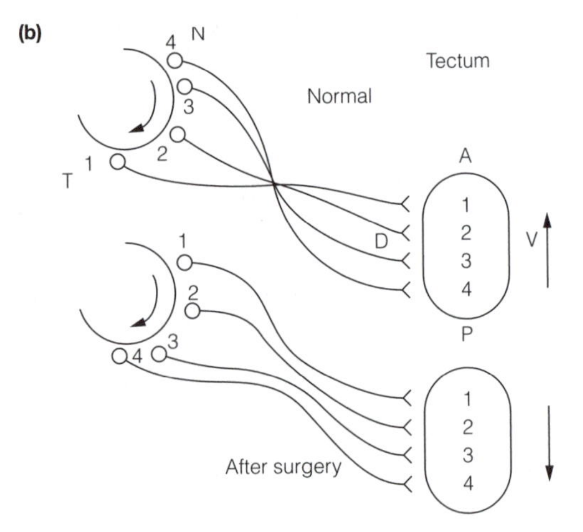

Fig. 2. Retinotectal specificity. (a) Behavior of a frog after cutting the optic nerve and 180° rotation of the eye. (b) rewiring of the tectum maintains the original position information. A retinal image, ↑, now has a tectal representation that is misaligned by 180° (i.e. it is upside down). N, nasal retina; T, temporal retina; A, anterior; P, posterior; D, dorsal; V, ventral. (c) Expansion of the topographic map after removal of half the retina. (d) Restoration of a complete map after removal of half of the tectum. In (c) and (d) the contours are isospatial lines at intervals 20° from the center of the retina.

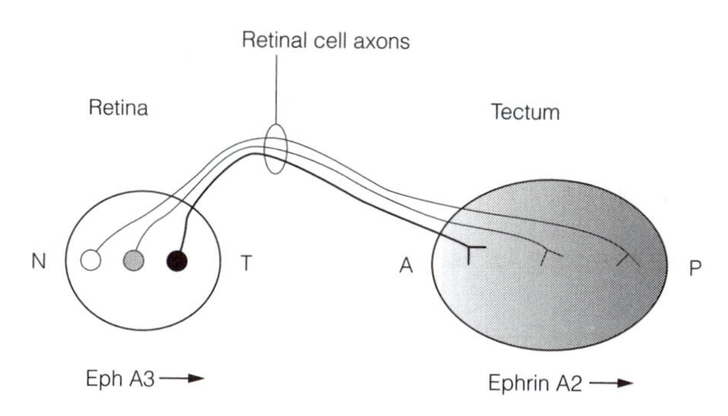

Fig. 3. Anteroposterior gradients for tectal ephrin A2 expression and retinal ephrin A3 receptor expression. These gradients provide positional information for specifying the retinotectal topographic map. N, Nasal retina; T, posterior retina; A, anterior tectum; P, posterior tectum.

ligand, **ephrin A2**, that is distributed as a gradient across the tectum, in highest concentration posteriorly. A corresponding gradient in the receptor for this ligand, the **ephrin A3 receptor** (Eph A3 receptor), exists on retinotectal axons with the greatest concentration on temporal retinal cells. Hence temporal cells will find increasingly posterior tectal regions repulsive. Eph A3 receptors are members of the large receptor tyrosine kinase (trk) superfamily (Topic P6).

P5 SYNAPTOGENESIS AND DEVELOPMENTAL PLASTICITY

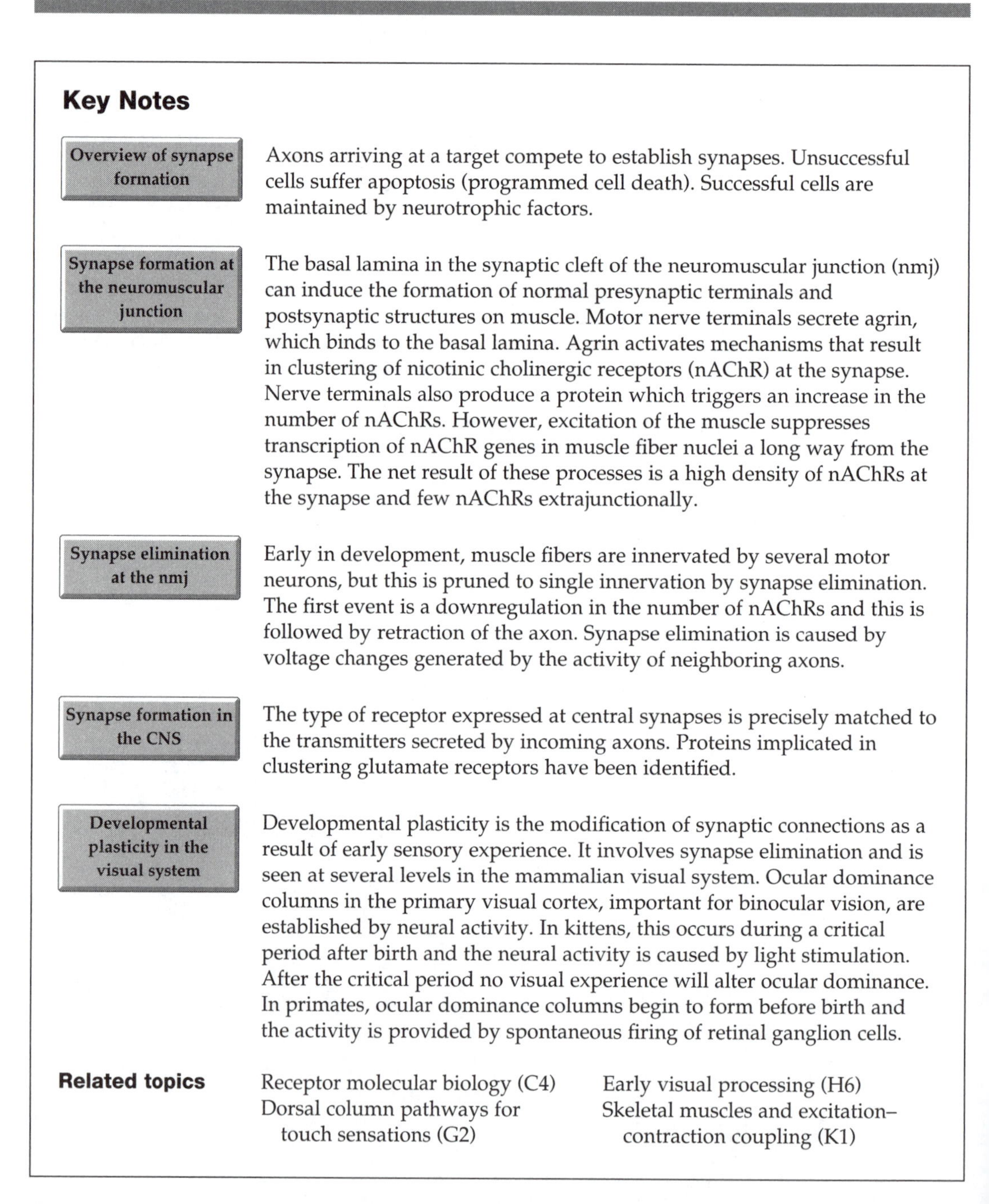

Key Notes

Overview of synapse formation

Axons arriving at a target compete to establish synapses. Unsuccessful cells suffer apoptosis (programmed cell death). Successful cells are maintained by neurotrophic factors.

Synapse formation at the neuromuscular junction

The basal lamina in the synaptic cleft of the neuromuscular junction (nmj) can induce the formation of normal presynaptic terminals and postsynaptic structures on muscle. Motor nerve terminals secrete agrin, which binds to the basal lamina. Agrin activates mechanisms that result in clustering of nicotinic cholinergic receptors (nAChR) at the synapse. Nerve terminals also produce a protein which triggers an increase in the number of nAChRs. However, excitation of the muscle suppresses transcription of nAChR genes in muscle fiber nuclei a long way from the synapse. The net result of these processes is a high density of nAChRs at the synapse and few nAChRs extrajunctionally.

Synapse elimination at the nmj

Early in development, muscle fibers are innervated by several motor neurons, but this is pruned to single innervation by synapse elimination. The first event is a downregulation in the number of nAChRs and this is followed by retraction of the axon. Synapse elimination is caused by voltage changes generated by the activity of neighboring axons.

Synapse formation in the CNS

The type of receptor expressed at central synapses is precisely matched to the transmitters secreted by incoming axons. Proteins implicated in clustering glutamate receptors have been identified.

Developmental plasticity in the visual system

Developmental plasticity is the modification of synaptic connections as a result of early sensory experience. It involves synapse elimination and is seen at several levels in the mammalian visual system. Ocular dominance columns in the primary visual cortex, important for binocular vision, are established by neural activity. In kittens, this occurs during a critical period after birth and the neural activity is caused by light stimulation. After the critical period no visual experience will alter ocular dominance. In primates, ocular dominance columns begin to form before birth and the activity is provided by spontaneous firing of retinal ganglion cells.

Related topics

Receptor molecular biology (C4)
Dorsal column pathways for touch sensations (G2)
Early visual processing (H6)
Skeletal muscles and excitation–contraction coupling (K1)

Overview of synapse formation

During nervous system development far more neurons are born than subsequently survive. When axons arrive at a target they compete with each other to form synapses – **synaptogenesis**. Cells which fail to establish synapses eventually suffer apoptosis. Neurotrophic factors released by the target tissue, and from other sources, ensure the survival of cells which successfully make contacts.

Synapse formation at the neuromuscular junction

Synapse formation has been most extensively studied at the nmj. (See Topic K1 for a review of the structure and function of the nmj.) Differentiation of both presynaptic and postsynaptic structures requires signals from the **synaptic basal lamina (bl)**, an extracellular matrix within the synaptic cleft which binds collagen, acetylcholinesterase, laminin and other proteins. The signaling role of the bl is revealed by experiments in which both muscle fibers and motor nerve axons are damaged in adult frogs. Both muscle fibers and axons degenerate to leave the bl and the Schwann cells that encapsulated the axon. After a few days, myoblasts invade and differentiate into new muscle fibers, the axon regrows and synapses are formed at precisely the same places as before. If muscle regeneration is prevented by X-irradiation, the axon still forms a normal presynaptic terminal with active zones exactly in register with the bl of the postjunctional folds, a feature of normal nmjs. If axon regrowth is prevented, the regenerating muscle fibers form normal postsynaptic structures at the original synaptic sites and nAChR cluster at the sites of the presumptive synapses as occurs normally.

Normally, developing muscle fibers express low levels of nAChR across their whole surface (about 1000 nAChR/μm^2). With the arrival of the nerve terminal, nAChRs cluster to a very high density immediately beneath the active zone (10 000 nAChR/μm^2), whereas in the extrajunctional region the density drops to 10 nAChR/μm^2. A relatively high density of nAChR is also seen over the entire surface of muscles that have been denervated for an appreciable time. This renders the muscle exquisitely sensitive to ACh, a state known as **denervation supersensitivity**.

Three mechanisms regulate nAChR at the developing synapse:

- Clustering of nAChR requires a large protein, **agrin**, that is secreted by motor neuron terminals and becomes bound to the synaptic basal lamina. The action of agrin is mediated by a receptor in the muscle membrane, one component of which is **MuSK (muscle specific kinase)**, a receptor tyrosine kinase that promotes clustering of nAChR by activating a membrane protein, **rapsyn** (*Fig. 1*). The agrin–MuSK–rapsyn cascade is also involved in the clustering of other synaptic proteins, e.g. acetylcholinesterase. Knockout mice lacking any of these proteins fail to form normal synapses, are immobile, cannot breathe and die after birth.
- Not only do nAChRs cluster in the postsynaptic membrane, there is also an upregulation of the receptor itself. This is due to a signaling mechanism that activates transcription of the nAChR subunit genes in nuclei that lie beneath the synapse. The nerve terminal expresses a transmembrane protein, **neuregulin** (also referred to as AChR-inducing activity, ARIA) which contains an epidermal growth factor (EGF)-like domain. It binds to EGF (ErbB) receptors to generate a transcriptional signal.
- ACh acting on nAChR causes depolarization of the muscle membrane. This activity-dependent change in membrane potential inhibits transcription of nAChR genes by nuclei that lie distant from the synaptic region.

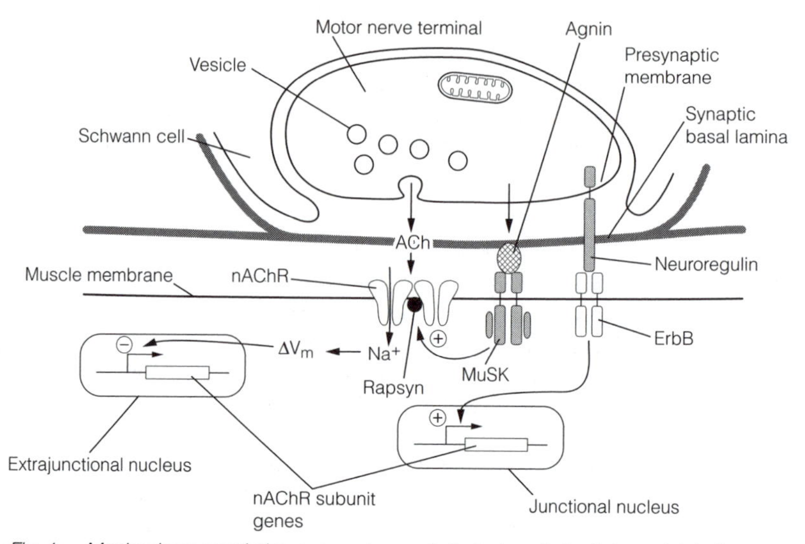

Fig. 1. Mechanisms regulating expression and clustering of nicotinic acetylcholine receptors (nAChR) at the neuromuscular junction. Erb B, Epidermal growth factor receptor; MuSK, muscle specific kinase.

The outcome of the processes above is that in normally innervated muscle the nicotinic receptors are clustered at the nmj. In denervated muscle, the lack of activity causes expression of nicotinic receptors across the whole muscle surface. This gives rise to the denervation supersensitivity.

Synapse elimination at the nmj

Each motor neuron has many branches by which it innervates numerous muscle fibers. Early in development, branches of several motor neurons converge to form synapses on each muscle fiber. All but the synapse from a single motor neuron on each fiber is eliminated in mammals, as motor neurons withdraw branches from multiply innervated fibers. Eventually, each motor neuron restricts its branches to a set of fibers over which it has sole control – the motor unit (*Fig. 2*). If a muscle is denervated, exactly the same phenomenon is observed, multiple innervation occurs initially and is pruned after 2–3 weeks to single innervation.

The earliest event in synapse elimination is a downregulation in the number of nAChRs. This is followed by a retraction of the presynaptic terminal. Clustering of nAChRs is important in maintaining the synapses, since mice lacking agrin or MuSK show loss of normal nerve terminals. The ultimate cause of synapse elimination is competition from other synapses, brought about by their activity. In adult animals, a motor neuron branches making several nmjs on a particular muscle fiber. Blocking one of these with the irreversible antagonist of nAChR, α-bungarotoxin (α-BTX), causes elimination of the blocked synapse. If several neighboring nmjs are blocked by α-BTX no synapses are eliminated. This implies that a local activity signal from unblocked synapses is responsible for elimination of blocked synapses. The molecular details of this process are not known.

Synapse formation in the CNS

Central synapses have poorly developed basal lamina and although agrin is present neither MuSK nor rapsyn are present in brain. However, appropriate receptors are clustered beneath the active zone, so mechanisms must exist which

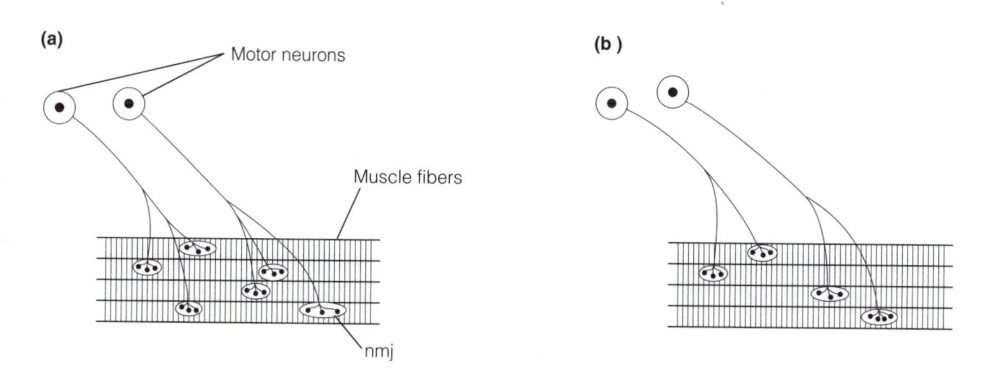

Fig. 2. Synapses elimination at the neuromuscular junction (nmj). (a) Normally each motor neuron would contact many muscle fibers. Initially, the territories of the two motor neurons overlap. Pruning of synapses leads to each neuron having exclusive control over a set of muscle fibers (b).

match transcription of the correct receptors in the postsynaptic cell to the arrival of terminals containing particular transmitters (i.e. glutamatergic terminals induce the expression of appropriate glutamate receptor subtypes). These receptors must also be clustered at the synapse. How this is achieved is not understood, but a family of proteins has been identified at glutamatergic synapses that is implicated in the clustering of receptors and anchoring them to cytoskeletal proteins

Developmental plasticity in the visual system

During development, synaptic connections can be modified as a result of early sensory experience. This is termed developmental plasticity. The rewiring usually involves synapse elimination and this comes about either because of apoptosis (programmed cell death) of neurons or because inappropriate axon branches are retracted. Developmental plasticity has been extensively studied in the visual system.

In lower vertebrates (e.g. frogs) the retinotectal pathways are not much refined after being laid down. By contrast in the rat, axons from nasal and temporal parts of the retina overlap considerably when they first reach the superior colliculus (tectum). However, by the end of the second postnatal week synapse elimination, largely by retraction of spurious axon terminals, results in temporal retinal axons being restricted to the anterior part, with nasal retinal axons being confined to the posterior part of the superior colliculus.

Plasticity in the primary visual cortex, V1, is responsible for the formation of ocular dominance columns and hence for the development of binocular vision. Kittens reared with one eye sutured closed have much smaller ocular dominance columns for the sutured eye. This occurs because the normal exuberant connections established early on are largely eliminated, if they are from the closed eye, by the excessively powerful competition exerted by the open eye. The competition exerted by the open eye is due to light evoked electrical activity. Injecting tetrodotoxin (TTX) into the open eye to block all action potentials in its retinal ganglion cells shifts ocular dominance to the sutured eye. Exactly the same shift is seen when activation of cortical cells by the open eye is blocked by enhancing GABAergic inhibition with GABA receptor agonists.

NMDA receptors are implicated in visual system plasticity. They are expressed in greater numbers and are activated at lower threshold in very young animals than in adults. The contribution they make to visual responses in deep layers of V1 decreases from 3 to 6 weeks and this decline can be delayed by rearing animals in the dark. This matches the **critical period** for developmental plasticity in the kitten visual system, the timing of which is delayed if animals are kept in darkness. A critical period is a narrow time interval during development in which experience can result in long-lasting changes in behavior. During the critical period for binocular wiring of the visual system, the effect of monocular deprivation can be partly reversed by restoring binocular vision. Outside the critical period, however, no amount of manipulation of visual experience will alter ocular dominance.

In primates, unlike kittens, ocular dominance columns are established before birth. Activity is still required, but in this case is provided by spontaneous firing of retinal ganglion cells, in the absence of light, that spreads in waves across the retina. The bursts of activity in the two retinae are out of phase and this is thought to be responsible for segregation of inputs from the left and right eyes. The burst firing may work by activating NMDA receptors, since the NMDA receptor antagonist, AP5, blocks neural activity in the visual system (much like TTX) and disrupts segregation of inputs. Although apparent before birth, ocular dominance columns in primates remain plastic until 6 weeks postnatally.

P6 Neurotrophic factors

Key Notes

Neurotrophic factors

Neurotrophic factors promote the survival of neurons by preventing apoptosis. They fall into three classes, neurotrophins, growth factors and cytokines.

Neurotrophins

Nerve growth factor (NGF) promotes the survival of sensory neurons in dorsal root ganglia and sympathetic neurons. Other neurotrophins, such as brain derived neurotrophic factor (BDNF) and neurotrophins 3–6 (NT3–6), promote the survival of central neurons. They have a common structure.

Neurotrophin signaling

Neurotrophins bind to low affinity and high affinity receptors. The high affinity receptors are tyrosine kinase receptors (trks). Binding of ligand causes the receptors to dimerize, activating tyrosine kinase domains on their cytoplasmic side. This forms binding sites for proteins, activating three signal transduction pathways that block apoptosis and promote growth by altering gene expression.

Actions of neurotrophins

Different classes of neuron require different neurotrophins for their survival. The requirement of a neuron for neurotrophins also changes during development. Neurotrophins may be secreted by the tissue through which axons must grow, their targets and even the neurons which depend on them. In the brain the highest expression of neurotrophins is in the hippocampus where they may be important in cellular mechanisms that underlie learning.

Apoptosis

Half of all neurons are estimated to suffer apoptosis (programmed cell death) during development. Apoptosis differs from necrosis, the mode of cell death that follows acute injury, by not stimulating immune responses. It is activated, in the absence of neurotrophic factors binding to trk receptors, by a biochemical cascade that switches on proteases.

Related topics

Slow neurotransmission (C3)
Axon pathfinding (P4)
Hippocampal learning (Q4)

Neurotrophic factors

In some instances, synaptic connections are eliminated by the death (apoptosis) of the neurons forming them. The remaining synapses are maintained by neurotrophic factors, molecules released from the target, or other source, which ensure the survival of neurons by preventing apoptosis. Neurotrophic factors fall into three major classes:

- **Neurotrophins**, first identified because of their role in promoting neuron differentiation and survival.
- **Growth factors**, originally recognized by their actions in stimulating proliferation and differentiation of numerous cell types.

- **Cytokines**, a large and diverse group of secreted molecules associated with regulation of the immune system.

Neurotrophins

The first neurotrophic factor to be discovered, **nerve growth factor** (**NGF**) promotes the survival of sensory neurons in the dorsal root ganglia (DRG) and sympathetic neurons. NGF treated chick embryos have greater numbers of neurons in their DRG, since fewer have undergone apoptosis, their cell bodies are larger and their axons are longer; similar changes are seen in sympathetic neurons and by some neurons. Chick embryos treated with anti-NGF antibodies lose most of their sympathetic neurons. NGF is secreted by the targets of the sensory and sympathetic neurons and by some neurons. Secretion of NGF from neurons is regulated by activity and depends on mobilization of Ca^{2+} from internal stores. Several other neurotrophins, such as **brain derived neurotrophic factor** (**BDNF**) and **neurotrophins 3–6** (**NT3–6**), promote the survival of central neurons. Neurotrophins share a common structure. Each consists of two identical peptides about 120 amino acids long, linked by several hydrophobic residues. Variable regions containing exposed basic amino acid residues provide for the specificity shown by individual neurotrophins.

Neurotrophin signaling

Neurotrophins bind two classes of receptor, high affinity receptors and low affinity receptors.

The low affinity ($p75^{LNTR}$) receptor ($K_D \approx 10^{-9}$ M) blocks apoptosis and enhances the binding of neurotrophins to the high affinity receptor.

High affinity receptors ($K_D \approx 10^{-11}$ M) are members of the **tyrosine kinase receptor** (trk; pronounced 'track') superfamily. These are large transmembrane receptors with single transmembrane segments, a tyrosine kinase domain on the intracellular face and immunoglobulin-like domains on the extracellular side. Three distinct trk receptors mediate the responses to neurotrophins, trkA–C, with the specificities shown in *Table 1*. Their expression is restricted to neurons responsive to the particular neurotrophins.

The trk superfamily of receptors is large and together with receptors for neurotrophins, includes receptors for growth factors (e.g. epidermal growth factor receptor, platelet-derived growth factor receptor), insulin and the Eph receptor for ephrin (see Topic P4). They share common signaling mechanisms. Binding of ligand causes receptors to dimerize (*Fig. 1*) which activates the intrinsic tyrosine kinase domains. This catalyzes autophosphorylation (each receptor phosphorylating its opposite number) of several tyrosine residues to

Table 1. Specificities of the tyrosine kinase receptors (trks)

Neurotrophin	Receptor	Principal neurotrophin targets
Nerve growth factor	trk A	Nociceptor afferents Sympathetic neurons Basal forebrain cholinergic neurons
Brain-derived neurotrophic factor	trk B	Mechanoreceptor afferents Retinal ganglion cells Hippocampus Cerebral cortex
Neurotrophin-2	trk A, B and C	Proprioceptor afferents Cochlear afferents Hippocampus

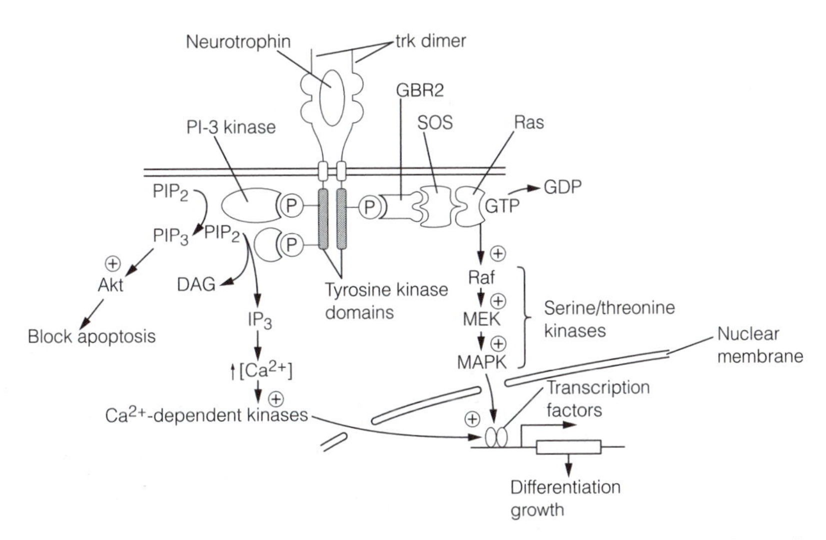

Fig. 1. Signaling by neurotrophins. Binding of SH2 domain proteins to phosphotyrosines on activated tyrosine kinase receptor (trk) dimers switches on three signaling pathways. DAG, Diacylglycerol; IP_3, inositol 1,4,5-trisphosphate; PI-3 kinase, phosphatidylinositol-3-OH kinase; PIP_2, phosphatidylinositol-4,5-bisphosphate; PIP_3, phosphatidylinositol-3,4,5-trisphosphate; SoS, Son of Sevenless.

create binding sites for **src homology domain 2 (SH2)** regions on three proteins. Binding of each of the SH2 proteins initiates three second messenger cascades.

(1) Binding and phosphorylation of **phospholipase C (PLC-γ)** generates diacylglycerol and inositol 1,4,5-trisphosphate from phosphatidylinositol-4,5-bisphosphate (PIP_2) (see Topic C3). This mobilizes Ca^{2+} from intracellular stores to have effects on assembly of the cytoskeleton and on gene transcription.

(2) Activation of **phosphotidylinositol-3-OH kinase (PI-3K)** generates phosphatidylinositol-3,4,5-trisphosphate from PIP_2 which activates a kinase, **akt**, that promotes neuron survival by inhibiting apoptosis.

(3) Binding of an adaptor protein, **GBR2**, couples the receptor to a complex containing a small guanosine 5′-triphosphate (GTP) binding protein, **ras**, which has some homology with the Gα subunit of G proteins, and a guanine nucleotide exchange protein called **Son of Sevenless (SoS)**. In its inactive state, ras binds guanosine 5′-diphosphate (GDP). SoS causes GDP to dissociate from ras so that GTP can spontaneously bind. In its GTP bound form, ras is activated and is released from the trk receptor to switch on the mitogen activated protein (MAP) kinase cascade, a set of serine/threonine protein kinases that are phosphorylated in succession. The end result is activation of transcription factors and expression of genes responsible for differentiation and growth.

Actions of neurotrophins

Different classes of sensory neurons in the DRG require different neurotrophins for their survival. This has been demonstrated in neurotrophin knockout mice which die at birth, but primary afferents differentiate and make connections before birth. Mutant mice lacking either the gene for NGF or trkA lose the small diameter primary afferents responsible for transmitting nociceptor and thermoreceptor information. Knockout mice lacking NT-3 or trkC genes lose large DRG cells that innervate muscle spindles and Golgi tendon organs. Mechanoreceptor afferents are unaffected by these gene deletions.

The neurotrophin requirements of neurons changes during development. Very early on, sensory neurons in the DRG and trigeminal ganglia are independent of neurotrophins, but subsequently require both BDNF and NT-3 for survival. On arrival at the target, sensory epithelium trigeminal afferents lose their dependence on BDNF and NT-3 and instead require NGF, which is secreted by the epithelium. The BDNF and NT-3 needed initially is secreted locally, either by the mesenchyme through which the axons grow or by the neurons themselves. The situation in which a cell produces a factor which stimulates its own growth is **autocrine** secretion. Many neurons express both the genes for BDNF and NT-3 and their receptors at the same time.

Genes for neurotrophins are expressed in the brain. Cholinergic neurons of the basal forebrain and in the striatum require NGF and trkA receptors are quite specifically restricted to these regions. The highest level of neurotrophin expression is in the hippocampus. Neurotransmission regulates neurotrophin expression; excitatory glutamate input to hippocampal neurons increases BDNF and NGF expression *in vivo*, while GABAergic input to the same cells decreases this expression. The link between neural excitation and activation of neurotrophic genes is Ca^{2+} entry. In long-term potentiation (LTP), a process thought to underlie some types of learning and which is triggered by calcium entry (see Topic Q4), there is upregulation of neurotrophin expression. Hence neurotrophins probably contribute to the morphological changes that occur at synapses in learning and memory.

Apoptosis

About half of all neurons that are born undergo programmed cell death, **apoptosis**, during development. They do so because, in competing with other neurons to innervate their targets, they fail to establish connections in the appropriate time and, insufficiently stimulated by neurotrophic factors, they die. Glial cells also experience apoptosis, though to what extent is not clear. It has been argued that apoptosis is the default mode for cells, and only the presence of appropriate trophic factors rescues them from apoptosis. Why nervous system development should proceed by overproliferation of cells followed by apoptosis, which seems uneconomic, is not clear. It may be to allow for optimal modeling of neural connections.

Apoptosis is very different from necrosis, the cell death that follows acute injury. In necrosis, cells swell, the cytoplasm vacuolates, subcellular organelles break down and the plasma membrane ruptures to release the cell contents. This activates immune cells, microglia and macrophages, triggering an inflammatory response. By contrast, in apoptosis (*Fig. 2*) cells initially shrink, the nucleus becomes **pycnotic** (that is, chromatin condenses onto the nuclear membrane and the nucleus fragments) and the cell blebs, breaking into membrane bound apoptotic bodies. Subcellular organelles (other then the nucleus) remain intact and because the cell contents are not released during apoptosis there is no inflammatory reaction. Local glial cells (e.g. Schwann cells) that are not normally phagocytotic, engulf the apoptotic bodies.

Numerous genes, many of which are involved in the regulation of the cell cycles, are implicated in the control of apoptosis. A simplified model of the process is illustrated in *Fig. 3*.

The sequence of events thought to be crucial to apoptosis is as follows: in the

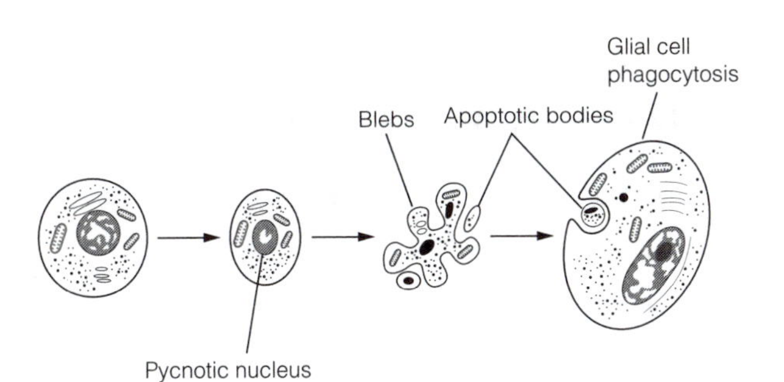

Fig. 2. Ultrastructural appearance of apoptosis. The nucleus becomes pycnotic, the cell blebs, fragments and is phagocytosed.

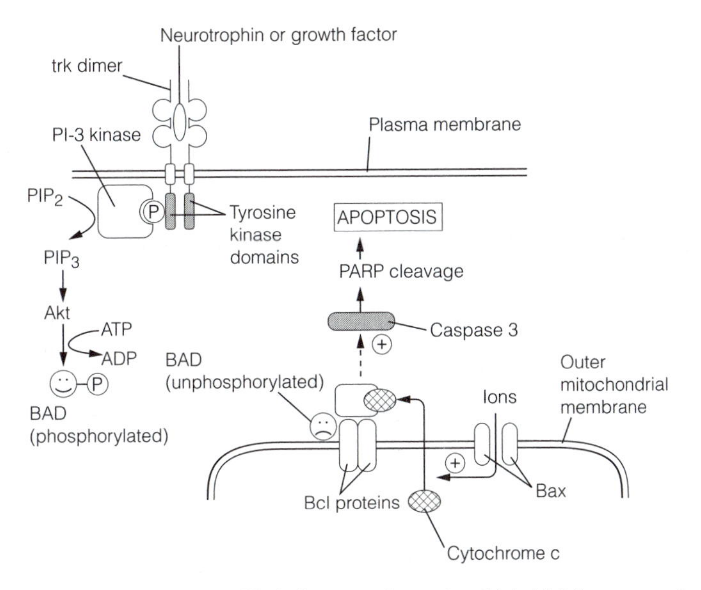

Fig. 3. A highly simplified diagram of events which inhibit or promote apoptosis. In the presence of neurotrophic factor, a soluble cytoplasmic protein, BAD, is phosphorylated and apoptosis inhibited. In the absence of neurotrophic factors BAD is unphosphorylated and initiates apoptosis (see the text for details).

absence of trophic factor, the cytoplasmic protein **BAD** is not phosphorylated and hence binds the anti-apoptotic **bcl** proteins in the outer mitochondrial membrane. This prevents the bcl proteins from interacting with **bax** proteins, which consequently form homomeric ion channels. The influx of ions through the bax channels translocates cytochrome c from the mitochondrion to the cytosol, where it activates cysteine proteases called **caspases**. One of these, caspase-3 cleaves **poly [ADP-ribose] polymerase (PARP)**, a key enzyme in DNA repair. Brains of caspase-3 knockout mice have many more cells (neurons and glia) than normal mice, and far fewer cells show apoptotic changes.

P7 Brain sexual differentiation

Key Notes

Sexual dimorphism

Differences in brain structure and physiology and in behavior or cognition between the sexes is sexual dimorphism. The rat hypothalamus is sexually dimorphic. The preoptic area (POA) of males is concerned with copulation and tonic output of gonadotrophin, but in females it is responsible for cyclical output of gonadotrophins that control the estrus cycle. The ventromedial nucleus in females organizes lordosis, a receptive sexual behavior that requires estrogens and progesterone for its expression. Other brain regions (e.g., amygdala, hippocampus and orbitofrontal cortex) show sexual dimorphism in the rat and might explain difference in cognitive skills between the sexes.

The rodent model

The rat brain is sexually differentiated by differences in hormone exposure of the two sexes during a critical period around the time of birth. High testosterone output by the male testis between embryonic day 15 and postnatal day 10 is responsible for the anatomical, physiological and behavioral masculinization of the male brain. Testosterone is converted to estradiol by neuronal aromatase and the estradiol acts on receptors in the hypothalamus, other limbic structures and the orbitofrontal cortex.

Human brain sexual differentiation

There is no clear evidence that human brain sexual differentiation results from early exposure to hormones. Naturally occurring mutations that expose the prenatal human brain to high concentrations of sex steroids have little effect on psychosexual development. Human male fetal testis secretes testosterone between 12 and 18 weeks of gestation but there are no estrogen or androgen receptors at this time. A second period of testosterone secretion occurs during the perinatal period when sexual differentiation might occur via androgen receptors.

Related topics

Neuroendocrine control of reproduction (M4)

Neurotrophic factors (P6)

Sexual dimorphism

Male and female brains differ in terms of structure and reproductive physiology and this is reflected in the distinctive reproductive behaviors and cognitive skills of the two sexes. This is sexual dimorphism. It has been studied particularly in the rat hypothalamus where it arises as a result of exposure to hormones during a critical period. A nucleus in the medial preoptic area (MPOA), called the **sexually dimorphic nucleus of the preoptic area (SDN–POA)** is larger in males than in females. This difference is established in the perinatal period (i.e around the time of birth), when male rats have higher concentrations of testosterone than females, and once established does not depend on the continued presence of gonadal hormones. The MPOA in male rats is involved in copulation and maintaining tonic output of reproductive hormones whereas in females it is concerned with regulating estrus cycles. In the rat, cells in the POA produce the gonadotrophin releasing hormone (GnRH) responsible for stimulating the secretion of luteinizing hormone and follicle stimulating hormone from the anterior pituitary (see Topic M4). In females, gonadotrophin release can be enhanced by the high concentrations of estrogens produced by mature follicles. This is mediated by neurons in the POA which express estrogen receptors and trigger GnRH secretion. In males, estrogens do not produce increased gonadotrophin release. In male rats and rhesus monkeys, MPOA lesions virtually abolish copulatory behavior, and firing of MPOA neurons correlates with copulation.

The ventromedial hypothalamus (VMH) is concerned with lordosis in female rats. **Lordosis** is receptive sexual behavior in which the rat raises her hindquarters and moves her tail out of the way to facilitate copulation. Ablation of the VMH abolishes lordosis, and recordings from behaving female animals show that VMH neuron activity correlates with lordosis. By contrast, in male rats VMH neurons are inhibited during copulation by the MPOA. Lordosis in female rats requires estrogen exposure for 24 h, followed by progesterone acting for 1 h. Estrogens cause the upregulation of progesterone receptors in the VMH in female rats but not in male rats.

Hypothalamic sexual dimorphism in rats is summarized in *Fig. 1*.

Other brain regions in the rat – amygdala, dorsal hippocampus and orbitofrontal cortex – are sexually dimorphic and might account for cognitive differences between the sexes. For example, while male rats are better at maze learning than females, the reverse is true for avoidance learning.

Brain sexual dimorphism occurs in primates, including humans. Reports that nuclei in the hypothalamus assumed to be equivalent to the SDN-POA of the rat are larger in human males than females have not been confirmed in other studies. The anterior commissure and another structure connecting the two hemispheres are larger in women than in men.

In rhesus monkeys, the orbitoprefrontal cortex is involved in certain spatial

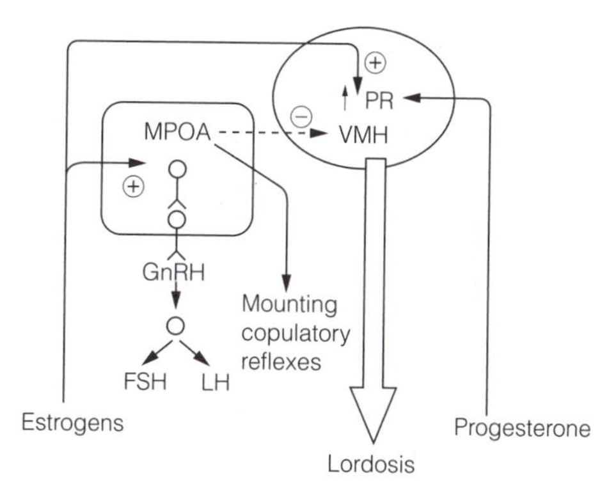

Fig. 1. A model for hypothalamic involvement in reproductive functions in the female rat. Typical female behavior, lordosis, requires progesterone and estrogens. The small medial preoptic area (MPOA) exerts only slight inhibition on the ventromedial hypothalamus (VMH). In the males the much larger MPOA organizes typical male sexual behavior and strongly inhibits the VMH.

discrimination tasks in which adult males outperform adult females. Lesioning the orbitofrontal cortex at different times shows that its ability to mediate these tasks arises earlier in male than female monkeys. In humans, lateralization of cognitive functions is seen in which in the great majority of people, the left hemisphere is specialized for language tasks, while the right hemisphere is specialized for nonverbal, visuospatial tasks. This functional asymmetry is weaker in females than in males. Positron emission tomographic scans during verbal language tasks show that women have some activity in the right as well as in the left hemisphere. In men there appears to be little involvement of the right hemisphere. Less hemispheric specialization of cognitive function in girls may allow their brains to retain greater plasticity for longer than boys, which confers advantages. Recovery of language function following left hemisphere damage in childhood is better in girls than boys, presumably because of the greater plasticity of the right hemisphere in the girls. Developmental dyslexia, aphasia and autism, conditions in which language deficits are predominant, are associated with left hemisphere dysfunction and are much more common in males.

The rodent model

Sexual differentiation of the rat brain is due to differences in hormone exposure during a critical period, the perinatal period. The testis secretes high concentrations of testosterone from embryonic day 15 (E15) to postnatal day 10 (P10) which is responsible for masculinizing the male brain; the gestation period in rats is 21 days.

This is confirmed in studies which exposed rats in the first 4 postnatal days (P1–P4) to inappropriate hormone environments. Female rats injected with testosterone during P1–P4, when adult, had anovulatory sterility, showed no enhanced gonadotrophin release in response to estrogens and had much reduced female typical sexual behavior (lordosis), together with male typical sexual behavior. These animals had large MPOAs. Castration of males on day P1 produced adult animals that would show enhanced gonadotrophin output when given estrogens. When given estrogens and progesterone they exhibited lordosis. Neonatal castration resulted in small (i.e. female-like) MPOAs. Hormone manipulation of rats beyond P10 had no effect on their brain sexual development.

Studies of hypothalamic cells in culture show that testosterone promotes neurite outgrowth, but it does so by first being converted by the enzyme **aromatase** to estradiol, which acts on estrogen receptors (*Fig. 2*). High levels of neuronal aromatase are expressed perinatally in the same cells that express estrogen receptors, that is the hypothalamus, amygdala and other limbic structures and orbitofrontal cortex.

Hence, masculinization of the male rat brain is brought about by aromatization of the testosterone (secreted between E15 and P10) to estradiol, which then

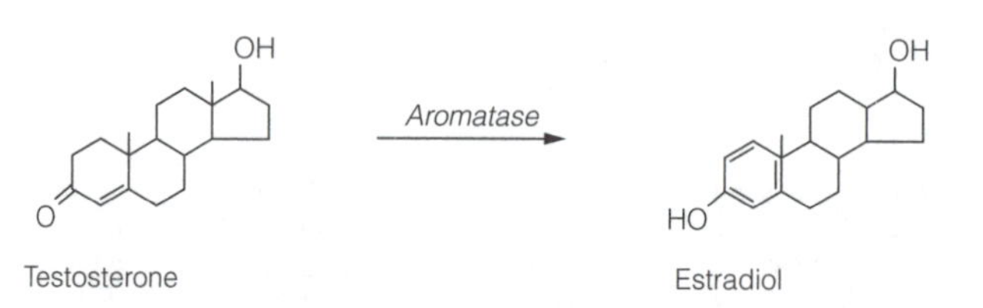

Fig. 2. Aromatization of testosterone to estradiol.

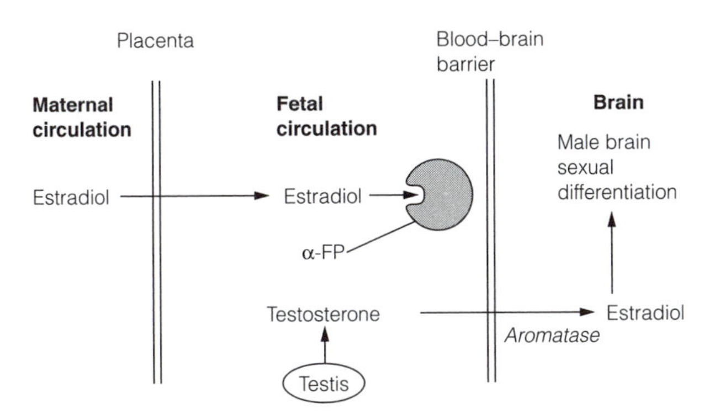

Fig. 3. A model for brain sexual differentiation in rats.

acts on estrogen receptors to promote neuron growth. Maternal blood concentrations of estradiol are high in late pregnancy. However, female fetal rats are not masculinized because the fetal circulation contains α-fetoprotein which binds estrogens so that they do not cross the blood–brain barrier (*Fig. 3*).

Human brain sexual differentiation

Although the human brain shows some sexual dimorphism, unlike the case with rats, there is no good evidence that this is due to early differences in hormone exposure. Girls exposed to very high androgen concentrations as a result of an inborn error of metabolism, **congenital adrenal hyperplasia**, do not generally show shifts towards male sexual behavior, although there is a slightly higher incidence of homosexuality in this group. **Androgen insensitivity syndrome** is due to loss of function mutations in the gene encoding androgen receptors. Affected genetic (XY) males have a short vagina and female secondary sexual characteristics because their testes secrete testosterone which is aromatized to estradiol. These individuals look like normal women and they behave as females, forming sexual relationships with men, despite their brain having been exposed to high concentrations of estrogens throughout their development. This implies that the rodent model is not applicable in humans.

In human male fetuses, the testis secretes testosterone between 12–18 weeks of gestation. Aromatase is present in the hypothalamus during this time. However, no estrogen or androgen receptors are found in the human brain between 12 and 24 weeks gestation. At two other times during development, in the perinatal period between 34 and 41 weeks of gestation, and at puberty, human males have much greater concentrations of testosterone. However, since human brain sexual dimorphism is apparent from about 2 years of age, if it is organized by hormone exposure sexual differentiation must occur during the perinatal period. Furthermore, the androgen insensitivity syndrome implies that if hormones are responsible it must be androgens acting at androgen receptors, not estrogens.

Q1 TYPES OF LEARNING

Key Notes

Definition of learning

Learning is the acquisition of altered behavior as a result of experience and occurs by rewiring of neural pathways (plasticity). The changes are stored (memory) as a trace, or engram. Previously learned behaviors are recalled by appropriate stimuli.

Declarative and procedural memory

There are two broad categories of memory, declarative and procedural. Declarative memory is memory for facts, is fast and consciously recalled. It can be for a set of facts that are associated together because they all relate to a single event (episodic memory), or for facts in isolation (semantic memory) that are stored in a manner that reflects similarities between them. Procedural memory is memory for motor skills. It is slow and not recalled consciously. Many learning situations include elements of both categories.

Short-term and long-term memory

Declarative memory has at least two temporal phases. Short-term memory (STM) is brief, of limited capacity and requires continual rehearsal of items to retain them in STM. STM (also referred to as working memory) has two independent subsystems. A phonological loop holds verbal information and requires the left cerebral hemisphere. A visuospatial sketch pad holds information about spatial relations and requires the right hemisphere. Long-term memory (LTM) is long lasting and apparently of unlimited capacity. Information may enter STM and LTM sequentially or in parallel. Consolidation allows information to be retained in LTM. Amnesias (loss of memory) following brain trauma usually afflict LTM and leave STM untouched. Loss of memory for events before the trauma is retrograde amnesia, the inability to form memories subsequent to the trauma is anterograde amnesia.

Nonassociative and associative learning

Procedural memory is either nonassociative or associative. Only a single type of stimulus is needed for nonassociative learning. In habituation, repetitive delivery of a weak stimulus causes the loss of a motor response. Sensitization is the enhancement of a response to innocuous stimuli seen after an unpleasant stimulus. Associative learning requires pairing of two events within a short time. In classical conditioning, animals learn an association between one stimulus (the conditioned stimulus) and the appearance of a second that may be rewarding or unpleasant (unconditioned stimulus), The conditioned stimulus must always be presented immediately before the unconditioned stimulus. In operant conditioning, animals learn an association between some action they perform and the arrival of a stimulus which may be either rewarding or aversive.

Related topics

Anatomy and connections of the hypothalamus (M1)
Invertebrate procedural learning (Q2)
Hippocampal learning (Q4)

Definition of learning

Some neural pathways establish connections during development that subsequently remain unaltered. These pathways are often said to be **hard-wired** and the generic term for these processes that bring it about is **specificity**. However, pathways subject to continual rewiring either during development or as a result of experience are referred to as **plastic**, and the rewiring processes are examples of **plasticity**. In general, across the animal kingdom, the less complex the nervous system (crudely speaking, the fewer the number of neurons) the more highly specified it is, and the less capable it is of making adaptive (plastic) changes in response to the environment.

The acquisition of reproducible alterations in behavior as a result of particular experiences is **learning**, which is a type of plasticity. **Memory** is the storage of the altered behavior over time. The biological substrate of memory is the **engram**, or memory trace. In animals, learning and memory can only be tested operationally by **recall**, in which the previously learned behavior is elicited by the appropriate stimuli.

Declarative and procedural memory

Learning occurs in a variety of distinct situations, differing in time course, stimulus requirements and outcomes, so is classified in several ways to reflect these features (*Fig. 1*). A major distinction is between declarative and procedural memory. Some authors argue for a third emotional category of learning.

Declarative (explicit) memory is memory for facts, for example Russian history. Declarative learning is fast, it requires few trials, requires conscious recall and may be readily forgotten. It has two components that are dissociable in patients with cortical damage. **Episodic memory** is memory for specific events, such as taking a holiday on Bali, in which associations are established at a specific time and place. The ability of rats to navigate through a maze in which they must learn to associate their positions in the maze with cues in their surroundings, **spatial navigation learning**, is a special case of episodic learning. In humans, positron emission tomographic (PET) scans reveal the involvement of the hippocampus, medial temporal lobe and prefrontal areas in episodic learning. The second component is **semantic memory** which is memory of facts unrelated to events; that Bali is part of Indonesia can be recalled without ever

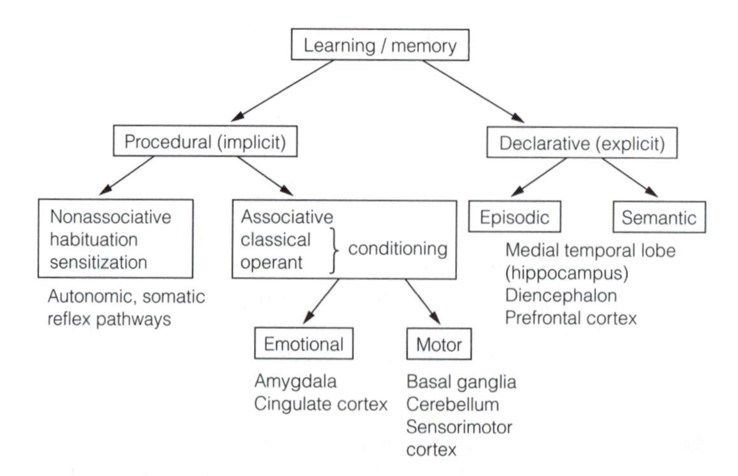

Fig. 1. Types of memory. Structures responsible for a particular category are shown in square brackets. Episodic learning is also associative.

having been there, so semantic memory is about 'knowing that'. PET scans of patients with defects of semantic memory show reduced metabolic activity in the anterior temporal lobes, particularly in the left hemisphere.

Studies of brain damaged patients show that semantic memories are sorted into a number of categories (sets of related objects), the engrams for which are located in different areas of brain. Interestingly, categories of living things (animals, plants, foods etc.) are quite separate from categories of nonliving things (stars, rocks, tools etc.) and recall of specific items seems to need activation of multiple brain sites, each of which codes for a given attribute (e.g. color, function, name) of the item. The principles by which knowledge is represented in the brain are poorly understood at present. Patients with deficits in single categories (e.g. the inability to name any fruit, but who could still name vegetables) have been reported.

Procedural (implicit, motor) memory is memory for skills, such as learning to walk, swim, ride a bike or play a musical instrument. It is 'knowing how' memory. Procedural memory is slow, it needs many trials – in other words a lot of rehearsal – and it is incremental in that improvement occurs gradually over time. Performance of procedural tasks does not involve conscious recall, and once established they are not forgotten even over many years without rehearsal.

Many tasks have both factual and skill memory components. Playing the flute requires declarative memory for the musical notation of the score and procedural memory for the sequence of finger movements and the breathing pattern needed to create the sounds.

Short-term and long-term memory

Declarative memory has at least two (and probably more) phases categorized by their time course. **Short-term (recent) memory (STM)** is temporary, limited in capacity, requires continuous rehearsal to keep it and is easily disrupted by conflicting input. It is continually decaying, as the oldest unrehearsed items are lost, and updated by addition of new input. **Long-term (remote) memory (LTM)** is, if not permanent, at least long lasting, has a capacity which is so great that there appears to be no obvious limit in a human lifetime, and does not require continual rehearsal. These two phases can be distinguished physiologically in that STM is disrupted by anesthesia or temporary cooling of the brain. LTM is unaffected by these procedures. In addition, **amnesias** (loss of memory) due to brain damage, show that STM and LTM are separable. Usually amnesia affects LTM but leaves STM untouched. A few patients, however, have intact LTM but impaired STM, showing that information can enter LTM even if STM is compromised. Amnesias of LTM are of two types, depending on whether memories are lost for events and facts acquired before, **retrograde amnesia**, or after, **anterograde amnesia**, the brain damage. Brain trauma often results in elements of both.

STM is often tested in humans by the ability to recall random strings of digits after a single presentation. If the subject successfully remembers five digits the next trial has six digits; each successive trial has one extra digit. If the subject makes an error, the previous sequence is repeated until the subject gets it right. The number of trials to success is plotted against the number of digits. Normal subjects show **primacy**, greater recall of material at the beginning of the list, because they are rehearsed most often, and **recency**, better recall for items at the end of the list, because STM for these has decayed the least.

Humans with cortical lesions show that there are at least two independent subsystems for STM. The **phonological loop (verbal sketch pad)** allows speech

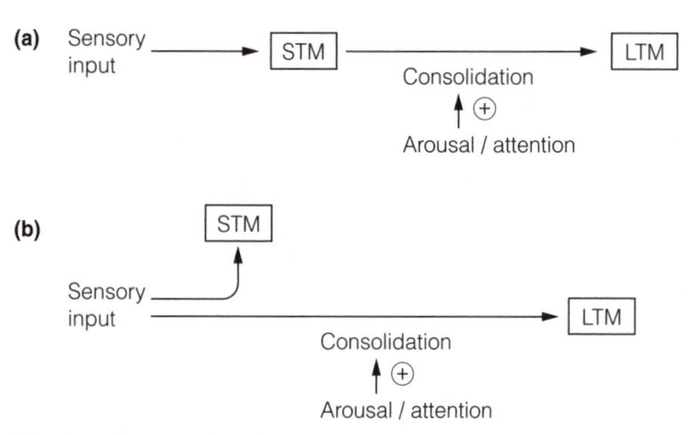

Fig. 2. Possible functional relationships between short-term memory (STM) and long-term memory (LTM): (a) serial model; (b) parallel model.

sounds to be held for long enough to give continuity to spoken language, so that phrases and sentences can be comprehended. It requires the left cerebral hemisphere. The **visuospatial sketch pad** is a temporary store for visual and spatial input that PET scans indicate involves several regions in the right hemisphere.

The exact relationship between STM and LTM is not absolutely clear. One model is that, although most perceptual input is lost as STM decays, some elements are selected by attention and arousal mechanisms for transfer to LTM. This is called **consolidation** (*Fig. 2*). A second possibility is that perceptual input goes in parallel to STM and LTM, with some elements selected to remain in LTM.

For complicated tasks (e.g. driving a car) information from several sources, ongoing sensory input, items in STM and material recalled from LTM, must be available for processing. This temporary store is **working memory** and is distributed in multiple sites in the brain. For example, when driving it is necessary simultaneously to hold in working memory the position and speed of other traffic (sensory input), the current state of road signals (STM) and the intended route (LTM), all of which are continually updated throughout the journey. Working memory is often taken to be synonymous with STM.

Nonassociative and associative learning

Procedural memory is learning to produce a motor response to a particular input. It is divided into two types: nonassociative and associative. **Nonassociative learning** occurs in response to only a single kind of stimulus. Two examples are **habituation**, in which repeated exposure to a weak stimulus results in a reduction or a loss of the response normally seen with occasional presentation of the stimulus, and **sensitization**, which is an exaggerated response to innocuous stimuli following a strong noxious (unpleasant) stimulus. The biochemistry underlying these in a marine **mollusc**, *Aplysia*, is well worked out (see Topic Q2).

Associative learning needs the pairing of two different types of stimulus within a short time and in the correct order. It enables animals to behave as if they can predict relationships of the kind, if A then B. **Classical conditioning** was first investigated in dogs that learned to associate a sound with a subsequent food reward. Hungry dogs salivate at the sight or smell of food. The food is the **unconditioned stimulus (US)** and the salivation an **unconditioned response**, so-called because the nervous system is hard-wired in such a way that salivation occurs as an autonomic reflex response to food; this is an example of specificity. If, in a series of training trials, a sound, the **conditioned stimulus (CS)**, is presented shortly before the arrival of the food, then in a subsequent test, presentation of the sound alone will elicit salivation. The salivation response is now a **conditioned response** since the animals have learnt to salivate when the CS (sound) is presented. Classical conditioning is characterized by **temporal contiguity** which refers to the requirement that the CS must be presented before the US, while **contingency** describes the fact that animals learn that a predictive relationship exists between the CS and the US. **Extinction** of the conditioned response occurs if the CS is repeatedly presented without the US or if the temporal pairing of the CS and US is disrupted (i.e. if either are presented randomly). Classical conditioning can allow perceptual capabilities to be explored. For example, if the CS is a red light, a response will be conditioned only if the animal can distinguish red light from light of other wavelengths. Classical conditioning in which the US is noxious and which results in fear responses to normally neutral stimuli is **aversive conditioning**.

In **operant (instrumental) conditioning** an animal learns an association between a motor activity it performs (e.g. pressing a lever) and the arrival of a stimulus, termed the **reinforcer** (e.g. a food pellet). Reinforcers may be positive, in which case they increase the probability that an animal will act to obtain it, or negative (an aversive stimulus, such as an electric foot shock) in which case the animal will work to avoid it. Operant conditioning is used to investigate motivated behaviors (see Topic M1).

Q2 INVERTEBRATE PROCEDURAL LEARNING

Key Notes

Aplysia

The marine mollusc, *Aplysia*, has a relatively simple nervous system and is used extensively to study the biochemistry of procedural learning. Tactile siphon stimulation stimulates the gill withdrawal reflex via a circuit in which a siphon sensory neuron excites motor neurons to the gill.

Habituation

Stimuli that repetitively activate synapses between sensory and motor neurons cause a reduction in neurotransmitter release which reduces the intensity of the withdrawal reflex (habituation) lasting minutes. Long-term habituation that lasts for several weeks is accompanied by protein synthesis and pruning of synapses.

Sensitization

Withdrawal reflexes are exaggerated in intensity (sensitized) by prior delivery of an unpleasant stimulus. In the short term, this is because the unpleasant stimulus causes release of serotonin onto sensory nerve terminals, causing them to release more transmitter whenever they are subsequently excited. Long-term sensitization involves transcription and translation, which brings about the formation of new synapses.

Associative learning

Pairing a mild tactile stimulus with an unpleasant stimulus produces classical conditioning in which the mild stimulus subsequently evokes a very powerful withdrawal reflex. It is mediated by an amplification of the biochemical events responsible for sensitization (set up by the unpleasant stimulus) by the mild stimulus.

Related topics

Slow neurotransmission (C3)
Voltage-dependent calcium channels (C6)
Hippocampal learning (Q4)

Aplysia

The marine snail, ***Aplysia*** (*Fig. 1a*), has served as a model invertebrate for studying the biochemistry that underpins procedural learning. It has the advantage of a relatively simple nervous system with 2×10^4 central neurons arranged in 10 ganglia. These have large neurons that are readily identified functionally and chemically since they appear consistently in all specimens. *Fig. 1b* summarizes the minimum circuitry needed to understand the procedural learning referred to below. Tactile stimulation (a squirt of seawater) of the siphon causes a brisk withdrawal response (a defensive reflex) of the siphon and gill into the mantle cavity.

Habituation

Ten repetitions of the tactile stimulus causes the withdrawal reflex to be progressively attenuated and this **habituation** lasts for several minutes.

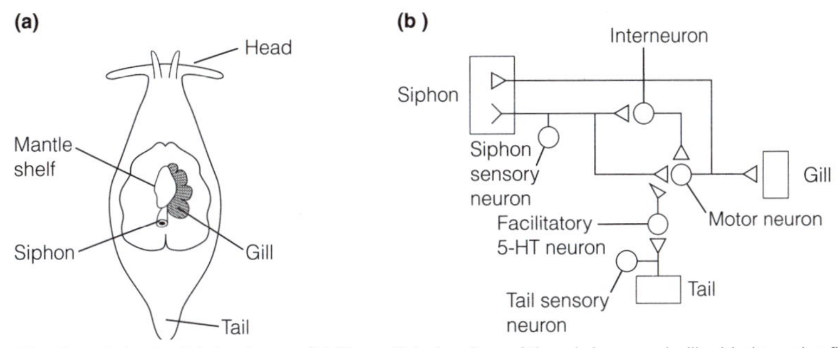

Fig. 1. Aplysia. *(a) Anatomy. (b) Essential circuitry of the siphon and gill withdrawal reflex used for procedural learning studies.*

Habituation reduces the release of neurotransmitter from the siphon sensory neuron. The repetitive stimulation brings about an initial increase in calcium entry into the sensory nerve terminals and produces a long-lasting inactivation of their N-type calcium channels, which consequently reduces Ca^{2+} influx. This habituation is an example of **homosynaptic depression** because repeated activation of a synapse decreases the strength of the same synapse. Long-term habituation, lasting for 3 weeks, is seen if the number of stimulus repetitions is increased. It is triggered by the same events initially but subsequently requires protein synthesis that brings about loss of synapses.

Sensitization

A single train of electric shocks (a noxious stimulus) applied to the tail sensitizes the mollusc to subsequent tactile (innocuous) stimulation of the siphon. The tactile stimulation now causes an exaggerated withdrawal reflex; the heightened responsiveness lasts for several minutes. However, giving several shock trains over 1.5 h causes long-term sensitization which can persist for several weeks. The initial biochemical events are the same for short- and long-term sensitization (*Fig. 2*). Noxious stimulation of the tail activates serotonergic facilitatory neurons which make axoaxonal synapses with the terminals of the sensory neuron. Serotonin acts on two populations of metabotropic receptor on the sensory terminal. One of these receptors is coupled to the cyclic adenosine monophosphate (cAMP) second messenger system and the activated protein kinase A (PKA) phosphorylates K^+ channels, decreasing a (serotonin-sensitive) component of the K^+ current ($I_{K,S}$). This broadens the action potential allowing greater Ca^{2+} entry into the terminal. In addition the PKA mobilizes additional synaptic vesicles from the releasable pool. The second serotonin receptor is linked to the second messenger diacylglycerol that activates protein kinase C with the result that the Ca^{2+} entry is prolonged. The overall effect of these actions is to increase the neurotransmitter release from the sensory terminal. Because sensitization involves strengthening of the synapse between the sensory and motor neuron by activity in a separate, modulatory synapse it is called **heterosynaptic facilitation**.

If sensitization is done after administering drugs which inhibit the synthesis of messenger RNA or proteins it does not persist beyond about 3 h. Hence long-term sensitization can be distinguished from short-term by its dependence on transcription and translation. One factor which triggers the switch from short- to long-term sensitization is that PKA can alter gene expression (*Fig. 3*). PKA phosphorylates a nuclear protein, **cAMP response element binding protein**

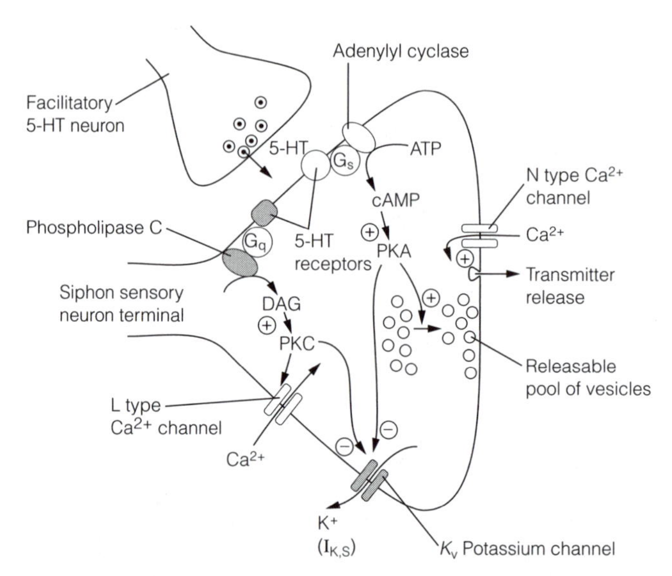

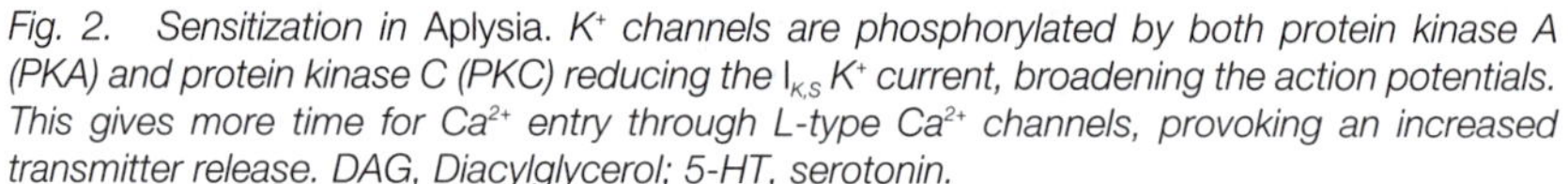
Fig. 2. Sensitization in Aplysia. *K^+ channels are phosphorylated by both protein kinase A (PKA) and protein kinase C (PKC) reducing the $I_{K,S}$ K^+ current, broadening the action potentials. This gives more time for Ca^{2+} entry through L-type Ca^{2+} channels, provoking an increased transmitter release. DAG, Diacylglycerol; 5-HT, serotonin.*

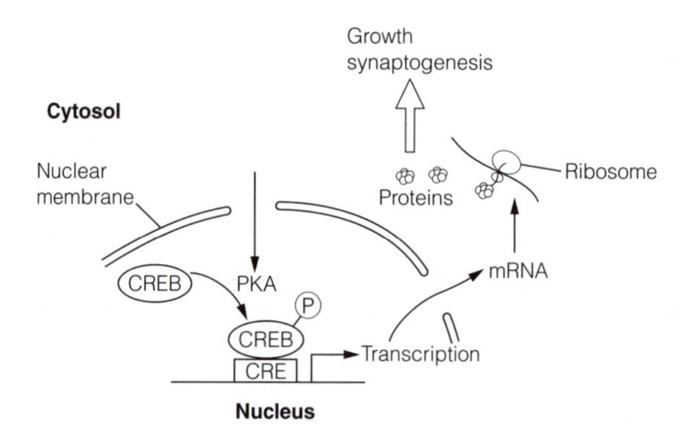

Fig. 3. In long-term sensitization, protein kinase A (PKA) activates transcription and translation producing proteins responsible for structural changes. CRE, cAMP response element; CREB, CRE binding protein.

(**CREB**), which engages with stretches of DNA (**cAMP response elements**, **CRE**) that control transcription of downstream genes. One gene activated in this way codes for a protease that degrades the regulatory subunits of PKA. Unrestrained by the regulatory subunits, the catalytic subunits of PKA are persistently activated and so maintain long-lasting phosphorylation of the K^+ channels. Transcription of other proteins responsible for sprouting and growth of the sensory neuron axon, and the formation of new synapses, are seen in long-term sensitization.

Associative learning

As well as habituation and sensitization, *Aplysia* is capable of the more complicated associative learning. For classical conditioning the unconditioned stimulus

(US) is a single shock to the tail which elicits the unconditioned response of a strong siphon withdrawal. The conditioned stimulus (CS) is a tactile stimulus to the siphon which produces a modest withdrawal response. After a few trials in which the CS is paired with the US (CS preceding by 0.5 s) the CS alone generates powerful siphon withdrawal. The cellular mechanism relies on the tail shock (US) setting up the usual sensitization cascade that results in elevated neurotransmitter release from the sensory terminals (*Fig. 2*). However, the tactile stimulus (CS) causes action potentials to invade the siphon sensory terminal, causing an influx of Ca^{2+}. Now the adenylyl cyclase activated by serotonin is Ca^{2+}-sensitive, so in response to the CS spike train, its activity is further enhanced by the Ca^{2+}. In other words, adenylyl cyclase is a coincidence detector, its activity being amplified by the near simultaneous pairing of serotonin release (US) and the rise in Ca^{2+} (CS). The conditioning is pathway specific since only in the siphon sensory terminal does the enzyme get both signals. The time interval between CS and US must be short, since the rise in terminal Ca^{2+} concentration is rapidly buffered; i.e. the Ca^{2+} signal is very short lived. The biochemical constraints which demand that the CS must precede the US are not known.

Q3 Memory circuitry in mammals

Key Notes

Medial temporal lobe

Damage to the structures in the medial temporal lobe, particularly the hippocampus in humans, monkeys and rats, causes both retrograde and anterograde amnesia. Procedural learning is unaffected. The hippocampus is particularly involved in the consolidation of new episodic memories into long-term memory (LTM). In primates, these memories are very general but in rats the hippocampus seems primarily concerned with place learning, by which animals are able to find their way around.

Diencephalon

Medial temporal lobe structures are connected to nuclei in the hypothalamus and thalamus. Lesions of these diencephalic structures in monkeys or humans (either through trauma or disease) cause severe amnesias.

The amygdala and aversive learning

Classical conditioning, in which a neutral conditioned stimulus is followed by an unpleasant unconditioned stimulus, causes fear responses to be elicited to the neutral stimulus. The amygdala is required for this aversive learning. Stimulation of the amygdala activates stress responses.

The amygdala and memory modulation

The degree of arousal determines the probability that specific memories will be consolidated. Arousal, signaled by release of catecholamines from the sympathetic nervous system, stimulates afferents in the vagus nerve. This activates the brain noradrenergic arousal system which projects to the amygdala and hippocampus to bring about consolidation. Hormones secreted by every level of the hypothalamic–pituitary–adrenal (HPA) axis also have actions on the amygdala and hippocampus. Optimal learning occurs with moderate catecholamine or glucocorticoid concentration. High levels of these hormones are detrimental. The amygdala may enhance consolidation by activating the cholinergic attentional system in the basal forebrain.

Prefrontal cortex

The connections of the prefrontal cortex with temporal lobe and diencephalic structures involved in learning and the effects of lesioning it, both in monkeys and humans, implies that it is concerned with tasks requiring working memory.

Related topics

Neuroendocrine control of metabolism and growth (M3)
Norepinephrine neurotransmission (N2)
Acetylcholine neurotransmission (N4)
Types of learning (Q1)
Hippocampal learning (Q4)

Medial temporal lobe

Much of the evidence implicating brain structures in memory is from studies of humans with brain damage or animals with lesions. Two clinical cases illustrate the involvement of the medial temporal lobe, particularly the hippocampus. The first, HM, had neurosurgery to treat intractable epilepsy. The surgery removed an 8 cm length of the medial temporal lobe including the amygdala, the anterior two-thirds of the hippocampus and the overlying cortex from both hemispheres (*Fig. 1*). Whilst the surgery successfully alleviated the seizures it produced devastating deficits in his declarative memory. While his short-term memory (STM) and very remote memory is normal he has a partial retrograde amnesia for some years before the surgery, and an anterograde amnesia so extreme he

(a)

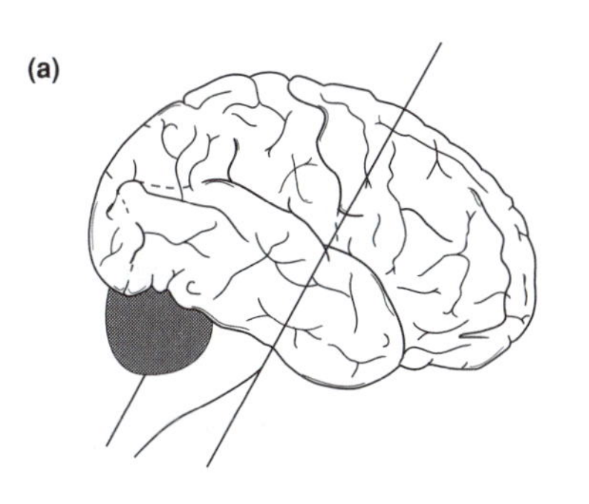

(b)

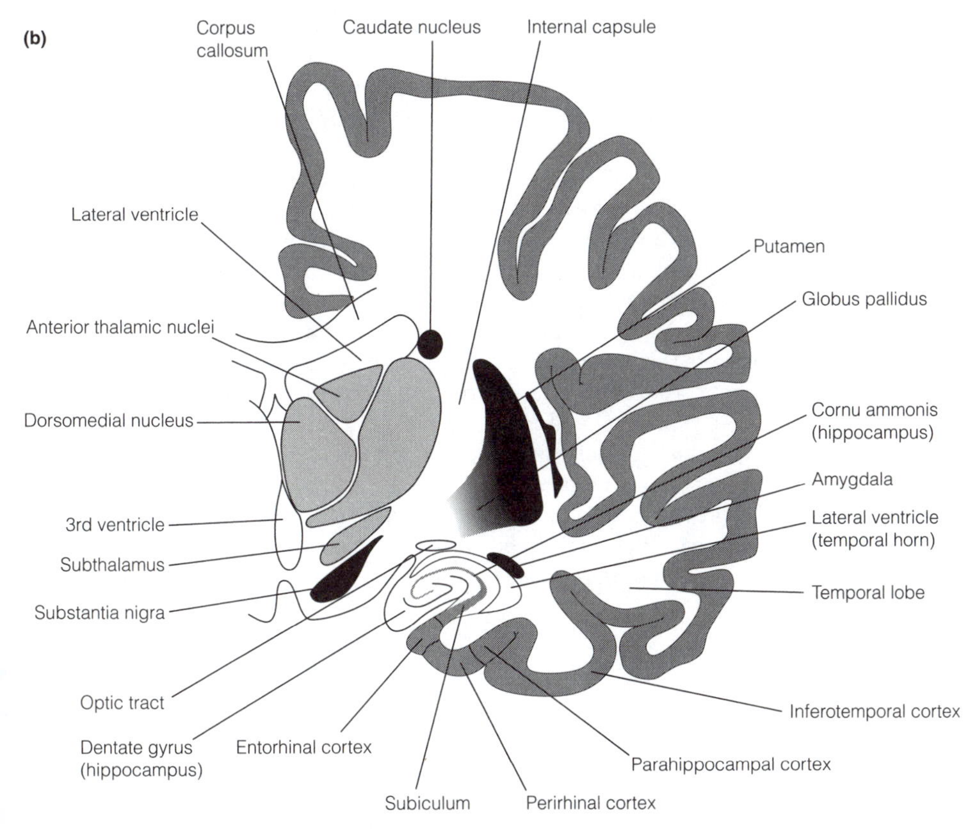

Fig. 1. Section (a) through the human brain to show (b) the gross anatomy of the medial temporal lobe.

cannot consolidate any new LTM; he is unable to retain the memory of events, places or people that he has experienced since his surgery for longer than the time he can hold them in STM. HM can learn new procedural tasks (e.g. tracing a pattern by looking in a mirror, a surprisingly difficult skill) though he cannot remember having acquired these skills. The second case, RB, suffered brain hypoxia following heart surgery, and subsequently had a memory deficit of the same character though not as severe as HM. Autopsy 5 years later revealed only a bilateral, highly selective, loss of pyramidal cells from one region of the hippocampus.

Bilateral medial temporal lobe lesions in macaque monkeys provide an animal model for human amnesias. Typically, the animals are trained on a **delayed nonmatching to sample task**. In this, the monkey is trained to select an object to get a food reward. Following a variable delay during which the animal cannot see any manipulations, it is given a choice between the same object and a novel object and is required to select the novel object to get the reward i.e. it needs to remember which object it saw first. Lesioned animals show an anterograde amnesia and selective lesioning shows that the most severe deficit occurs with damage to the perirhinal and parahippocampal cortex of the temporal lobe.

In rats, pure hippocampal lesions cause proportionally greater learning deficits than in primates. The hippocampus of rats is particularly important for **spatial navigation (place) learning**, by which animals acquire a memory for their location. One widely used way of investigating this is the **Morris water maze**. This is a circular pool (1.3 m in diameter) filled with opaque warm water. Hidden 1 cm below the surface is an 8 cm diameter platform. During learning trials rats swim in the pool, discover the platform and learn its position in the pool relative to cues in the laboratory. The motivation is that the platform provides an escape from the water. Learning takes several trials and can be tested by measuring the time taken for a rat to reach the platform or the length of the path it swims to reach it as recorded on a video camera. Cued control experiments in which the platform is raised just above the level of the water ensure that any differences in behavior are not attributable to any motivational, perceptual or locomotor factors.

Rats with hippocampal, but not selective neocortical, lesions are seriously compromised in the learning, but not the cued versions of this task. Rats given a microinjection of colchicine to destroy a specific population of cells (dentate granule cells) in their hippocampus either 1, 4, 8 or 12 weeks after learning the location of a submerged platform and tested 2 weeks later (*Fig. 2*) reveal that the hippocampus is not a permanent site for the spatial learning engram. The 12 week group remembered the location as well as control (uninjected) animals, but performance got progressively worse for 8, 4, and 1 week groups. This study shows that the hippocampus is needed for consolidation of spatial learning, but

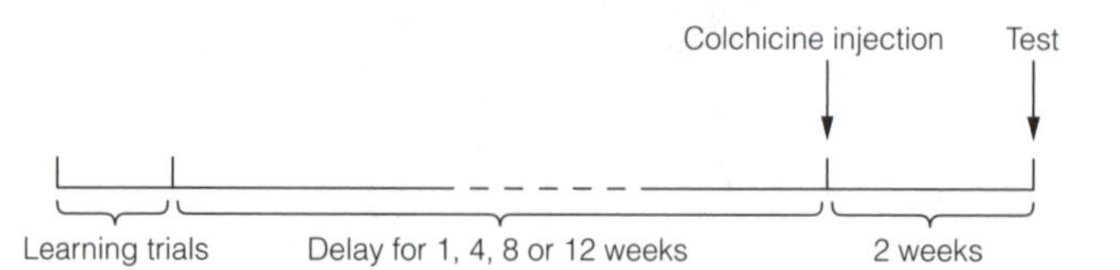

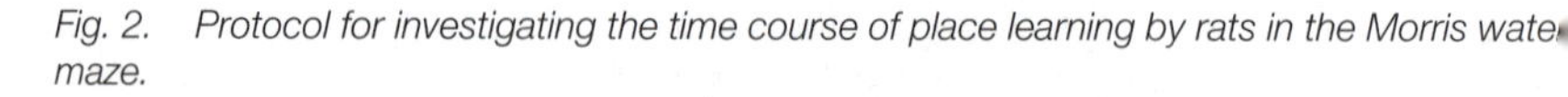
Fig. 2. Protocol for investigating the time course of place learning by rats in the Morris water maze.

that over successive weeks the site of the memory store is transferred elsewhere, probably the neocortex. Retrograde amnesia seen in humans with medial temporal lobe damage might result from the loss of memory not yet transferred out of the hippocampus and associated cortex to distant neocortex.

Although the hippocampus may be predominantly for spatial learning in rats, its role in primates is much broader in that it seems to make associations between multiple stimuli, of all modalities, so that discrete memories of events (specific occurrences in space and time) are formed; that is, in primates the hippocampus consolidates general episodic memories.

Diencephalon

Three diencephalon structures are extensively connected with the temporal lobe and play a role in memory. A major output of the hippocampus is the **fornix** which projects largely to the **mammillary bodies** of the hypothalamus, output from which goes to the **anterior thalamus**. Furthermore areas of the temporal cortex and amygdala make connections with the **dorsomedial nucleus** of the thalamus. Bilateral lesions confined to just one of these diencephalonic structures in monkeys modestly impairs performance in the delayed nonmatching to samples tasks, but larger lesions affecting all three produce very severe deficits. One human case, NA, sustained damage to his left dorsomedial thalamus when stabbed by a sword in a fencing accident. NA suffered amnesia very similar to that of HM, but it was less severe. Moreover, severe anterograde and retrograde amnesia are symptoms in **Korsakoff's syndrome**, in which damage occurs to the dorsomedial nucleus, mammillary bodies and other brain regions due to the thiamin (vitamin B1) deficiency of chronic alcoholism. Hence interconnected medial temporal lobe and diencephalon structures form components of a brain memory system.

The amygdala and aversive learning

Aversive learning or fear conditioning occurs when a neutral conditional stimulus, such as a tone, is paired with a noxious (unconditional) stimulus, such as a brief electric foot shock. After several tone–shock pairings the tone becomes a negative reinforcer and it elicits conditioned (fear) responses, including autonomic, endocrine and behavioral signs of fear. Neurons in the **central nucleus** of the amygdala fire in a way that correlates with the development of the fear responses. Lesions of the amygdala prevent acquisition of new conditioned fear responses or expression of preexisting ones in animals and humans. Electrical stimulation of the amygdala in humans during surgery evokes feelings of fear and anxiety. The functional connectivity of the amygdala (*Fig. 3*) supports its role in aversive learning since it activates the cholinergic attentional system, the sympathetic nervous system and the release of stress hormones.

The amygdala and memory modulation

Neural mechanisms that rely on the amygdala modulate the extent to which specific memories are consolidated. Events that are important trigger greater arousal, which increases the likelihood of consolidation. The arousal signals to which the amygdala responds are the stress hormones secreted by the adrenal glands and several CNS peptide neurotransmitters released in response to stress. Adrenalectomy causes learning deficits. Recall in emotionally neutral learning tasks is enhanced by norepinephrine and epinephrine given within a short time of the learning trials. The dose–response curve has an inverted 'U' shape; moderate concentrations of the catecholamines are more effective enhancers of memory than either high or low levels. As neither of these hormones crosses the blood–brain barrier, their actions on the CNS must be

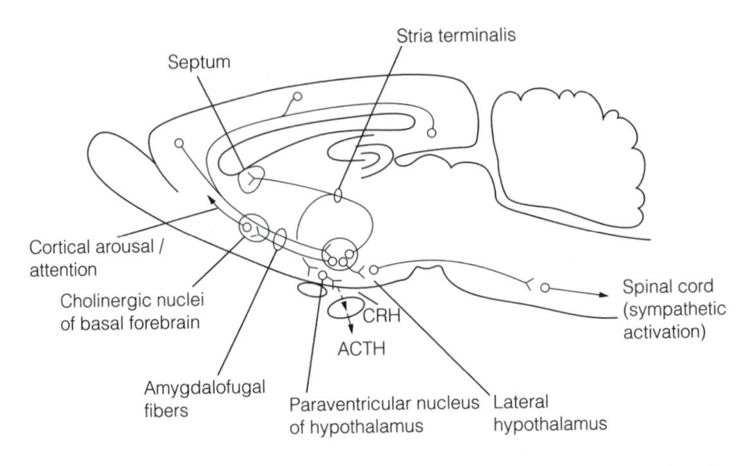

Fig. 3. Connections of the amygdala. CRH, corticotrophin releasing hormone; ACTH, adrenocorticotrophic hormone.

exerted peripherally. The catecholamines act on β adrenoceptors of visceral afferents that run in the vagus (X) nerve to the nucleus of the solitary tract (NTS). This results in activation of noradrenergic neurons of the locus ceruleus (see Topic N2) that constitute a brain arousal system. This system projects to the amygdala and hippocampus to modulate learning. Electrically stimulating the vagus nerve immediately after training improves recall in an inverted U relationship with firing frequency. Cutting the vagus nerves or lesioning the NTS blocks the effects of systematically administered catecholamines on memory.

Glucocorticoids released by activation of the HPA axis in stress (see Topic M3) also have effects on learning and memory. These hormones readily cross the blood–brain barrier to act on steroid receptors that are located in high density in the amygdala and hippocampus.

Low doses of glucocorticoids enhance, while high doses (or the chronic exposure that occurs in stress), impairs memory. Low concentrations occupy the high affinity mineralocorticoid receptors and this facilitates strengthening of synapses thought to be crucial for learning (Topic Q4). In contrast, high glucocorticoid concentrations fully saturate the low affinity glucocorticoid receptors and this blocks the synaptic strengthening necessary for learning.

Pituitary corticotrophs manufacture, from a single precursor, adrenocorticotrophic hormone (ACTH) and the opioid peptide, β endorphin, both of which impair learning by direct action on the CNS. Enkephalins, also opioid peptides, are coreleased from the adrenal medulla along with catecholamines and impair memory by a peripheral action. Naloxone, an antagonist of opioid receptors facilitates memory. Many drugs with actions on GABA, opioid or adrenergic receptors produce effects on learning, via the basolateral nucleus of the amygdala. Cutting the stria terminalis (*Fig.* 3) blocks many of the effects of drugs and stress hormones on memory. Noradrenergic transmission in the amygdala seems central to its role in memory modulation since microinjection of the β adrenoceptor antagonist, propranolol, blocks the effects of many drugs. Cholinergic enhancement of memory is well documented. Muscarinic receptor antagonists impair learning while inhibitors of acetylcholinesterase improve it. Acetylcholine modulation of memory is probably mediated by the septohippocampal pathway and the cholinergic nuclei of the basal forebrain, which are regulated by the amygdala.

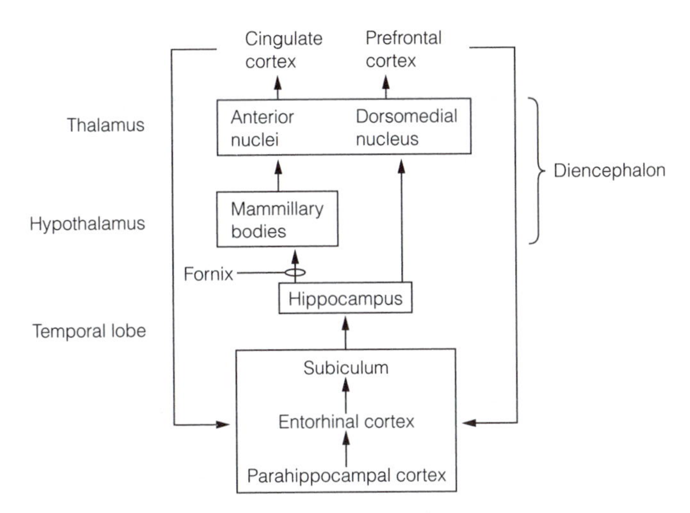

Fig. 4. Mammalian forebrain memory circuitry.

Prefrontal cortex

The **prefrontal cortex (PFC)** is involved in complex problem solving and planing future actions and there are good reasons for supposing that these executive tasks require working memory. The connectivity of the PFC argues for its role in working memory. Firstly, association fibers make reciprocal connections between the PFC and other cortical areas, so the PFC receives visual, auditory and somatosensory information. Secondly, the PFC is interconnected with the medial temporal lobe and dorsomedial thalamus that have a well documented role in learning and memory

In **spatial delayed response tasks**, monkeys see a food reward placed in one of several covered locations. After a delay, which can be varied over trials, the animals are tested to see if they remember the location of the food. Monkeys with lesions of the prefrontal cortex have deficits in these tasks, and performance degrades progressively as the delay is lengthened.

Recording from the PFC in alert behaving monkeys reveals cells that fire in predictable ways during delayed response tasks. For example, many cells fire throughout the delay period, others fire when the food is placed in the location and when the animal is allowed to choose the location. Particular regions of the PFC seem to be modality specific and so responsible for specific types of working memory.

Humans with prefrontal lesions also show deficits on working memory tasks in which they are required to use recent data to make correct decisions. Such individuals have great difficulty in tracing a path through a drawing of a maze. They will make the same errors repeatedly, and start right from the beginning of the maze after making a mistake, rather than from the position in the maze just before they made the error. The proposed circuitry for the brain memory system is summarized in *Fig. 4*.

Q4 Hippocampal learning

Key Notes

Cognitive map hypothesis

It is postulated that the rat hippocampus is the focus of a cognitive map which is continually updated by (episodic) learning as the animal explores its surroundings. The maps allow the animal to navigate its way through the world. Evidence for the cognitive map hypothesis comes from the presence in the hippocampus of place cells which fire only when the rat is in a particular location, which constitutes the place field of a cell. Place fields use a combination of sensory and locomotor cues to encode the spatial relations between objects in the rat's environment.

Hippocampal circuitry

Excitatory input to the hippocampus comes from the entorhinal cortex by way of the perforant pathway, axons of which synapse with granule cells of the dentate gyrus. Granule cell mossy fiber axons synapse with pyramidal cells in CA3. Branches of CA3 cell axons go to three possible destinations, the contralateral hippocampus, the hypothalamus, or to CA1 pyramidal cells via recurrent collaterals (Schaffer collaterals). Axons of CA1 cells go to the entorhinal cortex. All these principal neurons use glutamate and are excitatory. In addition, the hippocampus contains GABAergic inhibitory interneurons. Modulatory neurons using acetylcholine, norepinephrine or serotonin also provide input.

Long-term potentiation (LTP)

Learning is thought to be due to changes in the strength (weight) of synapses. Hebb's rule proposes that a synapse between two neurons is strengthened when the neurons are activated together. Synapses that obey this rule are said to be Hebbian. LTP is a long-term increase in the strength of synapses and may be associative (Hebbian) or nonassociative. LTP has been extensively studied at synapses between CA3 and CA1 cells in the hippocampus where it is Hebbian. LTP is produced by applying a high frequency (tetanic) stimulus to the CA3 axons. Subsequent single stimuli elicit a larger excitatory postsynaptic potential in the CA1 cell than before the tetanic stimulus. LTP can last for many hours in brain slices.

Cellular physiology of associative LTP

There are three phases to LTP, induction, expression and maintenance. Induction depends on the activation of NMDA receptors by the tetanic stimulation, which causes both increased glutamate release from CA3 axons and a large depolarization of the CA1 cell. This represents coactivation of pre- and postsynaptic cells (satisfying Hebb's rule) and fulfills the conditions for opening NMDA receptors. It is the Ca^{2+} entry through these glutamate receptors that is one of the necessary requirements for LTP induction. Expression of LTP involves second messenger mediated increases in the sensitivity of AMPA glutamate receptors, acquisition of AMPA receptors by previously silent synapses or perforation of synapses to give new active sites. In addition, an increase in presynaptic glutamate release may occur, triggered by a retrograde messenger liberated from the postsynaptic cell to affect the presynaptic terminal. Maintenance of LTP beyond 2 h requires transcription and translation to synthesize proteins that allow the synaptic modification to persist in the face of continual turnover of molecules.

Is LTP learning?

The optimal protocol for generating LTP is essentially identical to theta (θ) activity seen in rats learning a spatial task. The θ rhythm is caused by regular firing of hippocampal neurons driven by the cholinergic pathway from the septum. LTP *in vivo* may last for many months. Pharmacological or genetic engineering manipulations which impair LTP often produce deficits in spatial learning.

Related topics

Fast neurotransmission (C2)
Sleep (O4)
Types of learning (Q1)
Invertebrate procedural learning (Q2)
Memory circuitry in mammals (Q3)
Cerebellar motor learning (Q5)

Cognitive map hypothesis

The hippocampus of the rat and associated cortex is thought to provide the rat with a representation of the space around it and its location within it. This is the **cognitive map hypothesis** and it has several postulates. Firstly, the map allows the animal to find its way through the environment. Secondly, it is constructed by episodic (declarative) learning as specific locations come to be associated with particular sensory and motor cues. Thirdly, it does not require reinforcers and finally, the map is continually updated by exploration.

Evidence for the hypothesis is the existence of **place cells**, pyramidal cells in the hippocampus that fire when the rat is in a particular position in the environment. In a typical experiment, rats with electrodes chronically implanted in the hippocampus for extracellular recording, are allowed to explore a plus-shaped maze. The animals learn the spatial relationships between the maze and visual cues in the surrounding laboratory so as to find a food reward located in one of the four arms of the maze. The maze can be rotated with respect to the constellation of cues in the laboratory. The location in the maze which causes the place cell to fire is the cell's **place field** (analogous to the sensory field of sensory neurons, *Fig. 1*). The properties of place fields imply that they encode spatial relations between features of the rat's surroundings.

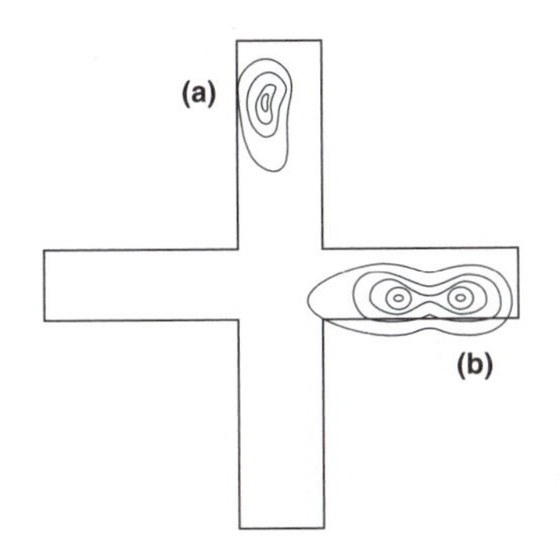

Fig. 1. Place fields of two cells (a, b) recorded as a rat explores a four arm maze. Contour lines reflect increased firing towards the centre of the field.

- An array of place fields represents a particular environment.
- There is no correspondence between the locations of place fields in the world and the positions of the place cells in the brain.
- A given place cell may have several place fields, each in a different context.
- Place fields move in concert with rotation of visual cues in the laboratory, but remain when the lights are turned off, so they are not *explicitly* coupled to sensory input.
- They enlarge or shrink with manipulations in the size or shape of an enclosed space in a way that implies the rat can use locomotor cues (e.g. the number of footsteps) to work out where it is.
- New place fields arise as an animal explores a novel environment.
- Altering familiar surroundings disrupts preexisting place fields.
- In maze tasks in which a rat must know its location in order to get a food reward, if the rat makes an error, the place fields correspond to the incorrect location (i.e. where the rat 'thinks' it is).
- Old rats, which consistently show deficits in spatial learning, have place fields with poorer spatial selectivity and reliability (stability from trial to trial).

Hippocampal circuitry

The hippocampal formation is folded archaecortex (ancient cortex) consisting of the **dentate gyrus** and the **cornu ammonis** (**CA**) – collectively termed the **hippocampus** – plus the subiculum. The cortex of the dentate gyrus and CA have three layers, while the subiculum is transitional cortex between the hippocampus proper and the six layered neocortex of the **entorhinal area**. A major input to the hippocampus from the entorhinal cortex comes via the **perforant pathway**, axons of which synapse with **granule cells** of the dentate gyrus or pyramidal cells in the CA3 region of the CA (*Fig. 2*). Axons of the granule cells (mossy fibers) also synapse with CA3 pyramidal cells.

The CA3 pyramidal cell axons branch, forming:

- **Commissural fibers** which pass to the opposite hippocampus.
- Efferents which leave the hippocampus via the **fornix** to terminate largely in the hypothalamus or thalamus.
- Collaterals which turn back to form synapses on the same and neighboring CA3 cells (**recurrent collaterals**), or which synapse with cells in the CA1 region of the CA (**Schaffer collaterals**).

CA1 cell axons go to the subiculum and entorhinal cortex. The perforant pathway, granule and pyramidal cells are glutamatergic and excitatory. The hippocampus also harbors inhibitory interneurons that are GABAergic. Other inputs to the hippocampus include a cholinergic pathway from the septum and noradrenergic and serotinergic axons from the brainstem reticular system. These inputs are modulatory.

Long-term potentiation (LTP)

A core idea of contemporary neuroscience is that learning occurs by changes in the strength of synapses (i.e. alterations in synaptic weights). A mechanism to account for how this might occur was proposed in 1949 and is called **Hebb's rule**. This states that all synapses between two neurons become stronger if both of the neurons are activated at the same time. Synapses which show this type of plasticity are said to be **Hebbian**, and can mediate associative learning since they act as coincidence detectors that associate firing of the presynaptic and postsynaptic cell. Hebb's rule is summarized by the aphorism 'what fires together, wires together'. Several mechanisms to bring about synaptic modifica-

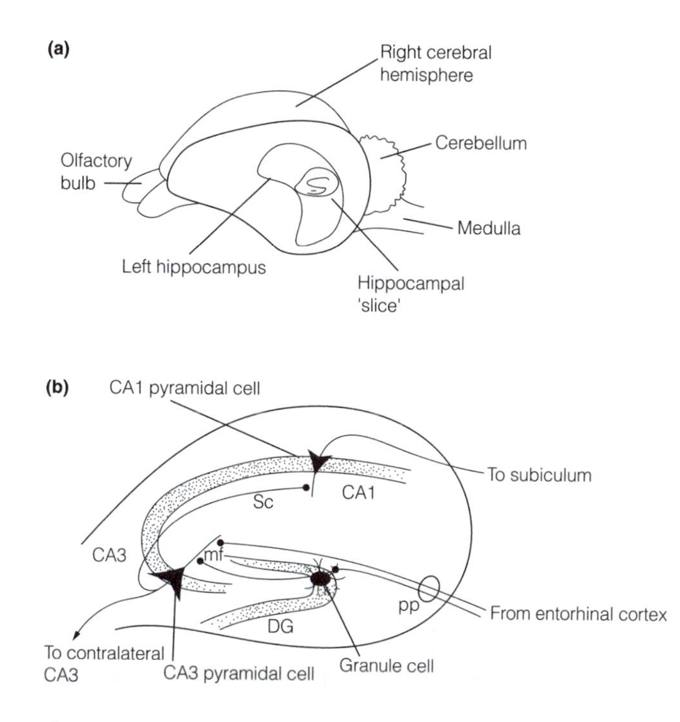

Fig. 2. Hippocampus. (a) Location of the left hippocampus in rat brain; a hippocampal slice is at right angles to the long axis of the hippocampus. (b) Structure of a hippocampal slice showing the principal excitatory neurons. From Revest, P. and Longstaff A. (1998) Molecular Neuroscience. *© BIOS Scientific Publishers, Oxford. DG, dentate gyrus; pp, perforant pathway; Sc, Schaffer collateral; mf, mossy fibers.*

tions are now known which either increase synaptic weighting, **long-term potentiation (LTP)**; or decrease synaptic weighting, **long-term depression (LTD)**. Both occur in the hippocampus, LTP is also seen in the neocortex, amygdala and at other sites in the nervous system, while LTD also occurs in the cerebellum. LTP and LTD are commonly thought of as cellular substrates of learning and memory.

LTP can be either associative (Hebbian) or nonassociative. LTP at the synapses between CA3 Schaffer collaterals (Scs) and CA1 cells in the hippocampus is Hebbian. It can be studied in hippocampal brain slices by intracellular or extracellular recording from CA1 neurons whilst electrically stimulating a bundle of Scs (*Fig. 3*). In response to brief, low frequency stimulation of the Scs, the CA1 cells show a brief excitatory postsynaptic potential (epsp) due to glutamate release. If a brief tetanic burst of high frequency stimulation is given (typically 100 Hz for 0.5 s), subsequent low frequency pulses now elicit a larger epsp. This is LTP, it may last for as long as the brain slice survives (many hours) and it has three properties.

(1) **Input specificity**. Delivery of low frequency stimuli to the CA1 cell via a different untetanized bundle of Scs does not elicit the enhanced epsp.
(2) **Cooperativity**. The probability of producing LTP increases with the number of afferent fibers (Scs) tetanically stimulated. While weak (i.e. low current) high frequency stimuli often fail to generate LTP because they excite only a few afferents, strong tetanic stimuli are successful because they recruit many afferents.

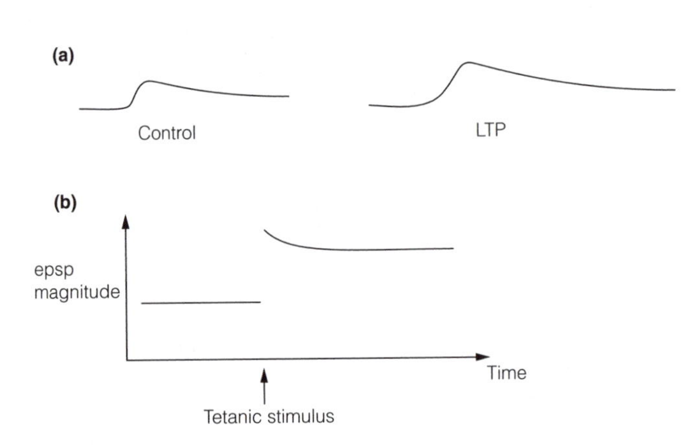

Fig. 3. LTP in a hippocampal slice. (a) Excitatory postsynaptic potentials (epsps) recorded from CA1 pyramidal cells before (control) and after tetanic stimulus (LTP). (b) Epsp magnitude remains elevated over several hours.

(3) **Associativity**. A given CA1 cell gets Scs from CA3 cells on the same side and commissural axons that come from CA3 cells in the contralateral hippocampus. A weak tetanic stimulus to either pathway that fails to generate LTP will do so if it is paired with strong tetanic stimulus in the other pathway. This is a case of **heterosynaptic LTP** since activity in one set of synapses causes potentiation of a second set of synapses on the same cell. The strongly tetanized pathway also shows LTP, in this case however it is **homosynaptic LTP**.

Cellular physiology of associative LTP

CA1 cells have AMPA and NMDA glutamate receptors. Recall (see Topic C4) that in order to be activated, NMDA receptors must bind glutamate and experience a depolarization sufficiently large to remove Mg^{2+} ions from the channel (voltage dependent blockade). This condition is not provided by low frequency Sc stimulation. The amount of glutamate released is low, few AMPA receptors are activated and the resulting epsp is too small to open NMDA receptors. However, high frequency stimulation opens numerous AMPA receptors and so depolarizes the cell sufficiently to activate NMDA receptors. Both cooperativity and associativity work by enhancing the depolarization of CA1 cells enough to open NMDA receptors. In fact, tetanic stimulation is not mandatory for LTP; any manipulation that depolarizes CA1 cells enough, including antagonizing the inhibitory actions of GABA, will ensure LTP to an input. NMDA receptors are permeable to Ca^{2+} as well as Na^+ and K^+ and it is Ca^{2+} entry that triggers the induction of LTP. Antagonists of NMDA receptors, for example the competitive antagonist, D-2-amino-5-phosphonovalerate (AP5, APV) or the open channel blocker, dizocilpine (MK801) prevent the induction of LTP.

There are several points to note regarding the **induction** of LTP.

- It is the properties of the NMDA receptor which makes the LTP Hebbian (associative), since to open its Ca^{2+} channel requires the near simultaneous activation of the presynaptic neurons (glutamate release) and activation of the postsynaptic neurons (depolarization).
- Back propagating calcium spikes generated in the CA1 cell by depolarization invade its dendritic spines adding to the NMDA receptor Ca^{2+} signal in the spine. The narrow spine neck means that Ca^{2+} is trapped within the spine enhancing its effectiveness. (Considerable evidence now exists for NMDA-

independent LTP that requires L-type voltage-dependent Ca^{2+} channels at other synapses.)

- Metabotropic glutamate receptors are also necessary for LTP induction, but only on the first occasion in which the synapse is tetanically stimulated; metabotropic glutamate receptors are said to act as 'molecular switches' for LTP.

The location of **expression** of LTP is a very controversial topic, both presynaptic and postsynaptic sites have been postulated. At the CA3–CA1 synapses a postsynaptic locus now seems quite well established. Several mechanisms for the postsynaptic expression of LTP have been proposed:

- That the postsynaptic membrane becomes more responsive to released glutamate because AMPA receptors become more sensitive.
- **Silent synapses** that previously harbored only NMDA receptors (and were silent because normal low frequency stimulation does not activate NMDA receptors) acquire AMPA receptors and so become responsive.
- Synapses perforate and eventually form two synapses where previously there was one.

There is good evidence for each of these mechanisms. For example, **calcium–calmodulin dependent protein kinase II (CaMKII)**, a major protein of the postsynaptic density, is thought to be activated by Ca^{2+} entry through NMDA receptors, and it then phosphorylates AMPA receptors (GluR1 subunits), enhancing their response to glutamate.

A presynaptic component to CA3–CA1 synapse LTP has been proposed in which there is increased glutamate release. This requires that the NMDA Ca^{2+} signal generates a **retrograde messenger** that travels the 'wrong' way across the synaptic cleft. One molecule proposed for this role is **nitric oxide (NO)**. In neurons NO is synthesized by a Ca^{2+}-dependent **nitric oxide synthase (NOS)**. As a small, freely diffusible molecule, NO rapidly diffuses out of the postsynaptic cell, across the cleft and into the presynaptic cell where it stimulates guanylyl cyclase and so enhances the probability of glutamate release.

With brief tetanic stimulation, LTP lasts for only about 2 h. However with several tetanic stimuli, LTP seems to last indefinitely. This long-lasting LTP depends on transcription and translation since it is blocked by drugs that inhibit messenger RNA or protein synthesis. Several mechanisms probably have a role in this **maintenance** of LTP. They must explain how synaptic alterations are retained, even while individual molecules are being turned over, long after the original signals that brought about the alterations have gone. Two well documented processes are as follows:

(1) CaMKII consists of four subunits. Ca^{2+} activation phosphorylates them and once the Ca^{2+} concentration has fallen to resting levels they remain phosphorylated. This is because if a subunit becomes dephosphorylated it will immediately become autonomously phosphorylated by one of the other subunits. In this way CaMKII remains persistently active.

(2) Ca^{2+} stimulates an isoform of adenylyl cyclase and consequently cyclic adenosine monophosphate concentrations increase in LTP. Protein kinase A becomes persistently activated and has effects on gene expression in much the same way as it does in *Aplysia* learning (see Topic Q2). Key events in LTP are summarized diagrammatically in *Fig. 4*. It is important to note that somewhat different mechanisms are responsible for nonassociative LTP that occur at some other synapses.

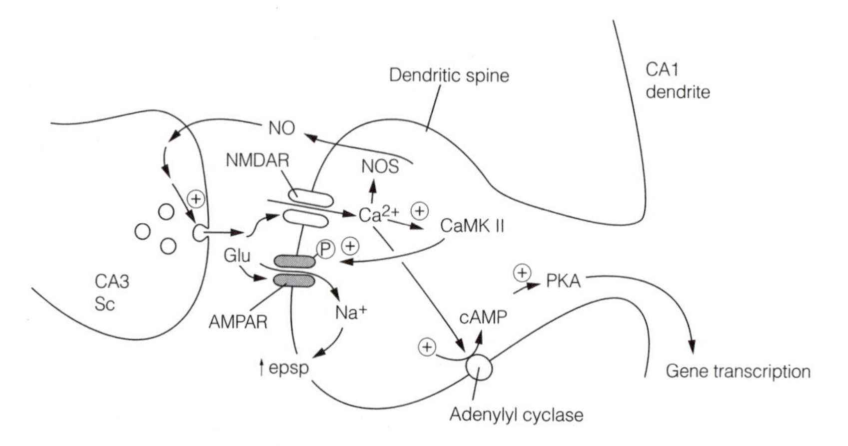

Fig. 4. Key events in LTP. Sc, Schaffer collateral; glu, glutamate; AMPAR, AMPA receptor, NMDAR, NMAD receptor; NO, nitric oxide; NOS, nitric oxide synthase; PKA, protein kinase A, CaMKII, Ca^{2+}–calmodulin-dependent kinase II.

Is LTP learning?

There is some dispute about whether LTP is a cellular substrate of learning. Against the view is that gene knockout experiments have been reported in which LTP is impaired specifically at synapses between the perforant pathway and the granule cells of the dentate gyrus (pp–gc synapses), but spatial navigation learning is unaffected. That the two processes are dissociable implies that there is no correspondence between LTP at the affected synapses and spatial navigation learning. The least radical interpretation of this result is that LTP at pp–gc synapses is not involved in this type of learning. The alternative is that LTP cannot explain how learning occurs.

Two types of evidence support the view that LTP is learning at the level of individual neurons. Firstly, tetanic stimulation arises physiologically; secondly, many manipulations that impair LTP do cause learning deficits and *vice versa*. When rats explore a maze, the electroencephalogram shows a theta (θ) rhythm with a frequency of 4–10 Hz. This reflects periodic firing of hippocampal neurons that is driven by the cholinergic septohippocampal pathway. During θ discharge, place cells fire in phase, but all other pyramidal cells are silenced by the increased discharge of inhibitory cells. Thus θ activity ensures that only cells involved in learning a particular environment are active. Moreover, one of the most effective stimulus protocols for producing LTP *in vitro* is just like θ discharge. Maybe the θ rhythm is the brain's 'natural' tetanic stimulus for learning. When LTP is generated in intact behaving rats via chronically implanted electrodes it may last for many months, so it has the timecourse for learning.

Preventing NMDA receptors from functioning not only blocks induction of LTP, it also prevents some types of learning. Blocking NMDA receptors with D-2-amino-5-phosphono-valerate (APV) or dizocilpine impairs learning of the Morris water maze by rats. In mice that are genetically engineered so that NMDA receptor subunits are deleted from the CA1 region in the third postnatal week (after hippocampal development is complete), no NMDA receptor-dependent LTP in CA1 is seen. These mice also show impaired spatial learning in the Morris water maze and degraded place fields, with a loss of the correlated firing of place cells that is seen in normal animals.

Q5 CEREBELLAR MOTOR LEARNING

Key Notes

The Marr–Albers–Ito model

Motor learning in the cerebellum comes about by a weakening of the strength of synapses between parallel fibers (pf) and Purkinje cells (Pc) that are active at the same time as error signals arrive at the Pc via climbing fibers. The synaptic weakening is long-term depression (LTD).

Classical conditioning of the eye blink reflex

A puff of air delivered to the eye normally causes a reflex eye blink. This can be classically conditioned by pairing a tone with the air puff. The tone activates pf–Pc synapses just before the air puff signal arrives at the Pc via the climbing fibers, and so the pf–Pc synapses suffer LTD. Subsequent occurrence of the tone causes reduced Pc excitation which translates into a larger cerebellar output to motor neurons driving the eye blink.

Long-term depression (LTD)

LTD is seen in the cortex and hippocampus as well as the cerebellum. In the cerebellar cortex, LTD requires simultaneous Ca^{2+} input into the Pc (caused by climbing fibers) and activation of glutamate receptors at pf–Pc synapses. The effect is to desensitize the AMPA receptors at the synapses.

Related topics

Cerebellar cortical circuitry (L2)
Cerebellar function (L4)
Types of learning (Q1)
Hippocampal learning (Q4)

The Marr–Albers–Ito model

Motor learning in the cerebellum involves alterations in the strengths of synapses between parallel fibers (pf) and Purkinje cells (Pc). Those synapses that are active at exactly the same time that there is climbing fiber input to the Purkinje cell, experience a reduction in the synaptic strength, a type of plasticity called **long-term depression (LTD)**.

In the **Marr–Albers–Ito model** of motor learning, the frontal cortex (via the corticopontine cerebellar tract) provides the mossy fiber-parallel fiber inputs and the climbing fibers from the inferior olive are thought to transmit error signals. All the pf–Pc synapses that happen to be activated by a pattern of mossy fiber inputs at the same time as climbing fiber error signals arrive will show LTD.

Synapses not concurrently active are unchanged. Subsequently, parallel fiber activity at the depressed synapses excites Pcs less, thereby reducing their inhibitory output on deep cerebellar nuclei. The overall effect is that synapses at which LTD occurs enhance cerebellar output.

Classical conditioning of the eye blink reflex

Motor learning that occurs during the classical conditioning of the eye blink reflex has been extensively studied. A puff of air delivered to the eye (unconditioned stimulus; US) will produce an eye blink (unconditioned response; UR). This eye blink reflex can be conditioned if the air puff is paired with a tone

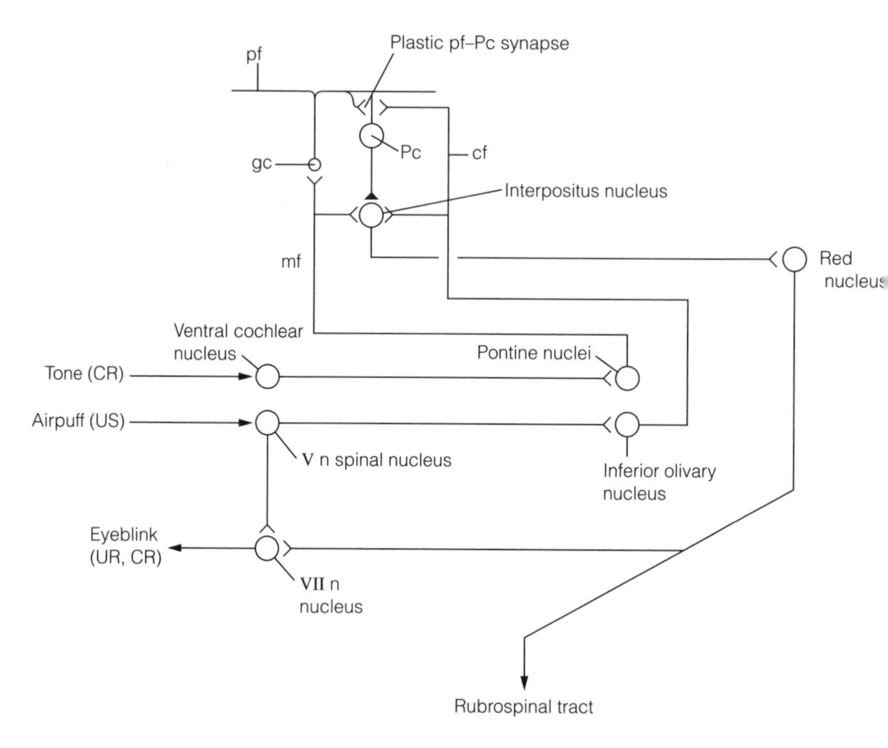

Fig. 1. Motor learning in the cerebellum; circuitry implicated in the conditioned eye blink reflex. pf, Parallel fiber; Pc, Purkinje cell; CR, conditioned response; UR, unconditioned response; US, unconditioned stimulus; cf, climbing fiber; gc, granule cell; mf, mossy fiber.

(conditioned stimulus; CS). The circuitry involved in motor learning in this reflex is shown in *Fig. 1*.

The air puff (US) is sensed by neurons in the spinal nucleus of the trigeminal (fifth) cranial nerve. The eye blink reflex (UR) is executed by connections between these cells and motor neurons in the facial (seventh cranial) nerve. Conditioning of the reflex requires the cerebellum. The US signal is transmitted via climbing fibers that arise from the inferior olivary nucleus. The tone (CS) signal goes by way of the ventral cochlear nucleus and pontine nucleus arriving at the cerebellum in mossy fibers. Activation of the pf–Pc synapse by the CS, 250 ms before the arrival of the US via the climbing fiber, results in LTD of the pf–Pc synapse. The effect of the LTD is that any subsequent arrival of the CS produces a smaller excitation of the Pc. Hence Pc inhibition of the interpositus neurons is diminished, so these cerebellar nucleus cells drive the eye blink via their connections with the red nucleus.

Long-term depression (LTD)

LTD is seen in the hippocampus and cerebral cortex where it can occur alongside long-term potentiation (LTP), and in the cerebellum (in which LTP is never seen). Induction of LTD in the cerebellum at the pf–Pc synapse requires coincident Ca^{2+} influx into the Pc and activation of AMPA and metabotropic glutamate receptors at the synapse. The Ca^{2+} influx is provided by the large depolarization due to climbing fiber activity which opens P type voltage-dependent Ca^{2+} channels. The receptors are activated by the release of glutamate from the parallel fibers. The final cause of the synaptic depression is desensitization of the AMPA receptors (*Fig. 2*) brought about by their phosphorylation by protein kinase C and possibly also by protein kinase G activated as a result of nitric oxide synthesis.

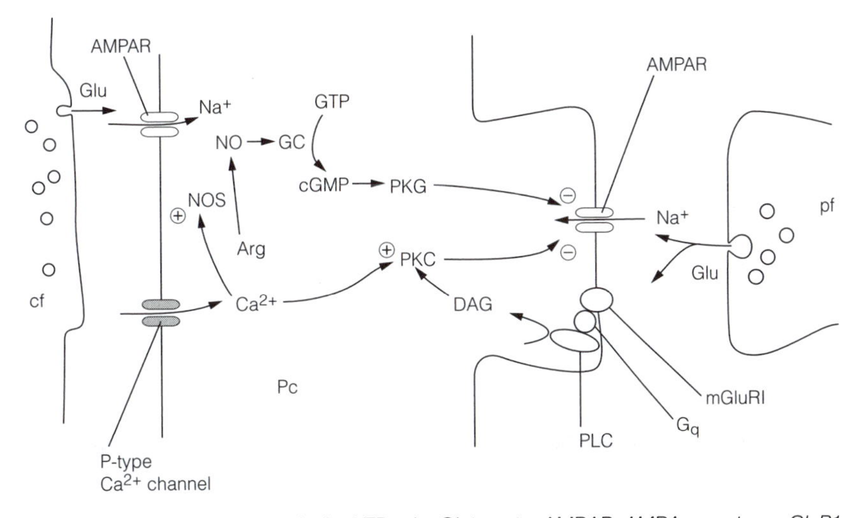

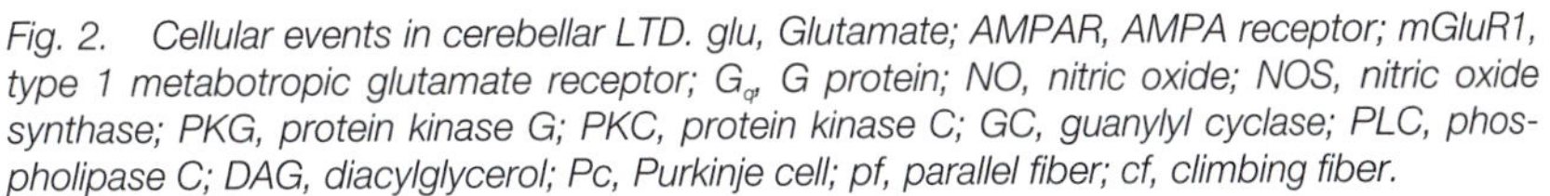
Fig. 2. Cellular events in cerebellar LTD. glu, Glutamate; AMPAR, AMPA receptor; mGluR1, type 1 metabotropic glutamate receptor; G_q, G protein; NO, nitric oxide; NOS, nitric oxide synthase; PKG, protein kinase G; PKC, protein kinase C; GC, guanylyl cyclase; PLC, phospholipase C; DAG, diacylglycerol; Pc, Purkinje cell; pf, parallel fiber; cf, climbing fiber.

R1 STROKES AND EXCITOTOXICITY

Key Notes

Cellular events in strokes

Most strokes are caused by blockage of cerebral arteries by blood clots produced locally or circulating from the heart. In the core of the ischemic region, cells starved of oxygen and glucose suffer rapid necrotic cell death. Death of cells in the surrounding penumbra is delayed and occurs because excessive glutamate release overexcites neurons, triggering apoptosis (programmed cell death).

Excitotoxic cell death

Deprived of oxygen, cellular ATP concentrations fall and the consequent reduced activity of the sodium pump facilitates Na^+ and Cl^- influx into cells (producing osmotic swelling) and K^+ efflux which depolarizes cells. Excessive glutamate release follows from the depolarization, stimulating NMDA, AMPA and metabotropic glutamate receptors to promote Ca^{2+} entry into neurons. Inappropriate Ca^{2+} entry can trigger either necrosis or (via free radical production or transcription) apoptosis. Excess secretion of Zn^{2+} from excitatory nerve terminals may act in a similar manner to Ca^{2+}. Excitotoxicity due to excessive glutamate is a final common path for cell death in a number of neurodegenerative conditions.

Possible treatments in stroke

Treatment strategies aim to rescue cells in the penumbra. Drug development involves using animal models of stroke, but some approaches that are successful in animals have not proven clinically useful. NMDA receptor antagonists, Ca^{2+} channel blockers and free radical scavengers have not proved efficacious in clinical trials. One successful approach in humans with acute ischemic stroke is the use of clot dissolving agents. AMPA receptor antagonists, apoptosis-inhibiting agents and Zn^{2+} chelators are currently being considered.

Related topics

Resting potentials (B1)
Slow neurotransmission (C3)
Receptor molecular biology (C4)
Neurotransmitter inactivation (C7)

Cellular events in strokes

Most strokes (cerebrovascular accidents) result from occlusion of a brain artery by a thrombus (blood clot) that forms *in situ* or travels there from the heart. The loss of blood flow (**ischemia**) starves brain tissue of oxygen and glucose. In the region of brain entirely dependent on the blocked blood supply, neurons and glia will die of **hypoxia** (lack of oxygen). This region is the **core** and the mode of cell death here is termed **necrosis**. Surrounding the core is the **penumbra**, this region suffers some ischemia but receives blood from other arteries. Cells in this region may survive or suffer delayed neuron death which occurs not directly from the lack of oxygen but from **excitotoxicity**. Here the reduced oxygen concentration causes neurons to secrete excessive amounts of glutamate, activating glutamate receptors to produce an influx of Ca^{2+} that is thought to kill cells by triggering **apoptosis** (programmed cell death).

Excitotoxic cell death

The cascade of events which leads to the death of cells from excitotoxicity is as follows. During ischemia, ATP concentrations are depleted so that energy requiring processes are compromised, including the Na^+/K^+-ATPase. Decreased activity of this cation pump increases the intracellular concentration of Na^+, and consequently Cl^- and water shift into the cells, which swell (*Fig. 1*). In addition, the extracellular K^+ concentration rises making the potassium equilibrium potential (E_K) more positive (see the Nernst equation, Topic B1), causing membrane depolarization. This depolarization activates voltage-dependent Ca^{2+} channels provoking neurotransmitter release. Since a large number of neurons in the brain use glutamate, this transmitter is released in excessive quantities. Glutamate release is exacerbated by the action of the Na^+/K^+ dependent glutamate transporter (see Topic C7) which normally removes the neurotransmitter from the cleft. This depends indirectly on metabolic energy since it is driven by the gradients for Na^+ and K^+ ions that are maintained by Na^+/K^+-ATPase. In ischemia the intracellular Na^+ and extracellular K^+ are thought to become so high as to force the transporter to work in reverse, extruding glutamate from the depolarized axons and astrocytes into the extracellular space.

The combination of a large depolarization and synaptic glutamate are precisely the conditions needed to activate NMDA receptors, resulting in Ca^{2+} influx. However, there are other routes by which intracellular Ca^{2+} is raised that may be as important:

- Ischemia evoked upregulation of the Ca^{2+} permeable subtypes of the AMPA receptor.
- Stimulation of metabotropic glutamate receptors (mGluRs) that are coupled to the phosphoinositide second messenger system (Type I mGluRs) since inositol trisphosphate causes the liberation of Ca^{2+} from internal stores.

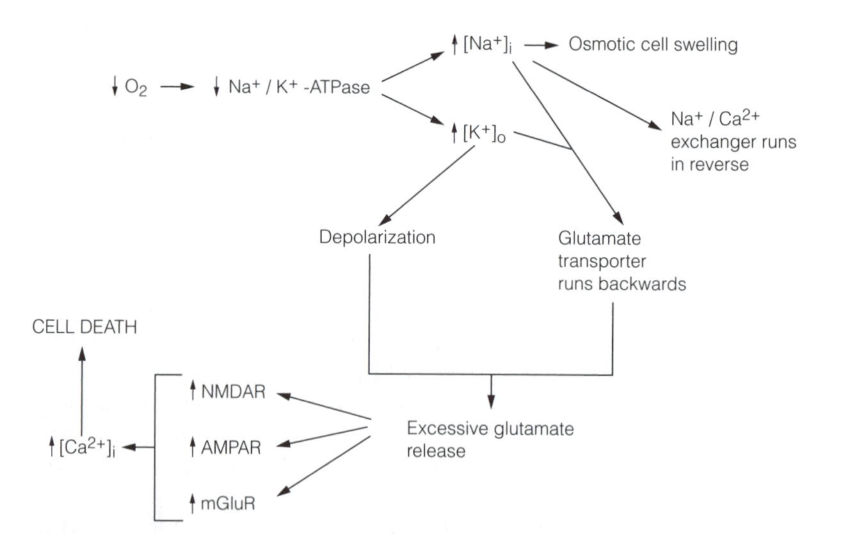

Fig. 1. Key events in ischemic stroke that lead to excitotoxic cell death. NMDAR, NMDA receptor; AMPAR, AMPA receptor; mGluR, metabotropic glutamate receptor.

- Operation of the Na^+/Ca^{2+} exchanger in reverse because of the high intracellular Na^+ concentration. Normally, this transport system uses sodium influx to expel calcium ions from the cell. If the sodium gradient falls, however, the transport occurs in the opposite direction.
- The uncontrolled Ca^{2+} influx overloads the transport systems and buffers which normally act to reduce the concentration of free Ca^{2+} in the cell. The Ca^{2+} triggers processes that eventually kill the cell.

Very high cytoplasmic Ca^{2+} concentrations are clearly cytotoxic and cause necrosis. The Ca^{2+} levels needed to trigger apoptosis are not well delineated. Although elevated Ca^{2+} concentrations seem to be associated with apoptosis in some cases, paradoxically, apoptosis of neurons in culture can be attenuated by modest increases in intracellular Ca^{2+} (e.g. by activating voltage-dependent Ca^{2+} channels).

Zinc ions (Zn^{2+}) which are released from some excitatory nerve terminals (e.g. mossy fiber terminals of the hippocampus) acts as a neurotransmitter or neuromodulator in the CNS, having actions at several types of receptor. Zn^{2+} is released from nerve terminals in ischemia, enters neurons via the same routes as Ca^{2+} (i.e. via NMDA and AMPA receptors and voltage-dependent Ca^{2+} channels) and is neurotoxic at the concentrations seen in ischemic brain. Administering agents which chelate Zn^{2+} before transient global ischemia in animals reduces subsequent cell death. Hence Zn^{2+} may, with Ca^{2+}, be an important contributor to ischemic brain cell death in strokes.

There are several ways in which Ca^{2+} may kill neurons. Firstly, Ca^{2+} activates a number of enzymes, endonucleases and proteases, the uncontrolled action of which would terminally disrupt cell function. Secondly, Ca^{2+} switches on processes that generate **free radicals**, including superoxide anions ($^{\bullet}O_2^-$), which are highly reactive, initiating extremely damaging chemical reactions, such as peroxidation of lipids in cell membranes. Free radical damage can lead to apoptosis. Thirdly, Ca^{2+} acting via a number of kinases can activate the transcription of genes that trigger apoptosis.

Excitotoxic cell death is also implicated in neurological disorders that arise as a result of genetic mutations that cause specific, regional cell loss; for example:

- **Huntington's disease** in which medium spiny neurons in the striatum which receive glutamatergic inputs from the cortex die.
- **Amylotropic lateral sclerosis** (**motor neuron disease**) which results from the death of motor neurons in the brain stem and spinal cord.

Possible treatments in stroke

Since cells in the core are killed within a very short time of the occlusion, short-term treatment strategies must focus on increasing the survival rate of cells in the penumbra and this means limiting excitotoxicity. Development of novel drugs for the treatment of stroke uses animal models in which ischemic lesions are generated by occluding the cerebral arteries. Unfortunately, some strategies that seemed promising in animal studies have not been vindicated in clinical trials. Possible approaches include:

(1) The endogenous thrombolytic (clot dissolving) protein, **tissue plasminogen activator** has proved effective in clinical trials and is now used for the treatment of acute ischemic stroke. This approach cannot be used in hemorrhagic strokes caused by the rupture of blood vessels.
(2) NMDA receptor antagonists (e.g. APV, dizocilpine, dextrorphan) in animal

studies restrict the size of the penumbra in focal ischemia but have not been effective in clinical trials in stroke, suggesting that NMDA receptor activation may not be such an important route for excitotoxicity in humans. In fact, AMPA receptor antagonists are now proving superior in protecting CA1 pyramidal cells from global (whole brain) ischemia. One promising approach, which is very effective in focal ischemia in rats, is the combination of glutamate receptor antagonists with drugs that block apoptosis (e.g cyclohexamide).

(3) Calcium channel antagonists, such as nimodipine, block Ca^{2+} entry through L type Ca^{2+} channels. Although these are not directly involved in neurotransmitter release, their presence in pyramidal cells means that they could act as a major route for Ca^{2+} influx into these cells. Unfortunately, these drugs have no efficacy in clinical trials.

(4) Free radical scavengers which curtail later stages in the excitotoxicity cascade have proved effective in reducing focal ischemic damage in animals, but so far have shown no significant effectiveness in clinical trials.

(5) An as yet untried approach would be to block Zn^{2+} entry into cells.

(6) The development of methods to prevent apoptotic cell death.

R2 EPILEPSY

Key Notes

Types of epilepsy

Epilepsy is characterized by recurrent seizures, brief periods of abnormal synchronized neural firing. Epilepsy can be acquired (head trauma, brain cancer and neurodegenerative disease are risk factors) or inherited. Generalized (tonic–clonic, absence) seizures are widespread and consciousness is always lost, whereas partial seizures originate from a single focus and may (complex) or may not (simple) be accompanied by loss of consciousness. A particular drug may not be effective in all types of epilepsy.

Neurobiology of epilepsy

Normally in the hippocampus, CA3 pyramidal cells fire bursts of action potentials and, via their Schaffer collaterals, drive CA1 cells. However extensive inhibition from GABAergic interneurons prevents burst firing of CA1 cells. In epileptic hippocampal slices however, CA3 cells do trigger bursting of CA1 cells, and burst firing of CA3 cells is itself exaggerated by paroxysmal depolarizing shifts (PDS) due to Ca^{2+} influx. PDSs cause abnormal (interictal) spikes seen in the EEG of the epileptic cortex between seizures. The development of the hyperexcitable state that predisposes to epilepsy seems to require NMDA receptors. However, individual seizures are initiated by AMPA receptors (NMDA receptors are subsequently activated) and terminated by adenosine produced from ATP as a result of the high neural activity in seizures.

Possible causes of hyperexcitability

Rare familial epilepsies caused by mutations in voltage-dependent Na^+ or K^+ channels or nicotinic cholinergic receptors have been discovered. Hyperexcitability of acquired epilepsy may be the result of an autoimmune response (e.g. antibodies generated to glutamate receptor subunits) or to excessive sprouting of damaged axons. In this case the hyperexcitability arises because of an increased number of recurrent excitatory connections. In an animal model of epilepsy, hyperexcitability is caused by increased sensitivity of NMDA receptors coupled to a reduction of inhibition mediated by $GABA_A$ receptors.

Pharmacology of epilepsy

Barbiturates and benzodiazepines enhance inhibition by GABA at $GABA_A$ receptors but may also be anticonvulsant by actions on voltage-dependent ion channels or AMPA receptors. Blockade of Na^+ channels is the mode of action of phenytoin and carbamazepine. Absence seizures are caused by the synchronized firing of thalamocortical circuits, caused by activation of T type Ca^{2+} channels. These channels are blocked by ethosuximide, a drug effective in absence seizures.

Related topics

Fast neurotransmission (C2)
Receptor molecular biology (C4)
Sleep (O4)
Hippocampal learning (Q4)

Types of epilepsy

Epileptic seizures are caused by an abnormal, synchronized firing of large populations of neurons that is usually self limiting. **Epilepsy** is defined as a disease in which such seizures recur. The incidence is about 1% and despite the variety of drugs currently available, control of seizures is acceptably good in only about 75% of individuals. Head trauma sufficient to cause coma, brain cancer, withdrawal from ethanol dependence and neurodegenerative brain disease are common risk factors for epilepsy. In addition, some epilepsies are inherited. Epilepsy is classified according to clinical presentation.

Generalized seizures are widespread. They include **tonic–clonic seizures** (*grand mal*), characterized by loss of consciousness and convulsions, and **absence seizures** (*petit mal*) in which the individual loses consciousness for just a few seconds, and which is accompanied by a 3 Hz EEG signal originating from the thalamus. **Partial (focal) seizures** are initiated from one region of the cortex, typically the motor cortex (in **Jacksonian** epilepsy), or the temporal lobe. In simple partial seizures consciousness remains; in complex partial seizures consciousness is lost. Drugs effective on some subtypes of epilepsy fail to be effective on others, which implies that different mechanisms are at work. However, some people have more than one type of fit, or progress from one type to another, which suggests a common underlying abnormality.

Neurobiology of epilepsy

Cellular and molecular mechanisms that might underlie epilepsy have been studied particularly in the hippocampus. Brief, high frequency electrical stimulation (1 s, 60 Hz) delivered to the hippocampus or amygdala of a rat once or twice each day, through chronically implanted electrodes, results (after about 2 weeks) in animals having seizures that resemble complex partial seizures in humans. This animal model is called **kindling** and epileptiform activity is investigated in brain slices removed from kindled animals. Epileptiform activity can also be provoked in hippocampal slices from normal (unkindled) animals by a variety of manipulations e.g. applying NMDA receptor agonists or $GABA_A$ receptor antagonists.

Spontaneous burst firing is part of the normal repertoire of CA3 pyramidal cells, but not of CA1 pyramidal cells (see *Fig. 2* of Topic Q4). When CA1 cells are driven physiologically by CA3 cells they are prevented from burst firing by inhibition delivered by GABAergic interneurons. However, epileptogenesis in hippocampal slices alters the firing behavior of pyramidal cells in several ways. Firstly, CA3 cells trigger burst firing in CA1 cells via Schaffer collaterals, which suggests a weakening of inhibition. Secondly, computer modeling of the hippocampus shows that if normal levels of inhibition are reduced, the connections between neighboring CA3 pyramidal cells would allow them to fire in synchrony. Thirdly, the burst activity of pyramidal neurons in slices made epileptic is itself abnormal. The cells display **paroxysmal depolarizing shifts (PDS)**, each a long-lasting depolarization, due largely to Ca^{2+} influx, which generates a burst of action potentials. These events underlie **interictal spikes**, abnormal EEG activity produced by epileptic cortex between seizures.

Epileptogenesis refers to the development of the hyperexcitable state that predisposes to seizures. It is blocked by **antiepileptic** drugs, and it is distinct from the processes that trigger individual seizures that can be inhibited by **anticonvulsants**. NMDA receptor antagonists are good antiepileptics in that they completely prevent hippocampal kindling, but they are not very effective at stopping epileptiform activity in slices already kindled (i.e. they are poor anticonvulsants). The implication is that NMDA receptors are necessary for epileptogenesis.

By contrast, AMPA receptor activity is thought to initiate the generation of individual seizures. NMDA receptors and L type Ca^{2+} channels are subsequently recruited which leads to prolonged burst firing. Adenosine is probably a key player in the termination of individual seizures. Neural activity is very high during seizures and adenosine derived from extensive ATP catabolism is transported across the plasma membrane to act on neural adenosine receptors. The rise in adenosine concentration that peaks about 30–60 s after the onset of a seizure is part of a normal physiological response that increases blood flow to match metabolic demand. Adenosine acts at A1 receptors that, by coupling to G_i proteins, opens K^+ channels and closes Ca^{2+} channels. Seizure termination is brought about by the resulting hyperpolarization. Adenosine A1 receptor agonists are anticonvulsants.

Possible causes of hyperexcitability

In three, rare, inherited epilepsies a single mutant gene is responsible in each case. In one, a point mutation occurs in a component of voltage-dependent Na^+ channels. In a second, the mutated gene codes for the subunit of a voltage-dependent K^+ channel. Those channels that include the mutant subunit have a 20–40% reduction in K^+ current and this accounts for the neuronal hyperexcitability in this epilepsy. The third familial epilepsy is caused by a mutation of the α4 nicotinic acetylcholine receptor (nAChR) subunit. Brain nAChRs have a significant permeability to Ca^{2+} and some are located presynaptically where they promote GABA release. The mutant nAChRs have a lower Ca^{2+} permeability and so may produce epilepsy by preventing proper synaptic inhibition.

Often, acquired epilepsy follows brain injury, but only after a delay ranging from weeks to years. Long-term processes postulated to be epileptogenic in acquired epilepsy are **autoimmune disease** and **axonal sprouting**. Antibodies raised to the GluR3 subunit of AMPA receptors (but not other subunits) have produced epilepsy in rabbits. Antibodies to GluR3 have also been seen in Rasmussen's encephalitis, a neurodegenerative disorder in humans, of unknown origin, characterized by seizures and progressive destruction of the cortex of one hemisphere.

Extensive sprouting of the axons (mossy fibers) of dentate gyrus granule cells (**mossy fiber sprouting, MFS**) occurs in numerous animal models of temporal lobe epilepsy and **Ammon's horn sclerosis**, in which extensive death of neurons and injury-evoked proliferation of glial cells occurs – a pathology that is commonly seen in the hippocampus in drug-resistant human epilepsies. Surgical resection of sclerotic hippocampus usually cures the epilepsy, indicating that MFS might cause the hyperexcitability. Normally it is extremely difficult to induce seizure-like activity in dentate granule cells, because they lack recurrent mossy fiber excitatory connections with neighboring granule cells. However in MFS, whole cell patch clamping of dentate granule cells shows that recurrent excitatory synapses are formed, and probably accounts for these cells becoming hyperexcitable.

At present it is difficult to decide the order in which events occur in epilepsy associated with MFS. A possible sequence of events is given in *Fig. 1*. The positive feedback component of this hypothesis, that seizures cause neuron death which worsens the situation by encouraging further MFS, almost certainly occurs in **status epilepticus**, in which seizures are prolonged over many minutes. Whether neurons die as a result of the more typical brief fits that last only tens of seconds is not known. It is the case that MFS can occur in the absence of neuron death.

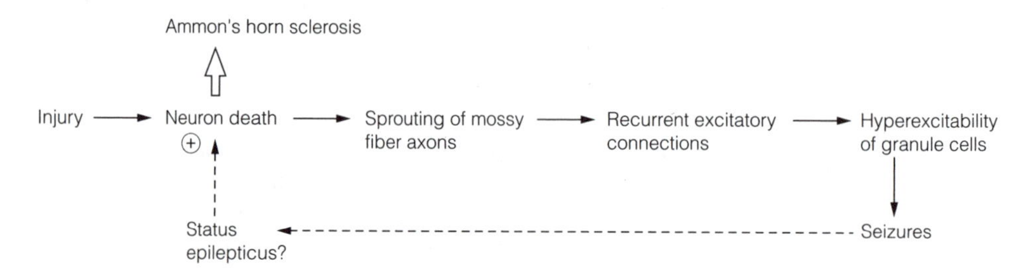

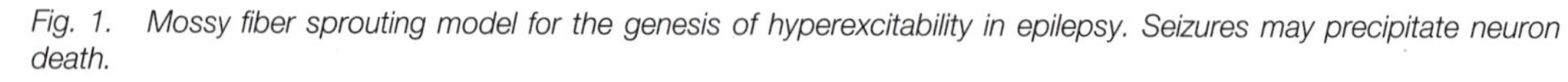
Fig. 1. Mossy fiber sprouting model for the genesis of hyperexcitability in epilepsy. Seizures may precipitate neuron death.

The modes of action of some drugs that successfully prevent seizures supports the long held contention that epilepsy is a matter of too much excitation and too little inhibition. In kindling, NMDA receptors become more sensitive to agonists and there is a reduction of inhibition by $GABA_A$ receptors that might be caused by NMDA receptor activation. The link between NMDA and $GABA_A$ receptors is Ca^{2+}. Influx of Ca^{2+} through the NMDA receptor activates a phosphatase, **calcineurin**. This dephosphorylates the $GABA_A$ receptor which consequently has a much reduced Cl^- current upon binding GABA. In summary, in this hypothesis, epileptogenesis results from greater NMDA receptor activation coupled with lower $GABA_A$-mediated inhibition. What causes the supersensitivity of the NMDA receptor is not known.

Pharmacology of epilepsy

Barbiturates and benzodiazepines act on $GABA_A$ receptors, increasing the Cl^- current caused when GABA binds. The anticonvulsive action of these drugs is usually attributed to this enhancement of GABA inhibition. However, in addition, barbiturates may owe some of their antiepileptic action to blocking L and N type Ca^{2+} channels and by noncompetitive inhibition of AMPA receptors. Moreover, blockade of adenosine uptake and block of voltage-dependent Na^+ channels may contribute to the anticonvulsant profile of benzodiazepines. Vigabatrin, an anticonvulsant useful in treating partial seizures, potentiates GABAergic inhibition indirectly by inhibiting the enzyme GABA transaminase, which normally breaks down GABA.

It has long been recognized that two widely used agents, phenytoin and carbamazepine, are anticonvulsant by binding inside the pores of voltage-dependent Na^+ channels, stabilizing them in an inactivated state.

During absence seizures, thalamic neurons go into the burst firing mode seen during slow wave sleep (see Topic O5). The bursting is synchronized and sustained by the reciprocal connections between the thalamus and cortex, and gives rise to the characteristic 3 Hz EEG signal. Burst firing is caused by activation of T type Ca^{2+} channels, and the brevity of absence seizures (just a few seconds) is presumably because T type Ca^{2+} channels are rapidly inactivated by depolarization. Ethosuximide and related drugs that are particularly effective in the treatment of absence seizures produce partial block of T type Ca^{2+} channels at therapeutic concentrations. Curiously another compound, valproate, used in absence seizures has no effect on T type Ca^{2+} channels; in fact it is one of several anticonvulsant drugs for which the mode of action is unknown.

R3 Parkinson's disease

Key Notes

Symptoms and etiology of Parkinson's disease (PD)

PD is a hypokinetic disorder characterized by tremor, rigidity, akinesia (difficulty in initiating movements) and bradykinesia (slowness of movements). Rare familial PD has been linked to mutations in α-synuclein, a component of Lewy bodies. Acquired PD is associated with head trauma, brain cancers and possibly environmental toxins.

Neuropathology

The major pathology in PD is the death of large numbers of dopaminergic neurons in the substantia nigra which project to the striatum. Other monoamine neurons are also lost. Lewy bodies are cytoplasmic inclusions containing α-synuclein, found in afflicted cells in PD, and other neurodegenerative diseases. Cell death in PD is caused by free radical reactions to which the substantia nigra is peculiarly vulnerable.

The MPTP model of PD

The pyridine 1-methyl-4-phenyl-1,2,3,6-tetrahydropyridin (MPTP), causes rapid, full blown PD in humans and monkeys. It has proved useful in studying how movement deficits arise in PD. MPTP crosses the blood–brain barrier and is oxidized to the toxic metabolite 1-methyl-4-phenyl pyridinium (MPP^+) which enters dopaminergic neurons, killing them by the production of free radicals. In monkeys, MPTP evoked PD reduces activity in the thalamocortical circuits that enable movement. The tremor arises from oscillations in the activity of neurons in the thalamus.

Treatment of PD

The key drug in the treatment of PD is L-DOPA, which crosses the blood–brain barrier and is converted to dopamine by dopamine-β-carboxylase. Although effective early on, after several years it becomes less useful and the majority of patients develop dyskinesia, a disabling hyperkinetic disorder.

Currently, dopamine receptor antagonists, monoamine oxidase (MAO) inhibitors and muscarinic cholinergic receptor antagonists also have a role in PD treatment. Future pharmacological strategies include the use of glutamate or adenosine receptor antagonists. Surgical approaches involve selective lesions of the thalamus or globus pallidus and the transplant of dopaminergic cells, harvested from the midbrain of human fetuses, into the striatum.

Related topics

Cortical control of voluntary movement (K6)
Anatomy of the basal ganglia (L5)
Basal ganglia function (L6)
Dopamine neurotransmission (N1)

Symptoms and etiology of Parkinson's disease (PD)

The most common hypokinetic disorder, **Parkinson's disease (PD)**, causes a 4–7 Hz **tremor**, especially of limbs, which reduces with intentional activity, an increase in muscle tone, **rigidity** of all limb muscles (unlike the selective rigidity of spasticity), a difficulty in initiating movements (**akinesia**) and movements that are made are slow (**bradykinesia**). The sufferer has a mask like facial expression with a very low blink rate, walks with a bent back and shuffling gait and if unbalanced may not recruit righting reflexes quickly enough to prevent themselves from falling.

Although PD is generally **idiopathic** (of unknown cause), rare familial cases occur. By studying family pedigrees several genes linked to familial PD have been identified. One of these families has point mutations in afflicted individuals in the gene coding for **α-synuclein**, a component of Lewy bodies (see below). There is also likely to be a genetic component to idiopathic PD. There is an increased incidence of the disease in close relatives of patients with PD, including a 53% concordance rate for monozygotic (identical) twins of PD patients showing dopaminergic dysfunction on PET scan. Sporadic PD occurs with head trauma or tumors that damage the midbrain and there is epidemiological evidence that it may result from exposure to environmental toxins; very severe PD follows ingestion of the pyridine compound, MPTP, which is chemically related to some herbicides.

Neuropathology

The defining pathology of PD is the death of large numbers of neurons in the substantia nigra pars compacta (SNpc) that give rise to the nigrostriatal pathway. Other dopaminergic neurons in the midbrain also die, but not to the same extent, and loss of noradrenergic cells in the locus ceruleus and cholinergic neurons in the basal forebrain is also seen. Bilateral destruction of the SNpc in monkeys with the neurotoxin 6-hydroxy dopamine causes rigidity and bradykinesia but not tremor. The cell death in PD is accompanied by the appearance of **Lewy bodies** in the cytoplasm of neurons, particularly in the SNpc. Lewy bodies are 5–25 μm in diameter and consist of a core of an abnormally folded protein, α-synuclein, surrounded by a halo of **ubiquitin**, a small protein present in all eukaryotic cells that tags proteins for destruction. Lewy bodies also feature in other neurodegenerative diseases and are probably a vehicle for sequestering abnormal proteins.

There is a normal age related loss of neurons from the substantia nigra, at a rate of about 5% per decade, but a 50% loss (associated with a 70–80% fall in striatal dopamine) is necessary to account for the onset of symptoms, so normal losses could only account for a very late onset of the disease. PET scans show that the rate of cell loss in PD is massively accelerated (up to 12% per year). This suggests that the disorder causing PD starts only about 5 years before symptoms appear.

The vulnerability of cells in PD correlates with their neuromelanin content (a dark pigment that accumulates with age). The SNpc has the greatest number of pigmented cells of any nucleus (about 90%) and hence suffers most. The significance of neuromelanin is that it binds iron and this metal contributes to the mechanism of cell death in PD.

Cell death in PD is caused by reactive oxygen species (*Fig. 1*). Normally the **superoxide anion** (**$^{\bullet}O_2^-$**) is converted by **superoxide dismutase** to hydrogen peroxide which is subsequently reduced to water by **glutathione peroxidase**. However, in the SNpc of PD, concentrations of glutathione are less than half those in the normal substantia nigra, while the amount of iron bound to neuromelanin is greater. This promotes the **Fenton reaction** which converts hydrogen peroxide to give the highly toxic **hydroxyl radical** (**$^{\bullet}OH$**).

The MPTP model of PD

The chance discovery in 1982 that an illicitly manufactured pyridine **MPTP** caused very severe PD in a group of heroin addicts who ingested it, has led to a

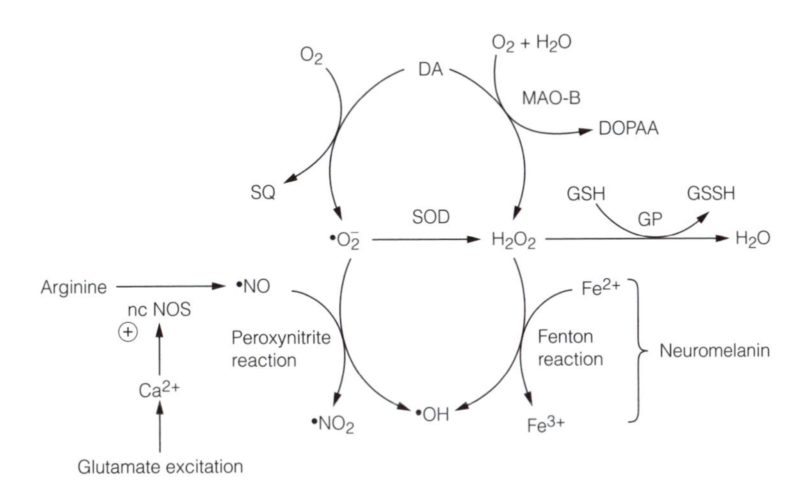

Fig. 1. Free radical reactions in Parkinson's disease. DA, dopamine; DOPAA, 3,4-dihydrophenyl-acetaldehyde; GP, glutathione peroxidase; GSH and GSSH, reduced and oxidized glutathione; ncNOS. neuronal isoform of nitric oxide synthase; SOD, superoxide dismutase; SQ, semiquinone. See text for details.

useful animal model of PD. In monkeys, MPTP causes a full blown PD. Following injection, MPTP crosses the blood–brain barrier and is taken up by astrocytes where it is oxidized to a metabolite MPP^+ by glial monoamine oxidase (MAO-B). MPP^+ is subsequently taken up via the specific dopamine transporter into dopaminergic neurons where it is responsible for the toxic effects of MPTP.

MPP^+ inhibits mitochondrial respiration, which depletes ATP levels, and generates superoxide anions, both of which kill the dopamine cells. Monkeys with MPTP induced PD show increased firing of neurons in the globus pallidus pars interna (GPi), and the subthalamic nuclei, and reduced firing of neurons in the globus pallidus pars externa (GPe) (see *Fig. 1*, Topic L6). The effect of this is to reduce the activation of the thalamocortical circuits which enable movement; this accounts for the akinesia and bradykinesia.

In monkeys rigidity is probably due to abnormally active long loop cortical stretch reflexes (see Topic K6). Patients with PD are unable to suppress stretch reflexes when attempting to change posture. For example, when instructed to sit they show inappropriate co-contraction of back and limb muscles, which is normally the required postural set for standing. The neurophysiological cause of the tremor is unclear, but the tremor correlates with a 3–6 Hz oscillation of neural activity in the ventrolateral thalamus. This provides the rationale for surgically lesioning this region of the thalamus to alleviate the tremor of PD (see below).

Treatment of PD

Both pharmacological and surgical treatments for PD exist. L-DOPA is the mainstay of drug therapy. It is the immediate precursor for dopamine (see Topic N1), crosses the blood–brain barrier and probably works by being taken up by the remaining dopaminergic terminals where it is converted to dopamine. Since L-DOPA is rapidly metabolized by DOPA decarboxylase in the periphery it is usual to administer it with a DOPA decarboxylase inhibitor that does not cross the blood–brain barrier. This improves uptake into the brain and reduces peripheral side-effects. L-DOPA is effective in the early years of treatment but becomes less so with long-term treatment and 80% of patients develop dyskinesia, a hyperkinetic disorder, which resembles the motor dysfunction of Huntington's disease (see Topic L6).

Other current pharmacological approaches include:

- dopamine receptor agonists;
- MAO-B inhibitors, which retard the catabolism of dopamine (see Topic N1);
- muscarinic acetylcholine receptor antagonists, which presumably work by reducing excitation of the GABAergic striatal neurons by the large aspiny cholinergic interneurons.

Other approaches that might be fruitful in the future are:

- Antioxidants have been used, on the premise that PD is due to free radical damage, though a large multicentre clinical trial has provided no evidence that high doses of vitamin E are beneficial.
- The finding of increased firing of glutamatergic neurons in the subthalamic nucleus in MPTP induced PD has prompted clinical trials of glutamate receptor antagonists.
- The activity of the GABA enkephalin containing striatal neurons of the indirect pathway is enhanced in PD by the lack of dopamine inhibition. However these cells also experience recurrent inhibition by GABAergic interneurons (*Fig. 2*). Release of GABA from these interneurons is reduced by presynaptic adenosine ($A2_A$) receptors. Hence antagonists of $A2_A$ receptors could have a role in treatment of PD by increasing the recurrent inhibition of the indirect pathway. An $A2_A$ antagonist improves the motor disability of marmosets with MPTP-induced PD.

Surgical approaches include lesions and tissue transplants. Discrete lesions of the globus pallidus, subthalamic nucleus or ventrolateral thalamus are performed. Lesions of the thalamus are particularly successful in reducing tremor but less effective in relieving rigidity or bradykinesia. In transplant therapy, human fetal midbrain dopaminergic cells are injected into the striatum. PET scans using [^{18}F]fluorodopa, which is taken up by dopaminergic cells, show that the transplanted cells survive and even increase their dopaminergic activity with time. The majority of patients show partial improvement.

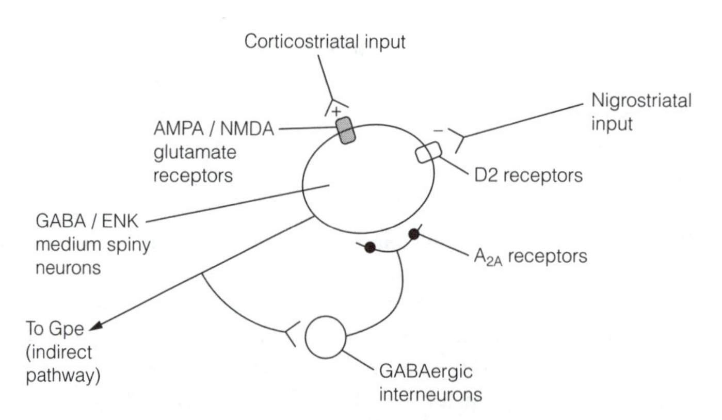

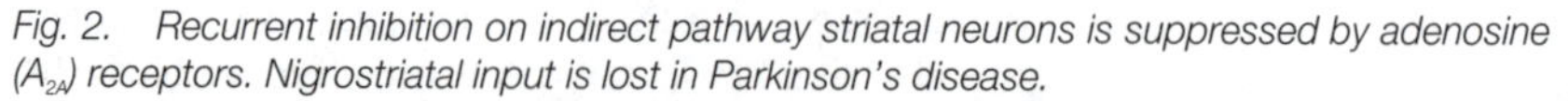
Fig. 2. Recurrent inhibition on indirect pathway striatal neurons is suppressed by adenosine (A_{2A}) receptors. Nigrostriatal input is lost in Parkinson's disease.

R4 ALZHEIMER'S DISEASE

Key Notes

Features of Alzheimer's disease

Alzheimer's disease (AD) is the most common dementia of the old and its incidence increases with age. AD is insidious, progressive and causes defects in memory, cognition, attention and motivation. Most AD sufferers also have symptoms of Parkinson's disease and neurodegeneration of the substantia nigra.

Neuropathology of AD

Patients with AD have massive shrinkage of the cortex and subcortical structures. Two characteristic lesions are found, particularly in the neocortex, hippocampus and amygdala. Neuritic plaques are extracellular deposits of β-amyloid peptide (βA) surrounded by degenerating neurites and glial cells activated by inflammatory processes. Diffuse plaques lacking damaged neurites and reactive glia are found in AD in brain regions not implicated in the disease, and in the brains of normal old people. Neurofibrillary tangles are cytoplasmic bundles of paired helical filaments (PHFs) made from an abnormal, hyperphosphorylated state of tau, a protein normally associated with microtubules. The density of tangles correlates with the severity of the dementia. Death of glutamatergic cells in the cortex and loss of several transmitter systems that project to the cortex (e.g. cholinergic pathways from the basal forebrain and septum to the hippocampus) dominate the pathology.

Familial AD

Rare, familial AD is linked to mutations in four genes and much idiopathic AD might be accounted for by polymorphisms of an apolipoprotein gene. Mutations of the gene for amyloid precursor protein (located on chromosome 21) which gives rise to βA are linked to early onset AD in several families and in Down's syndrome. Two genes that code for presenilins have numerous mutations linked to early onset AD, and *tau* gene mutations are seen in dementia with Parkinson's disease. There are three common alleles of apolipoprotein E (apoE), one of which is a clear risk factor for developing late onset idiopathic AD. Polymorphisms of the apoE gene could account for up to 90% of idiopathic AD.

β-amyloid and AD

Two pathways cleave amyloid precursor protein (APP), one of which leads to the formation of βA. βA is a normal secretory product found in human cerobrospinal fluid (CSF). APP is overexpressed in Down's syndrome and following brain ischemia or head injury. βA aggregates and is deposited in plaques where it is toxic, leading to the formation of neurofibrillary tangles. Mutations of the APP gene increase the production of βA and its deposition in plaques. Presenilin mutations may act by altering the regulation of the APP cleavage pathways.

Tau and AD

Tau is a protein needed for the assembly of microtubules. Mutations of the *tau* gene are linked to dementias in which neurofibrillary tangles are found extensively in the frontal and temporal cortex, but in which there are few plaques. Tau mutations may disrupt microtubule formation, and increase aggregation of tau into PHFs. Although tau in PHFs is hyperphosphorylated this is not the result of mutation and the consequence of this abnormal phosphorylation is unclear.

Pharmacological intervention in AD

Acetylcholinesterase inhibitors, which potentiate the effects of ACh released in the cortex from cholinergic neurons, can produce modest improvements in some patients. Anti-inflammatory drugs can arrest cognitive decline. However real progress will require novel strategies, for example, the discovery of drugs to inhibit the enzymes that process APP.

Related topics

Acetylcholine neurotransmission (N4)
Parkinson's disease (R3)

Features of Alzheimer's disease

Alzheimer's disease (**AD**) is the most common cause of **dementia** (insanity due to the loss of cognitive abilities) in the elderly. Its incidence increases with age so that while about 5% of people over 70 years old are sufferers, this increases to 20% of those over 80 years old. It is a disease predominantly of the nervous system (although deficits of gut function are also seen) and should be distinguished from **multi-infarct dementia**, the second most common cause of dementia, that is usually due to a succession of small cerebrovascular accidents.

AD is insidious, progresses unevenly, and at different rates in different individuals. Sudden deterioration may follow a stressful event. The earliest impairment is in semantic memory (particularly verbal), followed by failure to consolidate new long-term memories. Remote memory recall tends to be preserved until later. Deficits occur in cognition, attention and motivation (decreased appetite and libido) and AD sufferers are often depressed or frustrated. Two thirds of AD patients also have symptoms of Parkinson's disease, with neurodegeneration of the substantia nigra and Lewy bodies (Topic R3). Eventually, AD sufferers can no longer care for themselves. The average life expectancy after diagnosis is about 5 years.

Neuropathology of AD

Computer assisted tomographic scans of patients with AD show severe atrophy of cortical and subcortical regions with enlargement of the cerebral ventricles. Brain weight is reduced by 30–40%. Two lesions are characteristic of AD, found predominantly in the neocortex (often concentrated in frontal and temporal regions), hippocampus and amygdala.

(1) **Neuritic plaques** are spherical extracellular lesions 5–150 μm in diameter. They consist largely of deposits of an insoluble fibrillar form of **β-amyloid peptide** (**βA**), but also contain apoE and components of the complement cascade (proteins of the immune system). Surrounding and within this core are dystrophic (swollen, damaged and degenerating) neurites, together with microglia and astrocytes activated by cytokines resulting from inflammatory processes. In addition, there are diffuse plaques which are non-fibrillar βA deposits, lacking dystrophic neurites and reactive glial cells. These are localized in regions not clinically implicated in AD (e.g. the thalamus and cerebellum) and are seen in the brains of old, normal humans. The significance of these diffuse plaques to AD is not known.

(2) **Neurofibrillary tangles** are cytoplasmic bundles of paired 10 nm filaments that are twisted into a helix to form **paired helical filaments (PHFs)**. Tangles are found in large numbers in neurons (often pyramidal cells) in the entorhinal cortex, hippocampus, amygdala and in many areas that project to them. They are often seen in the dystrophic neurites of plaques. There is a correlation between the density of tangles and severity of the dementia. Tangles appear in other neurodegenerative diseases from which plaques are absent, so these two lesions are independent of each other. PHFs are composed of tau, normally a soluble cytoplasmic microtubule-associated protein, present in an abnormal highly phosphorylated, insoluble form, and associated with ubiquitin.

The cells that die in greatest numbers in AD are glutamatergic pyramidal cells of the cortex (including the hippocampus), cholinergic cells in the septo-hippocampal pathway and in the nucleus basalis of Meynert (see Topic N4) and noradrenergic and serotonergic projections from the locus ceruleus (LC) and raphe nuclei (RN) respectively. Nigrostriatal dopaminergic cell loss is responsible for the Parkinson's disease associated with AD. GABAergic and peptide neurotransmitter systems are spared. It has been suggested that the neurodegenerative changes start in the olfactory bulb (AD sufferers have a defective sense of smell) and spreads to the entorhinal cortex and hippocampus. Subsequently, cortico-cortical association axons die, so regions of cortex become disconnected from each other. Subcortical regions (nucleus basalis of Meynert, locus cerulus, raphe nuclei) deprived of their normal cortical targets degenerate.

Familial AD

The great majority of AD cases are idiopathic, but there are a number of rare familial types of AD linked to mutations of four genes. These provide important clues to causes of the disease. Idiopathic AD might also be genetic since it is linked to polymorphisms of an apolipoprotein gene.

Although the majority of AD cases are late onset (>60 years), rarely the illness may begin as early as 30 years old. This **early onset AD** is linked to mutations on three genes. **Amyloid precursor protein (APP)** is a membrane glycoprotein that gives rise to the β-amyloid peptides deposited in plaques. Its gene is on chromosome 21. Linkage between the APP gene and rare early onset AD in several families has been established. Notably, early onset AD is seen in **Down's syndrome**, which is caused by the presence of an extra copy of chromosome 21 (**trisomy 21**), so it is conceivable that an excess of APP can lead to AD. Transgenic mice engineered to have copies of mutant APP genes are proving useful animal models of AD since they express aspects of the abnormal pathology.

Two genes (on chromosome 14 and 1) linked to early onset AD code for the closely related **presenilin 1** and **presenilin 2**. These are large membrane proteins, in the endoplasmic reticulum (ER) and Golgi body that might be involved in the trafficking of newly synthesized proteins. Over 50 missense mutations have been identified in the presenilins that cause an aggressive early onset AD. Missense mutations in the gene (on chromosome 17) coding for tau is linked to dementia with PD in over 12 families.

Late onset idiopathic AD is linked with polymorphisms of apoE, a molecule involved in recycling cholesterol during membrane repair. The apoE gene, on chromosome 19, has three common alleles, ε2, ε3 and ε4. The ε4 allele is a risk factor for AD in that there is a high association between AD and possession of the ε4 allele. In contrast, the ε2 allele confers some protection. Individuals homozygous for ε4 are not only more likely to develop AD than other genotypes, but if they do they become sick earlier. It is estimated that between 60 and 90% of total AD cases can be accounted for by apoE genotype. The disease promoting effect of inheriting the ε4 isoform seems to be related to the fact that it enhances the aggregation of βA that accumulates in plaques (see below).

β-amyloid and AD

βA is derived from APP by the action of proteases. Since all the mutations in APP that give rise to AD are clustered around the cleavage sites in APP where these enzymes act, it is postulated that abnormal processing of the APP molecule leads to inappropriate βA production and that this underlies Alzheimer's disease.

APP is processed by two pathways, only one of which gives rise to the two βA peptides, with 40 and 42 amino acids, that are deposited in plaques. Only a minority of APP molecules are processed under normal circumstances. APP contains a 23 residue hydrophobic region near its C terminal that anchors it in membranes of the ER, Golgi body and the cell membrane. In the first pathway, a protease called **α-secretase** cuts APP (*Fig. 1*) producing a large soluble N-terminal product, sAPPα that is released into the lumens of the ER and Golgi body or from the cell surface. [This leaves behind a C terminal fragment containing 83 amino acids (CTF83) that is further cleaved by γ-secretase to liberate a p3 fragment.] In the second pathway, **β-secretase** cuts APP at a site closer to the N terminal than α-secretase to release sAPPβ. The remaining membrane bound C terminal component has 99 amino acids (CTF99) and is also acted on by **γ-secretase** to produce the βA peptides (βA_{40}, βA_{42}).

βA peptides are secreted by normal cells and are found in normal human CSF. At present only the β-secretase (also known as β-site APP-cleaving

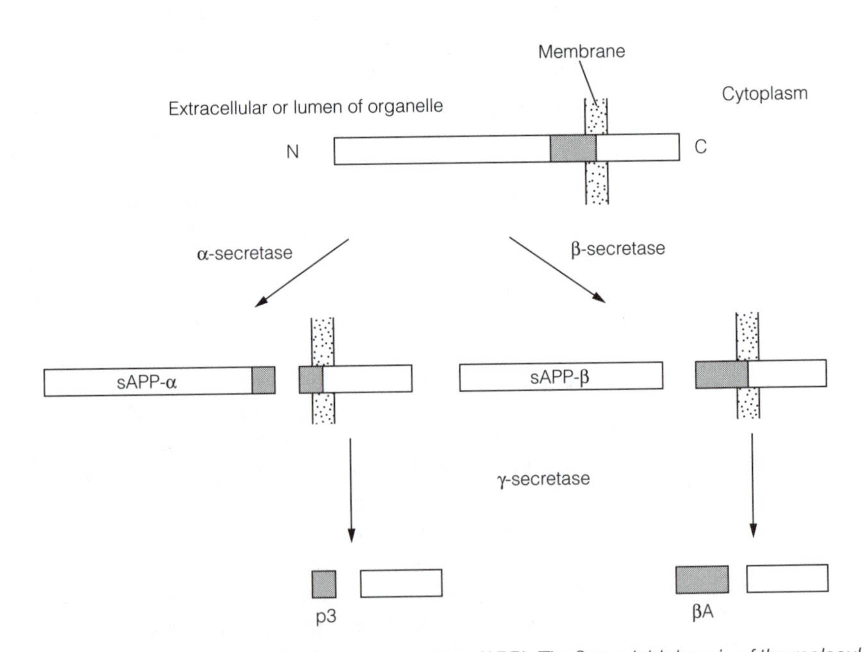

Fig. 1. Processing of amyloid precursor protein (APP). The β-amyloid domain of the molecule is hatched. sAPP, secreted amyloid precursor protein.

enzyme, BACE) has been identified and sequenced. It is a membrane bound aspartyl protease.

One way to understand what is amiss in AD is to know what roles APP and βA serve. The APP protein, or the products of its processing, may function as a protease inhibitor, a cell adhesion molecule or be neuroprotective against glutamate excitotoxicity. APP expression is increased during conditions that cause cell stress, such as ischemia and following head injury in humans.

The hypothesis that βA contributes to the pathogenesis of AD is based on the assumption that in AD, βA is produced in excessive amounts, converts into the fibrillar form that is deposited in neuritic plaques and that βA has well documented potentially adverse effects. Although soluble βA can spontaneously aggregate into the insoluble fibrillar form, numerous factors influence this process. Higher βA concentrations favor aggregation. Moreover, the ε4 isoform of apoE enhances the aggregation of βA. Crossing transgenic mice that have mutant forms of the APP gene with knockout mice that lack the apoE gene results in offspring which develop far less plaque formation than ordinary APP transgenic mice.

The mutations of the APP gene in familial early onset AD enhance the cleavage of APP by β-secretase or γ-secretase, but are also likely to increase the tendency for the mutant peptides to aggregate. Hence, APP mutations increase the production and deposition of βA.

The role of presenilins is uncertain, but brains of individuals with presenilin mutations, or animals genetically engineered to overexpress presenilins, have higher concentrations of βA_{42}. One hypothesis is that presenilin regulates the transport of either γ-secretase or APP so that cleavage to generate βA is facilitated. Other evidence suggests that presenilins might actually be γ-secretase enzymes.

How might abnormal amounts of βA lead to AD? Part of βA is very similar to the peptide transmitter, substance P and it binds **tachykinin** receptors, so it could act on neurons via such receptors. In cell culture, βA is neurotrophic in low doses, but it is toxic at high doses increasing Ca^{2+} entry into neurons, activating the processes that lead to the formation of paired helical filaments, and tangles which are eventually lethal.

Tau and AD

Tau, a cytoplasmic protein required for the proper formation of microtubules, and which forms the PHFs in tangles, is crucial to the neurodegeneration that leads to dementia. Indeed, it is the occurrence of neurofibrillary tangles, not plaques, which best correlates with the severity of AD. Malfunction of tau arises either as a direct consequence of βA neurotoxicity or as a result of mutations of the *tau* gene. The Ca^{2+} influx into cells caused by toxic amounts of βA activates a protease, **calpain**, which cleaves p35 protein to p25 protein. Normally, p35 controls the activity of **cyclin-dependent kinase 5 (cdk5)** which is needed for neurite outgrowth. Cleavage of p35 to p25 results in cdk5 being mislocated and permanently activated, and in this state it could hyperphosphorylate tau. Since hyperphosphorylation prevents binding of tau to microtubules, it disrupts the cytoskeleton, killing the cell. *Fig. 2* summarizes the progression of events that might occur in AD.

One problem for the amyloid cascade hypothesis of AD is that in **Pick's disease**, a dementia in which there is extensive tangle formation restricted to the frontotemporal cortex, there is no significant plaque formation. Numerous *tau* gene mutations are linked to frontotemporal dementias. Mutations in the

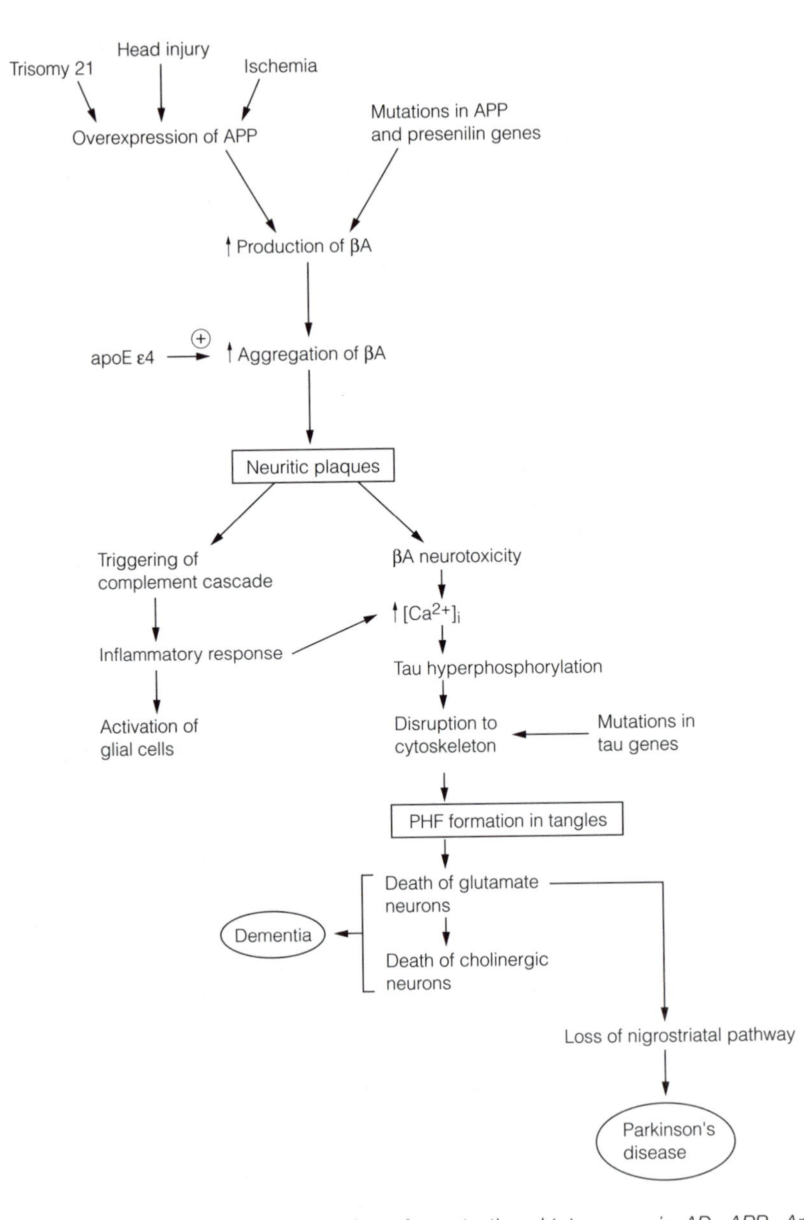

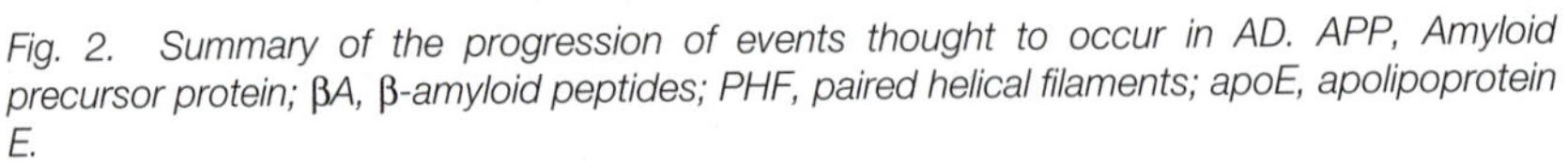
Fig. 2. Summary of the progression of events thought to occur in AD. APP, Amyloid precursor protein; βA, β-amyloid peptides; PHF, paired helical filaments; apoE, apolipoprotein E.

introns of the *tau* gene cause defects in the alternative splicing (see *Instant Notes in Biochemistry*, 2nd edn) of the *tau* mRNA, so an abnormal ratio of *tau* isoforms is generated. This mechanism is linked to a familial early onset AD in which twisted ribbon filaments are seen in glia, as well as in neurons. Mutations in the exons are located in positions which imply that they interfere with tau's ability to interact with microtubules. This could have two effects:

- destabilization of microtubules with disruption of axoplasmic transport; that is, loss of function;

- increases the probability of tau assembling into filaments which are toxic; a gain of (toxic) function.

What causes tau to assemble into paired helical filaments is not certain. It may require interactions with negatively charged sulfated glycosaminoglycans such as heparan sulfate. Heparan sulfate has been identified in nerve cells early in the formation of neurofibrillary tangles.

Pharmacological intervention in AD

Currently the only drugs licensed specifically for the treatment of AD are acetylcholinesterase inhibitors (e.g. donezepil). These are presumed to exert their therapeutic effect by potentiating the action of ACh released from cholinergic neurons in the basal forebrain and septum, and of course depends on their being some functioning cholinergic neurons remaining. A double blind trial of the nonsteroidal anti-inflammatory drug, indomethacin, showed that cognitive decline is halted in treated patients, which implies that the inflammatory component to AD is worth targeting. Other therapeutic approaches to try include:

- Ca^{2+} channel antagonists, antioxidants and free radical scavengers to target the downstream neurotoxic effects of βA accumulation;
- neurotrophins to promote the survival of injured cells;
- inhibitors of β- or γ-secretases, or of the aggregation of βA;
- inhibiting the calpain-mediated p35 cleavage pathway.

Further reading

There are many comprehensive textbooks of neuroscience and no single volume is likely to satisfy all needs. Different readers will prefer different textbooks depending, for example, on their prior learning, so I do not think it helpful to recommend one over another. Rather, I have listed some of the leading books which, experience shows, students find useful.

General reading

Bear, M.F., Connors, B.W., Paradiso, M.A. (1996) *Neuroscience: exploring the brain.* Williams & Wilkins, Baltimore.

Kandel, E.R., Schwartz, J.H., Jessel, T.M. (1991) *Principles of Neural Science*, 3rd Edn. Pearson Education, Harlow.

Nicholls, J.G., Martin, A.R., Wallace, B.G. (1992) *From Neuron to Brain*, 3rd Edn. Sinauer Associates, Sunderland, MA.

Levitan, I.B., Kaczmarek, L.K. (1997) *The Neuron; Cell and Molecular Biology*, 2nd Edn. Oxford University Press, Oxford.

Shepherd, G.M. (1994) *Neurobiology*, 3rd Edn. Oxford University Press, Oxford.

Zigmond, M.J., Bloom, F.E., Landis, S.C., Roberts, J.L., Squire, L.R. (1999) *Fundamental Neuroscience.* Academic Press, San Diego.

More advanced reading

The following selected articles are recommended for those who wish to know more about specific subjects. In many cases they are too advanced for first year students, but are very useful sources for students later in their studies.

Section A

Pardridge, W.M. (1998) *Introduction to the Blood–Brain Barrier.* Cambridge University Press, Cambridge.

Swanson, G (Ed.) (1996) Special issue: Glial signaling. *Trends Neurosci.* **19**, 305–369.

Walmsey, B., Alvarez, F.J., Fyffe, R.E.W. (1998) Diversity of structure and function at mammalian central synapses. *Trends Neurosci.* **21**, 81–88.

Section B

Hille, B. (1992) *Ionic Channels of Excitable Cells*, 2nd Edn. Sinauer Associates, Sunderland, MA.

Hodgkin, A.L. (1964) The ionic basis of nervous conduction. *Science.* **145**, 1148–1153.

Huxley, A.F. (1964) Excitation and conduction in nerve. *Science*, **145**, 1154–1159.

Matthews, G.G. (1998) *Cellular Physiology of Nerve and Muscle*, 3rd Edn. Blackwell Science, Malden, MA.

Ogden, D. (Ed.) (1994) *Microelectrode Techniques. The Plymouth Workshop Handbook.* The Company of Biologists Limited, Cambridge.

Section C

Buhl, E.H., Halasy, K., Somogyi, P. (1994) Diverse sources of hippocampal unitary inhibitory postsynaptic potentials and the number of synaptic release sites. *Nature* **368**, 823–828.

Nicholls, D.G. (1994) *Proteins, Transmitters and Synapses*. Blackwell Scientific Publications, Oxford.

Revest, P., Longstaff, A. (1998) *Molecular Neuroscience*. BIOS Scientific Publishers Ltd, Oxford.

Sudhof, T.S. (1995) The synaptic vesicle cycle: a cascade of protein–protein interactions. *Nature* **375**, 645–653.

Section D

Hoffman, D.A., Magee, J.C., Colbert, C.M., Johnston, D. (1997) K^+ channel regulation of signal propagation in dendrites of hippocampal pyramidal cells. *Nature* **387**, 869–875.

Midtgaard, J. (1994) Processing of information from different sources: spatial synaptic integration in the dendrites of vertebrate CNS neurons. *Trends Neurosci.* **17**, 166–173.

Stuart, G., Spruston, N., Sakmann, B., Hausser, M. (1997) Action potential initiation and backpropagation in neurons in the mammalian CNS. *Trends Neurosci.* **20**, 125–131.

Section E

Barr, M.L., Kiernan, J.A. (1983) *The Human Nervous System*, 4th Edn. Harper and Row Publishers, Philadelphia.

Berns, G.S. (1999) Functional neuroimaging. *Life Sciences* **65**, 2531–2540.

Fitzgerald, M.J.T. (1996) *Neuroanatomy: basic and clinical*, 3rd Edn. W.B. Saunders Company, London.

Tagamets, M.A., Horwitz,B. (1999) Functional brain imaging and modeling of brain disorders. *Prog. Brain Research* **121**, 185–200.

Section F

Konig, P., Engel, A.K., Singer, W. (1996) Integrator or coincidence detector? The role of the cortical neuron revisited. *Trends Neurosci.* **19**, 130–137.

Von der Malsburg, C. (1995) Binding in models of perception and brain function. *Curr. Opin. Neurobiol.* **5**, 520–526.

Section G

Berlucchi, G., Aglioti, S. (1997) The body in the brain: neural bases of corporeal awareness. *Trends Neurosci.* **20**, 560–564.

Melzack, R., Wall, P. (1993) *The Challenge of Pain*, 2nd Edn. Penguin, London.

Wall, P. (1999) *Pain: the science of suffering*. Weidenfeld and Nicholson, London.

Section H

Bullier, J., Novak, L.G. (1995) Parallel versus serial processing: new vistas on the distributed organization of the visual system. *Curr. Opin. Neurobiol.* **5**, 497–503.

Crick, F., Koch, C. (1995) Are we aware of neural activity in primary visual cortex? *Nature* **375**, 121–123.

Goodale, M.A., Milner, A.D. (1992) Separate visual pathways for perception and action. *Trends Neurosci.* **15**, 20–25.

Grossberg, S., Mingolla, E., Ross, W.D. (1997) Visual brain and visual perception: how does the cortex do perceptual grouping? *Trends Neurosci.* **20**, 106–111.

Hubel, D.H. (1982) Exploration of the primary visual cortex, 1955–78. *Nature* **299**, 515–524.

Livingstone, M.S. (1988) Art, illusion and the visual system. *Scientific American* **258**, 68–76.

Masland, R.H. (1986) The functional architecture of the retina. *Scientific American* **255**, 90–99.

Sharpe, L.T., Stockman, A. (1999) Rod pathways: the importance of seeing nothing. *Trends Neurosci.* **22**, 497–504.

Section I

Brainard, M.S. (1994) Neural substrates of sound localization. *Curr. Opin. Neurobiol.* **4**, 557–562.

Cohen, Y.E., Knudsen, E.I. (1999) Maps versus clusters: different representations of auditory space in the midbrain and forebrain. *Trends Neurosci.* **22**, 128–135.

Hudspeth, A.J. (1997) How hearing happens. *Neuron* **19**, 947–950.

Hudspeth, A.J. (1997) Mechanical amplification of stimuli by hair cells. *Curr. Opin. Neurobiol.* **7**, 480–486.

King, A.J. (1999) Sensory experience and the formation of a computational map of auditory space in the brain. *Bioessays* **21**, 900–911.

Section J

Freeman, W. (1991) The physiology of perception. *Scientific American* **264**, 34–41.

Mombaerts, P. (1999) 7TM proteins as odorant and chemosensory receptors. *Science* **286**, 707–711.

Mori, K., Nagao, H., Yoshihara, Y. (1999) The olfactory bulb: coding and processing of odor molecule information. *Science* **286**, 711–715.

Nakanishi, S. (1995) Second-order neurons and receptor mechanisms in visual- and olfactory-information processing. *Trends. Neurosci.* **18**, 359–364.

Smith, D.V., Margolis, F.L. (1999) Taste processing: wetting our appetites. *Curr. Biol.* **9**, 453–455.

Smith, D.V., St John, S.J. (1999) Neural coding of gustatory information. *Curr. Opin. Neurobiol.* **9**, 427–435.

Section K

Blake, D.J., Kroger, S. (2000) The neurobiology of Duchenne muscular dystrophy: learning lessons from muscle? *Trends Neurosci.* **23**, 92–99.

Clarac, F., Cattaert, D., Le ray, D. (2000) Central control components of a 'simple' stretch reflex. *Trends Neurosci.* **23**, 199–208.

Georgopoulos, A.P. (1995) Current issues in directional motor control. *Trends Neurosci.* **18**, 506–510.

Grillner, S. (1996) Neural networks for vertebrate locomotion. *Scientific American* **274**, 48–53.

Rowe, J.B., Frackowiak, R.S. (1999) The impact of brain imaging technology on our understanding of motor function and dysfunction. *Curr. Opin. Neurobiol.* **9**, 728–734.

Section L

Alexander, G.E., DeLong, M.R., Strick, P.L. (1986). Parallel organization of functionally segregated circuits linking basal ganglia and cortex. *Ann. Rev. Neurosci.* **9**, 357–381.

Chesselet, M-F., Delfs, J.M. (1996) Basal ganglia and movement disorders: an update. *Trends Neurosci.* **19**, 417–422.

Grieve, K.L., Acuna, C., Cudeiro, J. (2000) The primate pulvinar nuclei: vision and action. *Trends Neurosci.* **23**, 35–39.

Swanson, G. (Ed.). (1998) Special issue: cerebellum development, physiology and plasticity. *Trends Neurosci.* **21**, 367–418.

Section M

Hadley, M.E. (1992) *Endocrinology*, 3rd Edn. Prentice-Hall International, New Jersey.

Herman, J.P., Cullinan, W.E. (1997) Neurocircuitry of stress: central control of the hypothalamo–pituitary–adrenocortical axis. *Trends Neurosci.* **20**, 78–84.

Johnson, M., Everitt, B. (1988) *Essential Reproduction*, 3rd Edn. Blackwell Scientific Publications, Oxford.

Jordon, D. (Ed.). (1997) Central nervous control of autonomic function. Harwood Academic, Amsterdam.

Kalin, N.H. (1993) The neurobiology of fear. *Scientific American* **208**, 54–60.

LeVay, S. (1993) *The Sexual Brain*. MIT Press, Cambridge, MA.

Zakon, H.H. The effect of steroid hormones on electrical activity of excitable cells. *Trends Neurosci.* **21**, 202–207.

Section N

Cooper, J.R., Bloom, F.E., Roth, R.H. (1991) *The Biochemical Basis of Neuropharmacology*, 6th Edn. Oxford University Press, Oxford.

Nemeroff, C.B. (1998) The neurobiology of depression. *Scientific American* **278**, 28–35.

Perry, E., Walker, M., Grace, J., Perry, R. (1999) Acetylcholine in mind: a neurotransmitter correlate of consciousness. *Trends Neurosci.* **22**, 273–280.

Section O

Bergh, C., Sodersten, P. (1996) Anorexia nervosa, self-starvation and the reward of stress. *Nature Medicine* **2**, 21–22.

Elmkuist, J.K., Maratos-Flier, E., Saper, C.B., Flier, J.S. (1998) Unraveling the central nervous system pathways underlying responses to leptin. *Nature Neurosci.* **1**, 445–450.

Inui, A. (1999) Feeding and body weight regulation by hypothalamic neuropeptides-mediation of the actions of leptin. *Trends Neurosci.* **22**, 62–67.

Kalivas, P.W., Nakamura, M. (1999) Neural systems for behavioral activation and reward. *Curr. Opin. Neurobiol.* **9**, 223–227.

Spanagel, R. and Weiss, F. (1999) The dopamine hypothesis of reward: past and current status. *Trends Neurosci.* **22**, 521–527.

Steriade, M., Contreras, D., Amzica, F. (1994) Synchronized sleep oscillations and their paroxysmal developments. *Trends Neurosci.* **17**, 199–208.

Section P

Eisen, J.S. (1999) Patterning motoneurons in the vertebrate nervous system. *Trends Neurosci.* **22**, 321–326.

Jessen, K.R., Mirsky, R. (1999) Schwann cells and their precursors emerge as major regulators of nerve development. *Trends Neurosci.* **22**, 402–410.

Mehler, M.F., Mabie, P.C., Zhang, D., Kessler, J.A. (1997) Bone morphogenetic proteins in the nervous system. *Trends Neurosci.* **20**, 309–317.

Parnavelas, J.G. (2000) The origin and migration of cortical neurones: new vistas. *Trends Neurosci.* **23**, 126–131.

Rakic, P. (1988) Specification of cerebral cortical areas. *Science* **241**, 170–176.

Ruegg, M.A., Bixby, J.L. (1998) Agrin orchestrates synaptic differentiation at the vertebrate neuromuscular junction. *Trends Neurosci.* **21**, 22–27.

Shwaab, D.F., Hofman, M.A. (1995) Sexual differentiation of the human hypothalamus in relation to gender and sexual orientation. *Trends Neurosci.* **18**, 264–270.

Wiesel, T.N. (1982) Postnatal development of the visual cortex and the influence of the environment. *Nature* **299**, 583–591.

Section Q

Bliss, T.V.P., Collingridge, G.L. (1993) A synaptic model of memory: long-term potentiation in the hippocampus. *Nature* **361**, 31–39.

Buckner, R.L., Kelley, W.M., Petersen, S.E. (1999) Frontal cortex contributes to human memory formation. *Nature Neurosci.* **2**, 311–314.

Edwards, F. (1995) LTP-a structural model to explain the inconsistencies. *Trends Neurosci.* **18**, 250–255.

Fletcher, P.C., Frith, C.D., Rugg, M.D. (1997) The functional neuroanatomy of episodic memory. *Trends Neurosci.* **20**, 213–218.

Kim, J.J., Thompson, R.F. Cerebellar circuits and synaptic mechanisms involved in classical eyeblink conditioning. *Trends Neurosci.* **20**, 177–181.

Klintsova, A.Y., Greenough, W.T. (1999) Synaptic plasticity in cortical systems. *Curr. Opin. Neurobiol.* **9**, 203–208.

Linden, D.J. (1994) Long-term synaptic depression in the mammalian brain. *Neuron.* **12**, 457–472.

Morales, M., Goda, Y. (1999) Nomadic NMDA receptors and LTP. *Neuron.* **23**, 431–434.

Rose, S.P.R. (1995) Cell-adhesion molecules, glucocorticoids and long-term memory formation. *Trends Neurosci.* **18**, 502–506.

Silvia, A.J., Kogan, J.H., Frankland, P.W., Kida, S. (1998) Creb and memory. *Ann. Rev. Neurosci.* **21**, 127–148.

Soderling, T.R., Derkach, V.A. (2000) Postsynaptic protein phosphorylation and LTP. *Trends Neurosci.* **23**, 75–80.

Wilson, M.A., Tonagawa, S. (1997) Synaptic plasticity, place cells and spatial memory: study with second generation knockouts. *Trends Neurosci.* **20**, 102–106.

Section R

Campbell, P. (Ed.) (1999) Neurological disorders. *Nature.* **399** (suppl.), A3–A45.

Dirnagl, U., Iadecola, C., Moskowitz, M.A. (1999) Pathobiology of ischemic stroke: an integrated approach. *Trends Neurosci.* **22**, 391–397.

Kempermann, G., Gage, F. New nerve cells for the adult brain. *Scientific American* **280**, 38–43.

Hardy, J. (1997) Amyloid, the presenilins and Alzheimer's disease. *Trends Neurosci.* **20**, 154–159.

Schoepp, D.D., Conn, P.J. (1993) Metabotropic glutamate receptors in brain function and pathology. *Trends Pharmacol. Sci.* **14**, 13–20.

Wheal, H.V., Bernard, C., Chad, J.E., Cannon, R.C. (1998) Pro-epileptic changes in synaptic function can be accompanied by pro-epileptic changes in neuronal excitability. *Trends Neurosci.* **21**, 167–174.

INDEX